HANDBOOK OF MUTAGENICITY TEST PROCEDURES

Edited by

B.J. KILBEY, together with
M. LEGATOR
W. NICHOLS and
C. RAMEL

Department of Genetics, University of Edinburgh, Edinburgh, Scotland
Department of Medicine, University of Texas, Galveston, Tex., U.S.A.
Institute for Medical Research, Camden, N.J., U.S.A.
Wallenberg Laboratory, University of Stockholm, Sweden

ELSEVIER SCIENTIFIC PUBLISHING COMPANY
AMSTERDAM — NEW YORK — OXFORD

Second printing 1979

Library of Congress Cataloging in Publication Data

Main entry under title:

Handbook of mutagenicity test procedures.

Some of the chapters were originally published in Mutation research, v. 31 (1975) and v. 38 (1976).
Includes index.
1. Mutagenesis. 2. Genetics--Technique.
I. Kilbey, B. J.
QH465.A1H36 575.2'92 77-22531
ISBN 0-444-41338-3

Published by:

Elsevier North-Holland Biomedical Press
335 Jan van Galenstraat, P.O. Box 211
Amsterdam, The Netherlands

Sole distributors for the U.S.A. and Canada:

Elsevier/North-Holland Inc.
52 Vanderbilt Avenue
New York, N.Y. 10017, U.S.A.

Printed in The Netherlands

HANDBOOK OF MUTAGENICITY TEST PROCEDURES

LIST OF CONTRIBUTORS

J.W. Allen, Clinical Genetics Division, Mental Retardation Program, Children's Hospital Medical Center, Boston, Mass. 02115, U.S.A.

B.N. Ames, Biochemistry Department, University of California, Berkeley, Calif. 94720, U.S.A.

C.F. Arlett, MRC Cell Mutation Unit, University of Sussex, Falmer, Brighton BN1 9QG, England

A.J. Bateman, Paterson Labs, Christie Hospital and Holt Radium Institute, Manchester M20 9BX, England

M.C. Benge, Occupational Health and Medical Research, Dow Chemical U.S.A., Freeport, Tex., U.S.A.

R. von Borstel, Department of Genetics, University of Alberta, Edmonton, Alberta T6G 2E9, Canada

C. Bradt, Institute for Medical Research, Camden, N.J. 08103, U.S.A.

J.G. Brewen, Biology Division, Oak Ridge National Laboratory, Oak Ridge, Tenn. 37830, U.S.A.

U. Buckel, Institut für Anthropologie und Humangenetik, Universität Heidelberg, G.F.R.

J.E. Cleaver, Laboratory of Radiobiology, University of California, San Francisco, Calif. 94143, U.S.A.

D. Clive, Environmental Mutagenesis Branch, National Institute of Environmental Health Sciences, P.O. Box 12233, Research Triangle Park, N.C. 27709, U.S.A.

T.H. Connor, Division of Environmental Toxicology and Epidemiology, Department of Preventive Medicine and Community Health, University of Texas Medical Branch, Galveston, Tex. 77550, U.S.A.

R. DeMars, Laboratory of Genetics, University of Wisconsin, Madison, Wisc. 53706, U.S.A.

L. Ehrenberg, Stockholms Universitet, Wallenberglaboratoriet, Strålningsbiologiska Institute, S-106 91, Stockholm, Sweden

J. Ellenberger, Zentrallaboratorium für Mutagenitätsprüfung der Deutschen Forschungsgemeinschaft, Freiburg i.Br., G.F.R.

H.J. Evans, MRC, Clinical and Population Cytogenetics Unit, Western General Hospital, Edinburgh, Scotland

R. Fahrig, Zentrallaboratorium für Mutagenitätsprüfung der Deutschen Forschungsgemeinschaft, Freiburg i.Br., G.F.R.

C.N. Frantz, Environmental Mutagenesis Branch, National Institute of Environmental Health Sciences, Research Triangle Park, N.C. 27709, U.S.A.

M.H.L. Green, MRC Cell Mutation Unit, University of Sussex, Falmer, Brighton BN1 9QG, England

I. Hansmann, Institut für Humangenetik, Universität Göttingen, G.F.R.
L. Jacobs, Laboratory of Genetics, University of Wisconsin, Madison, Wisc. 53706, U.S.A.
L.A. Jeurgens, Department of Pediatrics, Harvard Medical School, Boston, Mass. 02115, U.S.A.
B.A. Kihlman, Department of Genetics and Plant Breeding, Agricultural College, S-75007 Uppsala, Sweden
D.J. Kilian, Occupational Health and Medical Research, Dow Chemical U.S.A., Freeport, Tex., U.S.A.
S.A. Latt, Clinical Genetics Division, Mental Retardation Program, Children's Hospital Medical Center, Boston, Mass., U.S.A.
A. Léonard, Laboratory of Genetics, Department of Radiobiology, C.E.N.-S.C.K., B-2400 Mol, Belgium
M.S. Legator, Division of Environmental Toxicology and Epidemiology, Dept. of Preventive Medicine and Community Health, University of Texas Medical Branch, Galveston, Tex. 77550, U.S.A.
H.V. Malling, Environmental Mutagenesis Branch, National Institute of Environmental Health Sciences, Research Triangle Park, N.C., 27709 U.S.A.
J. McCann, Biochemistry Department, University of California, Berkeley, Calif. 94720, U.S.A.
R.C. Miller, Institute for Medical Research, Camden, N.J. 08103, U.S.A.
G.R. Möhn, Zentrallaboratorium für Mutagenitätsprüfung der Deutschen Forschungsgemeinschaft, Freiburg i.Br., G.F.R.
F.M. Moreland, Genetic Toxicology Branch, FDA, Department of Health, Education and Welfare, Washington, D.C., U.S.A.
W.J. Muriel, MRC Cell Mutation Unit, University of Sussex, Falmer, Brighton BN1 9QG, England
H.B. Newcombe, Biology and Health Physics Division, Atomic Energy of Canada Ltd., Chalk River, Ont., Canada
W.W. Nichols, Institute for Medical Research, Camden, N.J. 08103, U.S.A.
M.L. O'Riordan, MRC, Clinical and Population Cytogenetics Unit, Western General Hospital, Edinburgh, Scotland
T.G. Pullin, Division of Environmental Toxicology and Epidemiology, Dept. of Preventive Medicine and Community Health, University of Texas Medical Branch, Galveston, Tex. 77550, U.S.A.
W.E. Rogers, Department of Pediatrics, Harvard Medical School, Boston, Mass. 02115, U.S.A.
R. Röhrborn, Institut für Anthropologie und Humangenetik, Universität Dusseldorf, Düsseldorf, G.F.R.
W. Schmid, Division of Medical Genetics, Dept. of Pediatrics, University of Zurich, Zurich, Switzerland
A.G. Searle, MRC Radiobiology Unit, Harwell, Oxon OX11 ORD, England
R.H. Smith, 902 Euclid Ave., Berkeley, Calif., U.S.A.
F.H. Sobels, Department of Radiation Genetics and Chemical Mutagenesis, University of Leiden, Leiden, The Netherlands
J.-A.F.S. Spector, Environmental Mutagenesis Branch, National Institute of Environmental Health Sciences, P.O. Box 12233, Research Triangle Park, N.C. 27709, U.S.A.

E. Vogel, Department of Radiation Genetics and Chemical Mutagenesis, University of Leiden, Leiden, The Netherlands

C.A. Wachtmeister, Stockholms Universitet, Wallenbergslaboratoriet, Strålningsbiologiska Institute, S-106 91 Stockholm, Sweden

E.B. Whorton, Jr., Division of Biometry, Department of Preventive Medicine and Community Health, University of Texas Medical Branch, Galveston, Tex., U.S.A.

F.E. Würgler, Department of Zoology, Swiss Federal Institute of Technology, Universitätstrasse 2, CH-8006 Zurich, Switzerland

E. Yamasaki, Biochemistry Department, University of California, Berkeley, Calif. 94720, U.S.A.

F.K. Zimmermann, Mykologie/Genetik, Technische Hochschule, 61 Darmstadt, Schnittspahnstrasse 10, G.F.R.

PREFACE

At a meeting on the Biological Activity of Fluorescent Whitening Agents which was held in Stockholm in 1973, a number of us became concerned over variations in procedures adopted by different laboratories using nominally the same test for mutagenesis. In view of the wide variety of compounds subjected to these routines, some degree of variation in method cannot, of course, be avoided. However, to facilitate comparisons, it seemed extremely important to keep procedural variation to a minimum. With this in mind, Marvin Legator, Claes Ramel, Warren Nichols and I invited some 25 people with special experience in particular techniques of mutagen testing to set these out in a form which could provide anyone attempting a specific test with all the necessary background information on materials, experimental design, pitfalls and difficulties needed to perform the test in an acceptable way with a minimal need for additional expert help. By combining this material in one volume we hoped to provide an 'on the bench' methodological text for most of the common assays for genetic damage apparently in use.

The present volume is the result of this attempt. Some articles are far more extensive than we had originally envisaged but this may well be to the advantage of the user. Furthermore, in addition to our first list of tests, we have obtained contributions on the use of data from population records, and the handling and safety aspects of mutagens and carcinogens. We have also included a chapter which discusses some of the basic statistical concepts which must be borne in mind when mutation experiments are designed. A negative response of a population to a suspected mutagen may well result from lack of mutagenic activity but it may also be an artifact resulting from the exposure of too small a population in the first place. In few areas of research is it so important to attach confidence limits to negative results.

Our choice of material may, and probably will, be criticised. The methods included here have been sufficiently used and tested to make us reasonably confident of their reliability. We were, of course, aware of a number of exciting developments which are taking place but which are still at the very important proving stages. Because of this, we have regretfully excluded them but we hope that we will be able to include them in a later, updated version of the book if, and when, it is prepared. This is also true of developments currently taking place in the methods used to measure the reaction mutagens with cellular components. Obviously these techniques will be of great significance for problems of risk evaluation.

The object of the book was not to suggest to industrial laboratories or any other section of the Scientific community which tests are the most satisfactory ones. We have, of course, attempted to persuade the authors to contribute information concerning the manpower and material requirements, the damage assayed and the level of expertise necessary for each test. However, when the experimenter is faced with making a choice between tests we would strongly advise him to seek the help and advice of both his National Regulatory Bodies and of those bodies which are able to provide a reservoir of scientific help and information such as the Environmental Mutagen Societies of America, Europe, India and Japan. It is also well worth reminding possible users of these tests that the Environmental Mutagen Information Center at the Biology Division of the Oak Ridge National Laboratory in Tennessee is an important source of documentation on much of the existing experimental data.

Finally, we would like to thank all those who have contributed encouragement or helpful advice and suggestions. Most of all, however, we must thank every contributor to what we hope will be a useful aid to those concerned with tests of potential mutagens.

Brian Kilbey
Edinburgh, 1977

ACKNOWLEDGEMENTS

The Editors would like to thank the Authors and the Publisher for permission to include 12 chapters in this volume that were previously published in Mutation Research. These are the papers by:

Ames et al. on pp. 1—17 of this volume; originally published in Mutation Res., 31 (1975) 347—364;

Frantz and Malling on pp. 49—63; originally published in Mutation Res., 31 (1975) 365—380;

Green and Muriel on pp. 65—94; originally published in Mutation Res., 38 (1976) 3—32;

Zimmermann on pp. 119—134; originally published in Mutation Res., 31 (1975) 71—86;

Fahrig on pp. 135—147; originally published in Mutation Res., 31 (1975) 381—394;

Clive and Spector on pp. 161—173; originally published in Mutation Res., 31 (1975) 17—29;

Brewen on pp. 221—224; originally published in Mutation Res., 31 (1975) 5—8;

Schmid on pp. 235—242; originally published in Mutation Res., 31 (1975) 9—15;

Evans and O'Riordan on pp. 261—274; originally published in Mutation Res., 31 (1975) 135—148;

Léonard on pp. 293—299; originally published in Mutation Res., 31 (1975) 291—297;

Searle on pp. 311—324; originally published in Mutation Res., 31 (1975) 277—290; and

Kihlman on pp. 389—400; originally published in Mutation Res., 31 (1975) 401—402.

CONTENTS

METHODS FOR DETECTING CARCINOGENS AND MUTAGENS WITH THE SALMONELLA/MAMMALIAN-MICROSOME MUTAGENICITY TEST*

BRUCE N. AMES, JOYCE MCCANN** AND EDITH YAMASAKI

Biochemistry Department, University of California, Berkeley, California 94720 (U.S.A.)

INTRODUCTION

We have previously described a very sensitive and simple bacterial test for detecting chemical mutagens[2–4,7,25] which has recently been reviewed[23]. The compounds are tested on petri plates with several specially constructed mutants of *Salmonella typhimurium*[4,7,25] selected for sensitivity and specificity in being reverted from a histidine requirement back to prototrophy by a wide variety of mutagens.

The test has been adapted for use in detecting chemicals which are potential human carcinogens or mutagens by adding homogenates of rat (or human) liver directly to the petri plates thus incorporating an important aspect of mammalian metabolism into the *in vitro* test. In this way, a wide variety of carcinogens requiring metabolic activation can be detected easily as mutagens[4,25,23].

There is considerable evidence, much of it obtained using this test[2–5,7,8,14,19,23,25], that with few exceptions carcinogens are mutagens. This supports the desirability of using this type of rapid and economical test system as a screening technique to pinpoint potentially dangerous chemicals among the thousands of chemicals to which humans are exposed.

The test is highly efficient in detecting carcinogens as mutagens. Many carcinogens and non-carcinogens have been tested and we are currently compiling results obtained using the test in this and many laboratories throughout the world (McCann *et al.*, in preparation). So far about 85% of the carcinogens tested (135/158) have been detected as mutagens. These include a wide variety of carcinogens such as direct alkylating agents, nitrosamines, polycyclic hydrocarbons, fungal toxins, aromatic amines, nitrofuran carcinogens, a variety of antineoplastic agents, and antibiotic carcinogens such as adriamycin, daunomycin, and mitomycin C. Also the known human chemical carcinogens which have been tested are positive. These include β-naphthylamine, benzidine, cigarette smoke condensates, *bis*-chloromethylether, aflatoxin B_1, vinyl chloride, 4-aminobiphenyl. This and other evidence that a high percentage of carcinogens are mutagens is most easily explained if these carcinogens cause cancer by somatic mutation[2–8].

---

* This work was supported by ERDA Contract AT(04-3)34 P.A. 156.

** Postdoctoral Fellow of the California Division of the American Cancer Society.

Abbreviations: MNNG, *N*-methyl-*N'*-nitro-*N*-nitrosoguanidine; furylfuramide, 2-(2-furyl)-3-(5-nitro-2-furyl/acrylamide.

To date 106 non-carcinogens (chemicals negative in tests for carcinogenicity) have been tested, many of which are close relatives of carcinogens. Extremely few ($< 10\%$) are mutagenic in the test, and for several of these carcinogenicity tests have been extremely limited and there is some doubt as to the classification of these chemicals as non-carcinogens. Many hundreds of common chemicals of unknown carcinogenicity have also been tested, and in general very few chemicals are positive. We feel that chemicals to which humans are exposed which are clearly positive in the test should be considered potential human health hazards, and should be thoroughly tested in animal systems, and where extensive human exposure has occurred, appropriate epidemiologic studies should be done.

The economy of the bacterial/mammalian-microsomal assay suggests its usefulness as a tool in rapidly obtaining information about the potential mutagenic/carcinogenic activity of uncharacterized compounds in complex mixtures. Because quantitative information is obtained from the test (linear dose–response curves are usually observed) the test is quite valuable as a bioassay in identifying and purifying mutagenic components in complex mixtures. This has been useful in a detailed study that has been made of the mutagenic activity of cigarette smoke condensate and 12 standard smoke condensate fractions[19]. (The condensate from less than 0.01 cigarette could easily be detected). Recently, considerable mutagenic activity has been found in most commercial hair dyes[6], and several of the individual mutagenic components in the hair dyes have been identified. We have also shown that there is considerable mutagenic activity in soot from city air (with D. STREITWIESER).

We describe in this paper the general methods for using the Salmonella/microsome test as a mutagenesis screening system. The methods described include the standard plate test, the use and storage of the bacterial tester strains, preparation and use of the liver homogenates (S-9), and the methods of inducing the rats for elevated microsomal enzyme activity. The application of this test system to screening large numbers of compounds, and the interpretation of test results is also discussed.

METHODS

The bacterial tester strains

The strains used for mutagenesis testing are shown in Table I. There are several standard bacterial tester strains containing different types of histidine mutations. One strain (TA1535) can be used to detect mutagens causing base-pair substitutions and two (TA1537 and TA1538) to detect various kinds of frameshift mutagens. The molecular basis of the frameshift mutations in these strains has been investigated. Frameshift mutations occur by shifted pairing in repetitive sequences of DNA and frameshift mutagens can be very specific for the particular sequences they mutate. TA1538 has a repetitive –C–G–C–G–C–G–C–G– sequence near the site of the histidine mutation[17], and is reverted particularly well by many carcinogens, such as 2-nitrosofluorene. The other frameshift tester strain, TA1537, appears to have a run of C's at the site of the mutation[7] and, is reverted well by carcinogens such as 9-aminoacridine. In addition to the histidine mutation, each tester strain contains two additional mutations that greatly increase its sensitivity to mutagens: one causes loss of the excision repair system and the other loss of the lipopolysaccharide barrier that coats the surface of the bacteria[7].

TABLE I

GENOTYPE OF THE TA STRAINS USED FOR MUTAGEN TESTING*

Histidine mutation			*Additional mutations*			*Use*
hisG46	*hisC3076*	*hisD3052*	*LPS*	*Repair*	*R factor*	
TA1535	TA1537**	TA1538	*rfa*	*ΔuvrB*	—	Standard tester strains and R factor strains; recommended for general mutagenesis screening
TA100		TA98	*rfa*	*ΔuvrB*	+R	
(TA1975)	(TA1977)	TA1978	*rfa*	+	—	Used in combination with the standard tester strains in the *repair test****
hisG46	*hisC3076*	*hisD3052*	+	+	—	Additional related strains available
TA92		TA2420	+	+	+R	
TA1950	TA1952	TA1534	+	*ΔuvrB*	—	
TA2410			+	*ΔuvrB*	+R	
TA1530	TA1532	TA1964	*Δgal*	*ΔuvrB*	—	
TA2631		TA2641	*Δgal*	*ΔuvrB*	+R	
	TA2637		*rfa*	*ΔuvrB*	+R	

*All strains were originally derived from *S. typhimurium* LT2. Wild-type genes are indicated by a +. The deletion (Δ) through *uvrB* also includes the nitrate reductase (*chl*) and biotin (*bio*) genes. The *Δgal* strains (and the *rfa uvrB* strains) have a single deletion through *gal chl bio uvrB*. The *rfa*, repair$^+$ strains have a mutation in *galE*. R = pKM101. The standard tester strain TA1536, originally included in the tester set[7] and all other strains containing the histidine mutation *hisC207* have been deleted as they are reverted by only very few mutagens and these can be detected well by the other tester strains.

**A new tester strain is under development which will, with its R factor derivative, replace TA-1537.

***The TA1538/TA1978 pair is recommended.

We have recently developed[25] two new tester strains (TA100 and TA98) by transferring a resistance transfer factor (R factor) to our standard tester strains TA1535 and TA1538 respectively. These new strains are extremely effective in detecting classes of carcinogens that we previously had not detected with our original strains, and they are much more sensitive to a number of carcinogens we had previously detected only weakly[25] (McCANN *et al.*, in preparation). These carcinogens include aflatoxin B_1, sterigmatocystin, furylfuramide (the nitrofuran food additive AF-2) and other nitrofuran carcinogens, acetylenic diarylcarbamates, methyl methanesulfonate, nitroquinoline-*N*-oxide, benzo(*a*)pyrene, 7,12-dimethylbenzanthracene, benzyl chloride, 1′-acetoxysafrole, and a variety of polycyclic hydrocarbons.

For general mutagenesis testing we recommend that the three standard tester strains, TA1535, TA1537, and TA1538 be used in combination with the two new R factor strains, TA100 and TA98. For screening purposes TA1538 may be deleted, as TA98 appears to be more sensitive than TA1538 for the detection of all mutagens we have tested so far on the two strains[25]. The other frameshift tester strain, TA1537, is still useful, though we are developing a considerably more sensitive strain and its R factor containing derivative as a replacement. We recommend that TA1535 be used in addition to TA100 because it has a considerably lower spontaneous mutation rate and is thus more convenient and sensitive for the detection of mutagens that do not preferentially revert TA100[25].

Strain TA1978 can be used in the *repair test* in combination with TA1538[3,7]. This test is not a mutagenicity test, but indicates whether an agent is killing bacteria by damage to DNA that can be repaired by the *uvrB* excision repair system[3,7].

We are also developing several new tester strains which will be useful for the detection of special classes of mutagens: *(1)* a new tester strain with normal excision repair designed to detect certain DNA cross-linking agents such as the carcinogen mitomycin C[25] (with *E. coli*); and *(2)* new tester strains lacking nitro reductase enzymes. Many nitro carcinogens, such as nitrofurazone and furylfuramide, are activated directly to mutagens by bacterial nitro reductases. Mammalian nitro reductases can also activate this class of carcinogens. Bacterial nitro reductase mutants have been isolated by others (Ref. 22; T. KADA, personal communication from T. MATSUSHIMA), and in collaboration with M. VORE we are isolating nitro reductase mutants in our tester strains. These new strains will be useful for comparisons between bacterial and mammalian nitro reductase activities.

We would prefer that tester strains be obtained directly from us. We keep a list of people who are using the strains so that new developments in the test can be communicated. We also request that results obtained using the strains be sent to us, as we keep a running compilation of the chemicals tested on the strains.

Checking out the tester strains

We recommend that upon receiving a strain that a nutrient broth (8 g Difco-Bacto nutrient broth, 5 g NaCl/l) culture be inoculated and grown up. (The culture should also be streaked on a nutrient agar plate for storage until the liquid culture is checked out and found to be satisfactory.) A frozen permanent should be made from the nutrient broth culture as described in the next section, and the culture should also be checked in the following ways:

1) Histidine requirement. Streak the cultures on minimal plates both with and without histidine (spread 0.1 ml sterile 0.1 *M* L-histidine on the agar surface). For strains containing the *uvrB* deletion, biotin (0.1 ml of 0.5 m*M* per plate) is also required for growth.

2) Strains having the deep rough (*rfa*) character should be tested for crystal violet (and/or deoxycholate) sensitivity[7]. A sterile filter paper disc containing crystal violet (10 μl of 1 mg/ml) (or 2 mg deoxycholate) is placed on a nutrient agar petri dish containing 0.1 ml (about 10^8 bacteria) of the nutrient broth culture to be tested in a thin overlay of agar (*top agar*). After 12 h incubation at 37°, a clear zone of inhibition around the disc (about 14 mm diameter) indicates the presence of the *rfa* mutation, which permits large molecules such as crystal violet to enter the bacteria and inhibit growth. Wild-type strains, or strains containing the *gal* deletion (Table I) are not inhibited.

3) The new tester strains with R factors (TA100 and TA98) should be checked routinely for the presence of the ampicillin resistant R factor, as R factors are somewhat unstable and can be lost from bacteria[25]. A simple way to do this is to streak a small amount (10 μl of 8 mg/ml in 0.02 *N* NaOH) of an ampicillin solution across the surface of a nutrient agar plate. After the streak is dry, cultures to be checked are cross-streaked against the ampicillin, and after incubation for 12–24 h at 37°, strains which do not contain the R factor will show a zone of growth inhibition around the ampicillin streak, whereas R factor containing strains will not. Experiments confirming increased mutagenesis (with methyl methanesulfonate and aflatoxin B_1) compared to TA1535 and TA1538 are also recommended[25]. We also routinely include such positive mutagenesis controls when using these strains.

4) The *uvrB* deletion is quite stable and is not easily lost from bacteria, however its presence can be easily confirmed by checking the UV sensitivity of strains containing the *uvrB* deletion[7]. Cross-streak the culture to be checked on a nutrient broth agar plate and irradiate half of the streak for 6 sec (8 sec for strains containing the R factor) with a G.E. 15 W germicidal lamp at a distance of 33 cm. After incubation at 37° for 12–24 h the UV-sensitive strains will grow only on the un-irradiated side of the plate.
5) The spontaneous reversion rate should be checked and should be approximately as indicated in the section on *Mutagenesis assays on plates*. An abnormally high spontaneous reversion rate may indicate contamination. We discuss diagnostic mutagens for each tester strain in a later section (*Interpretation of Results*). These can be used for confirming the reversion properties of each strain.

To reisolate a strain, streak out on a nutrient agar plate and after incubation pick 5–10 single colony isolates and grow them up in nutrient broth. A new permanent can be prepared (see below) from the isolate with a low spontaneous reversion rate and with a good growth rate which should then be checked further as described above.

Storage of R factor and standard tester strains

This is at −80° after freezing on dry ice a fresh nutrient broth culture (0.8 ml) with dimethylsulfoxide (0.07 ml) in 2-ml (0.5 dram) sterile, glass screw-capped vials with rubber-lined screw caps. We prepare duplicate frozen cultures of each tester strain. One is stored as a master copy and is only opened when we need to regenerate our frozen stocks. The other is used routinely to obtain fresh cultures for mutagenesis testing by scraping a sterile wooden applicator stick over the surface of the frozen culture (which is not thawed), inoculating nutrient broth (5 ml), and shaking overnight (16 h maximum) at 37°. Fresh cultures can be kept in the refrigerator for a week. It is not desirable to subculture the tester strains because of selection for *rfa* revertants to wild-type and loss of the R factor. Strains not carrying R factors or *rfa* mutations are usually stored at room temperature as permanent stab cultures in 2-ml glass vials containing 1.2 ml soft agar (Difco-Bacto nutrient broth, 2 g; agar, 1.5 g; distilled H_2O, 250 ml) sealed with paraffin. This method can also be used for the tester strains (where a freezer is not available) if the cultures are checked out frequently.

Induction of rat liver enzymes for carcinogen activation

For general mutagenesis screening we recommend using liver homogenates (S-9) (9000 *g* supernatant) from rats induced with a polychlorinated biphenyl (PCB) mixture (Aroclor 1254)[1,13,15,20,27]. The induction procedure[19] is similar to the method of CZYGAN *et al.*[13]. We use male rats of about 200 g each (Sprague-Dawley/Bio-1 Strain, Horton Animal Laboratories). A single i.p. injection of Aroclor 1254 (diluted in corn oil to a concentration of 200 mg/ml) at a dosage of 500 mg/kg is given to each rat five days before sacrifice. The rats are given drinking water *ad libitum* and Purina Laboratory Chow until 12 h before sacrifice when the food is removed. On the fifth day of induction the rats are stunned by a blow to the head and then decapitated. Liver homogenates (S-9) are prepared as described in the following section[4].

Although we have, on occasion used other tissues (see later section), we find that, in general, rat liver is the most convenient source of activating enzymes. It is essential for efficient detection of a wide variety of carcinogens requiring metabolic activation that S-9 be obtained from induced animals. We have used various induction procedures[4,19] and find that liver from Aroclor-induced rats is the most efficient for detecting different classes of carcinogens. To illustrate the relative efficiencies of the different induction procedures, we have compared directly uninduced, phenobarbital, 3-methylcholanthrene, and Aroclor 1254 induced rat liver S-9 fractions for the activation of two polycyclic hydrocarbons (benzo(*a*)pyrene and 3-methylcholanthrene), and 2-acetylaminofluorene, an aromatic amine carcinogen. The results (Table II) indicate

TABLE II

INDUCTION OF RAT LIVER ENZYMES

Induction method	*µl S-9/plate*	*Revertants of TA1538 per plate**		
		5 µg Benzo(a) pyrene	*50 µg 3-methylchol-anthrene*	*50 µg 2-acetyl-aminofluorene*
Aroclor 1254	10	127	37	750
	20	318	71	3320
	50	443	206	9600
	100	189	349	3420
	150	146	297	4280
Phenobarbital	10	—	85	1050
	20	—	136	3020
	50	45	256	13,700
	100	23	356	14,700
	150	17	207	12,200
3-Methyl-cholanthrene	10	15	—	59
	20	121	—	620
	50	296	49	2045
	100	287	35	7187
	150	223	52	5640
Uninduced** (corn oil)	50	31	10	2344
	100	43	17	5824
	150	3	16	7184

The assays were performed as described in *Mutagenesis assays on plates* using S-9 samples which were prepared from 200 g rats as shown in *Induction of rat liver enzymes*.
*The number of spontaneous revertants (<30) have been subtracted from the appropriate experimental values. Assay plates were incubated for two days at 37° and then scored.
**The controls were 100 g rats which were injected with the vehicle, corn oil, which was used for the Aroclor 1254 and 3-methylcholanthrene induction. Induction with sodium phenobarbital (0.1% in the drinking water) and 3-MC (80 mg/kg; i.p.) was as described previously[4].

that, for the three compounds tested, induced rats are superior to uninduced animals, and of the three induction procedures, Aroclor induced S-9 is the most efficient, overall, for the detection of all three carcinogens. Although phenobarbital is a generally effective inducer and can be used for the efficient detection of 2-acetylaminofluorene, and many other aromatic amine carcinogens[4], it is very inefficient for detection of certain polycyclic hydrocarbons.

It is also important to note that for optimum mutagenesis with a particular compound the amount of S-9 per plate is critical and can be different. Too much as well as too little S-9 can drastically lower the sensitivity. We routinely check this variable when optimizing mutagenesis with a particular compound. We also routinely check each new S-9 preparation with several compounds, such as benzo(*a*)pyrene and an aromatic amine, *e.g.* 2-acetylaminofluorene or 2-aminofluorene, and aflatoxin B_1 to find the optimum amount of S-9 for general screening. In general, we find from Aroclor induced rats that 20–60 μl of S-9 per plate (0.04–0.1 ml of S-9 per ml of S-9 mix) is a good amount for general screening.

Preparation of liver homogenate fraction ("S-9")

We have used basically the procedure of GARNER *et al.*[16]. All steps are at 0–4° using cold, sterile solutions and glassware. The livers (rat livers are about 10–15 g each) are placed in beakers (pre-weighed) containing 0.15 *M* KCl (approx. 1 ml/g wet liver). After weighing, livers are transferred to a beaker containing 3 vol. of 0.15 *M* KCl (3 ml/g wet liver), minced with sterile scissors, and homogenized in a Potter-Elvehjem apparatus with a Teflon pestle. The homogenate is centrifuged for 10 min at 9000 *g* (8700 rev./min in SS-34 head of Sorvall RC2-B) and the supernatant, which we call the S-9 fraction, is decanted and saved. 1 ml of S-9 fraction contains microsomes from about 250 mg of wet liver; protein concentrations were fairly constant from preparation to preparation (about 40 mg/ml: LOWRY). The fresh S-9 fractions are distributed in 2-ml portions in small plastic tubes (2-ml liquid nitrogen storage tubes/4-Shore-USA, La Jolla, Calif.), quickly frozen in dry ice, and stored at −80° in a Revco freezer. As required, sufficient S-9 fraction is thawed (at room temperature) and kept on ice; the unused portion is discarded at the end of the day. The extent of bacterial contamination of the S-9 should be determined: if necessary the S-9 mix may be filter sterilized (see following section).

S-9 from other tissues and animals

Microsomal preparations can be made from various tissues of different animals. We have compared rat and human autopsy liver[4,6] and rat liver and lung[19] and also made preparations from mouse liver and mammary gland (J. RICHARDS, unpublished). The preparations were similar to that from rat liver except for problems of sterility (with lungs) and homogenization difficulties with lungs and human liver. Because of the fibrous nature of lung tissue, the lungs were homogenized at 0–4° for 5 min in a Polytron tissue homogenizer (Brinkmann Instruments, Westbury, N.Y.) rather than the Potter-Elvehjem apparatus used for the liver[19]. (This homogenizer would probably be useful for any tissue that is fibrous or difficult to homogenize.) The S-9 fraction was then prepared from the homogenate in the same manner as the liver fraction. Although liver preparations are usually sterile, lung preparations contained bacterial contaminants. These were removed by passage of the lung S-9 mix through a sterile Swinex Filter unit (Millipore Corp. Bedford, Mass.) equipped with a Millipore filter (0.45-μm pore size). Nalgene disposable membrane filter units are used for larger amounts. (S-9, before dilution into the mix, clogs the filter and it is better to filter the S-9 mix.)

It should be emphasized that in order to compare quantitatively two different S-9 preparations for activation of a carcinogen (this can be for different tissues of the same animal, the same tissue from different animals, or different conditions of induction) a one point comparison is not sufficient (*e.g.* Table II). As in any enzyme assay the activity (in this case histidine revertants) must be proportional to the amount of enzyme (S-9) added in order to determine specific activity. Thus a multi-point comparison is necessary. It is also advisable to do the comparisons at several concentrations of carcinogen. As can be seen from Fig. 1A if one did a one point comparison of

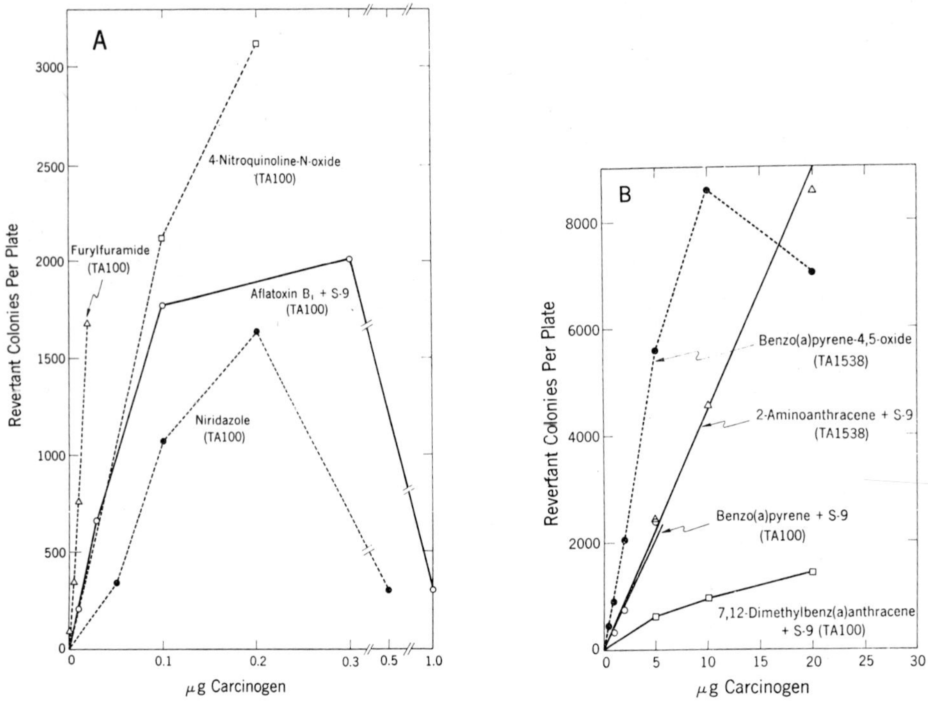

Fig. 1. A and B. The number of spontaneous revertants (varying from <30 for TA1538 to 160 for TA100) have been subtracted from the revertant values plotted against amount of carcinogen. Freshly prepared solutions of the carcinogens, and where indicated, S-9 mix, were incorporated into top agar as described in Methods.

aflatoxin activation at any concentration above 0.3 μg of aflatoxin the more active S-9 preparation might appear as less active.

There have been suggestions that in general screening of chemicals several S-9 preparations from different tissues or different animals should be used. We have seen no data so far that convinces us that the considerable extra work involved if this procedure is followed will in fact result in detection of mutagens that would not be detected using rat liver Aroclor-induced S-9 following the standard procedure we have outlined above.

Preparation of S-9 Mix[4]

S-9 mix contains per ml: S-9 (0.04–0.1 ml) (see Induction section), $MgCl_2$ (8 μmoles), KCl (33 μmoles), glucose-6-phosphate (5 μmoles), NADP (4 μmoles), and sodium phosphate, pH 7.4 (100 μmoles). Stock solutions of NADP (0.1 *M*) and glucose-6-phosphate (1 *M*) are prepared with sterile water in sterile tubes and stored at −20°. The stock salt solution (0.4 *M* $MgCl_2$, 1.65 *M* KCl) and phosphate buffer (0.2 *M*, pH 7.4) are autoclaved and stored in the refrigerator. S-9 mix is freshly prepared each day and may be kept on ice several hours before use.

Mutagens and carcinogens

Solutions are prepared in disposable sterile polystyrene screw-cap tubes. We generally dissolve compounds which are not water soluble in dimethylsulfoxide

(Schwarz-Mann, spectrophotometric grade, sterile as is). Less than 0.5 ml per plate does not interfere with mutagenesis, or with microsomal activity. We use fairly small bottles of dimethylsulfoxide (100 ml) which we only entei with sterile pipettes and which we keep dry by closing the top immediately after use. *p*-Dioxane or ethanol can also be used as solvents, and up to 100 μl of either solvent can be incorporated into the top agar overlay without inhibiting the lawn of bacterial growth. Most mutagen solutions can be stored at $-20°$ in the feezer or in the refrigerator for extended periods. Solutions of highly reactive compounds, such as alkylating agents, however, should be freshly prepared.

Top agar and petri plates

Top agar (0.6% Difco agar, 0.5% NaCl) is autoclaved and stored in bottles in volumes of 100 ml at room temperature. Before use the agar is melted by heating the bottle in a steam bath, and 10 ml of a sterile solution of 0.5 m*M* L-histidine·HCl–0.5 m*M* biotin is added to the molten top agar and mixed thoroughly by gentle swirling.

The trace of histidine in the top agar allows all the bacteria on the plate to undergo several divisions; *this growth is necessary in many cases for mutagenesis to occur*. The slight background that grows up also allows any inhibition by the compound to be seen. Further increase in the amount of histidine on the plate enhances mutagenesis, but also causes heavy growth of the background lawn that obscures the revertants.

We use plastic or metal caps (Bellco, Vineland, N.J.) for both the tubes and the agar bottles. These can be removed easily and are more convenient than cotton plugs or screw caps.

The petri plates (100×15 mm style, Falcon Plastics, Oxnard, Cal.) contain 30 ml of minimal-glucose agar medium. We use 1.5% Bacto-Difco agar in Vogel-Bonner Medium E[29] with 2% glucose, but other minimal media would presumably also serve. The plates can be stored at room temperature for several weeks. We have been informed (M. Nagao, personal communication) that plastic petri plates sterilized by the mutagen ethylene oxide can give rise to a high "spontaneous" rate for strains TA1535 and TA100 if there is residual ethylene oxide in the plates, though we have not had difficulty with this problem with the plates we have used.

Mutagenesis assay on plates[7,4]

The following are added (in order)to 2 ml of molten top agar at 45°: 0.1 ml of an overnight nutrient broth culture of the bacterial tester strain, the sample to be tested (usually in 0.1 ml or less), and 0.5 ml of the S-9 mix (if required). The bacteria can remain at 45° for a few minutes without harm, but the S-9 mix should not be left at this temperature for more than a few seconds. The contents are mixed (by rotating the tube between the palms) and poured on minimal glucose agar plates. Uniform distribution of the top agar on the surface of the plates is accomplished by gently tilting and rotating the uncovered plate and then setting the plates down to harden. The mixing, pouring and distribution should take less than 20 sec and the plates should be left to harden for several minutes. It is important to follow these time limits. If the top agar starts to harden in mid-operation a stippled plate surface will result which makes scoring of revertants difficult. We cover the plates promptly to avoid any light effects on photosensitive chemicals. Within an hour the plates should be put in a dark, 37° incubator. After 2 days the colonies (revertants to histidine prototrophy) in both test plates and controls are counted, and the presence of the light background lawn of growth (due to the trace amounts of histidine added) is confirmed. Spontaneous revertant colonies on control plates without mutagen and S-9 mix are about (at 48 h): 20 (TA1535), 7 (TA1537), 25 (TA1538), 160 (TA100), and 40 (TA98). Slightly larger numbers of spontaneous revertant colonies arise on plates containing the S-9 mix[4].

When mutagens are to be tested in a spot test the mutagen is left out of the top agar and instead is applied to the plate surface after the top agar containing the bacteria and S-9 mix is poured. The mutagen can be added to the agar surface as crystals, or a micro drop (about 10 μl). We also have used sterile 6-mm filter paper discs, though, some chemicals, such as daunomycin, are detected much less efficiently on discs, possibly due to adsorption. The plates can be stored at room temperature for up to an hour before applying the mutagen, but as the bacteria will begin to grow, it is best to add the mutagen promptly after the top agar hardens.

Testing of volatile compounds in the plate test

The testing of volatile, relatively water insoluble compounds, whether gas or liquid, cannot be done quantitatively in the manner described earlier for pour plate incorporation of solutions of chemicals. By a modification of the plate test, RANNUG *et al.*[26] were able to demonstrate the mutagenicity of vinyl chloride, a gas, by exposing petri plates having TA1535 and *S-9 Mix* in the soft agar overlay to known vinyl chloride/air mixtures in a 10 l desiccator for various lengths of time. BARTSCH *et al.*[10] also tested vinyl chloride using this method, and later showed the method applicable, to the testing of vinylidene chloride and 2-chlorobutadiene[11].

Establishing desired concentrations (v/v) of the compound in air can be accomplished in several ways:

(a) Evacuating the desiccator, introducing a known volume of the gas to be tested, then allowing air to enter until atmospheric pressure is reached.

(b) Using a manometer attached to the desiccator by a 3-way valve, remove a known volume of air, then allow the gas to enter until atmospheric pressure is reached.

(c) Using manometers, or flowmeters, measure the flow rates of the gas and compressed air into a mixing flask which then feeds the mixture into the desiccator.

(d) For liquids, the calculated weight to give a known volume of vapor is frozen and placed in the desiccator. The desiccator is partially evacuated and sealed, allowing the solid to vaporize. Then air is introduced until atmospheric pressure is reached.

A magnetic stirrer should be included in the sealed container to maintain a homogeneous mixture. The incubation is carried out at 37°. After the desired time of exposure, the compound is removed by three evacuations and refillings with air. Further quantitation can be attained by analyzing a sample of the gas/air mixture by GLC. After 48 h of total incubation time, the plates should be scored for revertants. (We are indebted to H. BARTSCH for contributions to this section.)

Interpretation of results: plate incorporation assay

In this method the mutagen is added directly to the molten top agar and is poured on the plate along with the S-9 mix and bacteria. This is the standard method that has been used for validating the test using hundreds of chemicals. For initial screening of a chemical we recommend testing concentrations over a wide dose range (say 0.2, 2, 20 and 500 μg per plate) both in the presence and absence of the standard S-9 mix (see *Induction and preparation of S-9 mix*), using the tester strains TA1535, TA100, TA1537 and TA98. A positive or questionable result should then be confirmed by demonstrating a dose-response effect using a narrower range of concentrations. A few dose–response curves are shown in Fig. 1. In general, we find that for most mutagens a concentration range exists when there is a linear dose–response, and the revertants per plate reported for any mutagen should be taken from this region of the curve. (We have occasionally obtained dose–response curves that are not linear: with 9-aminoacridine, MNNG, diethylsulfate and ethyl methanesulfonate[25]). Excessive bacterial killing by a mutagen causes a decrease in the number of revertants on the plate, and it can be seen (Fig. 1) that if one were to test just one concentration on the downhill part of a curve one could easily obtain a misleading result as to the quantitative activity of a particular mutagen.

Routine examination of the background lawn of bacterial growth resulting from the trace of histidine added to the top agar (see *Top agar and petri plates*) can be an aid in determining the presence of toxic effects. If massive cell death has occurred, the background lawn on the test plates will be sparse compared to control plates. In this case more histidine is available to the individual surviving bacteria and they undergo more cell divisions, consequently appearing as small colonies which can be mistaken for revertants if the absence of a normal background lawn is not noted.

Although the standard amount of S-9 recommended for general testing should permit detection of a wide variety of chemical mutagens requiring metabolic activation, to maximize reversion it is advisable to determine the optimum amount of S-9 in the S-9 mix required for activation, as this varies with the type of compound tested and its concentration on the petri plate (see *Induction* section).

In each experiment we routinely include positive mutagenesis controls using diagnostic mutagens to confirm the reversion properties of each strain. The characteristic reversion patterns of the standard strains to some diagnostic mutagens are shown in Table III. Positive controls using chemicals requiring metabolic activation confirm that the S-9 mix is active. For this purpose 2-aminofluorene or aflatoxin B_1 can be spot tested, as indicated above, but the polycyclic hydrocarbons, such as benzo(*a*)pyrene (5 μg should be sufficient for activation) which do not diffuse in the agar must be incorporated directly into the agar overlay. Sterility controls (solutions

TABLE III

DIAGNOSTIC MUTAGENS: FOR CONFIRMING REVERSION PROPERTIES OF TESTER STRAINS

Mutagen	*Amount spotted*	*S9*	*TA1535*	*TA100*	*TA1537*	*TA98*	*TA1538*
Methyl methanesulfonate	2 μl	−	−*	+++	−	−*	−
4-Nitroquinoline-*N*-oxide	10 μg	−	−*	+++	−	++	−*
MNNG	2 μg	−	+++	+++	−*	−	−*
9-Aminoacridine	10 μg	−	−	−	++	−	−
Daunomycin	5 μg	−	−	−	−	++	−
2-Aminofluorene	100 μg	+	−	+++	−	+++	+++
Aflatoxin B_1	1 μg	+	−	++	−	++	−*

All chemicals were dissolved in DMSO except for 9-aminoacridine which was in ethanol
−, <20; ++, >10; +++, >500 colonies in the spot test.
* We have combined slight positives with the negatives for the purpose of this table.

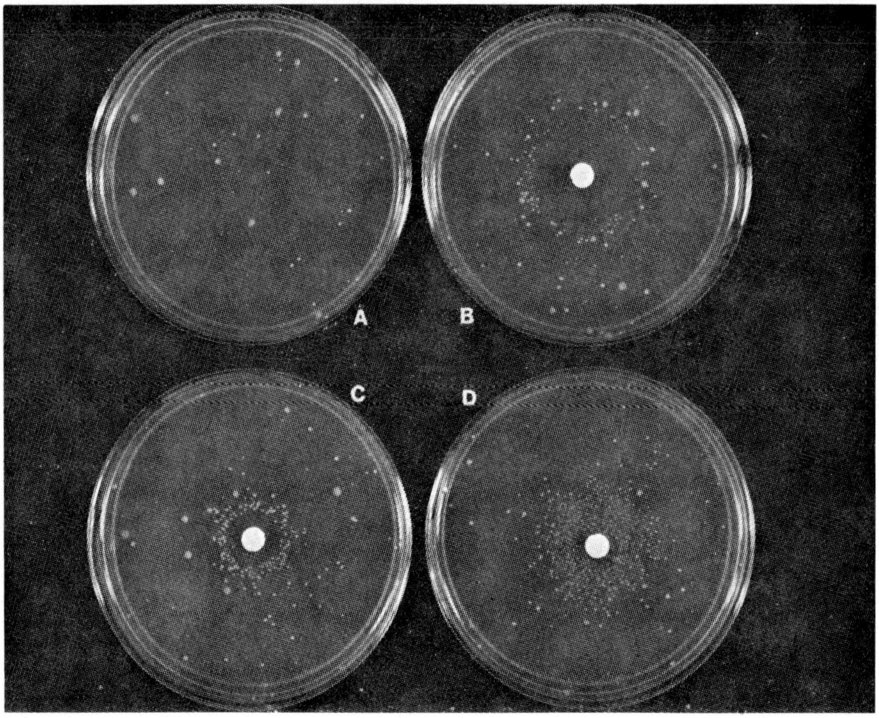

Fig. 2. The spot test. Each petri plate contains, in a thin overlay of top agar, the tester strain TA98, and in the case of plates C and D, a liver microsomal activation system (S-9 Mix) from aroclor-induced rats, (see text). Mutagens were applied to 6 mm filter paper discs which were then placed in the center of each plate: (A) spontaneous revertants, (B) furylfuramide (AF-2) (1 μg), (C) aflatoxin B_1 (1 μg), and (D) 2-aminofluorene (10 μg). Mutagen induced revertants appear as a ring of colonies around each disc.

of new test chemicals, and S-9 mix) are also routinely included in each experiment. It is important to do a control plate without the test chemical and with S-9 mix and bacteria as the small amount of histidine in the S-9 mix can influence the number of spontaneous revertants.

We would consider a chemical to have a negative response in the test if at least 500 μg had been tested on the plate (or the maximum non-inhibitory level) and if the number of induced revertants compared to the spontaneous was less than 2-fold. For compounds of low mutagenicity it is important to obtain reproducible dose -response curves.

Interpretation of results: spot tests

Spot tests are the simplest way to test compounds for mutagenicity and are particularly adaptable for the initial rapid screening of large numbers of compounds in a short period of time. A few examples of spot tests are shown in Fig. 2. We have tested 169 different hair dyes for mutagenicity using this method[6]. We have also used it in student laboratory experiments. There are several advantages of the spot test and it is often useful to test all new compounds by this method before doing the standard plate incorporation test. It is not necessary to prepare solutions of the chemicals to be tested, as a few crystals (or μl of a liquid) can be put directly on the agar surface; also, as the compound diffuses out from the central spot a range of concentrations are tested simultaneously. The spot test affords a preliminary indication of the toxicity of the chemical for the bacteria by the size of the zone of inhibition of the background lawn of bacterial growth around the spot and an indication whether the S-9 mix is necessary for mutagenicity, and in the case of a positive result, which tester strain should be used for the dose–response curve. The method can also be used to test samples which are not completely sterile, provided the colonies directly under the spot (or touching the paper disc) are not counted.

The spot test is primarily a qualitative test, and although very useful, has distinct limitations. It can only be used for the detection of chemicals which are diffusable in the agar, and thus most polycyclic hydrocarbons and other water insoluble chemicals are not easily detected by this procedure. It is also much less sensitive than the standard plate test (see above) as only relatively few bacteria on the plate are exposed to the chemical at any particular dose level.

A positive result in a spot test should be considered adequate evidence for mutagenicity only if there is a large increase in colonies (many times the number of spontaneous revertants) well away from the spot. In general, however, we believe that mutagenicity should be confirmed by demonstrating a dose-response effect using the standard plate incorporation assay.

Mutagenesis tests in liquid

The original experiments of MALLING[21] on dimethylnitrosamine activation and mutagenesis of the *Salmonella* strain *hisG46* (the mutation in TA1535) were in liquid suspension, and in fact detection of dimethyl- and diethylnitrosamine in the *Salmonella* test requires preincubation with the S-9 mix and the bacteria in liquid suspension before plating. An efficient and simple procedure is to follow the protocol described for the standard plate incorporation assay with the slight modification that the top agar is added *last*, after pre-incubating the other components 20 min at room temperature (T. MATSUSHIMA, D. R. STOLTZ, personal communications). It is important to add the carcinogen in the minimum amount of DMSO, or other organic solvent, so as not to inactivate the bacteria or S-9 during the 20 min. This slight modification of the standard plate test may be included as an addition to the screening protocol for completeness. However, at present liquid suspension procedures have several serious disadvantages for general mutagenesis testing, and should only be used as an adjunct to the standard plate test. Relatively few mutagens and carcinogens have been tested in liquid suspension tests, so the ability of this type of test to detect a wide variety of carcinogens is by no means clear. In fact, of the few chemicals tested dimethyl- and diethylnitrosamine are more efficiently detected in liquid suspension than in the standard plate assay[9, 17, 21], and all other chemicals tested, of which we are aware, including other *N*-nitrosamines such as dibutylnitrosamine and polycyclic hydrocarbons (H. BARTSCH, personal communication) which are efficiently detected in the plate test, are very inefficiently or not at all detected in a liquid suspension test.

It is not unreasonable that liquid suspension procedures would be relatively inefficient compared to a plate test for detection of many mutagens. Liquid mutagenesis procedures usually require incubation of the suspension at 37°, and at this temperature the microsomes are active for a relatively short period of time, whereas in the agar overlay in the plate test they remain active up to 9 h at 37° (ref. 10). We think this is due to stabilization of the immobilized enzymes in the agar layer. In addition, for many frameshift mutagens efficient mutagenesis requires growing bacteria for both induction and expression of the mutation. This is accomplished on the plates by having the bacteria undergo several generations growth on the trace of histidine incorporated in the top agar. Liquid suspension procedures designed to determine mutagenesis relative to killing usually involve exposure to mutagens in the absence of bacterial growth and plating of bacteria directly on medium lacking any histidine[16]. Under these conditions the degree of mutagenesis observed can be severely reduced.

Detection of mutagenic metabolites in urine

A wide variety of metabolites of drugs and other ingested compounds appear in the urine and we would like to expand the Salmonella test method as a general screening procedure for the detection of mutagenic metabolites in urine. We, and others, have reported procedures for the detection in urine of mutagenic metabolites of 2-acetylaminofluorene[12,14]. The addition of commercial β-glucuronidase to the petri plates along with the urine, liver homogenate, and bacteria allows detection of metabolites of these carcinogens which are excreted in urine primarily as β-glucuronide conjugates. By this method mutagenic activity is readily demonstrated with urine of rats administered as little as 200 μg (1.6 mg/kg) of 2-acetylaminofluorene[14]. We have recommended that testing urine for mutagenicity be done routinely in the case of experimental animals given a chemical in toxicology testing[14]. This would be a worthwhile addition to any mutagenesis screening program using the Salmonella test system. The urine test has been shown to detect several carcinogens requiring metabolic activation for mutagenic activity which are not efficiently detected using the standard *in vitro* liver microsomal method of activation. These are dimethylaminoazobenzene (butter yellow)[12], isoniazid[29], and safrole (H. ROSENKRANZ, personal communication). This method is potentially of great value and should be adaptable for the screening of human urines in order to detect mutagenic metabolites of drugs and of dietary components, and for testing urinary metabolites of drugs and food additives in experimental animals. However, as recently discussed[24], there are several technical problems which must be surmounted before any large scale application of the method for humans is justified. Of primary importance is the determination of an optimum extraction procedure for the detection of a variety of metabolites, and the need for, as in the case of liquid assay procedures (see above) validation for efficiency in detecting a large variety of carcinogens.

Disposal of carcinogen and mutagen waste

We recommend using disposable test tubes, petri plates and micropipettes in order to avoid re-circulation of these items with the rest of the laboratory apparatus. The disposal of these mutagen/carcinogen-contaminated materials is by contract with the firm which disposes of radioactive waste from our laboratory.

Safety precautions in the handling of carcinogens/mutagens

Every effort is made in this laboratory to isolate the areas where carcinogens/mutagens are used in order that the entire laboratory is not contaminated unknowingly by these compounds. A special area is designated in a well-ventilated hood where handling of solid and volatile liquid compounds is done during the preparation of solutions or during an experiment when the carcinogens/mutagens are incorporated in or spotted on the top agar overlay. The investigator wears disposable plastic gloves for these manipulations and all weighings are done by weight difference determination in order that the container of solid or liquid compound is not opened in the laboratory outside of the hood. A 37° incubator is used only for carcinogen/mutagen-containing petri plates. It has been adapted with a hose connection to a fan which draws a slight flow of air through the incubator to the hood. The fan motor is activated by a microswitch on the incubator door so that it operates whenever the petri plates are moved in or out of the incubator. This precaution is coupled with the use of desiccator jars in which the petri plates are incubated whenever dangerous volatile compounds are assayed.

Safety precautions in the handling of the Salmonella tester strains

Salmonella typhimurium can cause diarrhea and food poisoning. The particular *S. typhimurium* strain, LT2, the parent of all of our tester strains, is used by geneticists all over the world and is not very virulent. The deep rough mutation (*rfa*) present in the standard tester strains (see Table I) lowers virulence by orders of magnitude and these strains should be relatively harmless. The pKM101 plasmid, with one antibiotic resistance marker for ampicillin, and contained in the tester strains TA98 and TA100 should be a minimal hazard. Plasmids are extremely common in the enteric population in nature and most of the RTF plasmids isolated from Salmonella in hospitals contain many antibiotic resistance genes. Nevertheless, as a matter of routine with any Salmonella strain we use plugged pipettes and autoclave any material containing Salmonella before it is washed or disposed of. (The material containing carcinogens as well as Salmonella is not autoclaved and is disposed of as previously described.) It is also a laboratory rule not to keep any food material in the same refrigerators or area with carcinogens or Salmonella.

It may be prudent to keep the Salmonella away from mouse colonies.

Chemicals

Benzo(*a*)pyrene, MNNG, and 2-acetylaminofluorene were from Aldrich. 3-Methylcholanthrene was from Eastman, and Mallinckrodt was the source of sodium phenobarbital. 2-Amino-anthracene, 2-aminofluorene, and 4-nitroquinoline-1-oxide were from Schuchardt (Munich) and 7,12-dimethylbenz(*a*)anthracene, daunomycin, and 9-aminoacridine were from Sigma. Aflatoxin B_1 was from Calbiochem. The following were gifts: furylfuramide [2-(2-furyl)-3-(5-nitro-2-furyl)acrylamide: trade name AF-2] from T. SUGIMURA (National Cancer Center Research Institute, Tokyo); benzo(*a*)pyrene-4,5-oxide from P. L. Grover and P. Sims (Chester Beatty Research Institute, London); and niridazole from E. BUEDING (The Johns Hopkins University, Baltimore). Aroclor 1254 is available in 50 ml samples at no charge from W. B. PAPAGEORGE, Monsanto Chemical Co., St. Louis, Mo. 63166, and is also available commercially from Analabs, Inc. (North Haven, Conn).

Cost of determining mutagenicity

We estimate that once set up and working regularly one person ought to be able to test thoroughly several compounds a day and that the cost should be in the range of $100 to $300 a compound counting time and material but not overhead. This would include testing on four tester strains and, after the initial spot tests and set of concentrations, doing dose–response curves with those compounds that are positive. Spot testing only would miss some compounds but would of course be much cheaper.

ACKNOWLEDGEMENT

We would like to thank numerous collaborators and colleagues for their help with the material in this paper, in particular W. E. Durston, F. Lee, P. E. Hartman, D. Streitwieser, E. Choi and G. Jen

REFERENCES

1 Alvares, A. P., D. R. Bickers and A. Kappas, Polychlorinated biphenyls: a new type of inducer of cytochrome P-448 in the liver, *Proc. Natl. Acad. Sci. (U.S.A.)* 70 (1973) 1321–1325.
2 Ames, B. N., The detection of chemical mutagens with enteric bacteria, in A. Hollaender (Ed.), *Chemical Mutagens: Principles and Methods for their Detection*, Vol. 1, Plenum, New York (1971) 267–282.
3 Ames, B. N., A bacterial system for detecting mutagens and carcinogens, in H. E. Sutton and M. I. Harris (Eds.), *Mutagenic Effects of Environmental Contaminants*, Academic Press, New York (1972) 57–66.
4 Ames, B. N., W. E. Durston, E. Yamasaki and F. D. Lee, Carcinogens are mutagens: a simple test system combining liver homogenates for activation and bacteria for detection, *Proc. Natl. Acad. Sci. (U.S.A.)*, 70 (1973) 2281–2285.
5 Ames, B. N., E. G. Gurney, J. A. Miller and H. Bartsch, Carcinogens as frameshift mutagens: metabolites and derivatives of 2-acetylaminofluorene and other aromatic amine carcinogens, *Proc. Natl. Acad. Sci. (U.S.A.)*, 69 (1972) 3128–3132.
6 Ames, B. N., H. O. Kammen and E. Yamasaki, Hair dyes are mutagenic: identification of a variety of mutagenic ingredients, *Proc. Natl. Acad. Sci. (U.S.A.)* 72 (1975) 2423–2427.
7 Ames, B. N., F. D. Lee and W. E. Durston, An improved bacterial test system for the detection and classification of mutagens and carcinogens, *Proc. Natl. Acad. Sci. (U.S.A.)*, 70 (1973) 782–786.
8 Ames, B. N., P. Sims and P. L. Grover, Epoxides of polycyclic hydrocarbons are frameshift mutagens, *Science*, 176 (1972) 47–49.
9 Bartsch, H., C. Malaveille and R. Montesano, *In vitro* metabolism and microsome-mediated mutagenicity of dialkylnitrosamines in rat, hamster, and mouse tissues, *Cancer Res.*, 35 (1975) 644–651.
10 Bartsch, H., C. Malaveille and R. Montesano, Human, rat and mouse liver-mediated mutagenicity of vinyl chloride in *S. typhimurium* strains, *Int. J. Cancer*, 15 (1975) 429–437.
11 Bartsch, H., C. Malaveille, R. Montesano and L. Tomatis, Tissue- mediated mutagenicity of vinylidene chloride and 2-chlorobutadiene in *Salmonella typhimurium*, *Nature*, 255 (1975) 641–643.
12 Commoner, B., A. J. Vithayathil and J. I. Henry, Detection of metabolic carcinogen intermediates in urine of carcinogen-fed rats by means of bacterial mutagenesis, *Nature (London)*, 249 (1974) 850–852.
13 Czygan, P., H. Greim, A. J. Garro, F. Hutterer, F. Schaffner, H. Popper, P. Rosenthal and D. Y. Cooper, Microsomal metabolism of dimethylnitrosamine and the cytochrome P-450 dependency of its activation to a mutagen, *Cancer Res.*, 33 (1973) 2983–2986.
14 Durston, W. E. and B. N. Ames, A simple method for the detection of mutagens in urine: studies with the carcinogen 2-acetylaminofluorene, *Proc. Natl. Acad. Sci. (U.S.A.)*, 71 (1974) 737–741.
15 Ecobichon, D. J. and A. M. Comeau, Comparative effects of commercial aroclors on rat liver enzyme activities, *Chem-Biol. Interact.*, 9 (1974) 341–350.
16 Garner, R. C., E. C. Miller and J. A. Miller, Liver microsomal metabolism of aflatoxin B_1 to a reactive derivative toxic to *Salmonella typhimurium* TA1530, *Cancer Res.*, 32 (1972) 2058–2066.
17 Gletten, E., U. Weekes and D. Brusick, *In vitro* metabolic activation of chemical mutagens. I. Development of an *in vitro* mutagenicity assay using liver microsomal enzymes for the activation of dimethylnitrosamine to a mutagen, *Mutation Res.*, 28 (1975) 113–122.
18 Isono, K. and J. Yourno, Chemical carcinogens as frameshift mutagens: *Salmonella* DNA

sequence sensitive to mutagenesis by polycyclic carcinogens, *Proc. Natl. Acad. Sci., (U.S.A.)*, 71 (1974) 1612–1617.

19 KIER, L. D., E. YAMASAKI AND B. N. AMES, Detection of mutagenic activity in cigarette smoke condensates, *Proc. Natl. Acad. Sci. (U.S.A.)*, 71 (1974) 4159–4163.

20 LITTERST, C. L. AND E. J. VAN LOON, Time-course of induction of microsomal enzymes following treatment with polychlorinated biphenyl, *Bull Environ. Contam. Toxicol.*, 11 (1974) 206–212.

21 MALLING, H. V., Dimethylnitrosamine: formation of mutagenic compounds by interaction with mouse liver microsomes, *Mutation Res.*, 13 (1971) 425–429.

22 MCCALLA, D. R. AND D. VOUTSINOS, On the mutagenicity of nitrofurans, *Mutation Res.*, 26 (1974) 3–16.

23 MCCANN, J. AND B. N. AMES, A simple method for detecting environmental carcinogens as mutagens, *Ann. N.Y. Acad. Sci.*, in press.

24 MCCANN, J. AND B. N. AMES, The detection of mutagenic metabolites of carcinogens in urine using the Salmonella/microsome test, *Ann. N.Y. Acad. Sci.*, in press.

25 MCCANN, J., N. E. SPINGARN, J. KOBORI AND B. N. AMES, Detection of carcinogens as mutagens: bacterial tester strains with R factor plasmids, *Proc. Natl. Acad. Sci., (U.S.A.)* 72 (1975) 979–983.

26 RANNUG, U., A. JOHANSSON, C. RAMEL AND C. A. WACHTMEISTER, The mutagenicity of vinyl chloride after metabolic activation, *Ambio.*, *3* (1974) 194–197.

27 SCHMOLDT, A., H. F. BENTHE AND R. FRÜHLING, Induction of rat liver enzymes by polychlorinated biphenyls (PCBs) in dependence on the dose and chlorine content, *Arch. Toxicol.*, 32 (1974) 69–81.

28 STOLTZ, D. R. AND C. T. MILLER, Mutagenic activity in urine concentrates from rats orally dosed with a chemical which itself is inactive in the Ames test with microsomal activation, *Abstr. Environ. Mutagen Soc.* (1975) Ac-11.

29 VOGEL, H. J. AND D. M. BONNER, Acetylornithinase of *E. coli*: partial purification and some properties, *J. Biol. Chem.*, 218 (1956) 97–106.

METHODS FOR STUDYING EXCISION REPAIR OF DNA DAMAGED BY PHYSICAL AND CHEMICAL MUTAGENS *

JAMES E. CLEAVER

Laboratory of Radiobiology, University of California, San Francisco, Calif. 94143 (U.S.A.)

Contents

---

* This chapter is revised from "Methods for studying repair of DNA damaged by physical and chemical carcinogens", in H. Busch (ed.), Methods in Cancer Research, Vol. 9, Academic Press, New York, 1975, pp. 123–165.

Introduction

Exposure of cells to mutagens produces chemical changes (damaged sites) in most macromolecules, including DNA, RNA, and proteins; the damaged sites in DNA seem to be paramount in determining the mutagenic effect of the exposure. Once damage has been induced in DNA, cells experience an acute response in which many biochemical reactions are disturbed. After this acute response, surviving cells recover by virtue of repair processes that eliminate the damage or render it innocuous. Among the survivors some cells show mutations, which were generated by errors that occurred in DNA replication during the recovery period [104,105]. A simple but reasonable scheme by which exposure to mutagens induces mutations is the following sequence:

(a) Chemicals penetrate cells and become activated as electrophilic reagents; radiations produce reactive radicals or excited molecules.

(b) Activated chemicals react with numerous nucleophilic sites in DNA, RNA, and proteins; radiation-induced products lead to chemical changes in these molecules, thus damaging them.

(c) Damaged RNA and proteins are degraded and replaced by de novo synthesis; damaged DNA is repaired by excision of the damage and synthesis of replacement regions (excision repair).

(d) Damage that either cannot be repaired or has not yet been repaired interferes with semiconservative replication, stimulating post-replication repair, by which errors, and hence mutations, are induced.

Because replication errors in damaged DNA are likely to be directly involved in inducing mutations, it would be useful to have some rapid biochemical methods for measuring DNA damage and repair to help evaluate the possible mutagenic effect of radiations and chemicals. The most convenient biochemical measurements that can be made are measurements of the amount of excision repair. The nature of this repair process and methods by which it can be measured will therefore be discussed in detail.

Induction and classes of mutagen-induced damage to DNA

Chemical mutagens undergo many chemical and/or enzymatic reactions in cells and react as electrophilic reagents with nucleophilic sites in cellular components, especially DNA, RNA, and proteins. Some chemical mutagens, such as dimethyl sulfate and methyl methanesulfonate, are themselves electrophilic reagents, but others, such as benz(*a*)anthracene and acetylaminofluorene, undergo metabolic activation to become electrophilic derivatives. Nucleophilic sites are available in the DNA bases and the amino acid residues of proteins and include the O^6, N^3, N^7, and C^8 positions of guanine, the N^1 and N^3 positions of adenine, the N^3 position of cytosine, the N^3 and O^4 positions of thymine, and the S atoms of methionine and cysteine, the C^3 position of tyrosine, and the N^1 and N^3 positions of histidine [69]. Chemical mutagens can also react with the phosphate groups in polynucleotides to form phosphotriesters.

Ionizing radiations induce changes in DNA mainly through free radical reactions that produce modified deoxyribose residues, modified bases, and strand breaks. UV light is absorbed by unsaturated bonds in nucleic acids and proteins,

TABLE I

KINDS OF MUTAGEN-INDUCED MODIFICATION TO DNA

Nature of damage	Representative mutagens
Base damage	
Cyclobutane pyrimidine dimers	UV light
Alkylation (e.g., O^6-alkylguanine, 3-alkyladenine)	β-propiolactone alkylnitrosourea dialkylsulfate
Covalent adducts between a base and a mutagen	benz(*a*)anthracene *N*-acetylaminofluorene aflatoxin
Depurination	alkylating agents spontaneous hydrolysis
Strand breaks	
Breaks with sugar destruction and possible base loss	X-rays
Crosslinks	
Interstrand crosslinks	sulfur mustard mitomycin-C methoxypsoralen
DNA-protein crosslinks	UV light cyclohexylnitrosourea
Intercalation	ethidium bromide

and one of the major photoproducts is the cyclobutane pyrimidine dimer, although modifications to single bases and DNA-protein crosslinks can also be detected.

DNA damage induced by both radiations and chemical mutagens can be classified into three main groups: (a) base modification without associated strand breaks, (b) strand breaks (usually in one strand but occasionally in both strands) with or without associated base damage, and (c) crosslinks between strands or between DNA and protein (Table I). Each kind of damage is induced in characteristic relative proportions by various radiations and chemical mutagens. Modifications to DNA induced by intercalating agents (e.g. ethidium bromide) do not fit into any of these three groups. These agents bind electrostatically between adjacent base pairs, modify DNA functions, and are mutagenic, but whether the damage they cause is handled by repair processes is unknown at present.

DNA repair processes

Three major processes for the repair of base damage have been identified in mammalian cells: photoreactivation, excision repair, and post-replication repair. In addition, cell biochemistry is often perturbed and novel processes are sometimes induced after irradiation or chemical mutagen treatment that might imply the existence of other repair processes or variations on the major processes. One of particular current interest is the possibility of an inducible process, in contrast to the constitutive excision repair and post-replication repair processes, although current evidence for inducible processes is mostly confined to prokaryotic organisms [104].

Photoreactivation is a one-enzyme repair process specific for UV damage that reverts cyclobutane pyrimidine dimers to pyrimidines in situ. This process occurs in marsupials, amphibia, embryonic birds, and lower organisms [21] but has been difficult to detect in placental mammals. Although a photoreactivation enzyme has been isolated from human tissue [98] and monomerization of dimers in vivo by illumination of human cells with visible light has been described [99], the phenomenon appears to be demonstrable only under certain restricted experimental conditions in mammalian cells [20,99] and therefore is not of concern in general studies of repair of chemical mutagen damage.

Excision repair, which is the best known and characterized repair process, can repair most forms of base damage. It occurs in a large range of organisms, though it is very inefficient or missing in plants [73,106]. It occurs in cells at most stages of proliferation and differentiation, except the terminal differentiated states (e.g. sperm and nucleated erythrocytes). In this repair process, damaged bases are recognized and excised and a replacement region (patch) is synthesized and joined to the existing DNA. Two pathways for excision repair appear to exist, depending on the chemical form of the damage: nucleotide excision repair and base excision repair (Fig. 1).

For base damage involving several bases or covalent adducts between bases and activated chemical mutagens, specific recognition of the damaged bases is required, possibly involving cofactors that render the damage accessible to repair enzymes. The damage, in the form of nucleotide fragments, is then

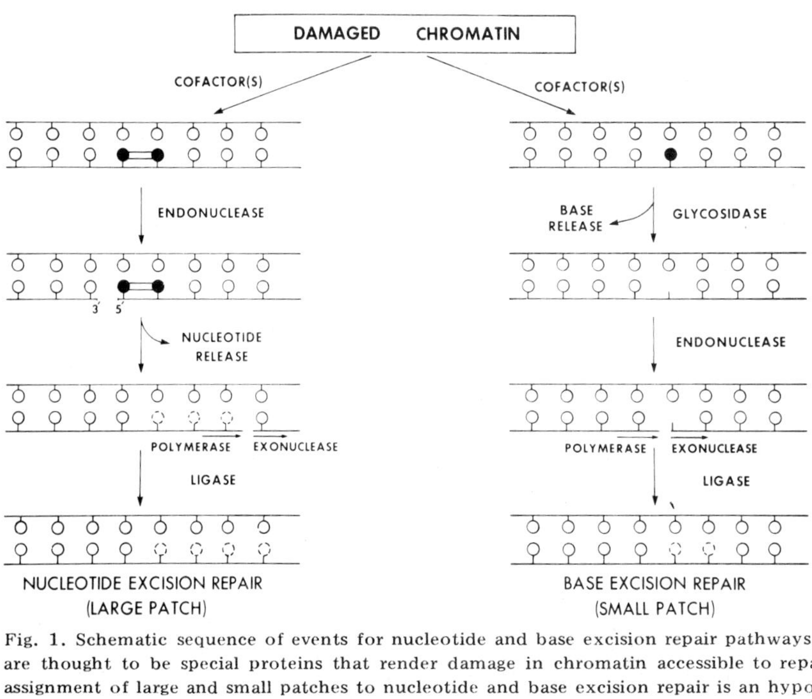

Fig. 1. Schematic sequence of events for nucleotide and base excision repair pathways. The cofactor(s) are thought to be special proteins that render damage in chromatin accessible to repair enzymes. The assignment of large and small patches to nucleotide and base excision repair is an hypothesis based on a variety of evidence from many sources (e.g. refs. 12 and 79).

excised by the sequential action of endonucleases and exonucleases, and the gap is filled by polymerases and ligases. For damage involving single bases (e.g. alkylated purines), different recognition factors and cofactors may be required. The damage is excised by a glycosidase followed by the action of apurinic/apyrimidinic endonucleases and exonucleases, and the gap is filled in by polymerases and ligases. Repair of strand breaks does not require endonuclease action, but it may require exonucleases for the damaged termini adjacent to breaks. Crosslinks between opposite strands of DNA are repaired by sequential excision of the bases at either end of the crosslink, and crosslinks between DNA and protein are repaired by either nucleotide or base excision repair.

Variations in these patterns occur in different organisms. For example, in *Escherichia coli* both the exonuclease and the polymerase functions are performed by DNA polymerase I [47], but in *Micrococcus luteus* [41] and human cells [37] separate exonuclease and polymerase molecules seem to be involved.

The early steps of nucleotide excision repair in man (performed by cofactors and an endonuclease) are determined by at least five genes [15], and defects in these genes are associated with the hereditary disease, xeroderma pigmentosum [70]. The early steps of base excision repair may also require multiple genes, and defects in these may be associated with the human disease, ataxia telangiectasia [75].

Excision repair is considered a highly accurate process, and wrong bases, leading to mutations, are thought to be inserted only rarely. Damaged regions that are not excised, and thus remain during subsequent DNA replication, are more likely to be associated with the incorporation of wrong bases and the generation of mutations.

A question for which there is conflicting evidence is whether excision-repair endonucleases are specific for particular damaged bases. Evidence supporting the idea of non-specific endonucleases is that mutation of single genes in prokaryotes [2,41] and man [8,11,79,86,95] renders cells defective in repair of a large range of mutagen-induced damage. Evidence suggesting the opposite idea is that the endonucleases that repair UV damage in prokaryotes and T_4 phage, which have recently been identified and purified, have a high degree of specificity for pyrimidine dimers [36,41].

Unlike excision repair, which occurs throughout a cell's life cycle, post-replication repair is confined to the DNA synthesis period. It can be considered a modification of normal DNA replication caused by the presence of damaged bases on parental DNA strands. Although there are differences in detail between prokaryotes and eukaryotes, the main features of this process are similar for all organisms. Immediately after organisms are exposed to radiation or chemical mutagens, the presence of damaged bases on parental strands blocks DNA replication [27] and appears to cause interruptions in the synthesis of daughter strands that produce gaps up to 1000 bases in length (Fig. 2), so that newly synthesized DNA has a lower single-strand molecular weight in damaged than in undamaged cells [56,58]. The gaps are sealed in the next few hours by both recombination [44,68,82] and chain elongation [56,57]. Recombination is detectable in both prokaryotes and eukaryotes, but it is less important than chain elongation as a mechanism for gap sealing in eukaryotes.

DNA that is synthesized many hours after DNA damage has been induced

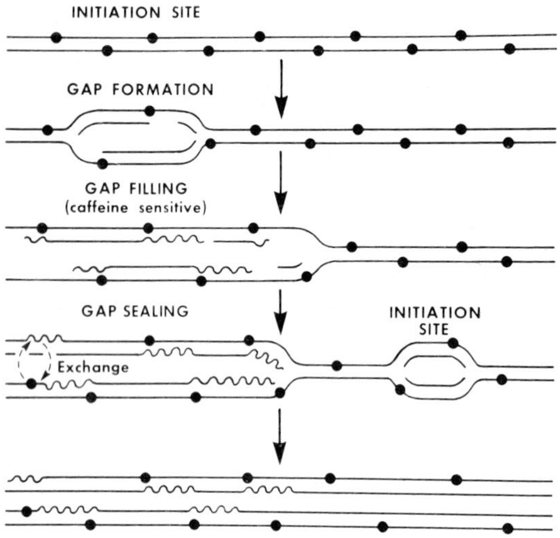

Fig. 2. Possible scheme for early events in DNA replication after UV irradiation (post-replication repair). Replication forks move in both directions from replication origins. Gaps appear in daughter strands in the vicinity of lesions on parental strands and occasional exchange events occur between parent and daughter strands.

does not appear to have large gaps [6]. This may be caused by various mechanisms. Many of the lesions may be excised or otherwise modified shortly after irradiation, sufficient adjustment may occur in the sites of initiation of normal replication and in the replication rate to minimize the effect of damage to the parental strand, or the phenomena associated with post-replication repair may be confined to replicons engaged in DNA replication at the time DNA is damaged.

The gap-filling step of post-replication repair is considered to be the main repair process by which wrong bases are incorporated into DNA and mutations are generated [105].

Because base damage seems to be the predominant damage caused by most mutagens, repair of base damage will be given major attention in this chapter.

Methods for studying excision repair

Excision repair is the predominant early response that cells make when they are damaged by mutagens and the one that has proven to be easiest to study. Each step of excision repair requires different methods for study, and ideally each step should be investigated for each mutagen and each cell type and tissue; but not all repair steps are equally easy to study in every tissue. The methods used to study excision repair, which are the same as those used to study post-replication repair, are detection of damaged sites, detection of strand breaks, and detection of incorporation of new bases.

A. Detection of damaged bases

Study of the excision of damaged sites from DNA is relatively straightforward once the chemical nature of the damage is defined. Studying excision then becomes an exercise in high-resolution chromatographic or enzymatic methods to measure very low levels of mutagen-induced products.

Chromatographic methods can be used to measure several types of damage. The number of pyrimidine photoproducts formed by UV irradiation is best assayed by digesting DNA to bases using formic acid or trifluoroacetic acid in sealed tubes for 30 min to 2 h at 180°C. The free bases and pyrimidine dimers can then be resolved by one-dimensional paper chromatography (*n*-butanol : water, 86 : 14) if moderate to high UV doses are used (i.e., above about 50 J/m^2). For lower UV doses it is necessary to use two-dimensional chromatography (*n*-butanol : water, 86 : 14 for the first dimension and saturated ammonium sulfate : 1 N sodium acetate : isopropanol, 80 : 18 : 2 for the second). Separation methods using thin-layer or column chromatography have also been successfully employed [39,42].

The number of alkylated purines in DNA can be studied by mild acid hydrolysis to release the purines (e.g. 0.1 N HCl at 80°C for 30 min or at 37°C for 20 h). The liberated purines can then be resolved by appropriate paper, thin-layer, or column chromatography, depending on the products under investigation. Alkylated adenine and guanine, for example, have been resolved using Sephadex G10 and Dowex column chromatography [40,55]. Many products of alkylation, however, are unstable in acid, so exclusive reliance on the products obtained from acid hydrolysis will overlook many potentially important products. For a comprehensive analysis of alkylated DNA it is therefore preferable to use an enzymatic method of digestion. A suitable procedure for digesting alkylated DNA to nucleosides is to digest purified DNA at pH 7.2 with an equal weight of deoxyribonuclease (4 h at 37°C) followed by equal weights of snake venom phosphodiesterase, alkaline phosphomonoesterase, and acid phosphomonoesterase (18 h at 37°C) [97]. The products of digestion are then analyzed by column or paper chromatography [97].

Many chemical mutagens, including bromomethylbenz(*a*)anthracene, acetylaminofluorene, and 4-nitroquinoline-1-oxide, form covalently bound adducts to DNA. If these chemicals can be radioactively labeled in stable positions, the extent of binding and excision can be followed by determining the amount of radioactivity bound to DNA at various times after administration of labeled mutagen (Fig. 3). Interpretations of these measurements can be difficult because the detailed biochemistry of many chemical mutagens is not well enough known to determine whether all radioactivity in DNA represents damaged sites in DNA alone. Additional technical problems occur if chemically unstable modifications or a large variety of different modifications of DNA is produced. Identification of the many products formed in DNA by X-rays, for example, has still not been completed [101].

In addition to these chromatographic methods, enzymatic studies of repair have progressed to the point that some excision enzymes (endonucleases) can be isolated from prokaryotes in a relatively pure form and used as tools for determining the numbers of damaged bases in eukaryotes. Thus far, the UV endonucleases isolated from T_4 phage-infected *E. coli* and from *M. luteus* have been used to assay the number of pyrimidine dimers in cultured cells [74] and *E. coli* endonuclease II has been used to assay the number of alkylated purines [48]. In contrast to the direct chromatographic methods for measuring exci-

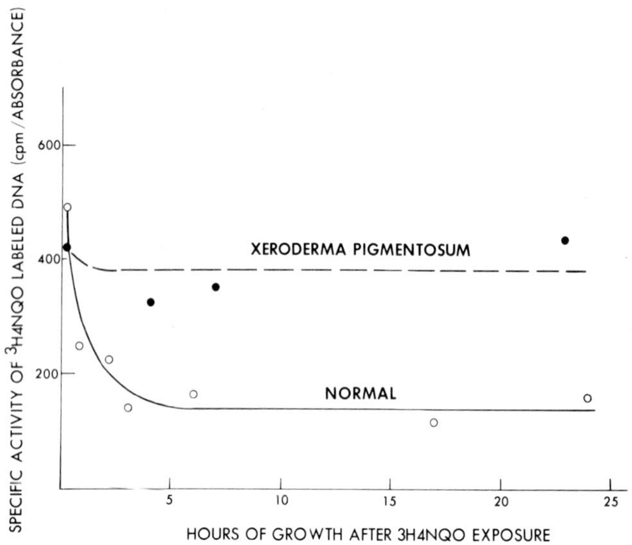

Fig. 3. Extent of excision of damaged bases as determined by specific activity (^{3}H cpm per A_{260}) of purified DNA from normal and xeroderma pigmentosum (XP12RO) human cells labeled for 1 h with 5 μM [^{3}H] 4-nitroquinoline-1-oxide and allowed to grow for various times before harvesting (Cleaver, unpublished results, 1974).

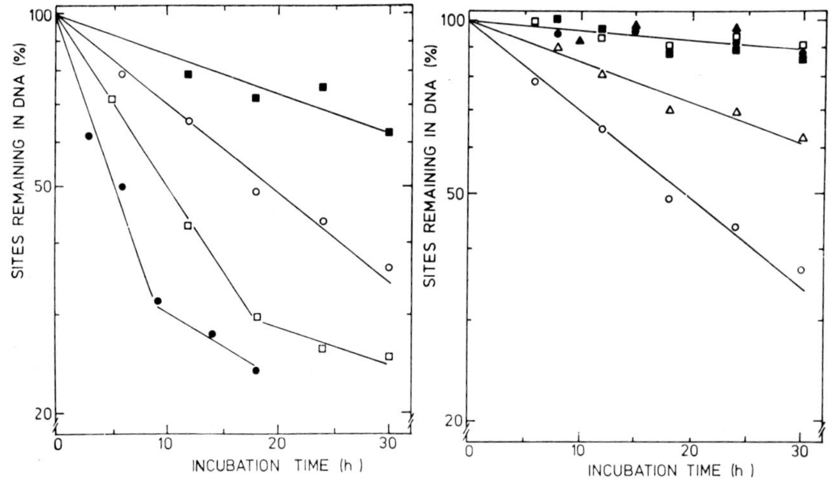

Fig. 4. Rate of disappearance of endonuclease-susceptible sites from DNA from human cells irradiated with 254-nm radiation. Left: DNA from normal human cells. After specified periods of post-UV incubation, cell samples were assayed to determine the number of nuclease-susceptible sites remaining in the extracted DNA. The percentages shown for the incubated samples are relative to that found for the parallel non-incubated ones. Each point is the arithmetic mean of two or more independent determinations. Incident UV dose (J/m^2): ●, 6.5; □, 12.5; ○, 25, and ■, 50. Right: DNA from several XP strains deficient in excision of pyrimidine dimers. For comparison, corresponding data for normal cells (○) are replotted here. Each point is the arithmetic mean of two independent determinations. All doses were 25 J/m^2. □, XP4; △, XP5; ●, XP12; ■, XP25; ▲, XP26, and ○, normal cells. (Reproduced from Paterson et al. [74] with permission from Elsevier Scientific Publishing Company.)

sion, which necessitate large numbers of cells or animals, the enzymatic method requires few cells for comparable levels of resolution.

In a typical protocol, labeled cells from irradiated or chemically treated cells either are rendered permeable to enzymes by osmotic shock [103] or are lysed and the DNA is isolated [74]. The cells or the DNA is then incubated with the appropriate endonuclease until all nuclease-susceptible sites (e.g. pyrimidine dimers, alkylated bases) are broken, and then the single-strand molecular weights are determined by alkaline sucrose gradient sedimentation. From the molecular weights the number of nuclease-sensitive sites, i.e. the number of damaged bases, can be determined (Fig. 4).

This procedure has been used in numerous studies of UV repair processes in tissue cultures and has considerable promise for more extensive applications. The limitation of the method resides in the availability of pure enzymes with clearly characterized specificities. Many enzyme preparations are not completely pure, and the presence of single-strand-specific nucleases can present a particular complication because many DNA modifications induced by mutagens either contain or resemble single-strand regions and are susceptible to attack by single-strand-specific nucleases that are not necessarily directly involved in DNA repair in vivo [87].

B. Detection of strand breaks

1. On alkaline sucrose gradients

The most widely used method for counting single-strand breaks is by determining the molecular weights of DNA single strands in alkaline sucrose gradients [67]. Strictly speaking, this method does not discriminate between true breaks and damaged sites that hydrolyze to become breaks at high pH, nor does it distinguish between breaks that are the direct result of damage and those transiently present during excision of damage.

The use of alkaline sucrose gradients for studying single-strand breaks in X-ray-damaged mammalian cells in culture was pioneered by Lett et al. [59] and was later extended to somatic tissues in vivo [23,102]. Although most major technical problems encountered in tissue culture studies now seem understood and under control, experience with cells derived from tissues in vivo is much more limited and many more investigations will be required to reach comparable levels of understanding. Most remaining technical problems involving alkaline sucrose gradients are associated with methods for detecting low concentrations of DNA and with the control of the extent of cell lysis and DNA strand separation.

a. The standard method. In a typical procedure (Fig. 5), single-cell suspensions from labeled tissue cultures or animal tissues that have been exposed to a mutagen are prepared in a physiological saline (perhaps also containing about 10^{-2} M EDTA), added to alkaline lysis layers on the top of preformed alkaline sucrose gradients, and then centrifuged in swinging bucket rotors. To minimize aggregation artefacts and concentration-dependent effects on DNA sedimentation, the cell concentration must be low (10^3—10^4 cells per 3.5-ml gradient or 10^5 per 30-ml gradient). Typical lysis layers consist of 0.1—0.5 ml of 0.5 N NaOH, 0.02—0.1 M EDTA, and possibly also 0.1% nonidet P40 or 1% sodium

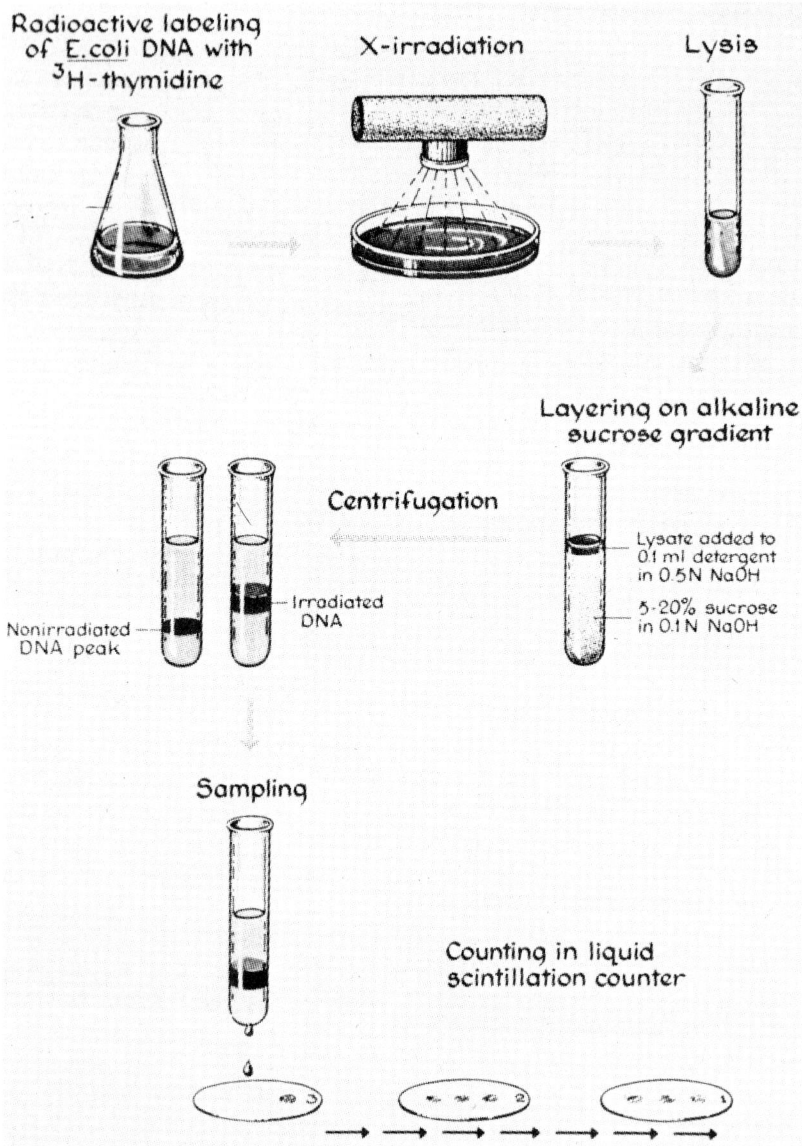

Fig. 5. Procedure for alkaline sucrose gradient sedimentation analysis of DNA from cells exposed to ionizing radiation to determine the number of breaks induced. (Adapted from Kaplan [46] with permission from the Radiological Society of North America.)

dodecyl sulfate detergents to dissolve cell membranes. Some experimenters have added either the same high salt concentration to the lysis layer as used in the gradient [29] or 5% sucrose to minimize transition zones between the lysis layer and the gradient itself [66]. Other experimenters add lysis solution gently to cells in a dish and, after several hours, transfer the resulting solution very gently to the top of the gradients [51]. Typical alkaline sucrose gradients are preformed linear or exponential gradients (5—20%) that contain, for example,

0.1 M NaOH, 0.9 M NaCl, and 0.01 M EDTA, with the pH ranging from 11.5 to 13. The bottom 1 ml or more of the gradient may also contain 50—60% sucrose to prevent the fastest sedimenting DNA molecules from hitting the tube bottom, but then molecular weight analyses must not include the bottom fractions because the gradient is distorted in this region. Most lysis layers and sucrose gradients give qualitatively similar results, but the merits and idiosyncrasies of each have to be evaluated by experience.

b. The method for detecting low concentrations of DNA. An essential requirement of sucrose gradient analysis is that DNA must sediment under controlled conditions at low concentrations, free of aggregation and intermolecular interaction (i.e., effectively as close to infinite dilution as possible). Sensitive methods of detecting low concentrations of DNA can be easily achieved for tissue culture cells by labeling the DNA with [^{3}H]- or [^{14}C]thymidine, the latter being preferable to avoid radiation damage from the labeling procedure itself [16,18]. For tissues in vivo the problem is more difficult. DNA in proliferating tissues or regenerating liver can be radioactively labeled in the adult animal, but non-dividing tissues require other methods. Either all tissues of the animals can be labeled by repeated injections or infusions of [^{3}H]thymidine in utero [33] or a fluorometric assay of unlabeled DNA must be used [49,102]. Fluorometric methods are difficult to employ in swinging bucket rotors: the amount of DNA required for accurate measurement necessitates either overloading the gradients or collecting large gradient fractions [71]. It is here that zonal rotors become invaluable [50,102].

c. Alkaline lysis and the production of various sizes of DNA. Lysis in the NaOH/EDTA lysis layer releases DNA, which progressively dissociates from other cell components and denatures to single strands. The lytic reaction depends on the cell type, cell concentration, time, and temperature and can be accelerated by irradiating cells with UV or X-rays immediately before they are added to the gradient. After lysis periods of several hours, depending on the cell type and the precise lytic conditions, the largest molecules that are stable in NaOH/EDTA for an extended period are single-stranded molecules of 120—165 S ((2—5) $\times 10^8$ daltons). For convenience these molecules are called "subunits" [60]. There seems to be general agreement that molecules in this size range represent single-stranded DNA that can be obtained reproducibly from all mammalian cell types. Whether or not the size represents a biological subunit of DNA or is due to some biophysical limit to the size of single-stranded DNA in sucrose at high pH has not been decided unequivocally at present.

There are therefore two major areas of sucrose gradient studies: (a) that of structures larger and either "anomalously sedimenting" or faster sedimenting than subunits, and (b) that of molecules ranging from subunit size downwards [12—14]. "Anomalously sedimenting" is a term that has been used for large DNA-containing structures whose sedimentation coefficients change with centrifugation speed (i.e. whose sedimentation coefficient is not a constant function of angular velocity squared times time, or $\omega^2 t$). These structures include Elkind's "complex," which contains double-stranded DNA [29,88], some structures with sedimentation coefficients greater than the subunit's [108, 109], and the molecules involved in the sharpening of the leading edge of the subunit profile that occurs at high centrifugal forces [72].

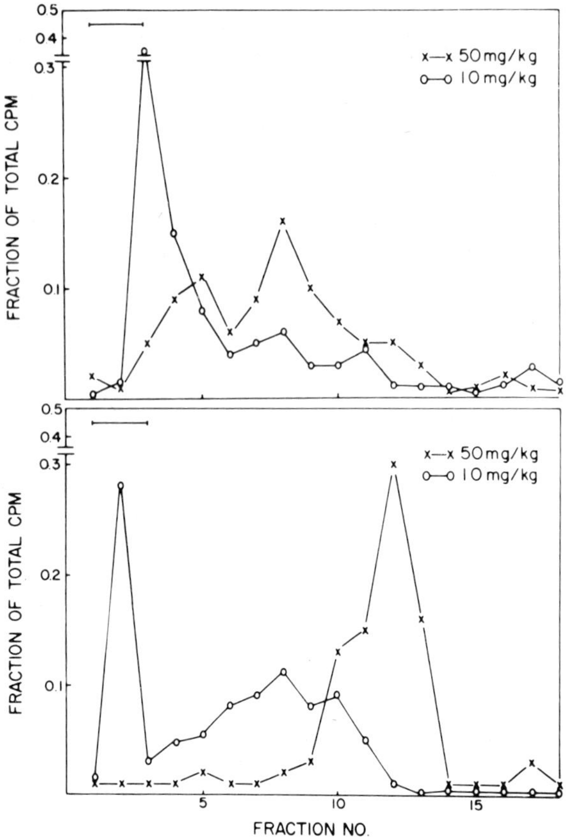

Fig. 6. Sedimentation profiles of rat liver DNA in alkaline sucrose gradients 4 h (top) and 24 h (bottom) after i.p. injection of 10 or 50 mg/kg of *N,N'*-dinitrosopiperazine. Bar (top left) indicates control DNA from rats not exposed to mutagen, which has a sedimentation coefficient in excess of 350 S (based on calibration by T_4 phage DNA). Sedimentation is from right to left. (Reproduced from Stewart and Farber [90] with permission from Cancer Research Inc.)

Faster sedimenting DNA often occurs in multi-subunit arrays, aggregated, only partially denatured, and possibly associated with other cellular material. These DNA-containing structures are often produced by brief lysis of cells from tissue cultures or tissues before sedimentation [13,14,24,60,90]. Small doses of mutagens produce large changes in their sedimentation properties (Fig. 6), but since adequate biophysical analysis of DNA in this size range has yet to be completed, quantitative analysis of strand breaks and other DNA conformation changes is difficult. Many of the apparent details of the time course and dose response of damage and repair may, therefore, be quite misleading because the sedimentation coefficients will be related in a complex way to the number of breaks and crosslinks and the nature and amount of base damage induced. Consequently, this area of study, although of great potential importance, is one that is as yet inadequately understood for use as a general tool in studying mutagen interactions.

d. Methods for measuring molecular weight of small DNA molecules ($\leqslant$*165 S).* At and below subunit size, DNA consists of single strands uncontaminated

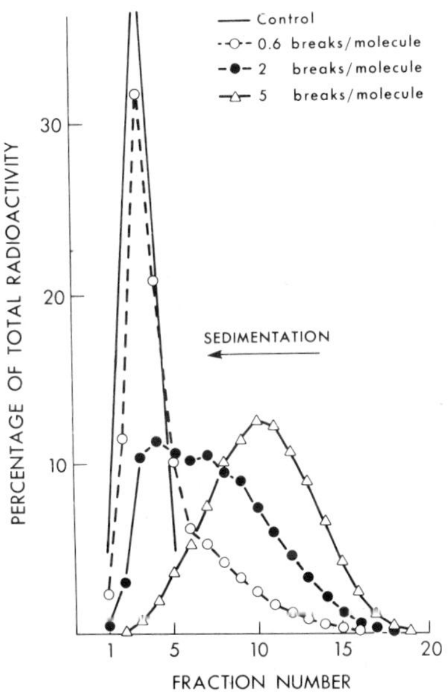

Fig. 7. Computer-simulated sedimentation profiles obtained by the introduction of increasing numbers of random single-strand breaks into the monodisperse distribution at 165 S of single-stranded DNA molecules. (Unpublished data from U.K. Ehmann, Stanford University, and J.T. Lett, Colorado State University, reproduced with their permission.)

by significant amounts of other material. These sediment according to predictable biophysical laws, so their molecular weights can be measured and used to determine numbers of single-strand breaks. For molecular weight calculations, gradients are calibrated with known viral DNA molecules such as λ phage DNA (43 S), SV40 DNA (53 S), or T_4 phage DNA (72.7 S). When exposure of cells to radiation or chemicals induces breaks in DNA, it is most likely that these breaks will be randomly distributed (a reasonable assumption for radiation and alkylating agents and a working assumption for most other DNA-damaging agents). The consequence of random breakage will be to produce smaller molecules with predictable distributions [28] (Fig. 7) that are described by a random distribution function. Any agent that causes random breaks will result in the formation of molecules with a random size distribution when enough breaks have been induced (about five or more per molecule) [64].

The molecular weight of DNA in any particular gradient fraction can be determined from Studier's equation [96], $S = 0.53\ M^{0.4}$. For high centrifugation speeds (about 15,000 rev/min and above) and large DNA molecules (about 10^8 daltons and above), this formula is not strictly applicable because drag and centrifugal forces distort the random coil configurations of DNA. Under these conditions the apparent S value of a molecule is less than predicted by Studier's equation and in extremes may actually *decrease* with increasing molecular weight [108,109].

Several molecular weight parameters can be calculated from gradient profiles, but the most useful are the number-average molecular weight (M_n), which is the mean of the distribution of the DNA molecules in the gradient, and the weight-average molecular weight (M_w), which is the mean of the distribution of the mass of the DNA. The equations are

$$M_n = \frac{\Sigma W_i}{\Sigma W_i/M_i} \quad \text{and} \quad M_w = \frac{\Sigma W_i M_i}{\Sigma W_i},$$

where W_i is the radioactivity in the ith fraction and M_i is the molecular weight of the ith fraction. In calculating M_n, the denominator is the radioactivity in the ith fraction divided by the molecular weight of the ith fraction, so if significant amounts of a gradient profile are present in the top fractions, calculation of M_n becomes unreliable. If the profiles are approximately random, then M_n can be derived from M_w. In addition, a median molecular weight, $M_{1/2}$, can be calculated. For a *random* distribution, the relationships between the various molecular weights are

$$M_n = 0.596\, M_{1/2} \quad \text{and} \quad M_n = M_w/2\,.$$

The number of breaks per original molecule is given by an equation based on number-average molecular weights calculated from the distributions of original and broken DNA:

$$\text{number of breaks per molecule} = \frac{M_n\ (\text{original})}{M_n\ (\text{broken})} - 1\,.$$

In all of these calculations it is assumed that control and mutagen-treated DNA sediment as single-stranded molecules effectively at infinite dilution. Calculations with these equations cannot be used in experiments in which control DNA is sedimented into one or two fractions at the bottom of the tube containing a concentrated (e.g. 50%) sucrose shelf because the sedimentation coefficients of DNA in the bottom fractions are unknown.

DNA breaks result in a gradual decrease in the amount of DNA at control size and a shift to lower sedimentation coefficients (Fig. 7). Repair of the breaks reverses these changes, and the number of breaks per molecule decreases steadily with time (Fig. 8). Large transitions of DNA peaks from one region of the gradient to another as a result of damage or repair or the appearance of multicomponent peaks can indicate either non-random breaks, non-random repair, or the presence of undenatured, collapsed, or entangled DNA whose changes in conformation as well as size alter the S values. The last possibility seems most likely because very little evidence has been obtained for non-random breakage from any agents in the area below subunit size or for non-random repair.

2. By the rate of strand separation in alkali

The slow rate of strand separation of large DNA molecules constitutes a source of difficulty in alkaline sucrose gradient studies and has limited their applicability to doses that cause breaks in subunit-sized DNA molecules. However, because the strand separation rate is a function of molecular weight [25], it can be be used as the basis of a method for determining molecular weight and, hence, the number of single-strand breaks induced at lower doses than possible in sucrose gradients [1,83].

If cells exposed to radiations or chemical mutagens are briefly exposed to

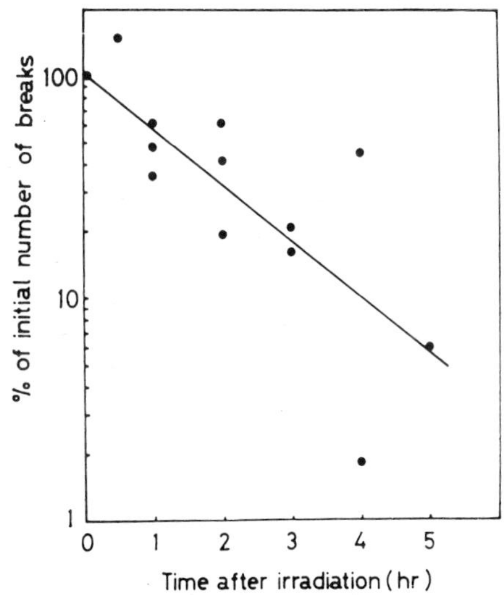

Fig. 8. In vivo rejoining of X-ray-induced single-strand breaks of DNA in mouse hepatocytes. The number of single-strand breaks obtained at the time mice were killed is expressed as a percentage of the number of breaks present immediately after irradiation. (Reproduced from Ono and Okada [71] with permission from Taylor and Francis.)

an alkaline pH and then neutralized, denaturation of DNA will proceed to a certain extent from each single-strand break. The amount of DNA that becomes single-stranded will depend on the frequency of single-strand breaks and the length of exposure to alkali. Thus, the fraction of DNA that becomes single-stranded during exposure to alkali for a fixed period of time will be a measure of the number of single-strand breaks.

In a typical experimental procedure [83], control or mutagen-treated cells (about 5×10^4 labeled cells in 50 μl of buffer) are squirted into 1 ml of alkaline phosphate buffer (0.03 M NaOH plus 0.01 M Na_2HPO_4 at pH 12.2 or 0.03 M NaOH, 0.01 M Na_2HPO_4, plus 0.9 M NaCl at pH 12.0) and allowed to stand in the dark at a measured temperature. The rate of strand separation increases with ionic strength and temperature, so these experimental conditions can be varied according to the resolution of single-strand breaks required. Strand separation is terminated by rapid addition of 1 ml of 0.034 M HCl followed by thorough mixing to achieve neutral pH (7.0 ± 0.2). The DNA is then sonicated to produce low molecular weights (2×10^6 daltons) and achieve permanent separation into single- and double-stranded fragments. The proportion of single-stranded DNA in the solutions can then be determined by one of many standard procedures available, such as hydroxyapatite chromatography [83] or nuclease S_1 susceptibility [100].

The proportion of DNA remaining double-stranded is a dose-dependent, non-linear function of time (Fig. 9). If it is assumed that breaks are present at random along the DNA, then strand separation is described by the relation [83],

$$F = -AM_n^{-1} t^b ,$$

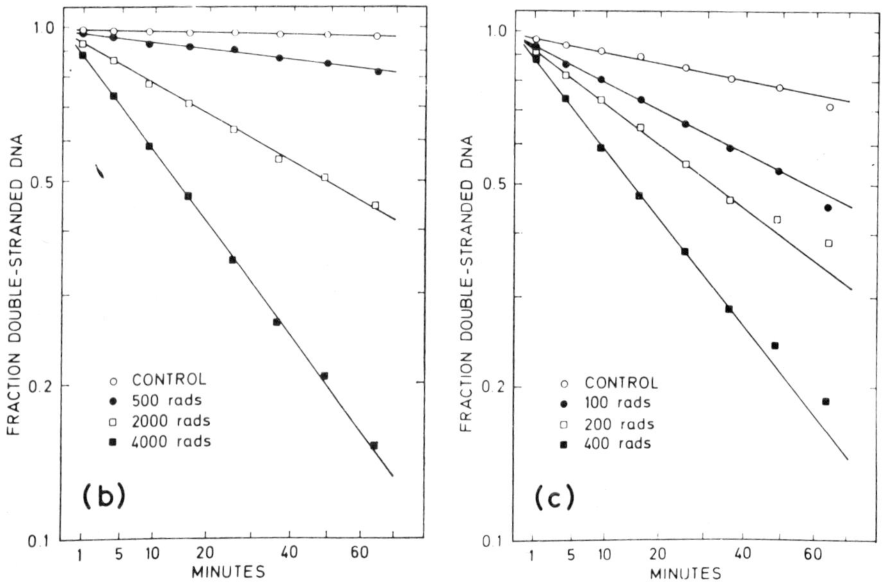

Fig. 9. Kinetics of DNA strand separation for two different alkaline conditions. Left (b): 0.03 M NaOH, 0.01 M Na_2HPO_4 at 21°C and right (c): 0.03 M NaOH, 0.01 M Na_2HPO_4, 0.9 M NaCl at 20°C. The cells were irradiated at 1100 rad/min. The ordinate, showing the fraction of the labeled DNA that is double-stranded after alkali treatment, is given in logarithmic scale. The non-linear abscissa, showing the time of alkali treatment, although graduated in minutes, is on a linear basis proportional to t^b, where t is time and b is 0.66. (Reproduced from Rydberg [83] with permission from Academic Press.)

where F is the fraction of the DNA that is double-stranded after alkali treatment for the time, t, A is a constant, M_n is the number-average molecular weight of the DNA, and b is a constant. The exponent, b, can be determined graphically and is usually in the range of 0.5—0.67, depending on salt concentration and temperature [83].

If a linear dose response is assumed for the induction of single-strand breaks along the DNA molecules, then $1/M_n = n = kD + n_0$, where n and n_0 are the number of single-strand breaks per unit molecular weight of DNA for the irradiated and unirradiated cells, respectively, k is a constant, and D is the dose given to the cells. For a constant alkali treatment time, the fraction, F, may then be written as

$$\ln F = -A'(kD+n_0).$$

This relationship holds well (Fig. 9), but shows slight deviation when extensive strand separation has occurred.

The method has considerable value in detecting single-strand breaks at low doses (Fig. 9) but the precise experimental conditions must be adjusted for each cell type and each mutagen to achieve reproducible results. Interpretation of the results can be complicated if the mutagens alter the rate of DNA denaturation by forming DNA-DNA or DNA-protein crosslinks or by other more obscure mechanisms. In the absence of an independent quantitative theory of DNA strand separation, this method is best calibrated with high doses that induce known numbers of breaks, which can be counted in alkaline sucrose gradients.

3. By alkaline elution

Another method of counting single-strand breaks based on the rate of strand separation of DNA containing these breaks is alkaline elution of DNA from cells lysed on filters [34,35,52]. Cells are collected on filters, lysed, and subjected to a steady flow of alkaline buffer through the filter. Then equal-volume fractions are collected and assayed for the fraction of the DNA that passed through the filters. The principle of the method is that single-strand breaks, which reduce molecular weights, increase the rate at which DNA passes through the filters, whereas crosslinks (DNA-DNA or DNA-protein) decrease the rate.

In a typical experimental procedure [34], 2×10^5 ^{14}C-labeled cells that have been exposed to a mutagen are mixed with 4×10^5 ^{3}H-labeled cells that have been X-irradiated with 150 rad at 0°C as an internal standard. They are then diluted in cold phosphate-buffered saline and filtered onto a polyvinylchloride filter and washed with cold saline. Next the cells are lysed at room temperature with 5 ml of 2 M NaCl, 0.02 M sodium EDTA, and 0.2% sarkosyl (pH 10.2), which is allowed to flow slowly through the filter without suction. The filter is then washed with 3 ml of 0.02 M sodium EDTA (pH 10.3) and eluted in the dark with 0.10 M tetrapropylammonium hydroxide and 0.02 M EDTA (acid form) (pH 12.2) at a pump speed of 0.04 ml/min. Fractions are collected at 90-min intervals for 15 h and counted in a liquid scintillation counter. The

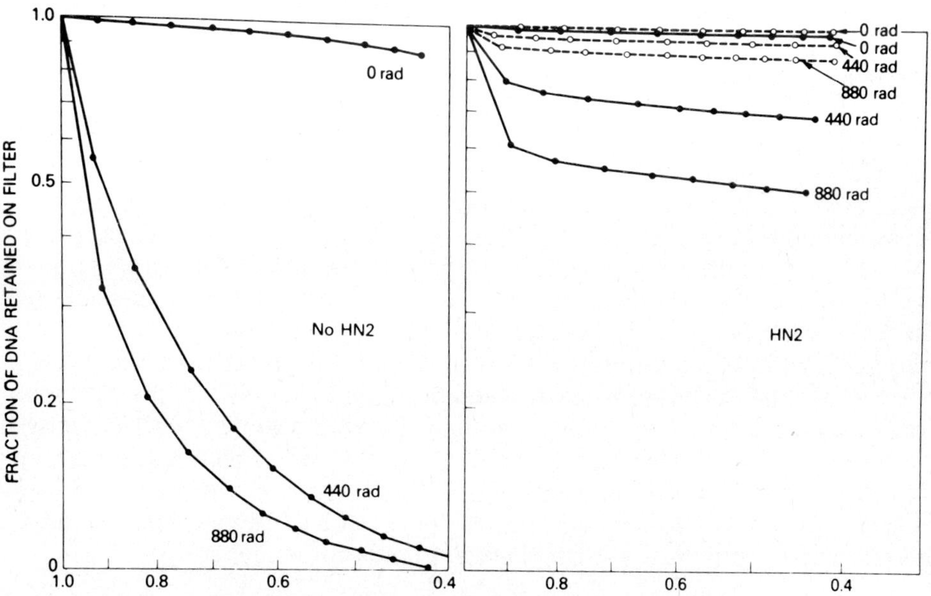

Fig. 10. Measurement of DNA breakage and crosslinkage in normal human fibroblasts by alkaline elution. Left: effect of strand breaks induced by X-irradiation on elution rates. Right: effect of strand breaks and crosslinks induced by X-irradiation and nitrogen mustard on elution rates. Confluent cultures labeled with [^{14}C]thymidine were treated for 16 min with either 10^{-5} M (●———●) or 4×10^{-5} M (○- - - - -○) nitrogen mustard; the cells were then suspended in cold phosphate-buffered saline, X-irradiated at 0°C, mixed with [^{3}H]thymidine-labeled cells that had received 150 rad as an internal standard, and analyzed by alkaline elution. (Reproduced from Fornace and Kohn [34] with permission from Elsevier Scientific Publishing Company.)

radioactivity remaining on the filter is determined by treating the filter with 0.4 ml of 1 M HCl at 70° C for 1 h followed by 2.5 ml of 0.4 M NaOH at room temperature for 30 min and 10 ml of Aquasol. The radioactivity remaining in the funnel is recovered by five washes with 3 ml of 0.4 M NaOH, and the counts are added to those remaining on the filter. If studies on crosslinking between DNA strands or between DNA and protein are to be done, the cells can be irradiated with X-rays immediately before lysis to determine the effect of breaking DNA between crosslinks on the elution rate, or they can be treated on the filters with proteinase K immediately after lysis at pH 10.2 [34] to determine the effect of removing protein on the DNA elution rate (Fig. 10). The fraction of DNA retained on the filter can be plotted as a function either of the elution time or of the fraction of internal standard DNA retained on the filter. This method at present has no independent quantitative theory for determining molecular weights from elution rates, so it must be calibrated by extrapolation from X-rays in a dose range where alkaline sucrose gradients are applicable. It is, however, a most sensitive method for detecting small numbers of breaks (<1 break in 5×10^9 daltons) and crosslinks.

C. Detection of incorporation of new bases

A major factor that has influenced the design of methods for detecting the incorporation of radioactively labeled bases into DNA after damaged bases have been excised has been the need to discriminate low levels of incorporation by excision repair from higher levels of incorporation by semiconservative replication. Semiconservative replication can occur in the same cells that are undergoing repair or in other cells of the population and often contributes several orders of magnitude more incorporation than does repair. The methods developed for detecting incorporation of new bases therefore distinguish between repairing cells and S-phase cells in autoradiographs and between repaired DNA and semiconservatively replicated DNA in density gradients. The proportion of S-phase cells can be minimized by using cell populations in which semiconservative replication is normally very low (e.g. lymphocytes, plateau-phase cultures, and differentiated tissues). The various names used to describe the incorporation of bases during excision repair have been adopted according to the experimental techniques used. Therefore, "unscheduled synthesis" (autoradiography), "stimulated [^{3}H]thymidine incorporation" (specific activity of populations with few S-phase cells), and "repair replication" (isopycnic gradients) all describe the same step in excision repair. Each of these methods is better suited to certain kinds of problems than to others, and the advantages of each depend on the particular cell population under study.

1. Unscheduled synthesis detected by autoradiography

Rasmussen and Painter [78] were the first to use autoradiography to detect the incorporation of [^{3}H]thymidine into the DNA of cultured cells during repair of UV damage. They observed that [^{3}H]thymidine was incorporated into DNA at all stages of the cell cycle within the first hours after irradiation. The term "unscheduled synthesis" was coined by Djordjevic and Tolmach [26] to describe this autoradiographically detected step of DNA repair. Un-

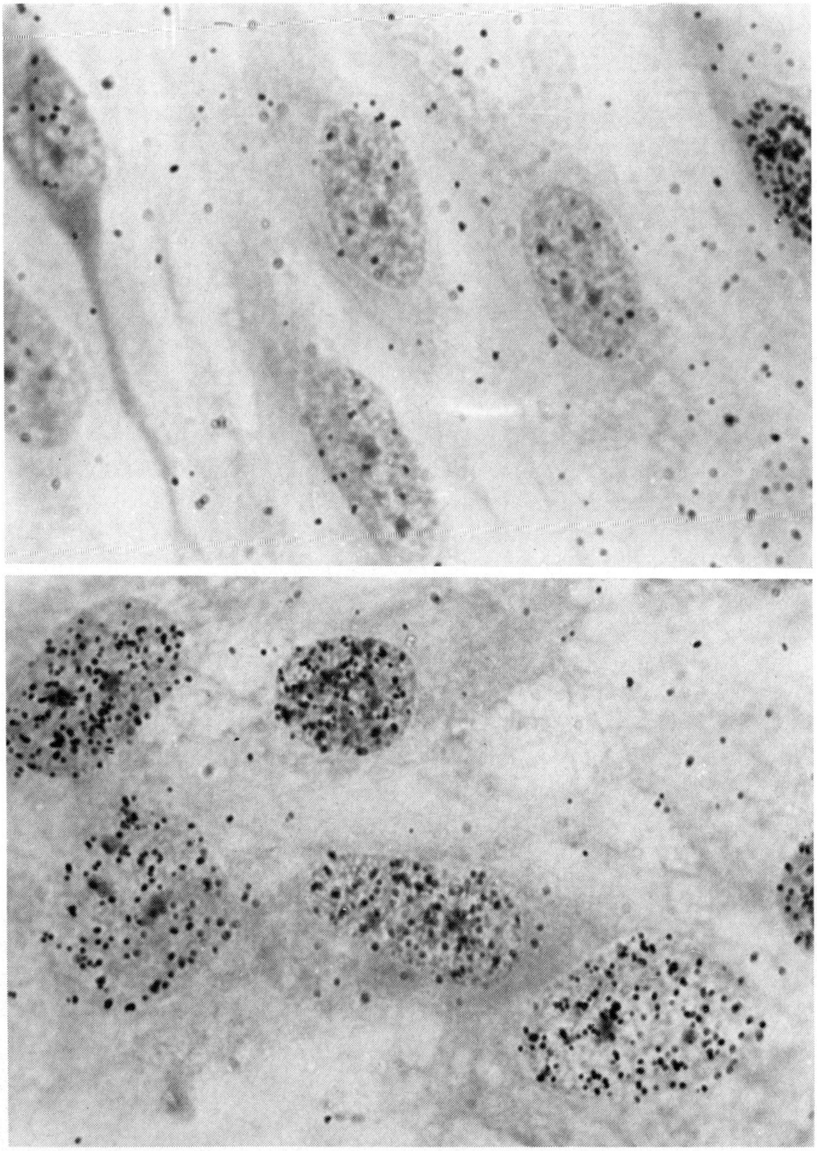

Fig. 11. Autoradiograph of cultured human fibroblasts labeled with [^{3}H]thymidine for 3 h. Top: control cultures, showing one heavily labeled S-phase cell. Bottom: cultures irradiated with 20 J/m^2 of UV light, showing all lightly labeled repairing cells.

scheduled synthesis is readily detectable in tissue cultures damaged by UV light (Fig. 11) or chemical mutagens such as 4-nitroquinoline-1-oxide, methyl nitrosourea, and *N*-methyl-*N*′-nitro-*N*-nitrosoguanidine, but only after high doses of X-rays (1—10 krad). The precise amount of incorporation depends on the particular mutagen and on such obvious factors as concentration and duration of [^{3}H]thymidine labeling and the ability of cells to use exogeneously supplied precursors.

A suitable protocol by which unscheduled synthesis can be detected in tissue cultures is to follow irradiation or exposure to a chemical with labeling in high specific activity [^{3}H]thymidine (e.g., 10 μCi/ml, 18—20 Ci/mmole). Repair begins immediately after damage is induced and continues for 24 h or longer, and labeling can be continued for as long as desired. Cultures are then fixed, processed for autoradiography, and developed, usually after about one week. Autoradiographs exhibit very heavily labeled S-phase cells and lightly labeled repairing cells (Fig. 11), and quantitative measurements of repair can be made by counting the grain numbers over lightly labeled cells. Repair during the S-phase cannot, of course, be detected by this means because of the overwhelming uptake of [^{3}H]thymidine by semiconservative replication.

Unscheduled synthesis in vivo can be detected either by injecting [^{3}H]thymidine into the animal after exposure to mutagens [30] or by removing the appropriate tissues and labeling them in vitro [61]. In either case, care is necessary in evaluating the significance of grain numbers because of variations in the availability of [^{3}H]thymidine to different tissues in vivo or different regions of a tissue fragment during in vitro labeling. In vitro labeling of tissue fragments often results in [^{3}H]thymidine incorporation only in the outer 50- to 100-μm layer of cells because of the limited diffusion of oxygen and [^{3}H]thymidine. There are generally fewer grains reflecting unscheduled synthesis in these systems because [^{3}H]thymidine is less available than in cultured cells.

Comparisons of grain numbers for different cell types in autoradiographs of unscheduled synthesis are fraught with potential sources of error, especially when attempts are made to relate grain numbers to biological endpoints such as cell death, chromosome aberrations, and mutation induction. Not only can there be a genuine difference in the ability of cells to perform unscheduled synthesis, but there may also be differences in the relative contributions of endogenous and exogenous sources of DNA thymine, differences in the rates of breakdown by thymidine phosphorylase, and differences in cell geometry that result in various amounts of self-absorption of ^{3}H β particles within the cell [7]. Differences in uptake of [^{3}H]thymidine can be minimized by blocking endogenous thymidine monophosphate synthesis by fluorodeoxyuridine (FdUrd) (about 2×10^{-6} M). Breakdown of thymidine could be minimized by blocking thymine degradation with diazouracil [22,32]. The variations in self-absorption among different cell types is unavoidable, but this and other sources of error can be compensated for by comparing grain numbers produced by unscheduled synthesis with those produced by S-phase incorporation under identical labeling conditions [19]. The amount of repair in terms of percentage of the diploid genome replaced (usually $<1\%$) can then be calculated. The most reliable comparisons between cell types are those in which the cell types under consideration are very similar, such as fibroblast cell lines (Fig. 11) or lymphocytes from various donors.

Despite these qualifications, unscheduled synthesis detected in autoradiographs is probably the most versatile method of observing one aspect of excision repair and has the fewest technical difficulties in concert with a high degree of resolution of small amounts of repair. Something that cannot be obtained by this method alone, however, is an indication of the biochemical details of the repair process.

2. Stimulated [^{3}H]thymidine incorporation

Because the amount of DNA replaced by excision repair in most cell types is less than 1% of the genome, incorporation of [^{3}H]thymidine into DNA by excision repair contributes insignificantly to the DNA specific activity in a proliferating cell population. But in the plateau phase, or under conditions of arginine starvation [92], the proportion of S-phase cells in tissue cultures becomes very low and the incorporation of [^{3}H]thymidine by excision repair results in a significant increase in the DNA specific activity [3,62,63,81]. Peripheral lymphocytes also incorporate very little [^{3}H]thymidine into DNA and can also be used. In all cases, improved resolution can be obtained by adding $(1-5) \times 10^{-3}$ M hydroxyurea to suppress the small amount of semiconservative replication that does occur. Measurement of excision repair by this means then simply becomes a matter of measuring the total amount of radioactivity incorporated per cell or per μg of DNA after labeling in 10 μCi/ml of high specific activity [^{3}H]thymidine with $(1-5) \times 10^{-3}$ M hydroxyurea. A small correction can be made for the amount of incorporation that occurs by semiconservative replication in control populations (Table II).

Some of the results obtained with peripheral lymphocytes, however, are different from those obtained with other cell types. For example, [^{3}H]thymidine incorporation into normal lymphocytes increases after irradiation [31] and chemical treatment [63] but incorporation of other pyrimidine or purine precursors does not [63,89], even though isopycnic gradient experiments show that all the DNA bases are involved in excision repair [11]. This is probably because of relative purine and pyrimidine pool sizes in various cell types.

Although this method has been extended to lymphocytes treated with a variety of chemicals [63], more care must be taken in interpretation than in experiments using UV light. The magnitude of the increased [^{3}H]thymidine incorporation is often much less than that attainable with UV light, and some chemical mutagens can interact with hydroxyurea to produce spurious results [5]. Also, the cell populations commonly used — plateau-phase cell cultures and peripheral lymphocytes — may have unique properties and therefore may not always respond the same as growing cell populations that are assessed for excision repair by other methods [84].

TABLE II

UV-INDUCED [^{3}H]THYMIDINE INCORPORATION INTO NORMAL HUMAN LYMPHOCYTES LABELED FOR 3 h WITH [^{3}H]THYMIDINE (10 μCi/ml, 15.8 Ci/mmole) PLUS 5×10^{-3} M HYDROXYUREA [a]

UV dose (J/m^2)	Specific activity (cpm per 0.5×10^6 lymphocytes ± S.E.)
0 (without hydroxyurea)	1281 ± 177
0 (with hydroxyurea)	222 ± 26
13	406 ± 49
50	552 ± 58

[a] J.E. Cleaver, unpublished experiments, 1974.

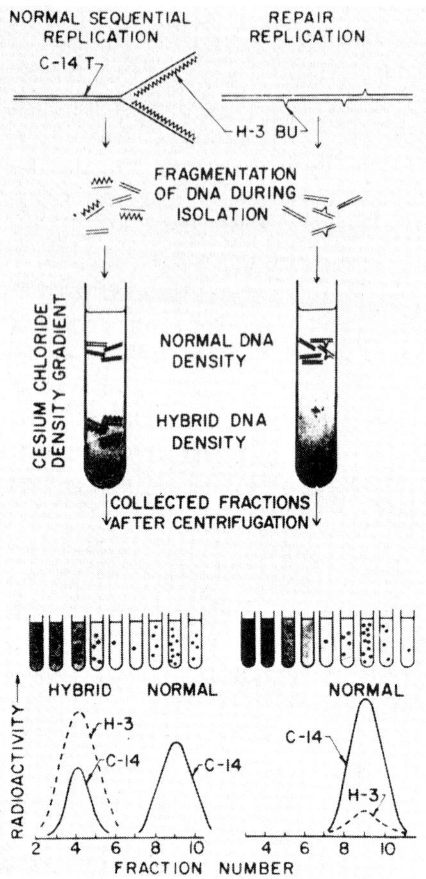

Fig. 12. Normal semiconservative DNA replication and repair replication demonstrated by growing cells in BrdUrd. After isolation of DNA, the long strands of newly synthesized DNA, which contain BrdUrd, are denser than normal in isopycnic gradients, whereas repair patches containing BrdUrd that are scattered throughout the molecule do not increase the density of DNA. (Reproduced from Hanawalt [43] with permission from Williams and Wilkins.)

3. Repair replication measured in isopycnic gradients

Demonstration that repair of damaged DNA involves insertion of small patches into DNA molecules (defined as "repair replication") was first achieved by using bromodeoxyuridine (BrdUrd) labeling and isopycnic gradient analysis [8,76,77] (Fig. 12). Proof that repair replication observed in this manner has functional meaning for mammalian cells was provided by the discovery that low amounts of repair replication in xeroderma pigmentosum cells correlate with clinical symptoms of high actinic carcinogenesis and increased UV sensitivity in cell cultures [8,10].

Demonstration of patches in DNA by BrdUrd labeling is possible because incorporation of small amounts of BrdUrd throughout the DNA has little effect on the density of the molecules (1.700 g/ml), but extensive replacement of thymine by bromouracil during semiconservative synthesis of new DNA strands significantly increases the density of the DNA up to 1.751 g/ml [9]. The patches can be detected either by using [^{3}H]BrdUrd as the label or by mixing a labeled nucleoside or base, such as thymidine, hypoxanthine, or adenine, with non-radioactive BrdUrd and detecting incorporation into normal-density DNA (Fig. 12). The latter method is less expensive, but it must be carefully performed, because labeling with mixtures of nucleosides or bases can lead to complex gradient profiles with molecules of various intermediate densities if the degree of bromouracil substitution along the DNA is not uniform. Repair replication seems to involve all four bases in DNA, even for UV light, which damages predominantly pyrimidines [11].

The protocol that has given the best data for observing repair replication involves a sequence of incubation in unlabeled BrdUrd (10^{-5} M) plus FdUrd (2×10^{-6} M) for 1 h; irradiation or exposure to chemicals in the continued presence of BrdUrd; labeling with 10^{-5} M [^{3}H]BrdUrd (20 μCi/ml) plus 2×10^{-6} M FdUrd and, if needed, 2×10^{-3} M hydroxyurea; harvesting; and analysis on isopycnic alkaline cesium chloride-cesium sulfate gradients. The incubation with unlabeled BrdUrd ensures that all labeled molecules made by semiconservative replication are fully substituted with BrdUrd in the new strands and that there are no labeled molecules with bromouracil at only one end that might give intermediate densities (Fig. 13).

Although this procedure has been described for cell cultures, a similar protocol could be followed for studying repair replication in vivo. Successive injec-

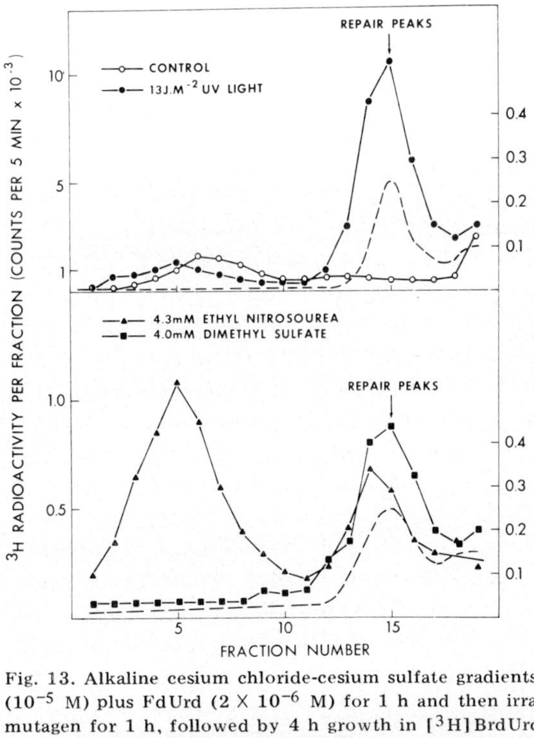

Fig. 13. Alkaline cesium chloride-cesium sulfate gradients of normal human fibroblasts grown in BrdUrd (10^{-5} M) plus FdUrd (2×10^{-6} M) for 1 h and then irradiated with UV light or exposed to a chemical mutagen for 1 h, followed by 4 h growth in [^{3}H]BrdUrd (10^{-5} M), FdUrd (2×10^{-6} M), and hydroxyurea (2×10^{-3} M) before harvesting. Replicated DNA containing [^{3}H]BrdUrd bands at fraction 5 and normal density (repaired) DNA bands at fraction 15. (Cleaver, unpublished observations, 1976).

tions of BrdUrd, a chemical mutagen, and BrdUrd plus a radioisotope, should enable detection of repair replication in vivo, but the lower efficiency of incorporation of precursors into tissues than into cell cultures may limit the resolution of low amounts of repair replication.

At the end of mutagen treatment and labeling procedures, DNA must be isolated from the cell cultures or tissues before isopycnic gradient analysis. A high degree of purification of DNA is not essential because isopycnic centrifugation is itself a purification that separates DNA from RNA and protein. The isolation is intended to (a) reduce the amount of UV-absorbing molecules (RNA and protein) so that an absorbance profile at 260 nm can be obtained to indicate the DNA distribution in the gradient and (b) reduce non-specific radioactivity, mainly in acid-soluble compounds, to improve resolution of peaks corresponding to radioactively labeled DNA. DNA has usually been isolated by suspending cells in saline-citrate buffer (e.g., 0.15 M NaCl, 0.015 M sodium citrate; 0.01 M Tris buffer, pH 7.4; 0.01 M EDTA), lysing by adding sodium dodecyl sulfate to a final concentration of 0.01—0.1%, and then degrading RNA and protein by adding RNAase (final strength, 0.5 mg/ml) and Pronase (final strength, 0.5 mg/ml) and digesting the RNA and protein at 37°C for several hours. The reaction mixture is then deproteinized by agitation with chloroform-amyl alcohol (24 : 1, v/v), centrifugation, and dialysis overnight against the saline-citrate buffer. The final dialyzed solution of labeled DNA is then mixed with cesium chloride (4.5 g of solution to 5.7 g of CsCl) or sodium iodide (5.0 g of solution containing 0.01 M sodium bisulfite to 4.65 g of NaI), and the solution is centrifuged at 33 000—39 000 rev/min in fixed-angle rotors for at least 35 h to allow the gradient to form and DNA to band at equilibrium densities. Gradients using CsCl or NaI [65] give adequate resolution of DNAs with different densities at neutral pH unless excess DNA is loaded onto the gradient. Overloading results in inadequate resolution and may give the illusion of [^{3}H]BrdUrd labeling in normal density regions.

When single-stranded DNA is to be studied, heat-denatured DNA at neutral pH can be centrifuged on CsCl and NaI gradients, but alkaline-denatured DNA at high pH must be studied on alkaline CsCl or alkaline CsCl plus Cs_2SO_4 gradients (0.1 N NaOH, 5 g of solution to 4.8 g of CsCl and 1 g of Cs_2SO_4 and centrifuged at 40 000 rev/min for 30—40 h), because NaI decomposes at high pH. The cleanest results, with least contamination between dense (i.e. semi-conservatively labeled) DNA and light (i.e. repaired) DNA are obtained with alkaline gradients (Fig. 13). Many early reports of repair phenomena in undamaged cells were unwittingly based on overloaded gradients or on heat-denatured DNA that had partially renatured [38]. These artefacts are at a minimum at high pH. If initial gradients do not reveal any distinct peaks of repair replication because of overlap from denser regions of the gradients, recentrifugation of normal density DNA in alkaline gradients will resolve repair replication. Recentrifugation is usually advisable to confirm the density of DNA that appears to be labeled because of repair replication and to estimate the amount of radioactivity incorporated by repair replication.

4. Repair replication measured by bromouracil photolysis

A method for assessing repair replication by bromouracil photolysis in tissue

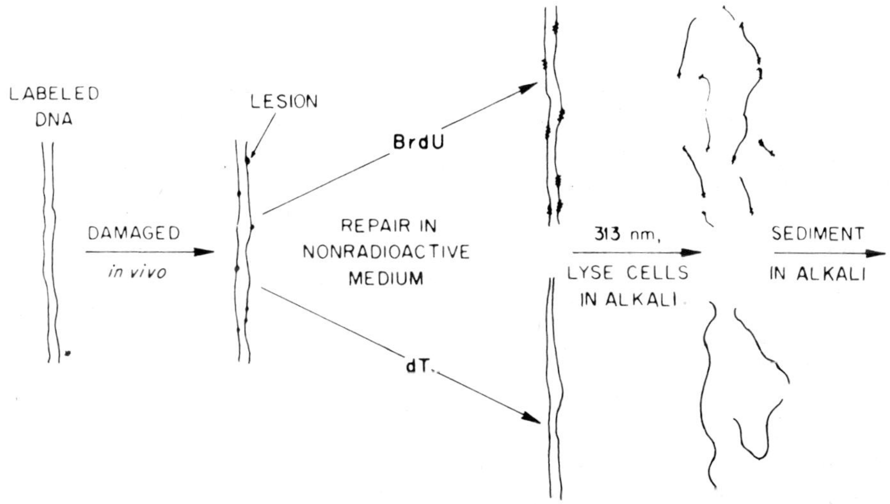

Fig. 14. General scheme for detecting the number and distribution of repaired regions (patches) in DNA by bromouracil photolysis. (Reproduced from Regan et al. [80] with permission from the author.)

cultures was introduced by Regan et al. [80]. It estimates the size of the patches involved in repair replication and the distances between them. These measurements are not readily obtained by any of the other methods used for studying insertion of new bases. The principle of the photolysis method depends on breakage of patches containing bromouracil molecules, but not normal DNA, by irradiation with 313-nm light (Fig. 14). At this wavelength there is a maximal difference between breakage of normal DNA and DNA containing bromouracil molecules [4,45,85]. The efficiency of breakage by 313-nm light depends on how many bromouracil molecules there are per patch, and the final molecular weights obtained when all patches are broken represents the average distance between patches. In a typical protocol, [^{3}H]- or [^{14}C]thymidine-labeled cells that have been damaged by radiations or chemical mutagens are allowed to grow in non-radioactive BrdUrd (10^{-5} M) and hydroxyurea (10^{-3} M) for a selected period and are then suspended in buffer, irradiated with 313-nm light to form alkali-labile bromouracil photoproducts, and analyzed immediately by lysis on alkaline sucrose gradients to determine the molecular weight of the labeled DNA [79,80]. Hydroxyurea must be used to suppress semiconservative replication because if replication has occurred in the presence of BrdUrd, photolysis of DNA in which one strand contains a large amount of bromouracil causes breakage of the opposite strand too [53,56], which could mask the effects of photolysis of repair patches.

This method appeals mostly to those working on photobiological problems and possessing sophisticated photobiological experience, but it has found increasing applications in studies of DNA repair. Its resolution is similar to that obtainable by the other methods of measuring incorporation of bases and it has been used in studies with radiations and various chemical mutagens [79, 80,86].

Significance of measurements of excision repair in indicating potential mutagenicity of chemicals

Detection of excision repair is prima facie evidence that exposure to a given radiation or chemical has caused reparable damage to DNA in the cell, but the next question to be asked is whether this has any significance for the mutagenic potential of the chemical [91,93]. This is especially important because measurements of excision repair are averages over all cells in a population, whereas mutation is a rare event in individual cells.

The amount of excision repair after exposure to a radiation or a chemical will depend on at least the following factors:

(a) the extent of reaction with DNA, i.e., the total number of damaged sites,
(b) the proportion of the damaged sites that are amenable to excision repair,
(c) the sizes of repaired regions,
(d) the kinetics of excision repair as functions of time and dose,
(e) the extent to which chemical interactions modify other sites, particularly in proteins, and possibly inhibit excision repair.

The amount of excision repair will therefore be greatest for mutagens that induce the greatest proportion of damaged sites that are reparable by the larger patches. The number of mutations, however, will depend on different factors, especially the amount of unrepaired or irreparable damage that is in DNA at the time of DNA replication and the probability that during post-replication repair this damage will induce mutations of any kind (e.g. point mutations, frameshifts, deletions, or insertions). In a very general and approximate sense, therefore, the relative amounts of excision repair and mutagenicity of a group of radiations or chemicals that react with DNA to similar extents will be inversely related to one another. An extreme example of this relationship may be the frameshift mutagens, such as acriflavine, which actually inhibit rather than induce excision repair [17].

A direct correlation between the amount of excision repair and the relative mutagenicity of a group of chemical mutagens can be expected only when the mutagens are closely related structurally and involve similar pathways of excision repair so that the same proportion of the total damage remains unrepaired for each mutagen. Thus, among the various derivatives of nitroquinoline-1-oxide, the more mutagenic also induce more excision repair, presumably because of their greater extent of reaction with DNA [94]. A similar relationship seems to hold for methyl and ethyl forms of various alkylating agents, although the relationship might be more complex because both the extent of reaction and the relative proportion of different alkylated sites vary with the individual alkylating agent. When a range of chemically disparate mutagens is used, therefore, although many or all may induce measurable amounts of excision repair, correlations between the relative mutagenicity and the relative amount of excision repair are poor.

Studies of the relationship between DNA damage, excision repair, post-replication repair, and mutagenesis must therefore be pursued with careful consideration of the numbers and varieties of lesions involved in mutagenesis and their modes of repair. Exclusive reliance on any one measurement would clearly be unwise, especially because there is no easy way to predict the amount of unrepaired damage from measurements of excision repair. A large variety of cellular measurements, including those of excision repair in conjunction with, for example, chromosome aberrations and sister chromatid exchanges [54] in both normal and repair-deficient mammalian cells (e.g. xeroderma pigmentosum and ataxia telangiectasia [107]) would seem advisable in assessing the possible mutagenicity of a chemical or radiation.

Acknowledgement

This work was performed under the auspices of the U.S. Energy Research and Development Administration.

References

1 Ahnström, G. and K. Erixon, Radiation induced strand breakage in DNA from mammalian cells: strand separation in alkaline solution, Int. J. Radiation Biol., 23 (1973) 285—289.
2 Ames, B.M., The detection of chemical mutagens with enteric bacteria, in A. Hollaender (ed.), Chemical Mutagens: Principles and Methods for their Detection, Vol. 1, Plenum Press, New York, 1971, pp. 267—282.
3 Baden, H.P., J.M. Parrington, J.D.A. Delhanty and M.A. Pathak, DNA synthesis in normal and xeroderma pigmentosum fibroblasts following treatment with 8-methoxypsoralen and long wave ultraviolet light, Biochim. Biophys. Acta, 262 (1972) 247—255.
4 Boyce, R., and R. Setlow, The action spectra for ultraviolet-light inactivation of systems containing 5-bromouracil-substituted deoxyribonucleic acid. I. *Escherichia coli* 15 $T^- A^- U^-$, Biochim. Biophys. Acta, 68 (1963) 446—454.
5 Brandt, W.N., W.G. Flamm and N.J. Bernheim, The value of hydroxyurea in assessing repair synthesis of DNA in HeLa cells, Chem.-Biol. Interact., 5 (1972) 327—339.
6 Buhl, S.N., R.B. Setlow and J.D. Regan, Recovery of the ability to synthesize DNA in segments of normal size at long times after ultraviolet irradiation of human cells, Biophys. J., 13 (1973) 1265—1275.
7 Cleaver, J.E., Thymidine Metabolism and Cell Kinetics, North-Holland Publ., Amsterdam, 1967.
8 Cleaver, J.E., Defective repair replication of DNA in xeroderma pigmentosum, Nature, 218 (1968) 652—656.
9 Cleaver, J.E., DNA damage and repair in light-sensitive human skin disease, J. Invest. Dermatol., 54 (1970) 181—195.
10 Cleaver, J.E., DNA repair and radiation sensitivity in human (xeroderma pigmentosum) cells, Int. J. Radiation Biol., 18 (1970) 557—565.
11 Cleaver, J.E., DNA repair with purines and pyrimidines in radiation- and carcinogen-damaged normal and xeroderma pigmentosum cells, Cancer Res., 33 (1973) 362—369.
12 Cleaver, J.E., Repair processes for photochemical damage in mammalian cells, in J.T. Lett, H. Adler, and M. Zelle (eds.), Advances in Radiation Biology, Vol. 4, Academic Press, New York, 1974, pp. 1—75.
13 Cleaver, J.E., Sedimentation of DNA from human fibroblasts irradiated with ultraviolet light: possible detection of excision breaks in normal and repair-deficient xeroderma pigmentosum cells, Radiation Res., 57 (1974) 207—227.
14 Cleaver, J.E., Conformation of DNA in alkaline sucrose: the subunit hypothesis in mammalian cells, Biochem. Biophys. Res. Commun., 59 (1974) 92—99.
15 Cleaver, J.E. and D. Bootsma, Xeroderma pigmentosum: biochemical and genetic characteristics, Ann. Rev. Genet., 9 (1975) 19—38.
16 Cleaver, J.E. and H.J. Burki, Biological damage from intranuclear carbon-14 decays: DNA single-strand breaks and repair in mammalian cells, Int. J. Radiation Biol., 26 (1974) 399—403.
17 Cleaver, J.E. and R.B. Painter, Absence of specificity in inhibition of DNA repair replication by DNA-binding agents, cocarcinogens, and steroids in human cells, Cancer Res., 35 (1975) 1773—1778.
18 Cleaver, J.E., G.H. Thomas and H.J. Burki, Biological damage from intranuclear tritium: DNA strand breaks and their repair, Science, 177 (1972) 996—998.
19 Cleaver, J.E., G.H. Thomas, J.E. Trosko and J.T. Lett, Excision repair (dimer excision, strand breakage and repair replication) in primary cultures of eukaryotic (bovine) cells, Exptl. Cell Res., 74 (1972) 67—80.
20 Cleaver, J.E., M. Paterson and E.C. Friedberg, Absence of photoenzymatic monomerization of pyrimidine dimers in normal and xeroderma pigmentosum cells, Biophys. J., 16 (1976) 185a.

21 Cook, J.S., Photoenzymatic repair in animal cells, in R.F. Beers, Jr., R.M. Herriott and R.C. Tilghman (eds.), Molecular and Cellular Repair Processes, Johns Hopkins University Press, Baltimore, Md., 1972, pp. 79—94.

22 Cooper, G.M., W.F. Dunning and S. Greer, Role of catabolism in pyrimidine utilization for nucleic acid synthesis in vivo, Cancer Res., 32 (1972) 390—397.

23 Cox, R., I. Damjanov, S.E. Abanobi and D.S.R. Sarma, A method for measuring DNA damage and repair in the liver in vivo, Cancer Res., 33 (1973) 2114—2121.

24 Damjanov, I., R. Cox, D.S.R. Sarma and E. Farber, Patterns of damage and repair of liver DNA induced by carcinogenic methylating agents in vivo, Cancer Res., 33 (1973) 2122—2128.

25 Davison, P.F., The rate of strand separation in alkali-treated DNA, J. Mol. Biol. 22 (1966) 97—108.

26 Djordjevic, B. and L.J. Tolmach, Responses of synchronous populations of HeLa cells to ultraviolet irradiation at selected stages of the generation cycle, Radiation Res., 32 (1967) 327—346.

27 Edenberg, H.J., Inhibition of DNA replication by ultraviolet light, Biophys. J. 16 (1976) 849—860.

28 Ehmann, U.K. and J.T. Lett, Review and evaluation of molecular weight calculations from the sedimentation profiles of irradiated DNA, Radiation Res., 54 (1973) 152—162.

29 Elkind, M.M. and C. Kamper, Two forms of repair of DNA in mammalian cells following irradiation, Biophys. J., 10 (1970) 237—245.

30 Epstein, J.H., K. Fukuyama, W.B. Reed and W.L. Epstein, Defect in DNA synthesis in skin of patients with xeroderma pigmentosum demonstrated in vivo, Science, 168 (1970) 1477—1478.

31 Evans, R.G. and A. Norman, Unscheduled incorporation of thymidine in ultraviolet-irradiated human lymphocytes, Radiation Res., 36 (1968) 287—298.

32 Ferdinandus, J.A. and G. Weber, Effect of 5-diazouracil on the synthetic and catabolic utilization of thymine in rat liver, Proc. Soc. Exptl. Biol. Med., 139 (1972) 592—596.

33 Fliedner, T.M., R.J. Haas, H. Stehle and A. Adams, Complete labeling of all cell nuclei in newborn rats with H^3-thymidine: a tool for the evaluation of rapidly and slowly proliferating cell systems, Lab. Invest., 18 (1968) 249—259.

34 Fornace, A.J., Jr. and K.W. Kohn, DNA-protein cross-linking by ultraviolet radiation in normal human and xeroderma pigmentosum fibroblasts, Biochim. Biophys. Acta, 435 (1976) 95—103.

35 Fornace, A.J., Jr., K.W. Kohn and H.E. Kann, Jr., DNA single-strand breaks during repair of UV damage in human fibroblasts and abnormalities of repair in xeroderma pigmentosum, Proc. Natl. Acad. Sci. (U.S.), 73 (1976) 39—43.

36 Friedberg, E.C., Studies on the substrate specificity of the T_4 excision repair endonuclease, Mutation Res., 15 (1972) 113—123.

37 Friedberg, E.C., J. Duncan and J.E. Cleaver, A thymine dimer excision nuclease in extracts of human cells, Radiation Res., 59 (1974) 98.

38 Gautschi, J.R., B.R. Young and R.B. Painter, Evidence for DNA repair replication in unirradiated mammalian cells — is it an artifact?, Biochim. Biophys. Acta, 281 (1972) 324—328.

39 Goldmann, K. and E.C. Friedberg, Measurement of thymine dimers in DNA by thin-layer chromatography, Analyt. Biochem., 53 (1973) 124—131.

40 Goth, R. and M.F. Rajewsky, Persistence of O^6-ethylguanine in rat-brain DNA: correlation with nervous system-specific carcinogenesis by ethylnitrosourea, Proc. Natl. Acad. Sci. (U.S.), 71 (1974) 639—643.

41 Grossman, L., Enzymes involved in the repair of DNA, in J.T. Lett, H. Adler and M. Zelle (eds.), Advances in Radiation Biology, Vol. 4, Academic Press, New York, 1974, pp. 77—129.

42 Günther, H.L. and W.H. Prusoff, Isolation of thymine dimers, in L. Grossman and K. Moldave (eds.) Methods in Enzymology, Vol. 12, Nucleic Acids, Part A, Academic Press, New York, 1967, pp. 19—30.

43 Hanawalt, P.C., Repair replication of damaged DNA in vivo, in Genetic Concepts and Neoplasia, Williams and Wilkins, Baltimore, Md., 1970, pp. 528—547.

44 Howard-Flanders, P. and W.D. Rupp, Recombinational repair in UV-irradiated *Escherichia coli*, in R.F. Beers, Jr., R.M. Herriott and R.C. Tilghman (eds.), Molecular and Cellular Repair Processes, Johns Hopkins University Press, Baltimore, Md., 1972, pp. 212—225.

45 Hutchinson, F., The lesions produced by ultraviolet light in DNA containing 5-bromouracil, Qtrly. Rev. Biophys., 6 (1973) 201—246.

46 Kaplan, H.S., Enzymatic repair of radiation-induced strand breakage in cellular DNA and its chemical inhibition, Radiology, 105 (1972) 121—126.

47 Kelly, R.B., M.R. Atkinson, J.A. Huberman and A. Kornberg, Excision of thymine dimers and other mismatched sequences by DNA polymerase of *Escherichia coli*, Nature, 224 (1969) 495—501.

48 Kirtikar, D.M. and D.A. Goldthwait, The enzymatic release of O^6-methylguanine and 3-methyladenine from DNA reacted with the carcinogen *N*-methyl-*N*-nitrosourea, Proc. Natl. Acad. Sci. (U.S.), 71 (1974) 2022—2026.

49 Kissane, J.M. and E. Robins, The fluorometric measurement of deoxyribonucleic acid in animal tissues with special reference to the central nervous system, J. Biol. Chem., 233 (1958) 184—188.

50 Klucis, E.S. and J.T. Lett, Zonal centrifugation of mammalian DNA, Analyt. Biochem., 35 (1970) 480—488.

51 Koch, C.J. and R.B. Painter, The effect of extreme hypoxia on the repair of single-strand breaks in mammalian cells, Radiation Res., 64 (1975) 256—269.

52 Kohn, K.W., C.A. Friedman, R.A.G. Ewig and Z.M. Iqbal, DNA chain growth during replication of L1210 cells. Alkaline elution of large DNA segments from cells lysed on filters, Biochemistry, 13 (1974) 4134—4139.

53 Köhnlein, W. and F. Mönkehaus, On the nature of the radiation damage in the thymine containing strand of hybrid BU-DNA after long wavelength U.V., Int. J. Radiation Biol., 22 (1972) 293—296.

54 Latt, S.A., J.W. Allen, W.E. Rogers and L.A. Juergens, In vitro and in vivo analysis of sister chromatid exchange formation (this volume).

55 Lawley, P.D. and C.J. Thatcher, Methylation of deoxyribonucleic acid in cultured mammalian cells by *N*-methyl-*N*′-nitro-*N*-nitrosoguanidine, Biochem. J., 116 (1970) 693—707.

56 Lehmann, A.R., Postreplication repair of DNA in ultraviolet-irradiated mammalian cells, J. Mol. Biol., 66 (1972) 319—337.

57 Lehmann, A.R., Post-replication repair of DNA in ultraviolet-irradiated mammalian cells. No gaps in DNA synthesized late after ultraviolet irradiation, Europ. J. Biochem., 31 (1972) 438—445.

58 Lehmann, A.R., S. Kirk-Bell, C.F. Arlett, M.C. Paterson, P.H.M. Lohman, E.A. de Weerd-Kastelein and D. Bootsma, Xeroderma pigmentosum cells with normal levels of excision repair have a defect in DNA synthesis after UV-irradiation, Proc. Natl. Acad. Sci. (U.S.), 72 (1975) 219—223.

59 Lett, J.T., I. Caldwell, C.J. Dean and P. Alexander, Rejoining of X-ray induced breaks in the DNA of leukaemia cells, Nature, 214 (1967) 790—792.

60 Lett, J.T., E.S. Klucis and C. Sun, On the size of the DNA in the mammalian chromosome: structural subunits, Biophys. J., 10 (1970) 277—292.

61 Lieberman, M.W. and P.D. Forbes, Demonstration of DNA repair in normal and neoplastic tissues after treatment with proximate chemical carcinogens and ultraviolet radiation, Nature New Biol., 241 (1973) 199—201.

62 Lieberman, M.W. and M.C. Poirier, Deoxyribonucleoside incorporation during DNA repair of carcinogen-induced damage in human diploid fibroblasts, Cancer Res., 33 (1973) 2097—2103.

63 Lieberman, M.W., S. Sell and E. Farber, Deoxyribonucleoside incorporation and the role of hydroxyurea in a model lymphocyte system for studying DNA repair in carcinogenesis, Cancer Res., 31 (1971) 1307—1312.

64 Litwin, S., E. Shahn and A.W. Kozinski, Interpretation of sucrose gradient sedimentation patterns of deoxyribonucleic acid fragments resulting from random breaks, J. Virol., 4 (1969) 24—30.

65 Lohman, P.H.M., M.L. Sluyter and I.A.A. Matthijs, Repair replication in human cells studied by sodium iodide isopycnic centrifugation of DNA in a fixed-angle rotor, Analyt. Biochem., 54 (1973) 178—187.

66 McBurney, M.W., F.L. Graham and G.F. Whitmore, Sedimentation analysis of DNA from irradiated and unirradiated L-cells, Biophys. J., 12 (1972) 369—383.

67 McGrath, R.A. and R.W. Williams, Reconstruction in vivo of irradiated *Escherichia coli* deoxyribonucleic acid; the rejoining of broken pieces, Nature, 212 (1966) 534—535.

68 Meneghini, R. and P. Hanawalt, T4-endonuclease V-sensitive sites in DNA from ultraviolet-irradiated human cells, Biochim. Biophys. Acta, 425 (1976) 428—437.

69 Miller, E.C. and J.A. Miller, Biochemical mechanisms of chemical carcinogenesis, in H. Busch (ed.), The Molecular Biology of Cancer, Academic Press, New York, 1974, pp. 377—402.

70 Mortelmans, K., E.C. Friedberg, H. Slor, G. Thomas and J.E. Cleaver, Defective thymine dimer excision by cell-free extracts of xeroderma pigmentosum cells, Proc. Natl. Acad. Sci (U.S.), 73 (1976) 2757—2761.

71 Ono, T. and S. Okada, Estimation in vivo of DNA strand breaks and their rejoining in thymus and liver of mouse, Int. J. Radiation Biol., 25 (1974) 291—301.

72 Ormerod, M.G. and A.R. Lehmann, Artefacts arising from the sedimentation of high molecular weight DNA on sucrose gradients, Biochim. Biophys. Acta, 247 (1971) 369—372.

73 Painter, R.B. and S. Wolff, Apparent absence of repair replication in *Vicia faba* after X-irradiation, Mutation Res., 19 (1973) 133—137.

74 Paterson, M.C., P.H.M. Lohman and M.L. Sluyter, Use of a UV endonuclease from *Micrococcus luteus* to monitor the progress of DNA repair in UV-irradiated human cells, Mutation Res., 19 (1973) 245—256.

75 Paterson, M.C., B.P. Smith, P.H.M. Lohman, A.K. Anderson and L. Fishman, Defective excision repair of γ-ray-damaged DNA in human (ataxia telangiectasia) fibroblasts, Nature, 260 (1976) 444—447.

76 Pettijohn, D. and P. Hanawalt, Evidence for repair-replication of ultraviolet damaged DNA in bacteria, J. Mol. Biol., 9 (1964) 395—410.

77 Rasmussen, R.E. and R.B. Painter, Evidence for repair of ultra-violet damaged deoxyribonucleic acid in cultured mammalian cells, Nature, 203 (1964) 1360—1362.

78 Rasmussen, R.E. and R.B. Painter, Radiation-stimulated DNA synthesis in cultured mammalian cells, J. Cell Biol., 29 (1966) 11—19.

79 Regan, J.D. and R.B. Setlow, Two forms of repair in the DNA of human cells damaged by chemical carcinogens and mutagens, Cancer Res., 34 (1974) 3318—3325.

80 Regan, J.D., R.B. Setlow and R.D. Ley, Normal and defective repair of damaged DNA in human

cells: a sensitive assay utilizing the photolysis of bromodeoxyuridine, Proc. Natl. Acad. Sci. (U.S.), 68 (1971) 708—712.

81 Robbins, J.H. and K.H. Kraemer, Abnormal rate and duration of ultraviolet-induced thymidine incorporation into lymphocytes from patients with xeroderma pigmentosum and associated neurological complications, Mutation Res., 15 (1972) 92—97.

82 Rupp, W.D. and P. Howard-Flanders, Discontinuities in the DNA synthesized in an excision-defective strain of *Escherichia coli* following ultraviolet irradiation, J. Mol. Biol., 31 (1968) 291—304.

83 Rydberg, B., The rate of strand separation in alkali of DNA of irradiated mammalian cells, Radiation Res., 61 (1975) 274—287.

84 Scudiero, D., A. Norin, P. Karran and B. Strauss, DNA excision-repair deficiency of human peripheral blood lymphocytes treated with chemical carcinogens, Cancer Res., 36 (1976) 1397—1403.

85 Setlow, R. and R. Boyce, The action spectra for ultraviolet-light inactivation of systems containing 5-bromouracil-substituted deoxyribonucleic acid, II. Bacteriophage T4, Biochim. Biophys. Acta, 68 (1963) 455—461.

86 Setlow, R.B. and J.D. Regan, Defective repair of *N*-acetoxy-2-acetylaminofluorene-induced lesions in the DNA of xeroderma pigmentosum cells, Biochem. Biophys. Res. Commun., 46 (1972) 1019—1024.

87 Shishido, K. and T. Ando, Cleavage of ultraviolet light-irradiated DNA by single strand-specific S1 endonuclease, Biochem. Biophys. Res. Commun., 59 (1974) 1380—1388.

88 Simpson, J.R., W.A. Nagle, M.D. Bick and J.A. Belli, Molecular nature of mammalian cell DNA in alkaline sucrose gradients, Proc. Natl. Acad. Sci. (U.S.), 70 (1973) 3660—3664.

89 Smets, L.A., Ultra-violet-light-enhanced precursor incorporation and the repair of DNA damage, Int. J. Radiation Biol., 16 (1969) 407—418.

90 Stewart, B.W. and E. Farber, Strand breakage in rat liver DNA and its repair following administration of cyclic nitrosamines, Cancer Res., 33 (1973) 3209—3215.

91 Stich, H.F., The link between oncogenicity of 4NQO and 4NPO derivatives, induction of DNA lesions and enhancement of viral transformation, in W. Nakahara, S. Takayama, T. Sugimura and S. Odashima (eds.), Topics in Chemical Carcinogenesis, University Park Press, Baltimore, Md., 1972, pp. 17—28.

92 Stich, H.F. and R.H.C. San, DNA repair and chromatid anomalies in mammalian cells exposed to 4-nitroquinoline-1-oxide, Mutation Res., 10 (1970) 389—404.

93 Stich, H.F., P. Lam, L.W. Lo, D.J. Koropatnick and R.H.C. San, The search for relevant short term bioassays for chemical carcinogens: the tribulation of a modern Sisyphus, Canad. J. Genet. Cytol., 17 (1975) 471—492.

94 Stich, H.F., R.H.C. San and Y. Kawazoe, DNA repair synthesis in mammalian cells exposed to a series of oncogenic and non-oncogenic derivatives of 4-nitroquinoline-1-oxide, Nature, 229 (1971) 416—419.

95 Stich, H.F., R.H.C. San, J.A. Miller and E.C. Miller, Various levels of DNA repair synthesis in xeroderma pigmentosum cells exposed to the carcinogens *N*-hydroxy- and *N*-acetoxy-2-acetylaminofluorene, Nature New Biol., 238 (1972) 9—10.

96 Studier, F.W., Sedimentation studies of the size and shape of DNA, J. Mol. Biol., 11 (1965) 373—390.

97 Sun, L. and B. Singer, The specificity of different classes of ethylating agents toward various sites of HeLa cell DNA in vitro and in vivo, Biochemistry, 14 (1975) 1795—1802.

98 Sutherland, B.M., Photoreactivating enzyme from human leukocytes, Nature, 248 (1974) 109—112.

99 Sutherland, B M., M. Rice and E.K. Wagner, Xeroderma pigmentosum cells contain low levels of photoreactivating enzyme, Proc. Natl. Acad. Sci. (U.S.), 72 (1975) 103—107.

100 Sutton, W.D., A crude nuclease preparation suitable for use in DNA reassociation experiments, Biochim, Biophys. Acta, 240 (1971) 522—531.

101 Teoule, R., A. Bonicel, C. Bert, J. Cadet and M. Polverelli, Identification of radioproducts resulting from the breakage of thymine moiety by gamma irradiation of *E. coli* DNA in an aerated aqueous solution, Radiation Res., 57 (1974) 46—58.

102 Wheeler, K.T., R.E. Sheridan, E.L. Pautler and J.T. Lett, In vivo restitution of the DNA structure in gamma irradiated rabbit retinas, Radiation Res., 53 (1973) 414—427.

103 Wilkins, R.J. and R.W. Hart, Preferential DNA repair in human cells, Nature, 247 (1974) 35—36.

104 Witkin, E.M., Ultraviolet-induced mutation and DNA repair, Ann. Rev. Genet., 3 (1969) 525—552.

105 Witkin, E.M. and D.L. George, Ultraviolet mutagenesis in *POLA* and *UVRA POLA* derivatives of *Escherichia coli* B/R: evidence for an inducible error-prone repair system, Genetics, Suppl. 73 (1973) 91—108.

106 Wolff, S. and J.E. Cleaver, Absence of DNA repair replication after chemical mutagen damage in *Vicia faba*, Mutation Res., 20 (1973) 71—76.

107 Wolff, S., B. Rodin and J.E. Cleaver, Sister chromatid exchanges induced by carcinogens in normal and xeroderma pigmentosum cells, Nature, 265 (1977) 347—349.

108 Zimm, B.H., Anomalies in sedimentation. IV. Decrease in sedimentation coefficients of chains at high fields, Biophys. Chem., 1 (1974) 279—291.

109 Zimm, B.H., V.N. Schumaker and C.B. Zimm, Anomalies in sedimentation. V. Chains at high fields, practical consequences, Biophys. Chem., 5 (1976) 265—270.

THE QUANTITATIVE MICROSOMAL MUTAGENESIS ASSAY METHOD

C. N. FRANTZ AND H. V. MALLING

Environmental Mutagenesis Branch, National Institute of Environmental Health Sciences, Research Triangle Park, North Carolina 27709 (U.S.A.)

INTRODUCTION

Mammals can convert some non-mutagenic compounds (promutagens) to highly mutagenic metabolites. Such promutagens will not be detected by mutagenicity screening techniques which use microorganisms to detect genetic damage unless mammalian metabolism is first allowed to act on the chemicals. Also, the active mutagen metabolites may be short-lived, such as alkylating agents which combine with many common chemical groups, so that the organism must be in close spatial and temporal contact with the metabolism of the promutagen in order to detect mutagenic activity[10].

The microsomal mutagenicity assay is an *in vitro* technique which allows tissues to activate promutagens in the presence of a test organism in which induced mutation frequencies are simply determined. Basically, test organisms such as *Salmonella typhimurium* histidine auxotrophs are incubated at body temperature with mammalian tissue (usually liver microsomes), the promutagen to be tested, and whatever cofactors are necessary for biotransformation of the promutagen to mutagenic metabolites. The specific system described in this paper utilizes fresh mouse or rat liver microsomes in the presence of oxygen, Mg^{2+} and NADPH (cofactors) to dealkylate dimethylnitrosamine (DMN) and diethylnitrosamine (DEN), producing alkylating agents which revert *S. typhimurium* His G46.

Any microorganism mutation detection system can be used so long as it is not seriously impaired by the requirements of *in vitro* metabolism. Test organisms which have been used include histidine auxotrophic strains of *S. typhimurium*, which can detect a wide range of mutagenic activity by means of the various specific reversion mechanisms built into these strains[1,2,8]. *Bacillus subtilis* 168 *ilv* leucine and valine auxotroph reversion[17], *Escherichia coli* reversion[15], forward mutations at *ad-3A* and *ad-3B* in *Neurospora crassa*, a fungus[16], and induction of gene conversion and mitotic recombination in the yeast *Saccharomyces cerevisiae*[3,12].

The range of genetic alterations detectable is limited by the type of mutation system used. Therefore, insofar as microorganisms cannot detect all types of genetic damage, *e.g.*, chromosome translocations, a negative result in a microsomal assay does not rule out mutagenic activity of the compound in mammals. The microsomal mutagenesis assay has not yet been successfully performed with cultured mammalian cells.

Any tissue preparation can theoretically be used to activate a promutagen. However, the vast majority of compounds that should and will be tested for metabolically activated mutagenicity are foreign compounds not closely resembling those normally found in mammals. The microsomal fraction is where such compounds would most likely be metabolized. Microsomes are derived from endoplasmic reticulum, are found in the 9000 *g* supernatant of mammalian tissue homogenates, and can be prepared in pure, active form by centrifugation of the 9000 *g* supernatant at 105,000 *g* (or at 1500 *g* in the presence of calcium).

Liver is the tissue richest by far in the microsomal enzyme activity, yet there are marked age, species, sex and strain differences. Also, some reactions which barely proceed in the liver may occur readily in lung, kidney, or gut epithelium. The major microsomal enzyme reaction mechanisms, known as "mixed function oxidases", typically involve enzymatic hydroxylation; for example, DMN demethylation probably occurs when one methyl group is oxidized, forming a highly unstable hydroxylated compound, which dissociates into formaldehyde and monomethylnitrosamine, which in turn is highly unstable and spontaneously degrades to release a methonium ion which will methylate many common groups[4]. Many commonly used drugs are metabolized by this system.

The essential make-up of this membrane bound enzyme complex is not completely characterized, but one component enzyme of all the reactions which can be easily measured is cytochrome P-450, the terminal electron acceptor of the system. Energy for the reaction is generally provided by NADH or NADPH, which are therefore necessary (and labile) cofactors and which must be provided for *in vitro* reactions. Other cofactors, such as calcium or magnesium ions, vary with the specific enzyme activity studied.

Some types of metabolic activation are not readily reproducible *in vitro*. For example, activation of the highly carcinogenic compounds 2-acetylaminofluorene, and subsequently of its microsomal metabolic product, *N*-hydroxyacetylaminofluorene, probably takes place *in vitro* and to some extent *in vivo* by a transacetylation reaction, and thus is mutagenic in the basic microsomal test system[1]. However, by far the most potent metabolic product in terms of binding to DNA is the sulfate ester, which can only be found in the presence of sulfotransferases (found in the cytosol) and the coenzyme adenosine-5′-phosphosulfate (PAPS) or a similar source of high energy sulfate[13]. If little or nothing is known of the metabolic activation of the promutagen to be tested, negative results are likely to occur due to lack of provision of proper cofactors (diverse metal ions, various phosphorylated nucleotides, or soluble enzymes).

Some reactions which can be produced *in vitro* may not occur to any significant extent *in vivo*. For example, nitrofurans, metabolized in the presence of oxygen by some oxidases yield different quantities of mutagenic products than when metabolized anaerobically (by different oxidases)[5,11,18].

Some mixed function oxidase reactions can be performed without enzymes using ascorbic acid[9]. Microsomes also perform reductions, hydrolyses and conjugations when appropriate cofactors are available. Many of the same type of reactions occur in the cytosol, even in mitochondria, but usually are specific for substrates occurring naturally in mammalian tissue.

In summary, use of microsomes and NADH and NADPH to screen *in vitro* for mutagenic metabolites when the specific biochemistry is unknown is a very limited screening test because (a) only mixed-function-oxidase reactions will be seen in quantity; (b) some mixed-function-oxidase reactions will not be seen since other needed cofactors (such as metal ions) will not be present in sufficient amounts, (c) although conjugation reactions generally detoxify compounds and allow excretion[8], conjugations (glucuronide formation, methylation, acetylation, etc.), reductions, and hydrolyses will probably not be seen because proper cofactors will not be present in significant quantity. Finally, some reactions which do not occur significantly *in vivo* may be seen *in vitro*.

Thus, although an intraperitoneal host-mediated assay may be much less sensitive, it is more appropriate for screening compounds the possible metabolism of which is unknown. Furthermore, the intrahepatic host-mediated assay is the most sensitive mutation detection technique yet devised employing relatively inexpensive mammalian metabolism and microorganisms[10,14].

The microsomal mutagenesis assay is appropriate as a screening technique only where a specific type of metabolic activation to a mutagenic product is highly suspected. However, it is a very useful technique for determining the character and kinetics of the biochemistry involved in activation or deactivation. Essentially, a biochemical enzyme assay is performed .The reaction product is measured as mutagen activity. To determine the exact activation pathway, both mutagenicity and biochemical product must be measured.

SPECIAL APPARATUS

The microsomal mutagenesis assay may be divided into five essential parts, (a) preparation of microsomes, (b) preparation of microorganisms, (c) the actual incubation of microorganisms, microsomes, and cofactors, (d) plating and incubation of test organisms, and (e) counting of test organisms.

Preparation of microsomes

Tissue homogenizer of the Potter-Elvehjem type in which a teflon pestle fits snugly into a glass tube containing tissue and solution. An elaborate foot pedal operated homogenizer kit with tubes is available from Tri-R-Instruments (Stock No. S-200). Alternatively, a grounded heavy duty 1/4" electric hand drill may be obtained at a local hardware store and two pestles (24.54 mm diameter), (Stock No. S-23) and four tubes to fit (Stock No. S-37) (tube capacity about 40 ml) are adequate. Also, many find the hand drill an easier way to homogenize in ice to keep the preparation cooled.

A high speed refrigerated centrifuge is needed for the first centrifugation of the tissue homogenate at between 9000 *g* and 20,000 *g*. Models adequate for the job start at around $3,500.00. Alternatively, if a good quality refrigerated centrifuge is already available, converter kits for high speed angled rotor operation are available.

Extra equipment. If biochemical quality work is considered, purification of microsomes is required in order to get a reliable measurement of microsomal protein for standardization. Purified microsomes also may be less damaging to delicate test organisms than a simple liver homogenate 9000 *g* supernatant. Purification may be performed on a preparative ultra-centrifuge at 105,000 *g*. Alternatively, the calcium precipitation technique allows shorter centrifugation time at 1500 *g*, so only a higher capacity head for a high speed (up to 9000 *g*) refrigerated centrifuge would be needed.

Incubation of reaction mixtures

Mechanical timer and water bath shaker. Microsomal enzyme reactions generally proceed optimally at around 37° in the presence of oxygen. If the reaction is performed in unstoppered flasks on a water bath shaker, exact temperature control is maintained and sufficient oxygen reaches the microsomes during shaking. A water bath shaker with a rack holding 9 250-ml erlenmayer flasks is adequate.

However, if there is a chance that active metabolites are volatile, flasks should be stoppered sterilely after gassing with pure oxygen. A small oxygen tank, holder, and regulator with gauge is then needed. Stoppering will also prevent accidental contamination during incubation. Ideally, reactions producing volatile carcinogens or mutagens should be done inside an exhaust hood which filters exhausted air.

Extra equipment. If biochemical reactions are to be monitored precisely, a footpedal on/off switch for the shaker-incubator and a stop watch are desirable.

Plating and incubation of test organisms

With *S. typhimurium*, test organisms are most easily plated in agar overlays which can be conveniently melted either by autoclave or by boiling water bath. However, a small autoclave is necessary for preparation of sterile solutions.

45° water bath. Bacterial plating overlays must be kept melted until ready for use.

Test organism colony counting

Extra equipment. A single button hand-held tabulator aids in counting colonies. More elaborate devices include counters that operate when the plate is touched or automatic computerized colony counters (about $5000 from Fisher Scientific).

Other equipment. Precision balance to weigh out NADPH, media components, mutagens, etc. Trip balance to balance centrifuge tubes. A pH meter is convenient for adjusting pH of stock solutions. A small refrigerator is necessary for storage of media, solutions, and bacteria on agar slants.

Glassware and routine nonchemical supplies for preparation of media and solutions

6 1000-ml erlenmeyer flasks; 11 250-ml erlenmeyer flasks; 2 125-ml screw top bottles; 10 125-ml erlenmeyer flasks; 20 14 × 120 mm screw cap tubes.

Glassware and routine nonchemical supplies needed for a 9 sample microsomal mutagenesis assay with S. typhimurium histidine auxotroph

Reusable. 3 125-ml erlenmayer flasks with stoppers; 10 250-ml erlenmeyer flasks with stoppers; 1 500-ml erlenmeyer flask with stopper; 40 20 × 150-mm test tubes with caps; 60 13 × 100-mm test tubes with caps; 2 50-ml capped centrifuge tubes; 4 50-ml beakers; 2 50-ml capped high speed centrifuge tubes; 1 small ice bucket; 1 large ice bucket or tray; 1 hand pipettor; 12 10-ml pipettes; 18 0.1-ml pipettes; 1 autoclavable fiberglass air filter; 2 large surgical scissors; 2 blunt surgical forceps; 1 fine surgical forceps; 1 250-ml beaker; 1 9-inch section of wooden gutter; 2 push pins; 1 plastic squeeze bottle (200 ml).

Disposable. 60 100 × 15 mm plastic petri dishes; 3 mice.

Chemical supplies for media and solutions prepared before experimentation

Distilled water; $MgSO_4 \cdot 7H_2O$ 10g; citric acid 100g; KH_2PO_4 500g; $NaHNH_4HPO_4 \cdot$

$4H_2O$ 175 g; agar (Bacto) 22 g; glucose 10 g; l-histidine 21 g; biotin 12.2 mg; NaCl 18.5 g; Difco-Bacto nutrient agar nutrient broth dehydrate; $MgCl_2 \cdot 6H_2O$ 610 mg; Tris–HCl (Sigma 7–9) 3.03 g; 1 N HCl 20 ml; sucrose 21.2 g.

Chemicals for 9 sample assay

TPNH (NADPH) 40 mg (alternatively, a TPNH generating system may be used which is less expensive[12]); DPN (NADH) 10 mg; mutagen.

Costing

Once the technique has been established in a given laboratory, a single technician, efficient but not necessarily with an extensive background, should be able to prepare microsomes and run 36 samples at 20-min incubation each, dilute, and plate the samples in an 8-h day. Hand counting the plates should take about 4 h, while preparation of media for this many tests should take about 4 man h. Thus, one experimental point could be determined on an average every 1/2 man h.

PRELIMINARY PREPARATION

Test organism

Many scientists now have the *S. typhimurium* histidine auxotroph tester strains developed by BRUCE AMES, University of California, Berkeley. Tester strains should be chosen carefully with the possible mechanism of mutagenic action in mind[4]. Grow up overnight at 37°. Dip a sterile stick into the broth and streak onto a number of nutrient agar slants. After 24 h of 37° incubation, bacteria should be visible on slant. Use these slants to inoculate, with a sterile stick, nutrient broth flasks for each experiment. To check for contamination and spontaneous reversions of your stock bacteria, simply grow up bacteria at appropriate concentrations on minimal media plates in overlays with and without l-histidine and check colony morphology.

Solutions for microsome preparation and buffer

Solution A. $MgCl_2 \cdot 6H_2O$ 0.61 g; Tris (Sigma 7–9) 3.03 g.

Dissolve in approx. 60 ml water. Adjust to pH 7.5 by adding approx. 20 ml 1 *N* HCl then fill up to 100 ml. Adjust pH after autoclaving. Store in refrigerator.

Solution B (0.62 *M* sucrose). Sucrose 21.2 g.

Add water until approx. 70 ml in a volumetric flask. Can be autoclaved. After autoclaving, add sterile distilled water to 100 ml. Store in refrigerator.

Media and stock solutions for Salmonella

Earl's salts stock solution (50×). Distilled H_2O 670 ml; $MgSO_4 \cdot 7H_2O$ 10 g; citric acid 100 g; KH_2PO_4 500 g; $NaHNH_4HPO_4 \cdot 4H_2O$ 175 g.

Important: add each chemical separately and in order. Dissolve each one before adding the next. No more than 50 g of KH_2PO_4 should be added at one time.

After chemicals are dissolved, pour approx. 100 ml in each 250-ml flask, then autoclave for 15 min.

Plating medium. Distilled H_2O 500 ml; agar 8.0 g (agar must be washed three times before use, as unwashed agar may be contaminated with histidine); glucose 10 g.

Wash agar 3 times; add 10 g glucose. Add distilled H_2O up to 500-ml mark.

Place magnetic stir bar in the flask. Autoclave for 15 min (sterilized). It is best not to make up smaller batches as washing may deplete agar concentration. Immediately after autoclaving, add the following to the flask.

salt solution (50×) (No. 1) 10 ml; histidine stock solution (No. 3) 1.25 ml; biotin stock solution (No. 4) 1.0 ml.

Shake the medium gently. Keep in 45° water bath for 1 h. Pour approx. 20 ml in each 100×15 mm petri dish. Refrigerate petri dishes.

l-histidine stock solution. Add 21 mg l-histidine to 100 ml H_2O (use screw top bottle). Autoclave for 15 min. Keep in refrigerator.

Biotin stock solution. Add 12.2 mg biotin to 100 ml H_2O (use screw top bottle). Autoclave for 15 min. Keep in refrigerator.

l-histidine soft agar. Distilled H_2O 1000 ml; NaCl 5 g; l-histidine 39 mg; agar (Bacto) 7 g (agar should be washed three times).

Melt agar. Dispense 2 ml in each 13×100 mm-test tube. Plug with metal cap. Autoclave for 10 min. Store in refrigerator.

Soft agar minus histidine. Distilled H_2O 1000 ml; NaCl 5 g; agar (Bacto) 7 g (Agar should be washed three times)

Melt agar. Dispense 2 ml in each 13×100-mm test tube. Plug with metal cap. Autoclave for 10 min. Store in refrigerator.

Salmonella growth flasks. Nutrient broth: follow the directions on bottle. Use Difco-Bacto nutrient broth; dehydrate.

Add 50 ml of the medium to 125 ml flask. Seal with a luminum foil. Autoclave for 15 min. Refrigerate.

Saline. Add 8.5 g NaCl in 1000 ml H_2O. Dispense approx. 150 ml in 250-ml flask. Cover with metal cap. Autoclave for 15 min. Refrigerate.

Nutrient agar slants. Follow directions for nutrient agar. Place in screw-cap tubes. Autoclave. Slant (approx. 45°). Refrigerate.

Sterile distilled water. Autoclave in 500 ml aliquots. Refrigerate.

Autoclave glassware, pipets, stoppers, oxygen filter, etc.

Bacterial concentration

It is preferable to consistently treat the same number of bacteria. When the bacteria are grown up overnight, they are in a non-log growth phase, as many nutrients are used up. Usually about the same number will be consistently grown up. However, in order to determine more exactly the number of bacteria treated, the logarithmic relationship of optical density to bacterial concentration should be determined separately for each test organism (and spectrophotometer) used. A graph of this relationship should be constructed by resuspending the centrifuged pellet of *S. typhimurium* grown up overnight in about 7 ml of iced sterile saline and making small step-wise dilutions. Determine optical density by adding 0.25 ml of each dilution to a separate spectrophotometer cuvette containing 4.75 ml iced saline. Read O.D. at 420 nm. To determine the number of bacteria, make a further dilution by a factor of $5 \cdot 10^4$ in iced sterile saline of any one of the above original dilutions, plate in triplicate 0.5 or 1.0 ml in histidine-containing overlay, incubate 48 h, and count the colonies. Calculate the number of bacteria in the original dilutions and plot against O.D. (The graph will be accurate in the range of 60–80% O.D. Always read O.D. at the same sample temperature.)

STEP-BY-STEP

The microsomal mutagenesis assay (9 samples) using liver homogenate supernatant and S. typhimurium His G46.

(1) Inoculate His G46 *S. typhimurium* bacteria from agar slant into 50 ml nutrient broth the day before experiment is planned, grow up overnight (16–18 h) at 37°. Bacteria will be in stationary growth phase.
(2) Take NADPH (TPNH), other cofactors, mutagens, etc. from freezer to warm to room temperature before opening containers. (TPNH should be stored in freezer in desiccant.)
(3) Make up Solution C on ice by adding 20 ml Solution A, 20 ml Solution B, 10 ml cold sterile H_2O. Mix. Keep on ice.
(4) Make up a set of 0.85% saline dilution tubes (capped, 20×150 nm) on ice for each of the nine mutagenesis flasks as follows: 9.5 ml, 9.9 ml, 9.9 ml, 4 ml.
(5) Refrigerate or place on ice all solutions and equipment. Prewarm petri dishes containing minimal medium in incubator.
(6) Centrifuge bacterial broth at 2000 rpm for 20–35 min at 4°. Decant supernate, and resuspend bacteria pellet in 10–20 ml (see above) of cold normal saline by gentle shaking. Do not vortex. Keep on ice until use.

(7) *Microsome preparation.*
(a) Kill 3 animals. Allow to bleed by decapitation in order to have as little blood in livers as possible. Pooling livers from three animals will reduce individual variation in enzyme activity.
(b) Remove livers using sterile technique. (With mice, it is perhaps easiest after bleeding to first place the mouse on its back on a 9″ section of wooden gutter, rinse the abdomen with alcohol from a squeeze bottle, push pin its tail to the board, and remove skin from abdomen (not necessary to use sterile instruments). Strip the skin over the mouse's head and push pin it to the wooden gutter. Rinse abdominal wall with alcohol. Using sterile forceps and scissors (kept in alcohol in a 250-ml beaker), open abdominal wall. It is best to remove the gall bladder, which in poorly kept mice may be infected with *E. coli,* and probably is infected with anaerobes. Using sterile fine scissors and forceps (kept in sterile H_2O or saline in a 50 ml beaker) grasp liver, pull and cut away fine attachments to diaphragm and cut hepatic veins. Pull liver away from animal and cut attachment at ligament of Treitz. Place liver in 50 ml sterile beaker containing 10 ml Solution C. Alternatively, cut portions from each lobe of the liver avoiding the gall bladder, and do not try to remove the entire liver.
(c) Rinse twice in ice cold Solution C (enough to cover livers) in 50-ml beakers by transferring livers.
(d) Weigh livers in tared 50 ml beaker containing a known amount of iced Solution C and adjust volume as desired. (4–5 ml/g liver is standard, while 2 ml/g liver is quite concentrated.)
(e) Homogenize on ice in a precooled glass tube with cold teflon pistil (tight-fitting); two or three passes through suspension is sufficient.
(f) Rinse centrifuge tubes with alcohol to sterilize, and rinse with iced sterile distilled water or saline. Keep on ice.

(g) Centrifuge at 20,000 *g* for 10 min at 4°, or at 9000 *g* for 10–20 min.
(h) Decant completely the supernatant into a cooled container. Mix well before each use.
(8) To force DMN demethylation, 0.8 mg TPNH/ml of reaction mixture or even more is necessary. However, many biochemical reactions proceed at maximum rates with less than 100 μg/ml of TPNH. Weigh out TPNH and mutagen. Do not dissolve TPNH or DPNH until just prior to use.
Note. It is simpler to prepare all solutions to be added to mutagenesis flasks so that 1 ml of each is added. Thus a stock solution of mutagen would be five times the desired concentration, as five items (1 ml each) are added to each flask.
(9) Place a 250 ml sterile flask on ice for each sample to be run. Add 1 ml each of Solution C, liver homogenate supernate, and bacteria to flasks. Add mutagen unless is extremely short-lived. Dissolve TPNH in iced sterile distilled water. To start reaction place flasks on shaker water bath at 37° for 1 min (to warm up) then add 1 ml TPNH to each at timed intervals. Through a sterile filter, gas each flask 1–2 sec; stopper. Time reaction from addition of TPNH. After 15–30 min microsomal reactions generally become non-linear with respect to time.
(10) To stop reaction remove 0.5 (or 1.0) ml aliquot from flask, place in 9.5 (or 9.0) ml first saline dilution tube (iced).
(11) Dilute for viable counts, mixing well at each step.

			0.1 ml		0.1 ml		1.0 ml	
0.5 flask aliquot	→	9.5 first dilution tube	→	9.9 ml	→	9.9 ml	→	4.0 ml last dilution tube

(12) Melt agar overlays by autoclave or boiling water, keep at 45°.

(13) *Plating.*
(a) Add 0.5 ml of last dilution tube to histidine-containing overlay, mix, spread evenly on plate. Repeat once for duplicate plates of viable count for each sample.
(b) Add 0.5 ml or 1.0 ml (depending on expected number of revertants) of first saline dilution tube to histidine-free overlay, mix, spread evenly on plate. Repeat 3 times for quadruplicate plates of the revertant count for each sample.

Swirl overlay around on plate while still molten to distribute bacteria evenly for easier counting.

(14) Incubate plates at 37°. Turn plate over after overlay has hardened to avoid condensation dripping onto plates. After 36 h colonies are visible. By 48 h they are easily counted.
(15) Count colonies. For a balance between ease of counting and significant, reliable results, aim for about 200 colonies per plate. Resuspension in 15 ml saline of the bacteria grown to stationary phase in 50 ml nutrient broth should give about 400 organisms/ml in the last dilution tube if there is no significant killing.

If contamination is seen, it is best to discard the experiment. An easy way to demonstrate that Salmonella are what you have recovered (personally tried so far only with G46 and TA1535) is to demonstrate that the bacteria agglutinate on a slide in the presence of Salmonella group B antisera. This technique does not rule out contamination with other group B Salmonella.

(16) *Calculations.* The possibility of differential survival between revertants and auxotrophs is never ruled out absolutely without a reconstruction experiment for each mutagen; it is best to calculate % survival and to base calculations on the number of revertants per surviving bacteria only with that possibility in mind, and to ideally perform a reconstruction experiment for each mutagen tested. When the ratio between induced and spontaneous mutants is low and test organism survival is low, a reconstruction experiment demonstrating equal survival of auxotrophs and revertants is mandatory.

(a) surviving organisms:

survivors/ml first dilution tube =

$$\frac{\text{histidine requiring colony count total}}{\text{(ml of last dilution tube/plate) (number of plates)}} \times \text{dilution factor}$$

(dilution factor is $5 \cdot 10^4$ in above instructions, *i.e.* the dilutions between the tube from which revertants and the tube from which survivors are determined.).

$$\text{(b) \% survival} = \frac{\text{experimental survivors/ml 1st dilution tube}}{\text{control survivors/ml 1st dilution tube}}$$

(c) revertants:

$$\text{revertants/ml 1st dilution tube} = \frac{\text{histidine non-requiring colony count total}}{\text{(ml of first dilution tube/plate) (number of plates)}}$$

$$\text{(d) revertants}/10^6\text{ survivors} = \frac{\text{(c)}}{\text{(a)}} = \frac{\text{revertants/ml first dilution tube}}{10^6\text{ survivors/ml first dilution tube}}$$

(17) The basic statistical approach is a chi square comparing number of bacteria and number of revertants in experimental and control flasks.

APPENDIX A

Preparation of Ca^{2+} precipitated microsomes[6]

Refer to 9000 *g* liver homogenate supernatant technique for completeness.

(1) (a) Make up solutions for liver preparation (sterile) MC1, MC2, etc. denote different solutions. The composition of these solutions are given at the end of the appendix.

Homogenizing solution, 1 part MC1 : 1 part MC2; Diluting solution, 1 part MC2 : 19 parts H_2O : 20 parts MC3; Sucrose solution, 1 part MC2 : 19 parts H_2O; KCl resuspension = MC4; Keep on ice or at 4°.

(b) Using sterile technique, remove livers, place immediately in iced Homogenizing Solution, rinse 2 times in Homogenizing Solution. Weigh livers. Make up to 4 ml Homogenizing Solution/g liver. Homogenize on ice.

(c) Spin at 4°, 10,000 *g* for 20 min (ss34 Sorval 9500 rpm). Pour off supernatant. Discard pellet.

(d) Dilute supernatant 1:6 (by volume) with Diluting Solution in large centrifuge tube. Spin at 1500 *g* for 10–20 min at 4°. Decant supernatant.

(e) If washed microsomes are desired, as for cytochrome P-450 determination, loosen pellet in a few ml of Sucrose Solution by vortex. Place mixture in sterile cold homogenizing tube and homogenize by hand on ice. Rinse back into large centrifuge tube with cold Sucrose Solution and make up to equal previous volume. Spin at 4°, 1500 *g*

for 10–20 min. Decant supernatant by suction through sterile pipet, discard. Loosen pellet by vortexing in a few ml of 1.15% KCl (MC4). Re-homogenize by hand in sterile cold homogenizing tube with teflon pistil. Rinse into sterile, cold, capped graduated cylinder. If a high concentration of enzyme is desired, use very little KCl.
(2) Protein determination (Lowry)[7]. Make up stock solutions prior to experiment.
(a) Add 2.0 ml H_2O to each 18 × 175 mm test tube needed.
(b) Standards, add 0, 50, 100, 150, 200, and 250 μl of bovine serum albumin (1 mg/ml) to each of 6 test tubes in duplicate.
(c) Add 5, 10, 15, 20 μl of well mixed sample (or of 1.15ä% KCl suspension for control) to 4 respective test tubes.
(d) Add 2.5 ml $CuSO_4$ solution and 2.5 ml tartrate solution to 250 ml graduated cylinder, mix well. Then fill up to 250 ml with Lowry A Solution. Add 10 ml to each tube. Mix. (This solution must be made just prior to use.)
(e) Wait 10 min.
(f) Dilute Folin-phenol reagent 1:1 with distilled H_2O; add 1 ml to each tube. Mix.
(g) Wait 20 min.
(h) Mix. Read O.D. at 660 nm.
(i) Construct graph of standards.
(j) Substract buffer reading from each sample, read protein concentration from net O.D. of sample on straight line portion of graph. Average results from the different concentrations.
(3) Adjust microsomal protein to desired concentration with cold 1.15% KCl.

Microsomal preparation solutions

MC1 Tris–HCl 1.21 g; $MgCl_2 \cdot 6H_2O$ 0.204 g; KCl 0.371 g; $CaCl_2$ 0.178 g.
Dissolve the above in 60 ml of distilled water. Autoclave. Adjust pH to 7.5 with 1 *N* HCl. Make up to 100 ml with sterile distilled water.

MC2 Sucrose 17.1 g.
Dissolve in about 70 ml distilled water. Autoclave. Add sterile distilled water to 100 ml.

MC3 $MgCl_2 \cdot 6H_2O$ 0.408 g; $CaCl_2$ 0.356 g.
Dissolve in 150 ml distilled water in a 200-ml volumetric flask. Autoclave. Make up to 200 ml with sterile distilled water.

MC4 KCl 1.15 g.
Dissolve in 90 ml distilled water. Autoclave. Make up to 100 ml with sterile distilled water.

Reagents for Lowry determination

2% $CuSO_4 \cdot 5H_2O$ (stock solution); 4% K–Na tartrate (stock solution). Lowry "A": NaOH 8 g; Na_2CO_3 60 g.

Dissolve in 2000 ml distilled water (stock solution). Folin-Cocteau Phenol reagent (commercial); Bovine serum albumin 1 mg/ml (stock solution). Refrigerate.

APPENDIX B

Factors affecting the microsomal activation of DMN and DEN

(1) Reaction mixture components. Microsomes, NADPH, and oxygen are absolute

TABLE I

EFFECTS OF TREATMENT CONDITIONS ON THE FREQUENCY OF HISTIDINE REVERSION INDUCED IN *S. typhimurium* STRAIN TA1530

Treatment	*% Survival*	*Number of revertants*	*Revertants per 10^6 survivors*
Complete system	114	916	8.16
Complete system plus N_2 minus O_2	95	2	0.02
Minus NADPH	121	1	0.01
Minus microsomal fraction	126	0	0
Minus $MgCl_2$	109	20	0.19
Minus DMN (control)	100	0	0

20,000 *g* supernatant of male B6C3F1 mouse livers prepared at 5 ml buffer/g liver was incubated 20 min with 3.2 mg NADPH and 45 m*M* DMN, and gassed with either 100% N_2 or 100% O_2.

requirements for the induction of mutations with DMN (Table I). When any one of these components were deleted from the reaction mixture, no reversions were obtained. When the ionic cofactor magnesium was left out of the reaction mixture, significant mutagenic activity was obtained, but almost two orders of magnitude less than with $MgCl_2$ (see ref. 8). The activation of DMN without added $MgCl_2$ may be due to Mg^{2+} already present in the liver homogenate supernatant.

(2) Comparison of buffers. The activation of DMN to a mutagen was compared in two different buffers (Table II). No significant difference in survival or reversion frequency with DMN was found between HEPES and Tris buffers, both at pH 7.4.

TABLE II

COMPARISON OF THE EFFECT OF TRIS AND HEPES BUFFERS ON THE FREQUENCY OF HISTIDINE REVERSION INDUCED IN *S. typhimurium* STRAIN TA1530 INCUBATED WITH MICROSOMES

Buffer	*DMN*	*Percent Survival*	*Revertants counted*	*Revertants/per 10^6 survivors*
Tris–HCl 200 μmoles	–	100	3	0.03
Tris-HCl 200 μmoles	90 m*M*	87	1168	11.6
HEPES 200 μmoles	–	84	3	0.03
HEPES 200 μmoles	90 m*M*	81	1389	14.8

The microsomes were prepared as the 9000 *g* supernatant of B6C3F1 male mouse livers homogenized in 5 ml 0.25 *M* sucrose per gram liver. The 5 ml mixture, incubated in a 250-ml flask for 20 min, contained 200 μmoles Tris–HCl or HEPES buffer pH 7.4, 24 μmoles $MgCl_2$, 1.8 μmoles NADPH, 500 μmoles sucrose and 1 ml of mouse liver homogenate 20,000 *g* supernatant.

(3) Duration of incubation. The reaction mixture was incubated at 37° for varying lengths of time in the presence of DMN (Table III). Although the bacteria were in stationary growth phase at the start of the experiment, incubation clearly allowed them to multiply. Induction of reversions per 10^6 surviving bacteria increased in a linear fashion for 30 min, but had decreased dramatically when measured after 60 min of incubation. This finding suggests a difference between revertants and histidine requiring cells in ability to survive or multiply in a 60-min incubation.

(4) The frequency of reversions increased with increasing concentrations of DMN in the reaction mixture. Even at the lowest concentration, 5.6 m*M* DMN, there was about a 60-fold increase in mutant frequency over control (Table IV).

(5) SKF 525-A inhibition of DMN mutagenicity. SKF 525-A was added to the incubation mixture in varying concentrations (Table V). At a high concentration, a signifi-

TABLE III

EFFECT OF DURATION OF TREATMENT ON MUTAGENICITY OF DMN

Duration in min	*DMN concentration*	*% Bacterial survival*	*Number of revertants counted in 1.0 ml*	*Revertants/ 10^6 survivors*
0	180 m*M*	100	2	0.10
5	180 m*M*	106	42	2.0
10	180 m*M*	113	365	17
20	180 m*M*	166	540	34
30	180 m*M*	136	659	50
60	180 m*M*	217	761	35
60	0	204	1	0.03

A 20,000 *g* supernatant of B6D2F1 male mouse livers homogenized in 4 ml buffer per gram liver and a final concentration of 0.8 mg/ml TPNH were used.

TABLE IV

EFFECT OF DMN CONCENTRATION ON MUTANT FREQUENCY

DMN concentration (mM)	*% Bacterial survival*	*Number of revertants counted in 0.25 ml*	*Revertants/ 10^6 survivors*
0	100	1	0.02
5.6	87.5	60	1.27
11.3	85.8	152	3.28
22.5	87.5	189	4.73
45	89.3	432	9.00

A 20,000 *g* supernatant of C3H×C57B/6 male mouse livers homogenized in 5 ml buffer per gram liver and a final concentration of 0.8 mg/ml TPNH were used.

TABLE V

SKF 525-A INHIBITION OF DMN MUTAGENICITY IN THE MICROSOMAL ASSAY

SKF 525-A concentration	*DMN concentration*	*% Bacterial survival*	*Number of revertants counted in 1.0 ml*	*Revertants/ 10^6 survivors*
0	0	100	1.0	0.03
10 m*M*	0	65	1.3	0.11
0	180 m*M*	100	664	38
10 m*M*	180 m*M*	39	16	2.3
1 m*M*	180 m*M*	>100	1.5	0.07
0.1 m*M*	180 m*M*	>100	24	1.2

The 20,000 *g* supernatant of B6D2F1 male mouse livers homogenized in 4 ml buffered sucrose per gram liver was incubated with 180 m*M* DMN and 0.8 mg/ml TPNH for 20 min.

cant percentage of bacteria were killed. At 1 m*M* SKF 525-A, a concentration frequently found to inhibit mixed function oxidase reactions, no DMN induced reversions were observed and no effect on bacterial survival was seen.

(6) Mouse strain differences in the activation of DMN and DEN to mutagens. Liver homogenate supernatant was prepared from three different strains of mice and were incubated with DMN or DEN (Table VI). Significant differences were seen between the strains in reversions per 10^6 surviving bacteria when incubated with DMN, but no

TABLE VI

STRAIN DIFFERENCES IN THE METABOLISM BY MOUSE LIVER OF DMN AND DEN TO MUTAGENS

Mouse strain	*Mutagenic activity—revertants/10⁶ survivors*		
	Control	*200 mM DEN*	*90 mM DMN*
C3H (Oak Ridge)	0.02 (100%)	0.60 (36%)[a]	41 (93%)
B6C3F1	—	0.51 (29%)	15 (89%)
DBA/2J	—	0.56 (29%)	10 (97%)

20,000 *g* supernatant of 13-week-old male mouse livers homogenized in 4 ml buffer/g liver and incubated with 0.8 mg/ml TPNH for 20 min.
[a]Figures in parentheses are the percentage of bacteria surviving treatment.

differences were seen when incubated with DEN. The differences in liver metabolic activity seen are therefore probably specific for the site of DMN metabolism, and not an effect of microsomal protein or cytochrome P-450 content per gram of liver. Alternatively, DEN activation may be limited in this incubation by the absence of some unknown cofactor such that quantitative data in this particular incubation system are not meaningful.

(7) Mouse and rat livers were compared for their ability to metabolize DMN and DEN to mutagens. Dramatic differences between the two species were found for DMN, but none for DEN (Table VII). When microsomes were prepared to the same protein concentration by the calcium precipitation technique, differences between mouse and rat were much smaller (C. N. FRANTZ AND H. V. MALLING, unpublished).

TABLE VII

SPECIES DIFFERENCES IN THE METABOLISM BY LIVER OF DMN AND DEN TO MUTAGENS

Species	*Compound*	*Concentration (mM)*	*% Bacterial survival*	*Number of revertants counted*	*Revertants/ 10⁶ survivors*
	Control	0	100	1	0.01
Mouse	DMN	45	93	967	20.3
Rat	DMN	45	87	30	0.40
Mouse	DEN	45	100	7	0.14
Rat	DEN	45	70	13	0.22

20,000*g* supernatants of male 12-week-old Sprague-Dawley random bred rat livers and C3H × C57B/6 male mouse livers homogenized in 5 ml buffer per gram liver and a final concentration of 0.64 mg/ml TPNH and 0.56 mg/ml DPNH were used.

(8) Microsomal protein concentration. In two separate experiments, microsomes were prepared and washed, and incubated with DMN and TPNH (Table VIII). Induction of reversions of *S. typhimurium* His G46 increased dramatically as amount of enzyme in the incubation mixture was increased.

(9) Effect of 3-methylcholanthrene pretreatment. Mice of two different strains were treated with 3-methylcholanthrene, and 24 h later microsomes were prepared from the livers of treated mice and matched controls. The microsomes from all four groups were washed and adjusted to identical protein concentrations, and incubated with DMN and TPNH (Table IX). Microsomes from C3H mice activated DMN to a mutagen more than did those of C57BL6 mice. Pretreatment with 3-methylcholanthrene, which is known to induce activity of mixed function oxidase microsomal enzymes,

TABLE VIII

THE EFFECT OF MICROSOMAL PROTEIN CONCENTRATION ON THE METABOLISM OF DMN TO A MUTAGEN

mg microsomal protein/ml in incubation	*revertants per 10^6 survivors*
1.6	41
1.2	17
1.0	9
0.8	2.7
0.6	0.68
0	0.04
0.75	1.61
0.60	0.62
0.45	0.42
0.30	0.34
0.15	0.28
0	0.10

The upper experiment was done with microsomes prepared in the traditional way, homogenized in 4 ml HEPES buffer/g B6D2F1 male mouse liver centrifuged at 10,000 *g*, the supernatant of which was centrifuged at 105,000 *g* twice for 1 h. 180 m*M* DMN and 0.8 mg/ml TPNH were used. The lower experiment was performed with C57BL/6 mouse liver microsomes prepared by calcium precipitation and treated with 100 m*M* DMN and 1.7 mg/ml TPNH. The results nevertheless correspond closely.

TABLE IX

EFFECT OF 3-METHYLCHOLANTHRENE ENZYME INDUCTION ON LIVERS OF C3H AND C57BL/6 MICE IN TERMS OF MICROSOMAL PROTEIN YIELD, DMN DEMETHYLASE ACTIVITY, AND METABOLISM BY LIVER MICROSOMES OF DMN TO A MUTAGEN

Mouse strain	*Inducer*	*mg liver*	*mg purified microsomes*	*mg microsomes per g liver*	*Demethylase[a] activity*	*% Survival*	*Number of revertants in 1.0 ml*	*Revertants per 10^6 survivors*
C57BL/6	Control	4.6	40	8.7	2.66	126	300	13.8
C57BL/6	3-MC	3.9	65	16.6	4.66	138	1030	42.9
C3H	Control	4.9	69	14.1	3.53	132	465	20.3
C3H	3-MC	4.6	75	16.3	3.88	136	757	31.8

[a] n moles HCHO/mg protein/min.

All mice were males matched for age and raised in the same environment for at least two weeks prior to sacrifice. 0.5 mg 3-methylcholanthrene in 0.25 ml olive oil was injected i.p. 24 h prior to sacrifice; controls received 0.25 ml olive oil i.p. Pools of 3 animals were used to determine each data point. Microsomes were prepared according to the method of KUPFER AND LEVIN[6]; protein was measured by the Lowry technique[7]. Identical 15-min incubations were done for formaldehyde production and bacterial mutagenesis except no bacteria were present in the demethylase assay. Incubation contained 100 m*M* DMN, 1.6 mg microsomal protein per ml, 1.6 mg TPNH per ml, pH 7.5 buffer, and either water or *S. typhimurium* in 0.9% saline.

increased activation of DMN by both strains, but to different degrees. Formaldehyde, a side product of DMN activation, was also measured[10], and correlated well with the induction of histidine reversions (Fig. 1 and Table IX). Therefore, the initial demethylation of DMN is proportional to its activation to a mutagen, and DMN demethylation is probably the rate controlling step in the activation of DMN to a mutagen. However, alternative metabolic pathways proportional to demethylation have not been ruled out.

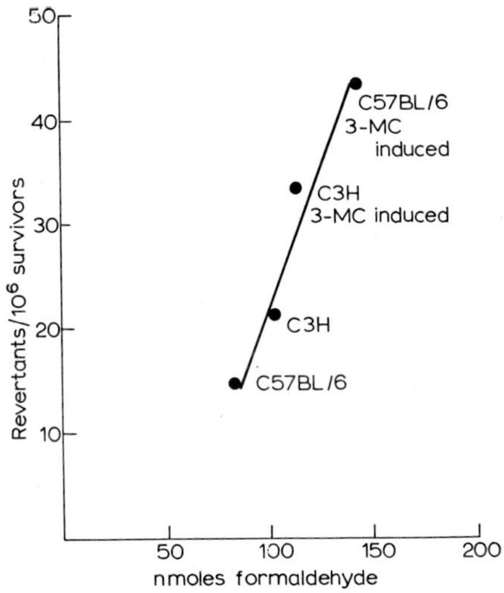

Fig. 1. Relationship of DMN demethylase activity to DMN mutagenicity in two inbred mouse strains with and without 3-methylcholanthrene induction. Data are from Table VII.

REFERENCES

1 Ames, B. N., et al., Carcinogens are mutagens. A simple test system combining liver homogenates for activation and bacteria for detection, *Proc. Natl. Acad. Sci. USA*, 70 (1973) 2281.

2 Ames, B. N., F. D. Lee and W. E. Durston, An improved bacterial test system for the detection and classification of mutagens and carcinogens, *Proc. Natl. Acad. Sci. USA*, 70 (1973) 782-786.

3 Brusick, D., and H. Andrews, A comparison of the genetic activity of chemical mutagens in *Saccharomyces cerevisiae* strains D3, D4 and D5 utilizing *in vitro* assays with and without hepatic enzyme activation. Environmental Mutagen Society, Washington, D.C. (1974).

4 Druckrey, H., et al., Chemische Konstitution und carcinogene Wirkung bei Nitrosaminen, *Naturwiss.*, 48 (1961) 134.

5 Feller, D. R., M. Morita and J. P. Gillette, Enzymatic reduction of niridazole by rat liver microsomes., *Biochem. Pharmacol.*, 20 (1971) 203.

6 Kupfer, D., and E. Levine, Monoxygenase drug metabolizing activity in $CaCl_2$-aggregated hepatic microsomes from rat liver, *Biochem. Biophys. Res. Commun.*, 47 (1972) 611.

7 Lowry, O. H., et al., Protein measurement with the Folin phenol reagent, *J. Biol. Chem.*, 193 (1951) 265.

8 Malling, H. V., Dimethylnitrosaimne: Formation of mutagenic compounds by interaction with mouse liver microsomes, *Mutation Res.*, 13 (1971) 425.

9 Malling, H. V., Mutagenicity of two potent carcinogens, dimethylnitrosamine and diethylnitrosamine, in *Neurospora crassa.*, *Mutation Res.*, 3 (1966) 537.

10 Malling, H. V., and C. N. Frantz. *In vitro* versus *in vivo* metabolic activation of mutagens, *Environmental Health Perspectives*, 6 (1971) 71.

11 McCalla, D. R., and D. Voutsinos, On the mutagenicity of nitrofurans, *Mutation Res.*, 26 (1974) 3.

12 McGregor, D. B., Enhancement by rat liver preparations of the gene conversion frequency induced by ethylmethanesulfonate in *Saccharomyces cerevisiae, Mutation Res.*, 23 (1974) in press

13 Miller, J. A., and E. C. Miller, Chemical carcinogenesis: mechanisms and approaches to its control, *J. Natl. Cancer Inst.*, 47 (1971) v.

14 Mohn, G., and J. Ellenberger, Mammalian blood-mediated mutagenicity tests using a multipurpose strain of *Escherichia coli* K-12, *Mutation Res.*, 19 (1973) 257–260.

15 Nakajima, T., and S. Iwahara, Mutagenicity of dimethylnitrosamine in the metabolic process by rat liver microsomes, *Mutation Res.*, 18 (1973) 121.

16 Ong, Tong-man, (NIEHS, Research Triangle Park, N.C.) Personal Communication.

17 Popper, H., et al., Mutagenicity of primary and secondary carcinogens altered by normal and induced hepatic microsomes, *Proc. Soc. Exp. Biol. Med.*, 142 (1973) 727.

18 Wolpert, M. K., J. R. Althause and D. G. Johns, Nitroreductase activity of mammalian liver aldehyde oxidase, *J. Pharm. Exp. Ther.*, 185 (1973) 202.

MUTAGEN TESTING USING TRP$^+$ REVERSION IN *ESCHERICHIA COLI*

M.H.L. GREEN and W.J. MURIEL

MRC Cell Mutation Unit, University of Sussex, Falmer, Brighton BNI 9QG (England)

Contents

Summary

Escherichia coli strain WP2 and its repair-deficient derivatives are suitable strains for mutagen screening. In these strains, agents which cause base substitution mutations can be shown to increase the frequency of Trp$^+$ revertants. In addition, agents causing many types of DNA damage can be detected through increased killing of the repair deficient derivatives.

Four ways of performing tests are described: (a) *Spot tests* in which a small amount of the agent under test is placed directly on a selective agar plate. Trp$^+$ revertants are counted and increased sensitivity of repair-deficient strains determined from the size of the zone of inhibition of cell growth. (b) *Treat and plate tests*, where a strain is treated with the agent under test and subsequently plated to determine survival or frequency of Trp$^+$ revertants. (c) *A simplified fluctuation test* which shows exceptional sensitivity in measuring mutation with low levels of mutagens. (d) *Use of a liver microsomal fraction* in conjunction with treat and plate tests to detect metabolically activated mutagens.

The merits and defects of these systems are discussed. Common pitfalls in evaluating tests and procedures for avoiding them are described.

Introduction

Trp$^+$ reversion in E. coli

In view of the popularity of *Escherichia coli* as a laboratory organism, it is not surprising that several *E. coli* mutation systems have been adapted for mutagen screening [3, 12, 17]. Such tests have, of course, advantages of simplicity, speed and sensitivity. The one which will be described here uses a number of derivatives of WP2, a tryptophan requiring strain of *E. coli*, and measures reversion to tryptophan independence. The WP2 mutation system has probably received more attention than any other and its use for mutagen screening has been described by Bridges [3].

Strain WP2 and its derivatives all carry the same defect in one of the genes for tryptophan biosynthesis. Tryptophan independent (Trp$^+$) mutants (revertants) can arise either by a base change at the site of the original alteration or by a base change elsewhere in the chromosome so that the original defect is suppressed [5, 18, 19]. This second possibility can occur in several different ways so that the system seems capable of detecting all types of mutagen which substitute one base for another.

The derivatives of strain WP2 described here are deficient in different DNA repair processes. Such repair deficient bacterial strains may either (a) show enhanced mutability if the repair system lost would normally remove the damaged part of the DNA molecule, accurately repairing it afterwards, or (b) show reduced mutability if the repair process is necessary for the conversion of the DNA damage into the final mutated base sequence. Strains of both types are very useful, the (a) type because lower doses can be used to obtain the same mutagenic effect, and the (b) type because they are hypersensitive to agents damaging DNA. Thus although they are not mutated, they are more readily inactivated by such agents than repair proficient strains [9, 11, 22, 25]. An additional advantage is that it is possible to partly characterise the type of DNA damage caused by an agent from the pattern of response of repair deficient strains (Appendix C).

For these reasons, and because Trp^+ reversion can in fact arise through mutation at several loci, our preference is to use one mutation system in several repair deficient strains, rather than several mutation systems in a single strain.

Disadvantages

(a) A number of important mutagens and carcinogens require to undergo metabolic activation in order to produce their effect and the enzymes necessary are not present in bacteria. The problem can partly be alleviated by incorporating in the test system a liver microsomal fraction which will perform the necessary activation [1]. This technique works as well with *E. coli* as with *Salmonella typhimurium*. *E. coli* strains can also be used in the host-mediated assay [16].

(b) The Trp^+ reversion system does not detect agents which cause frameshift mutations (addition or deletion of a base causing the reading of the genetic code to go out of phase). Systems for detecting this type of mutation exist in *E. coli* [7] but the best characterised are those in *S. typhimurium* [2]. We would suggest that the *E. coli* system described here is complementary to the *S. typhimurium* system of Ames et al. [2] because it is more versatile and informative in the characterization of base substitution mutagens.

Conclusion

No bacterial mutagen test system will be ideal, bacteria are too different from people for one thing. Nevertheless, the simplicity and sensitivity of such tests make them useful for mass screening. Some compounds may give a positive response in a bacterial test, but be negative in a mammalian system. This may be because of the greater sensitivity of bacterial systems or, more rarely, because of differences in mechanism. However, in our view, a positive in a properly conducted bacterial test should always be taken seriously, and the compound referred to a higher level of the 3-tier system [4]. The possibility that a mammalian mutagen may fail to be detected in a bacterial test must always be borne in mind.

Mutagen screening

General

(a) All operations are carried out using aseptic techniques and clean sterile glassware, solutions etc. Where necessary solutions of test substances can be sterilised by filtration (0.22 μm pore size) though in many cases it is sufficient to make up in sterile water in a sterile container.

(b) Contaminated glassware can be soaked in Decon or a similar solution. This will probably not inactivate a mutagen, so that a dilute mutagenic solution must still be carefully washed down a sink. 10% sodium thiosulphate will inactivate some alkylating agents.

(c) It should never be necessary to pipette mutagens or mutagenic solutions by mouth. Finnpipettes and Pi-pumps (see section Equipment) or other remote pipetting aids should be used in all the procedures described here.

(d) Where necessary, an efficient fume cupboard should be used and plates can be incubated in sealed canisters. Carcinogenic solids can be dispensed by pre-weighing a small closed bottle, adding the agent in the fume cupboard, (if necessary switching off extraction momentarily), cleaning the bottle carefully, and reweighing.

(e) Never open a plate with a fungal contaminant. Otherwise the next experiment will also be contaminated. Seal a contaminated plastic plate by gently heating the edge in a bunsen flame and then count it.

(f) *E. coli* WP2 and its derivatives are non-pathogenic and do not carry transmissible plasmids.

Spot tests

We suggest using strains WP2, WP2 *uvrA*, WP67 *uvrA polA* and strain CM611 *uvrA exrA* in spot tests. One is looking for either an increased number of mutants in the presence of the agent or greater killing of a sensitive strain.

Required to test 4 compounds

Growing cultures of strains WP2, WP2 *uvrA*, WP67 and CM611 (see section Equipment).

Bottles, pipettes etc., to prepare a stock solution of each agent under test.
Finnpipette (5—50 μl) + 50 sterile tips.
60 SA2 plates (see section Special methods).
At least 60 sterile antibiotic test discs.
Spreader and forceps in beaker of methylated spirit.
Bunsen burner.
1-ml pipettes.
37°C incubator.

Protocol

(a) The 4 strains are each inoculated into 10 ml growth medium from either overnight or refrigerated cultures (see section Special methods).

(b) The cultures are incubated with aeration for about 2.5 h at 37°C. This should give about 2×10^8 cells/ml in each case.

(c) A series of SA2 plates are spread with 0.1 ml of a culture. For each strain

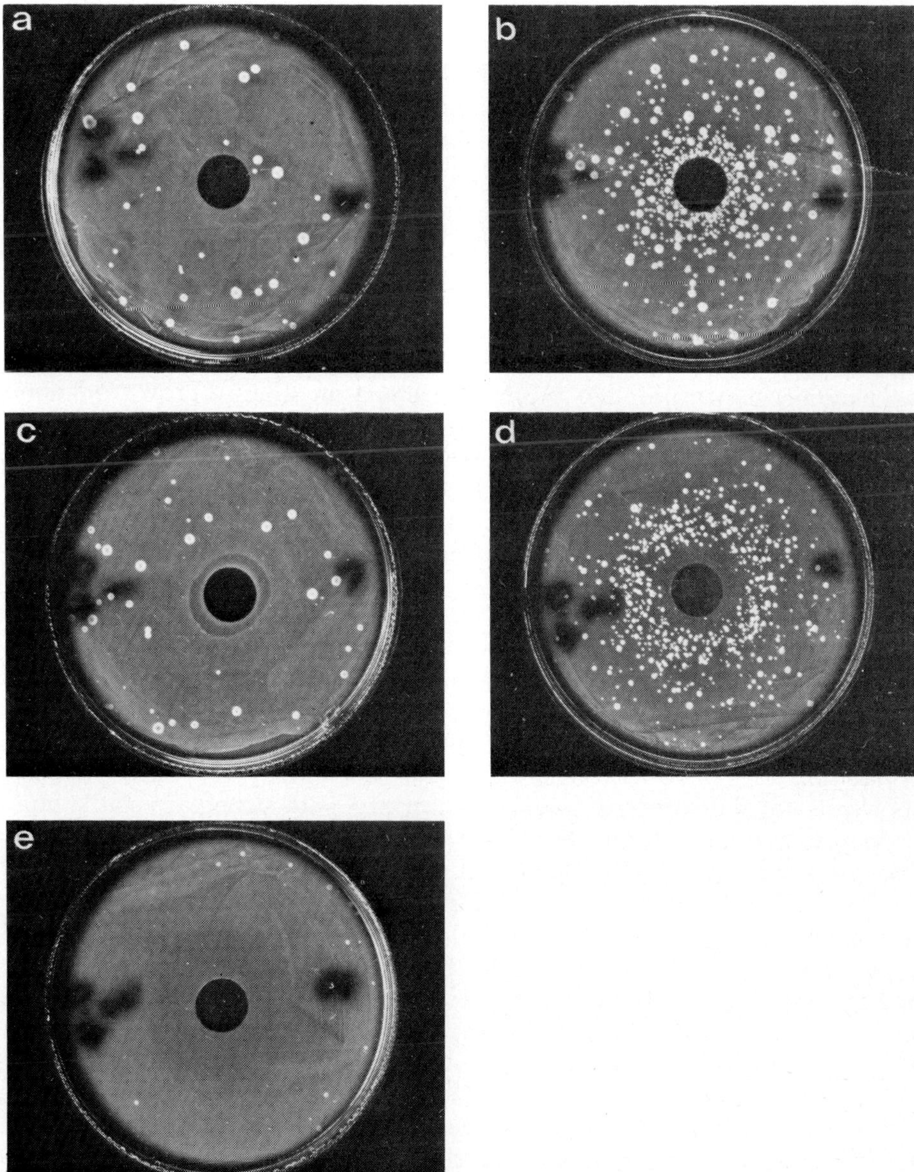

Plate 1. Typical plates from spot tests. (a) A control. Strain WP2. No zone of inhibition and spontaneous mutants only. The lawn of Trp bacteria and Trp$^+$ revertant colonies are readily visible. (b) A positive mutation plate. Strain WP2 treated with 2.5% v/v methyl methanesulphonate. The mutants tend to be clustered round the test disc. No zone of inhibition in this instance. (c) A zone of inhibition. Strain WP2 treated with savlon. The clear area round the test disc is readily visible. No more mutants than in the spontaneous control. (d) A positive mutation test and a zone of inhibition. Strain WP67 *uvrA polA* treated with 2.5% methyl methanesulphonate. (e) An indistinct zone of inhibition. Strain WP67 *uvrA polA* treated with mitomycin C (500 μg/ml). This strain should show increased sensitivity to mitomycin C (Table X) but there is an indeterminate thinning of the lawn round the test disc rather than a distinct zone.

there should be at least 3 control plates, and 3 plates for each test agent.

(d) The compounds under test are added. With solids, it may be possible to add a few crystals directly to the plate. With liquids, antibiotic test discs (or pieces of filter paper) are placed on the spread plates, and 25 μl of the test substances spotted on the disc with a Finnpipette within a minute or two.

(e) Solids can be dissolved in water, ethanol or dimethyl-sulphoxide. Solutions in ethanol or dimethylsuphoxide should be diluted *at least* 20-fold in water. Again 25 μl is pipetted on to an antibiotic test disc.

Finally it is possible to dissolve a compound in a toxic solvent (e.g. chloroform), add a measured amount to an antibiotic test disc, allow the solvent to evaporate, and add the disc to a plate.

(f) The plates are carefully inverted to prevent condensation running on to the agar surface and incubated for 48 h at 37°C.

(g) The plates are scored. Since SA2 agar contains a small amount of tryptophan (1 μg/ml) the tryptophan-requiring bacteria are able to grow from about 2×10^7 to about 10^{10} cells per plate before the tryptophan is exhausted, and form a thin but readily visible lawn. When the tryptophan is exhausted the Trp^+ bacteria continue to grow through the lawn and form larger distinct colonies (typical plates are shown in Plate 1). The supplement of 1 μg/ml tryptophan in the selective agar will allow sufficient growth for about 30—60 spontaneous mutant colonies per plate to arise.

The Trp^+ mutant colonies on the control and treated plates are counted. If there is a zone where growth of the lawn is inhibited round the test spot, this is measured. Use of 1 μg/ml tryptophan in the selective agar rather than 0.25 μg/ml, makes zones easier to distinguish.

Evaluation

Spot tests for mutagenicity are most successful with strong or volatile mutagens (methyl methanesulphonate is both). In other cases the technique is very insensitive, since only a minority of cells on the plate are in contact with an effective level of the mutagen. A test for mutation is a potential positive if there are more mutants per plate on the treated than on the control plates. It is a definite positive if the difference is statistically significant using a Student's t test. A negative result may simply mean that the mutagenic effect is below the level of detection in this system. Rather than using statistics therefore, the result should be corroborated with a treat and plate experiment.

Differences between strains in the size of the zone of inhibition are more sensitive but may generate false positives with weaker growing strains. Once again, any consistent differences in the size of the zone of inhibition between strains should be corroborated with a survival test treating in liquid suspension and plating.

The value of spot tests is debatable. Where a large number of compounds are to be tested with limited resources they offer a major advantage since compounds can be tested directly, without first having to determine an effective concentration, and this greatly reduces the work involved. In rare cases (e.g. the volatile breakdown product of captan) they are the only method of detection. On the other hand, spot tests are not very reliable, and for this reason treating in liquid suspension is usually preferable wherever practicable.

Illustration

The compounds tested were methyl methanesulphonate (2.5% v/v) mitomycin C (500 μg/ml) and Savlon (chlorhexidine gluconate 0.3%, cetrimide 3%). The normal Protocol was followed and typical plates obtained are shown in Plate 1. Results are summarised in Tables I and II. With respect to mutation, methyl methanesulphonate clearly gives a positive result, and the other two compounds are negative. Mitomycin C does mutate strain WP2 and the negative result reflects insensitivity of the spot test. Methyl methanesulphonate only gives a zone with the *polA* strain WP67. It can be seen that mitomycin C tends to give a general thinning of the lawn rather than a zone (which is why we employ a different technique when using mitomycin C to test for repair deficiency). Savlon gives an apparent positive response, since the repair deficiency strains show larger zones.

Most of the commoner difficulties are shown in this test. Methyl methanesulphonate is positive, both for mutation and zone of inhibition. Mitomycin C, a weak mutagen is apparently negative for mutation, and gives uninterpretable zones of inhibition. Savlon, which should give a clear negative, gives an apparent positive for zones of inhibition (though this is not repeatable). The danger always exists that a slower-growing strain will give a larger zone of inhibition.

Treat and plate test

For a more sensitive test, and especially in order to test a strain with a known concentration of an agent, it is necessary to treat the cells in liquid suspension and plate on selective agar. Once again, a test is positive if there is an increase in mutants per plate in the treated sample compared to the control. Also, a test is positive if one or more of the repair deficient strains shows more killing with an agent than does a repair proficient strain.

This method of testing is more sensitive than spot testing because it enables a large number of cells to be treated with an effective level of a mutagen. The main difficulty is not the actual test, but determining what is an effective level. If the concentration of a mutagen is too low, too few mutants will be induced to be detected; if the concentration is too high, too many mutants will be killed for any to be detected.

TABLE I

TRP^+ MUTANTS PER PLATE IN SPOT TESTS USING METHYL METHANESULPHONATE, MITOMYCIN C AND SAVLON

Agent	Strain	Average number Trp^+ mutant colonies per plate			
		WP2 (UV resistant)	WP2 *uvrA*	WP67 *uvrA polA*	CM611 *uvrA exrA*
Control		42	48	62	23
Methyl methane-sulphonate		409	456	576	60
Mitomycin C		38	23	23	5
Savlon (0.3% chlorhexidine gluconate 3% cetrimide)		38	51	67	19

TABLE II

ZONES OF INHIBITION IN SPOT TESTS WITH METHYL METHANESULPHONATE, MITOMYCIN C AND SAVLON

	Diameter of zone of inhibition			
	WP2 UV resistant	WP2 *uvrA*	WP67 *uvrA polA*	CM611 *uvrA polA*
Control	—	—	—	—
Methyl methanesulphonate	—	—	20-23 mm	—
Mitomycin C	20 mm	?	?	?
Savlon	16 mm	18 mm	20 mm	22 mm

The highest number of induced mutants per plate and hence the most sensitive test for mutation will be with a treatment that causes slight killing (the optimum may be from 0—90% and will depend on the exact kinetics of killing and mutation induction in any particular instance). To determine a dose response curve it is commonest to treat with different concentrations of an agent for a fixed time, although in particular cases it may be preferable to treat with one concentration for different times.

The protocol given here is to determine a dose response curve for mutation. For a survival curve, mutation plates would be omitted and levels giving higher killing used. It is important in this case to dilute each sample at least 10—100 fold before plating so that killing does not continue on the plate.

If multiple mutation tests are being carried out, it is still desirable to test 2—3 different levels of an agent (giving say 100%, 70% and 10% survival) in order that one will be near the optimum level for detecting mutation.

Strains WP2 *uvrA*, WP67 *(uvrA polA)* and WP2 (UV resistant) are the most likely strains to provide a positive in a preliminary test for mutation. Strain WP2 should be compared with strains WP67 *(uvrA polA)*, CM611 *(uvrA exrA)* and WP100 *(uvrA recA)* in preliminary survival experiments.

Preliminary test to determine an effective concentration

Unfortunately, it is not possible to predict killing in liquid suspension from a spot test. We suggest as a preliminary test:

(a) Take a growing culture of 2×10^8 cells/ml.
(b) Dilute to obtain about 30 ml of 10^7/cells ml in growth medium.
(c) Dispense seven 3-ml samples in universal bottles.
(d) Make up a high concentration of the test agent in 4 ml of the diluted culture.
(e) Serially dilute 1 ml : 3 ml into the other universal bottles, leaving an untreated control (use a fresh pipette for each step).
(f) Incubate the bottles all day or overnight.
(g) Look for the level at which growth is not inhibited. This may give a rough indication of a level of the agent, suitable for testing for mutation.

Full test

Requirements for a mutation experiment with two strains.

Growing cultures of 2 strains;
stock solution of agent under test;
12 empty universal bottles;
27 universal bottles with 10 ml buffer;
100 ml dilution buffer;
centrifuge + sterile tubes (or membrane filtration equipment);
10 ml, 1-ml pipettes;
finnpipette 5—50 μl + 20 sterile tips;
shaking water bath;
40 SA1 plates (0.25 μg/ml tryptophan);
(giving 3 mutation and 2 survival plates for 3 doses + control and using 2 strains);
spreader, Bunsen, 37°C incubator, etc.

Protocol

(a) Logarithmically growing cultures of 2×10^8 cells per ml are prepared as described in section Special methods.

(b) The cultures are membrane filtered (or centrifuged) washed and resuspended in an equal volume of buffer.

(c) It is often simplest to dilute and measure the correct amount of the test agent into a series of universal bottles and then add the appropriate volume of cells suspended in buffer. Where possible the concentrations of test agent should be sufficiently low that they do not interfere with growth of the cells on subsequent plating.

(d) A control of cells treated similarly but without the test agent is included.

(e) The cells are incubated with the agent for an appropriate period (30 min is enough in many cases).

(f) It may be desirable to concentrate the cells by filtering after treatment. With dry plates it is possible to plate up to 0.2 ml, and up to 2×10^8 cells per plate. This may be necessary with a weak mutagen (e.g. γ-irradiation). It may also be necessary to filter to remove a toxic level of a mutagen. More than about 2×10^8 cells cannot be plated because in many cases there will not be sufficient growth for induced mutation to be expressed.

(g) The cell suspension is spread (0.1 or 0.2 ml per plate). At least 3 plates are needed for each treatment.

(h) Samples of the treated suspension are diluted and plated on the same agar (SA1) to determine viability. At high dilutions on the selective agar Trp^- bacteria are able to grow to form small but readily visible colonies (Plate 2). We suggest dilution steps of 0.1 ml : 10 ml and 0.1 ml to 0.9 ml. A fresh tip must be used for each dilution. A total dilution of 10^5- or 10^6-fold will be required.

(i) Plates are incubated at 37°C for 48 h.

(j) The number of colonies on the mutation plates is determined as described earlier.

(k) The number of colonies on the viability plates is determined.

Evaluation and statistical treatment

A test is positive if the number of mutant colonies on treated plates is significantly greater than the number on control plates. This can be tested statistical-

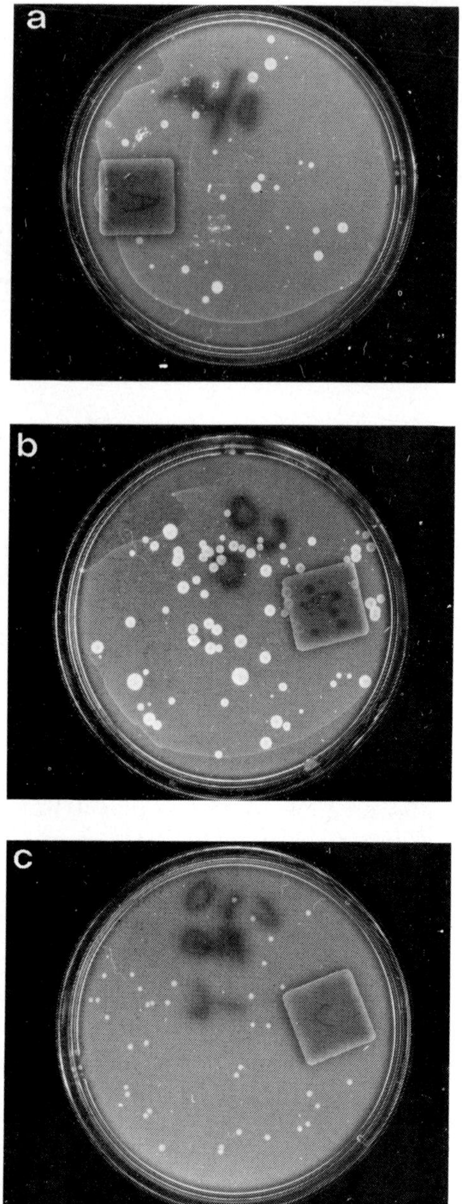

Plate 2. (a) Typical control mutation plate with strain WP2 on SA1 agar. The Trp^+ revertants and the lawn of Trp bacteria are clearly visible. (b) Typical mutation plate. Obtained after strain WP2 received 5 J m^{-2} UV irradiation (SA1 agar). (c) Typical survival plate. When a diluted sample is plated individual Trp bacteria grow to form small but clearly visible colonies on SA1 agar.

ly by Student's t test, or over a series of experiments by an analysis of variance. However, simple rules of thumb may be applied. If a substance is to be considered positive, it should be possible to adjust the conditions of treatment to consistently double the number of mutants per plate, over the spontaneous level (10—20 per plate) in a series of experiments.

If a substance is to be considered negative it should give *no* increase over the spontaneous level in at least 3 separate experiments. These should be under optimum conditions, i.e., with slight killing and testing 2×10^8 cells/plate. Substances which give a non significant increase in mutants should be retested until a consistent picture emerges. The conditions of treatment should be modified as thought necessary to increase the sensitivity of the test, remembering that it is possible to conceal even a potent mutagenic effect by use of inappropriate conditions (Appendix A).

In survival experiments a test is positive if one or more repair deficient strains show increased killing by an agent compared to the wild type. Differences in viability between strains in such cases are usually so great, especially at higher levels of killing (10—100-fold or more) that statistics are unnecessary. Nevertheless, any test should be repeated at least 3 times, or until a consistent pattern emerges. Small differences in surviving fraction (e.g. 2-fold) may be trivial in origin even if statistically significant. It may be possible to corroborate a small effect by testing the remaining strains (CM561, CM571 and WP100), otherwise it should be treated with caution.

To determine the induced mutation frequency for a given treatment the following calculation can be performed:

$$\frac{\text{number of mutants per treated plate} - \text{number of mutants per control plate}}{\text{number of viable cells per treated plate}}$$

It is a slight under-estimate and an exact method is given in Appendix B. The exact calculation requires that the untreated control be plated for mutation on SA3 agar as well as SA1. Some popular but erroneous methods of calculation are given in Appendix A, in a rather extreme form, in the hope of discouraging their use.

Induced mutation frequencies can be plotted directly against dose, or as a logarithm of mutation frequency against a logarithm of dose. Both plots are shown in the illustration. The log/log plot may give a straight line plot when mutation frequency does not increase linearly with dose. Survival is generally plotted as the logarithm of the surviving fraction against dose.

Illustration

Plate 2 shows a typical control and treated mutation plate and a typical survival plate. Dose response curves are shown for mutation (Fig. 2) and for survival (Fig. 1) for strain WP2 treated with methyl methanesulphonate (MMS). Protocol was as described. From Fig. 2(a) it can be seen that mutation does not increase linearly with dose. The log/log plot in Fig. 2(b) is a straight line and suggests that over the range studied mutation increases with the 1.65th power of dose. Results of this type (slope steeper than 1.0) are fairly common with misrepair mutagens.

A modified fluctuation test for detection of weak mutagens

Although the treat and plate protocol for testing mutagens is more sensitive than a spot test, it is still less than ideal. The tryptophan supplement in the selective agar is essential to allow a period of growth during which initial DNA damage is converted into a mutation, and the mutation is expressed as a func-

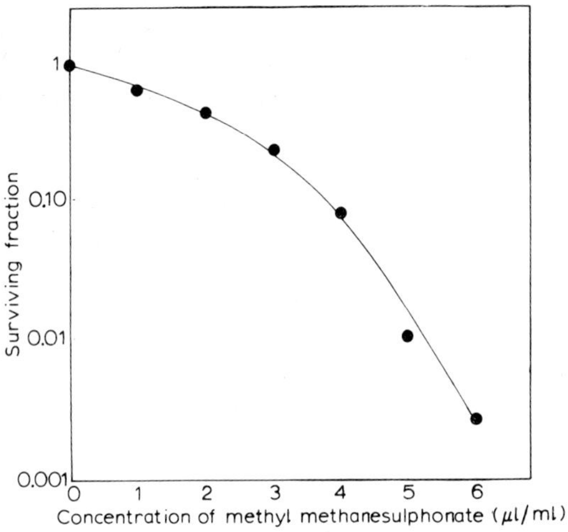

Fig. 1. Survival of strain WP2 treated for 30 min with different concentrations of methyl methanesulphonate.

tional gene product. Only then can a mutant grow to form a colony. However, the supplement also limits the sensitivity of the test. If cells are treated and plated, and grow for 4—5 generations on the plate, the comparison will be between induced mutation in 10^8 cells and spontaneous mutation in 2.5×10^9 cells. Only a large increase over the spontaneous mutation rate will be detected, even under optimal conditions. To overcome this difficulty it is necessary to start with a method that measures spontaneous mutation rate and we have therefore used a modification of the Luria and Delbruck fluctuation test [14]. This system has been used in its classic form by Voogd et al. [23], but we have adopted a modification by Ryan [20], which gives an attractively simple test. The method is suitable for use with reasonably stable agents which are mutagenic at non-toxic concentrations, and it can be remarkably sensitive in detect-

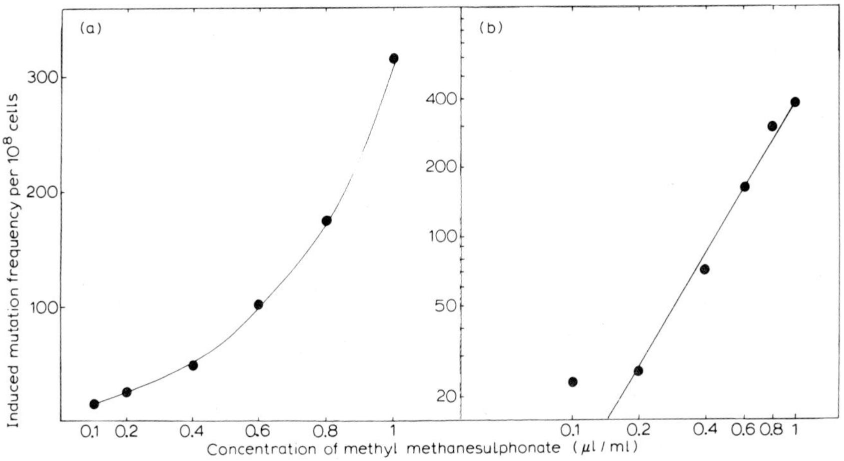

Fig. 2. Induced mutation frequency in strain WP2 treated with methyl methanesulphonate. (a) Linear plot; (b) log/log plot.

ing low levels of mutagen. It will give a false positive with any agent which allows extra growth of tryptophan requiring cells. It offers no advantage with unstable mutagens, or agents which are mutagenic only at toxic levels.

Requirements

A growing culture of the strain under test;
4 racks of 50 small test tubes;
400 ml solution F, 0.1 ml solution E, 8 ml solution C;
4 × 115 ml bottles;
100 ml sterile measuring cylinder;
20-ml pipettes + Pipump (remote pipetting aid);
stock solution of agent under test;
37°C incubator, etc.

Protocol

(a) A preliminary test (as described above) will be necessary in order to estimate suitable concentrations of the test agent.

(b) A growing culture of strain WP2 or one of its derivatives is prepared as described previously.

(c) To 400 ml of solution F is added 8 ml of solution C, 0.1 ml solution E and about 0.1 ml of growing culture (to give 1—2 × 10^5 cells/ml). The mixture is divided into 4 × 102 ml samples.

(d) To three samples, low concentrations of the test agent are added (the fourth sample is a control). The concentration of the test agent used should not inhibit growth.

(e) Each 102 ml is dispensed in 2 ml aliquots in 50 small test tubes (12 mm × 75 mm are suitable), preferably in a rack where all the tubes are directly visible. This step in the procedure should be capable of automation if the method is being used on a large scale.

(f) The rack is covered and incubated at 37°C.

(g) The auxotrophic bacteria grow to exhaust the tryptophan, at which stage only Trp^+ revertants can grow.

(h) From 48 h onwards tubes containing Trp^+ revertants become unequivocally turbid. More tubes become positive on subsequent days.

(i) We suggest that the experiment should be scored after 72 h (and possibly re-scored at 96 h). The number of turbid tubes (i.e., containing Trp^+ revertants) in the control and treated racks is compared.

Evaluation

For a test to be considered negative we would suggest that a compound ought to give no increase in the number of turbid tubes in 3 separate experiments. These experiments should test 3 different concentrations of the agent from just below an inhibitory concentration down (any compound will appear negative at an inhibitory concentration, and if there are significantly fewer tubes positive in the treated than in the control series, the concentration of agent is almost certainly too high).

For a test to be considered positive, there should be significantly more turbid tubes in the treated than in the control sample in 3 separate experiments.

TABLE III

SIGNIFICANCE OF 50 TUBE FLUCTUATION TESTS (CALCULATED USING χ^2)

Tubes turbid in control	Tubes turbid in treated sample for significance at probability		Tubes turbid in control	Tubes turbid in treated sample for significance at probability	
	5%	1%		5%	1%
0	4	7	26	36	39
1	6	9	27	37	40
2	8	11	28	38	41
3	10	13	29	39	41
4	11	14	30	40	42
5	13	16	31	40	43
6	14	17	32	41	44
7	16	19	33	42	44
8	17	20	34	43	45
9	18	21	35	44	46
10	19	23	36	44	46
11	21	24	37	45	47
12	22	25	38	46	48
13	23	26	39	46	48
14	24	27	40	47	49
15	25	28	41	48	49
16	26	29	42	48	50
17	27	30	43	49	50
18	28	31	44	49	—
19	29	32	45	50	—
20	30	33	46	50	—
21	31	34	47	—	—
22	32	35	48	—	—
23	33	36	49	—	—
24	34	37	50	—	—
25	35	38			

Significance is estimated by χ^2. For 50 tube experiments, if a specified number of tubes are turbid in the control, Table III gives the minimum number which must be turbid in the treated sample for significance at the 5% or 1% probability levels.

If a compound gives a non-significant positive response, it should be re-tested until a consistent picture emerges. In this instance if there are equal numbers of tubes in the treated and control series and, t = total number of tubes turbid in the whole treated series, c = total number of tubes turbid in the whole control series, n = total number of tubes in the whole treated or control series.

Then

$$\chi^2 = \frac{2n(t-c)^2}{(t+c)(2n-t-c)} \text{ with 1 degree of freedom,}$$

if χ^2 is greater than 3.84 it is significant at the 5% probability level;
if χ^2 is greater than 6.64 it is significant at the 1% probability level;
if χ^2 is greater than 10.83 it is significant at the 0.1% probability level.

This type of fluctuation test as it stands is not a universal panacea, but in particular instances it can be remarkably effective. It can be applied equally well to other organisms. J.M. Parry (pers. comm.) has successfully used the method for mitotic conversion in yeast, and it should be equally applicable to the histidine requiring *S. typhimurium* strains of Ames (2). Its particular merit is that it measures mutation frequencies with levels of mutagens which are likely to be found in the environment.

Illustration

Table IV shows the number of tubes positive after 3 days in a typical series of experiments. In a treat and plate test methyl methanesulphonate can only be detected as a mutagen at concentrations at least 20-fold higher and mitomycin C is too weak a mutagen to be detected at all. With K_2CrO_4, the effect of using too high a concentration can be seen and $NiCl_2$ is an example of an agent giving a negative result. However, extremely low concentrations of methyl methanesulphonate, mitomycin C and K_2CrO_4 are all significantly mutagenic at the 1% probability level.

Use of E. coli with liver microsomal preparations

With *E. coli* we have found the following adaptation of the treat and plate method convenient.

(a) Cultures are grown to 2×10^8 cells/ml as described previously.

(b) For a survival experiment, the cultures are diluted 10^4-fold in buffer.

(c) To a universal bottle is added 1 ml of liver microsomal preparation with the appropriate co-factors [8] 0.1 ml of the diluted cells and an aliquot of the substance under test. A control without the agent is included for each strain.

(d) The bottle is closed and incubated at 37°C in a shaking water bath for 30 min.

(e) 0.1 ml aliquots of the mixture are plated directly on nutrient agar.

(f) Plates are incubated for 24 h.

(g) Plates are counted.

(h) For a mutation experiment, a 50 ml growing culture is centrifuged, resus-

TABLE IV

MUTAGENESIS BY METHYL METHANESULPHONATE, MITOMYCIN C, K_2CrO_4, and $NiCl_2$, TYPICAL DATA FROM FLUCTUATION TESTS

Agent	Concentration	Control Tubes turbid (after 72 h)	Treated Tubes turbid (after 72 h)	Significance (probability)
Methyl methanesulphonate	0.5 μg/ml	8	32	$P < 0.01$
	0.5 μg/ml	12	28	$P < 0.01$
Mitomycin	0.05 μg/ml	6	20	$P < 0.01$
K_2CrO_4	2.5 μg/ml	10	1	$P < 0.01$[a]
	0.5 μg/ml	17	39	$P < 0.01$
$NiCl_2$	25.0 μg/ml	8	4	$P > 0.05$

[a]Significantly fewer in treated sample.

The standard protocol was used with 50 tubes for each treatment.

pended in 1/10 or 1/20 volume and 0.1 ml added to the liver preparation. Following treatment 0.2 ml aliquots are plated on 3 SA1 plates. A further sample is diluted 10^6 -fold and plated for survival on 2 SA1 plates.

Evaluation

The results are similar in form to a treat and plate test, though rather more variable. Similar calculations and statistical analyses can be performed.

Illustration

Table V gives the result of a survival experiment with the pyrrolizidine alkaloid heliotrine. It can be seen that the doubly repair deficient strains show increased killing. The pattern of sensitivities (WP2 *uvrA* more sensitive than WP2, CM611 and WP100 most sensitive) is similar to that for the DNA cross-linking agent mitomycin C (see Appendix C) which agrees with evidence from in vitro studies that heliotrine is a metabolically activated DNA cross-linking agent [24].

Modifications and snags

(a) If it is desired to perform a treat and plate experiment on growing cultures, the initial filtration (transferring the cells from growth medium to buffer) is simply omitted. Solution D (casamino acids) can be omitted from the growth medium if it is suspected that it would inactivate the test agent.

(b) Solution D can be omitted from SA1 or SA2, if it is suspected that it is inactivating a potential mutagen (e.g., an alkylating agent). With SA1, the level of tryptophan (Solution E) should then be increased to 0.75 μg/ml since with some agents more generations of growth on the plate may be required for induced mutations to be expressed.

(c) If a compound is positive in a test with one or more of the strains described here, it can be tested with all the strains, in order to characterise it as fully as possible.

(d) If the number of Trp^+ colonies on control plates (selective agar with 0.25 μg/ml tryptophan) exceeds an average of approx. 25 per plate (60 for WP67)

TABLE V

SURVIVAL OF STRAIN WP2 AND ITS DERIVATIVES FOLLOWING TREATMENT WITH 10 MG/ML HELIOTRINE IN THE PRESENCE OF A LIVER MICROSOMAL FRACTION

Strain	Viability (counts per plate)				Survival
	Control		Treated		(%)
WP2 UV resistant	48	44	40	66	115
WP2 *uvrA*	48	75	42	39	66
WP67 *uvrA polA*	43	52	28	37	68
CM561 *exrA*	52	42	48	52	106
CM611 *uvrA exrA*	53	43	6	7	13.5
WP100 *uvrA recA*	111	96	1	3	1.9

The liver microsomal fraction and co-factors were prepared according to Ames et al. [1].

there may be an excessive background of Trp$^+$ revertants in the stock culture. This can be tested as described in section Special methods. An experiment with an excessive background should be treated with reserve.

If a high Trp$^+$ background is not responsible for an excessive number of spontaneous Trp$^+$ revertants, there may be a small amount of tryptophan in the casamino acids or liver microsomal preparation. The amount of Solution E added can be reduced to compensate for this.

(e) The sensitivity of the fluctuation test can be increased slightly by using 50 × 8 ml aliquots instead of 50 × 2 ml. On statistical grounds the test is most sensitive with 50—80% of the tubes positive in the control. However with 8-ml samples all the tubes are faintly turbid and the experiment is much harder to score.

Future developments

(a) MacPhee [15] has recently demonstrated the use of strains of *S. typhimurium* containing a plasmid which enhances mutability. He is currently developing similar strains of *E. coli*. Such strains will show greatly increased mutation with certain types of agent.

(b) We do not currently possess an isogenic derivative of strain WP2, lacking the endonuclease which recognises 0—6 alkyl guanine. Such a strain might be extremely useful.

(c) We are currently trying to construct a *uvrA exrA recA* triple mutant. A *recA exrA* double mutant in a different genetic background shows extreme repair deficiency with excellent growth and if this were true for the triple mutant, it would be a good substitute for strains CM611 and WP100.

(d) Finally, we are attempting to develop the fluctuation test to deal with unstable and liver-activated mutagens. If we are successful it should greatly extend the usefulness of this technique.

Equipment

Apparatus

All the apparatus required for these tests is likely to be present in any laboratory that routinely handles micro-organisms. A partial check-list of some more expensive items is given in Table VI. Useful but non-essential items have been bracketed. Quantities of glassware etc., will depend on the work being done, for instance fluctuation tests require thousands of small test tubes and no petri dishes. Finnpipettes are available from Jencons Ltd., Hemel Hempstead, England. Pipumps from Schuco Ltd., London W12, England.

Chemicals

These are listed in Table VII.

Bacterial strains

These are described in Table VIII. They are nearly isogenic derivatives of *E. coli* strain WP2, which is tryptophan-requiring, radiation resistant derivative of *E. coli* B (the strain used by Luria and Delbruck [14] in their original demonstration of mutation in bacteria). These strains have been deposited with the

TABLE VI

MAJOR EQUIPMENT FOR MUTAGEN TESTING WITH *E.COLI*

Autoclave

Oven

Incubator for 37°C incubation, about 6 cu. ft. needed per person.

Shaking water bath with rack for universal bottles. A simple water bath + aquarium pump to bubble air through cultures is also useful to aerate larger volumes.

Small refrigerator with freezer compartment. 6 cu. ft. for 1—2 people.

(Microscope) × 400 magnification dry field with phase contrast (or dark field) + Thoma ruled counting chambers.

Centrifuge with autoclavable tubes (50 ml are convenient). Benchtop will suffice, though a larger model is convenient.

Membrane filtration equipment. Filter holders and suction source, can be used instead of a centrifuge for most purposes.

Balance. To weigh 1—50 g.

Balance. To weigh 5 mg—1 g (or more accurate).

Fume cupboard (Hood) efficient enough to handle voltaile mutagens safely.

Source of distilled or at least deionised water.

TABLE VII

CHEMICALS REQUIRED

K_2HPO_4	For media
KH_2PO_4	
$(NH_4)_2SO_4$	
Trisodium citrate	
$MgSO_4 7H_2O$	
NaCl	

New Zealand Agar (Davis, Christchurch, New Zealand) (or a higher purity agar)

Difco Casamino Acids (Vitamin free) (Difco Detroit, Michigan, U.S.A.)

Nutrient Broth No. 2 (Oxoid, London England)

D-Glucose

L-Tryptophan

Chloroform

Dimethylsulphoxide

Analar ethanol

Industrial methylated spirit

Mitomycin C (Sigma, London)

TABLE VIII

BACTERIAL STRAINS

Strain		Relevant genes		Constructed by
Primary test strains	WP2	*trp*	UV resistant	E.M. Witkin
	WP2 *uvrA*	*trp*	*uvrA*	R.F. Hill
	WP67	*trp*	*uvrA polA*	E.M. Witkin
	CM611	*trp*	*uvrA exrA*	M.A. Rothwell
Secondary test strains	CM561	*trp*	*exrA*	M.A. Rothwell
	CM571	*trp*	*recA*	M.A. Rothwell
	WP100	*trp*	*uvrA recA*	E.M. Witkin

National Collection of Industrial Bacteria, Torrey Research Station, P.O. Box 31, 135 Abbey Road, Aberdeen, AB9 8DG, Scotland, U.K.

We have suggested that strains WP2, WP2 *uvrA*, WP67 and CM611 be used in an initial test. Strain WP2 is the repair-proficient parent strain, strain WP2 *uvrA* lacks excision repair, so that it is more readily mutated by agents such as UV. Strain WP67 contains in addition to *uvrA* a *polA* mutation. A *polA* strain lacks DNA polymerase I and shows increased mutability, and also increased sensitivity to a particularly wide spectrum of agents including those such as methyl methanesulphonate which cause DNA strand breakage [22]. Finally strain CM611 is deficient in both excision repair *(uvrA)* and error-prone rec-dependent repair *(exrA)*. This means that it is almost completely unable to repair its DNA, and is extremely easily killed by most mutagens.

Of the remaining strains, CM561 and CM571 are excision proficient but *exrA* and *recA* respectively. Both mutations block the error-prone repair process, but *recA* strains tend to be more sensitive, since *exrA* strains seem able to perform a limited amount of repair without mutation. Strain WP100 is *uvrA recA* and thus the most sensitive strain of all. The reason that we prefer to use strain CM611 instead of WP100 for primary screening, is that WP100 grows poorly, with a low plating efficiency, which makes simple tests of lethality hard to evaluate.

Costing

For one person performing tests, over half their time would be spent preparing media, counting experiments, tabulating data etc., leaving about 3 h per day to carry out experiments. With spot tests, about 15 compounds could be dealt with in this time. If each test were repeated at least 3 times, and allowing for extra repetition of dubious results, devising better procedures for awkward compounds and ordinary natural disasters, the final estimate would be 2 compounds per person per day. With the treat and plate method, about 6 survival or mutation curves of 3 points + control can be performed in a similar period, and correcting for repetition and difficulties, this gives one compound per person per 9 days. Output with liver microsome tests is similar.

The fluctuation test does not require a survival curve, so that up to 2 tests of 3 concentrations of a compound + control can be performed per day. Assuming that 2 strains are tested per compound, and correcting as before, this again gives one compound per person per nine days.

Special methods

Media

SA1 *selective agar for general use* (0.25 μg/ml L-tryptophan)
300 ml solution B;
100 ml solution A;
8 ml solution C;
10 ml solution D;
0.1 ml solution E.

Constituents are given in Table IX. To pour plates we melt a bottle of agar (B) by re-autoclaving, allow to cool for about 20 min (to 60—70°C), pour in A, C,

TABLE IX

CONSTITUENTS OF MEDIA USED FOR MUTAGEN TESTING

Solution A:	(Davis Mingioli salts)	
	K_2HPO_4	28 g
	KH_2PO_4	8 g
	$(NH_4)_2SO_4$	4 g
	Trisodium citrate	1 g
	$MgSO_4 \cdot 7H_2O$	0.4 g
	Distilled water	1 l
Dispense in 100 ml quantities and sterilize by autoclaving. 15 lbs/sq in. 15 min.		
Solution B:	Davis New Zealand agar	6 g
	Distilled water	300 ml
Make up in 5 50 ml bottle. Autoclave. 15 lbs/sq in. 15 min.		
Solution C:	D-Glucose 20 g	
Dissolve and make up to 100 ml in distilled water, dispense in 8 ml quantities. Autoclave. 5 lb sq in. 10 min.		
Solution D:	Difo casamino acids	50 g
	Distilled water	500 ml
Dissolve, dispense in 10 ml quantities, autoclave. 15 lbs/sq in. 15 min.		
Solution E:	L-Tryptophan	100 mg
	Distilled water	100 ml
Dissolve, dispense in 10 ml quantities, autoclave, 5 lbs/sq in. 10 min. add a few drops chloroform and stored in refrigerator.		
Solution F:	Solution A	100 ml
	Distilled water	300 ml
Make up in 5 50 ml bottles, autoclave.		

and D, pipette in E, swirl well (shaking makes too many bubbles) and pour. This gives enough for about 16 × 9 cm plastic petri dishes. Plates must be dried before use either by storing 3—4 days at room temperature or just before use placing them upside-down with the lids off at up to 50°C in a sterilising oven. Plates can be used for 2—3 weeks after pouring. Plates should normally be kept upside down to prevent condensation from the lid falling on to the agar surface.

SA2 *Selective agar for spot tests* (1 μg/ml L-tryptophan)
Prepared as SA1, but with 0.4 ml solution E.

SA3 *Unsupplemented agar*
Prepared as SA1, but omitting solutions D and E.

Growth medium
To 400 ml solution F (Table IX) add aseptically;
8 ml solution C;
10 ml solution D;
4 ml solution E (giving 10 μg/ml L-tryptophan).

Nutrient agar

Oxoid nutrient broth No. 2	12.5 g;
Davis New Zealand Agar	7.5 g;
Distilled water	500 ml.

Make up in 20 oz bottles and autoclave (15 lbs/sq in 20 min). Dispensed by melting (re-autoclaving), cooling to 50°C and pouring about 20 plates from 1 bottle.

Stab agar

Davis New Zealand Agar	0.6 g
Oxoid Nutrient Broth No. 2	0.75 g;
NaCl	0.3 g;
Distilled H_2O	100 ml.

The constituents are made up and melted in a beaker. 2—3 ml aliquots are dispensed into small test tubes (e.g. 3″ × 0.375″) in a rack. The rack of tubes is covered with foil and autoclaved. Suitable corks are sterilized, coated with wax and the tubes sealed. The tubes are incubated at 37°C for several days to check that they are completely sealed and that there is no contamination. An important point about this medium is that the concentration of broth is less than half the normal level.

Dilution buffer

Solution F. (Table IX). This can be dispensed in 10-ml aliquots in universal bottles and re-autoclaved. When autoclaving 10-ml aliquots it may help to place a 1 litre beaker of distilled water in the centre of the autoclave, to reduce loss of volume of the buffer. Dispensing 0.9 ml volumes for 10-fold dilutions is best done aseptically just before the experiment.

Essential points concerning media

On irradiation with UV light tryptophan forms a toxic and mutagenic photoproduct. We have never found this phenomenon to affect results but it would seem advisable to store tryptophan containing media away from direct light.

If the media and brand names described here are not available others may often be substituted. The minimal salts used need not necessarily be Davis-Mingioli, though if another formula is used, it is essential that it contain trisodium citrate (at least 0.25 mg/ml). Any good quality agar can be used. Any standard bacterial dilution buffer can be used. Any standard nutrient broth can be used at the appropriate strength provided that it contains tryptophan, but casamino acids and similar acid hydrolysed media do not contain significant amounts. This lack of tryptophan instead means that casamino acids can conveniently be incorporated in the selective agar where they tend to make zones of inhibition easier to distinguish, and increase the mutation yield by certain mutagens. The casamino acids used should be tested for the absence of tryptophan. Plates prepared including Solution D but omitting Solution E should show no visible lawn and not more than 3—5 spontaneous mutants per 10^8 cells plated.

The growth medium must contain sufficient (10 μg/ml) tryptophan to prevent overgrowth by Trp^+ revertants. The level of tryptophan in the selective agar plates (0.25 μg/ml) is even more critical and Solution E must be dispensed as accurately as possible. Any inaccuracy will reduce the sensitivity of the test, and could even lead to false positives. The values given are for L-tryptophan, with DL-tryptophan, quantities should be doubled. Dilutions can be carried out in test tubes rather than universal bottles, but we would advocate the use of closed containers such as universal bottles for treating cells. (In fact in cost and storage space test tubes and universal bottles are very comparable.)

Maintenance of bacterial stocks

(a) Cultures are streaked out on nutrient agar plates, incubated at 37°C overnight and stored in the refrigerator for up to a month before restreaking.

(b) For long-term storage, after a streaked plate is incubated overnight, a single colony is touched with a sterile inoculating needle and a small inoculum is stabbed into a stab agar tube. The tube is carefully sealed and stored at room temperature. Several tubes are prepared for each strain. After a few days it is necessary to check that the stab has grown and that the tube is sealed. Cultures can survive for several years stored in this way.

(c) For a test, a single colony on a plate is inoculated into 10 ml of growth medium and incubated at 37°C with aeration overnight. Next day it is diluted into fresh growth medium and incubated until it reaches about 2×10^8 cells/ml (about 2½ h).

Dilute strains WP2, WP2 *uvrA* and CM561 0.1 ml to 10 ml; WP67, CM611 and CM571 0.2 ml to 10 ml; WP100 0.4 ml to 10 ml.

(d) For larger volumes of cells, air can be bubbled through the culture with an aquarium pump.

(e) If a series of tests are planned over a period of 2—3 weeks, it is convenient to store an overnight culture in the refrigerator. Provided the first experiments show an acceptable background of spontaneous mutants, this stock can be used instead of an overnight culture in subsequent experiments.

TABLE X

TREATMENT OF RADIATION SENSITIVE DERIVATIVES OF STRAIN WP2 WITH UV RADIATION AND MITOMYCIN C

UV irradiation Strain	Dose Jm^{-2} 0	1	4	8	20
WP2 UV resistant	+++	+++	+++	+++	+++
WP2 *uvrA*	+++	+++	+++	++	+
WP67 *uvrA polA*	+++	+++	++	+	o
CM611 *uvrA exrA*	+++	++	+	o	o
CM561 *exrA*	+++	+++	+++	++	+
CM571 *recA*	+++	++	+	o	o
WP100 *uvrA recA*	+++	o	o	o	o

Mitomycin C Strain	Dose μg/ml (0.1 ml spread on plate) 0	20	40	100	250
WP2 UV resistant	+++	+++	+++	+++	++
WP2 *uvrA*	+++	+++	++	++	+
WP67 *uvrA polA*	+++	+++	++	+	o
CM611 *uvrA exrA*	+++	+++	+	+	o
CM561 *exrA*	+++	+++	++	++	+
CM571 *recA*	+++	++	+	+	o
WP100 *uvrA recA*	+++	+	o	o	o

+++ No visible lethality.
++ Some lethality visible.
+ Little survival.
o No survival (0—3 colonies).
Our UV lamp has been recalibrated and the revised doses are 0.6, 2.4, 4.8 and 12 J m^{-2}.

WP2 UV resistant +++

WP2 *uvrA* +++

WP67 *uvrA polA* +++

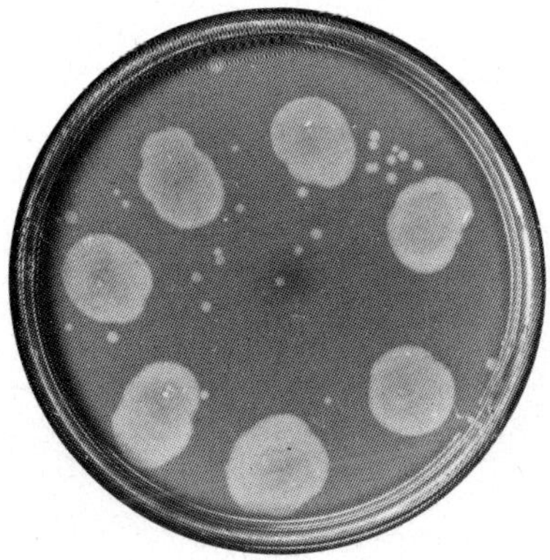

CM 571 *recA* +++

CM611 *uvrA exrA* +++

CM561 *exrA* +++

WP100 *uvrA recA* +++

Plate 3a. Pre-testing of WP2 derivatives for repair deficiency. 0 J m^{-2}.

WP2 UV resistant +++

WP2 *uvrA* +++

WP67 *uvrA polA* +++

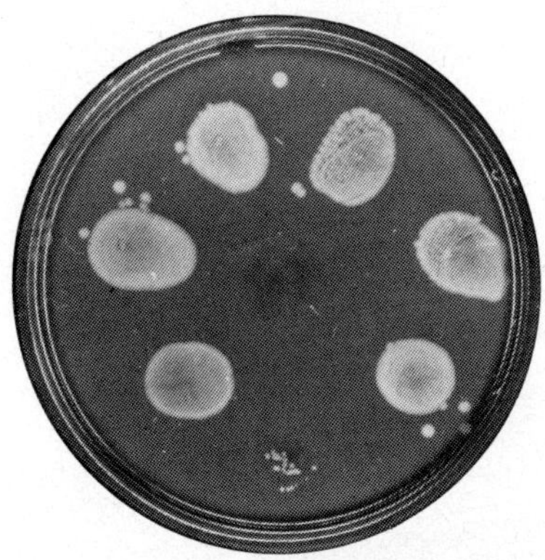

CM571 *recA* ++

CM611 *uvrA exrA* ++

CM561 *exrA* +++

WP100 *uvrA recA* +

Plate 3b. 1 J m^{-2}.

WP2 UV resistant +++

WP2 *uvrA* +++

WP67 *uvrA polA* +++

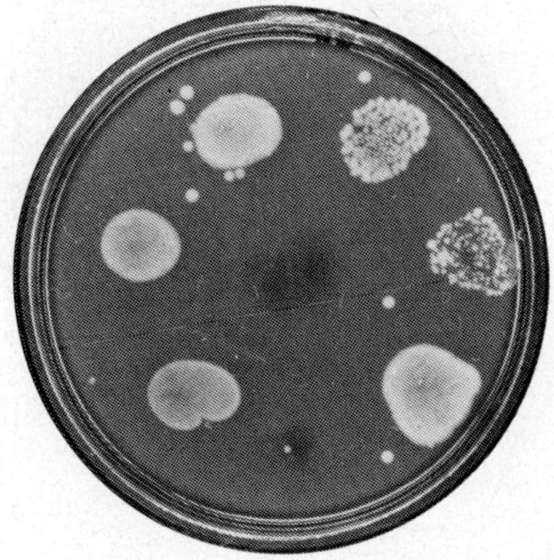

CM571 *recA* ++

CM611 *uvrA exrA* ++

CM561 *exrA* +++

WP100 *uvrA recA* 0

Plate 3c. 4 J m^{-2}.

WP2 UV resistant
+++

WP2 *uvrA*
+++

WP67 *uvrA polA*
++

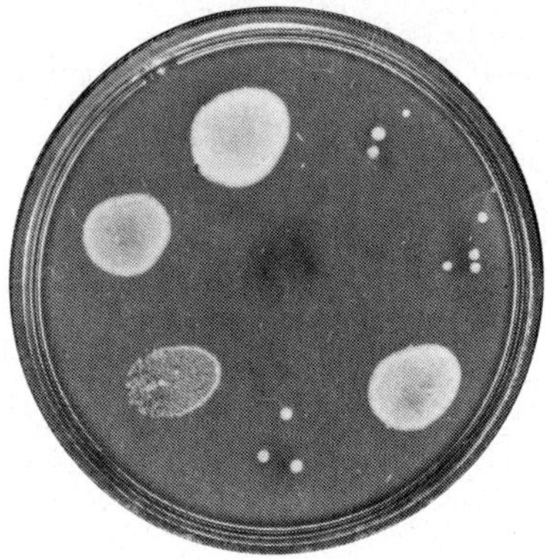

CM571 *recA*
0

CM611 *uvrA exrA*
0 +

CM561 *exrA*
+++

WP100 *uvrA recA*
0

Plate 3d. 8 J m^{-2}.

WP2 UV resistant
+++

WP2 *uvrA*
++

WP67 *uvrA polA*
+

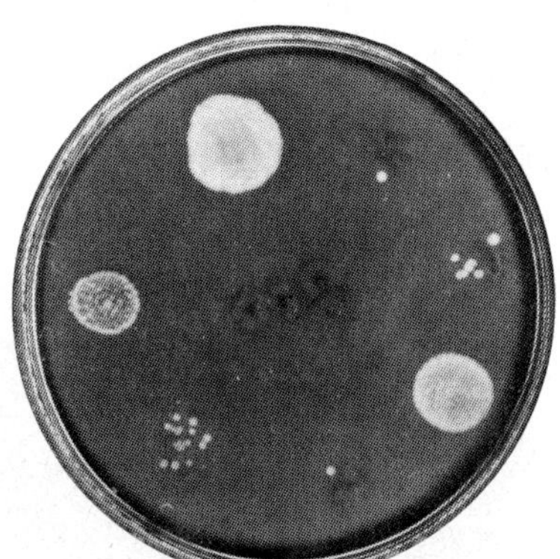

CM571 *recA*
0

CM611 *uvrA exrA*
0 +

CM561 *exrA*
++

WP100 *uvrA recA*
0

Plate 3e. 20 J m^{-2}. The colonies at high doses in WP100 and CM611 are almost certainly UV resistant revertants.

(f) To test the repair deficiency of the strains it is best to use UV irradiation as a reference DNA damaging agent, though mitomycin C can be used as an alternative. The strains are grown to 2×10^8/ml as described in (c) above. Using a 0.1-ml pipette 0.01-ml spots of the strains are made on six nutrient agar plates so that each plate has a spot of each strain. Each plate is irradiated with a 254 mm UV lamp and given one of the doses shown in Table X. The plates are incubated overnight and the spots examined next day and classified according to Table X. Typical results are shown in Plate 3. If the UV lamp is not calibrated, it will be necessary to try different doses and compare the relative sensitivities of the strains.

(g) A series of solutions of mitomycin C, of the strengths indicated in Table X are prepared by diluting in buffer. For each concentration, 3 nutrient agar plates are carefully spread with 0.1 ml of the mitomycin C solution. Immediately after the mitomycin is spread, each plate is inoculated with a 0.01-ml spot of each strain using a 0.1-ml pipette and starting with plates containing least mitomycin. Plates are incubated at 37°C overnight and scored. Results are less reproducible than with UV but it should be possible to determine relative sensitivities according to Table X.

(h) To test whether a stock has an excessive background of spontaneous Trp^+ revertants, about 1 ml of the stock is filtered, resuspended in the same volume of buffer, undiluted 0.2-ml samples are plated on SA3 agar, and diluted samples (approx. 10^6-fold) are plated on nutrient agar to determine viability. More than about 2 mutants per 10^8 viable cells plated (3 for WP2 *uvrA*, 5 for WP67) suggests an excessive background of revertants and the stock should be discarded.

Acknowledgement

We would like to thank Professor B.A. Bridges for his help and advice.

Appendix A

The distilled water effect or how to make every experiment a positive

Recent work in environmental mutagenesis has shown that many substances, long thought harmless, are in fact mutagenic. However, other compounds, common in the environment, remain to be tested and among these is water. Lack of interest in water as a potential mutagen is surprising since it is a common ingredient of foodstuffs, and drinks, moreover it is a major atmospheric pollutant, especially in countries such as Britain. We have, therefore, performed the following experiment. A growing culture of strain WP2 was filtered and resuspended in distilled water. Over a period of 12 days samples were plated on selective agar to determine viability and the number of mutant colonies per plate. Fig. 3a shows that the viability declined steadily during the 12-day period, but Fig. 3b shows that the number of mutant colonies on the selective agar plates remained constant.

If we ignore the calculation given in section Mutagen screening and deter-

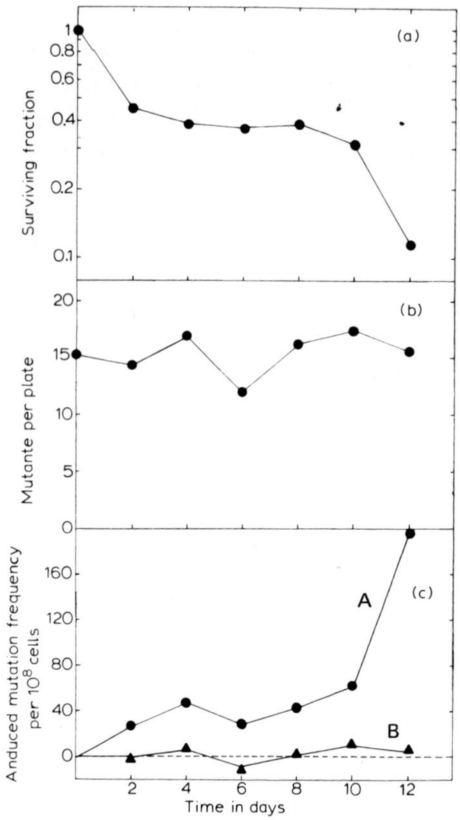

Fig. 3. Survival and mutation of strain WP2 incubated in distilled water. (a) Survival; (b) mutants per plate; (c) induced mutation frequency per 10^8 cells, (●———●) calculated erroneously, (▲———▲) induced mutation frequency calculated correctly.

mine induced mutation frequency as

$$\frac{\text{mutants/treated plate}}{\text{viable cells/treated plate}} - \frac{\text{mutants/control plate}}{\text{viable cells/control plate}}$$

Line A in Fig. 3c shows that the apparent induced mutation frequency steadily increases.

The experiment seems to indicate that distilled water is a potent mutagen. What is the catch? The experiment is a perfectly valid example of the treat and plate method and the only departure from our recommended procedure is in the use of an invalid calculation. The simplest valid calculation of the induced mutation frequency is

$$\frac{\text{mutants/treated plate} - \text{mutants/control plate}}{\text{survivors/treated plate}}$$

The reason for this calculation is that there are two classes of spontaneous mutant, (a) pre-existing mutants, present in the population of cells at the time of plating, and (b) plate mutants, which arise during the period of growth on the plate. In practice nearly all the spontaneous mutants are in the second category of plate mutants. The number of plate mutants is dependent on the final number of auxotrophic cells which grow on the selective agar plate and mutate. This is in turn dependent on the supplement of tryptophan in the selective agar. *It is completely independent of the number of cells plated.* However many cells are plated, (within limits of between perhaps 5×10^4 and 5×10^8) the same number of cells will grow on a selective agar plate containing a given amount of tryptophan and give rise to approximately the same number of plate mutants.

Using a proper calculation, the induced mutation frequency is given by line B in Fig. 3c. This is less sensational but considerably more reassuring. We make this point at such length, because although it has been known for over 20 years, it is widely ignored, and the miscalculation given here is still frequently used. The miscalculation is equally applicable to His^+ reversion in Salmonella, and any similar system where growth after treatment is to a fixed number of cells, rather than for a fixed number of generations. A significant amount of mutation data is calculated using a method that would show distilled water, or anything that kills bacteria, to be a mutagen.

How to make every experiment a negative

Innumerable ways of failing to detect mutation exist. One can fail to activate a mutagen, one can successfully inactivate it, one can use the wrong strain or treat under wrong conditions. One can simply add too little mutagen and, especially with fluctuation tests, there is the extra subtlety that negative results can be generated by adding too much mutagen.

However, one of the simplest ways not to detect mutation is to plate too few cells. Using the treat and plate method described here, the lowest induced mutation frequency which can be detected is about 1 in 2×10^7 cells, an increase of about 20-fold over the spontaneous mutation rate. This maximum sensitivity is obtained plating 2×10^8 cells/plate. If, instead, 10^7, 10^6 or 10^5 cells are plated, the controls will have a quite normal number of spontaneous mutants (as explained above) but an increase of 400, 4000 or 40,000 fold over the spontaneous mutation frequency will be required for an agent to be detected as a mutagen. Hence when declaring agents to be non-mutagenic it is important to consider the induced mutation frequency which could have been detected in the test.

Appendix B

Exact calculation of induced mutation frequency [3, 21]

M = induced mutation frequency per survivor;
a = total number mutants per SA1 plate (treated cells);
b = total number mutants per SA1 plate (untreated cells);
c = total number mutants per SA3 plate (untreated cells);
d = number of viable cells per mutation plate (treated cells);
e = number of viable cells per mutation plate (untreated cells).

Spontaneous mutants will comprise pre-existing mutants, present when the population was plated and plate mutants arising during growth on the plate. Only the pre-existing mutants can grow on unsupplemented agar (SA3), so that

TABLE XI

TYPES OF DNA DAMAGE INDUCED BY CERTAIN AGENTS

	Type of damage	Examples of agents	Effect on survival	Effect on mutation
A	Excisable misreplication damage	Volatile breakdown product of Captan monofunctional sulphur mustards	—	More mutation with WP2 *uvrA* than WP2, CM611 *uvrA exrA* than CM561 *exrA*, WP100 *uvrA recA* than CM571 *recA*
B	Excisable misrepair damage	UV radiation, Captan (8-methoxypsoralen)	WP2 *uvrA* more sensitive than WP2	More mutation with WP2 *uvrA* than WP2
			CM611 *uvrA exrA*, and WP100 *uvrA recA* more sensitive than WP2 *uvrA*	No mutation with CM561 *exrA*, CM571 *recA*, CM611 *uvrA exrA*, WP100 *uvrA recA*
C	Non-excisable misreplication damage	*N*-methyl-*N*′-nitro-*N*-nitroso guanidine, *N*-nitroso-*N*-methylurea (ethyl methanesulphonate)	WP2 *uvrA* as resistant as WP2	Equal mutation with WP2 and WP2 *uvrA*
			WP67 *uvrA polA*, CM561 *exrA*, CM571 *recA*, CM611 *uvrA exrA*, WP100 *uvrA recA* all more sensitive than WP2	Mutation with CM561 *exrA*, CM571 *recA*, CM611 *uvrA exrA*, WP100 *uvrA recA*
			CM611 *uvrA exrA* as resistant as CM561 *exrA*	
D	Non-excisable misrepair damage (strand breaks?)	γ-radiation, methyl methanesulphonate, dichlorvos	WP2 *uvrA* as resistant as WP2	Equal mutation with WP2 and WP2 *uvrA*
			WP67 *uvrA polA*, CM561 *exrA*, CM571 *recA*, CM611 *uvrA exrA*, WP100 *uvrA recA* more sensitive than WP2	No mutation with CM561 *exrA*, CM571 *recA*, CM611 *uvrA exrA* or WP100 *uvrA recA*
			CM611 *uvrA exrA*, as resistant as CM561 *exrA*	
E	DNA crosslinks	Mitomycin C, heliotrine	WP2 *uvrA* more sensitive than WP2	Little mutation with WP2
			CM611 *uvrA exrA* more sensitive than WP2 *uvrA*	Less with WP2 *uvrA* than WP2
				No mutation with CM561 *exrA*, CM571 *recA*, CM611 *uvrA exrA*, WP100 *uvrA recA*
F	Strand breaks?	Iodoacetamide, nalidixic acid	WP2 *uvrA* as resistant as WP2	
			WP67 *uvrA polA* more sensitive than CM611 *uvrA exrA*	

1. More killing with strain WP2 *uvrA* than strain WP2, CM611 *uvrA exrA* than CM561 *exrA*, WP100 *uvrA recA* than CM571 *recA* — B,E

2. Equal killing with WP2 and WP2 *uvrA*, CM561 *exrA* and CM611 *uvrA exrA*, WP100 *uvrA recA* and CM571 *recA* — C,D

3. No mutation with CM561 *exrA*, CM571 *recA* CM611 *uvrA exrA* and WP100 *uvrA recA* — B,D

4. Mutation with CM561 *exrA*, CM571 *recA* CM611 *uvrA exrA* and WP100 *uvrA recA* — A,C

%. More mutation with WP2 *uvrA* than WP2 — A,B

6. Killing as in 1, little mutation — E

7. WP67 *uvrA polA* more sensitive than CM611 *uvrA exrA* — F(C,D)

there are c pre-existing mutants in the untreated sample and therefore b-c plate mutants on SA1 plates. Pre-existing mutants will be killed during treatment, just as untreated cells, so that cd/e will be present on the treated plate. The plate mutants will not decrease (Appendix A). The treated plate will contain induced mutants, pre-existing spontaneous mutants and plate spontaneous mutants. Therefore after correcting for both classes of spontaneous mutants the frequency of induced mutants

$$M = \left(a - (b - c) - \frac{cd}{e}\right)\frac{1}{d}$$

$$= \left(a - b + c\left(1 - \frac{d}{e}\right)\right)\frac{1}{d}$$

Appendix C

Use of repair deficient strains to characterise mutagens

Table XI attempts a classification of mutagens on the basis of their effect on repair deficient strains. Damage may lead to mutation by a direct alteration of the genetic code (misreplication) or by an error during its repair (misrepair). Such damage may or may not be subject to excision repair. In this way, groups A to D are formed. Not all DNA crosslinking agents fall into group E. 8-methoxypsoralen, for instance, is believed to crosslink DNA and falls into class B. Agents in group F are characterised by their effect on the *polA* strain WP67. It is uncertain how such agents cause DNA damage or whether they are significant environmental mutagens.

Some agents fall into more than one category, methyl methanesulphonate for instance causes a small amount of type C, as well as type D damage. Other agents may not fit any category exactly, and new agents with different patterns of response may be found in the future. Nevertheless, it is likely that many agents may fall into one of the classes given here.

Note added in proof

It has recently been agreed the *exrA* locus should be renamed *lexA*.

In the fluctuation test, we have found that addition of 5 μg/ml bromcresol purple to the medium simplifies scoring of the assay. With this pH indicator tubes containing a mutation are yellow and tubes without a mutation are purple.

References

1 Ames, B.N., W.E. Durston, E. Yamasaki and F.D. Lee, Carcinogens are mutagens: a simple test system combining liver homogenates for activation and bacteria for detection, Proc. Natl. Acad. Sci. (US), 70 (1973) 2281—2285.

2 Ames, B.N., F.D. Lee, W.E. Durston, An improved bacterial test system for the detection and classification of mutagens and carcinogens, Proc. Natl. Acad. Sci. (US), 70 (1973) 782—786.

3 Bridges, B.A., Simple bacterial systems for detecting mutagenic agents, Laboratory practice, 21 (1972) 413—419.

4 Bridges, B.A., The three-tier approach to mutagenicity screening and the concept of radiation equivalent dose, Mutation Res., 26 (1974) 335—340.

5 Bridges, B.A., R.E. Dennis and R.J. Munson, Differential induction and repair of ultraviolet damage leading to true reversions and external suppressor mutations of an ochre codon in *Escherichia coli* B/r WP2, Genetics, 57 (1967) 897—908.

6 Bridges, B.A., R.P. Mottershead, C. Colella, Induction of forward mutations to colicin E2 resistance in repair deficient strains of *Escherichia coli:* experiments with ultraviolet light and captan, Mutation Res., 21 (1973) 303—313.

7 Clarke, C.H., M.J. Wade, Evidence that caffeine, 8-methoxypsoralen and steroidal diamines are frameshift mutagens for *E. coli* K12, Mutation Res. (in press).

8 Frantz, C.N. and H.V. Malling, The quantitative microsomal mutagenesis assay method, Mutation Res., 31 (1975) 365—380.

9 Howard-Flanders, P., DNA repair and recombination, Brit. Med. Bull., 29 (1973) 226—235.

10 Ishii, Y. and S. Kondo, Comparative analysis of deletion and base-change mutabilities of *Escherichia coli* B strains differing in DNA repair capacity (wild-type, *uvrA, polA, recA*) by various mutagens, Mutation Res., 27 (1975) 27—44.

11 Kada, T., K. Tutikawa and Y. Sadaie, In vitro and host mediated "rec-assay" procedures for screening chemical mutagens; and phloxine, a mutagenic red dye detected, Mutation Res., 16 (1972) 165—174.

12 Kondo, S., Evidence that mutations are induced by errors in repair and replication, Genetics, 73 (1973) 109—122.

13 Loveless, A., Possible relevance of 0—6 alkylation of deoxyguanosine to mutagenicity and carcinogenicity of nitrosamines and nitrosamides, Nature, 223 (1969) 206—207.

14 Luria, S.E., M. Delbruck, Mutations of bacteria from virus sensitivity to virus resistance, Genetics, 28 (1943) 491—511.

15 MacPhee, D.G., *Salmonella typhimurium his* G46 (R-Utrecht) — possible use in screening mutagens and carcinogens, Appl. Microbiol. 26 (1973) 1004.

16 Mohn, G., J. Ellenberger, Mammalian blood-mediated mutagenicity tests using a multipurpose strain of *Escherichia coli* K 12, Mutation Res., 19 (1973) 257—260.

17 Mohn, G., J. Ellenberger, D. McGregor, Development of mutagenicity tests using *Escherichia coli* K 12 as indicator organism, Mutation Res., 25 (1974) 187—196.

18 Osborn, M. and S. Person, Characterisation of revertants of *E. Coli* WU 36-10 and WP2 using *amber* mutants and an *ochre* mutant of bacteriophage T4, Mutation Res., 4 (1967) 504—507.

19 Osborn, M., S. Person, S. Phillips and F. Funk, A determination of mutagen specificity in bacteria using nonsense mutants of bacteriophage T4, J. Mol. Biol., 26 (1967) 437—447.

20 Ryan, F.J., Spontaneous mutations in non-dividing bacteria, Genetics, 40 (1955) 726—738.

21 Sedgwick, S.G., and B.A. Bridges, Survival, mutation and capacity to repair single-strand DNA breaks after gamma-irradiation in different exr$^-$ strains of *Escherichia coli,* Molec. Gen. Genet., 119 (1972) 93—102.

22 Slater, E.E., M.D. Anderson and H.S. Rosenkranz, Rapid detection of mutagens and carcinogens, Cancer Research, 31 (1971) 970—973.

23 Voogd, C.E., J.J. van der Stel and J.J.J.A.A. Jacobs, The mutagenic action of nitroimidazoles. I metronidazole, nimorazole, dimetridazole and ronidazole, Mutation Res., 26 (1974) 483—490.

24 White, I.N.H., and A.R. Mattocks, Reactions of dihydropyrrolizines with deoxyribonucleic acids in vitro, Biochem. J., 128 (1972) 291—297.

25 Witkin, E.M., Ultraviolet-induced mutation and DNA repair, Annual Rev. of Genetics, 3 (1969) 525—552.

THE USE OF *ESCHERICHIA COLI* K12/343/113(λ) AS A MULTI-PURPOSE INDICATOR STRAIN IN VARIOUS MUTAGENICITY TESTING PROCEDURES

GEORGES R. MOHN * and JÜRGEN ELLENBERGER

Zentrallaboratorium für Mutagenitätsprüfung der Deutschen Forschungsgemeinschaft, Freiburg i.Br. (G.F.R.)

Contents

Summary

Escherichia coli K12 was selected initially as an indicator organism for mutagenicity testing because it is best characterized genetically and physiologically,

---

Abbreviations: DENA, diethylnitrosamine; DMSO, dimethyl sulfoxide; EMS, ethyl methanesulfonate; ICR-191, 2-methoxy-6-chloro-9-(2-chloroethylaminopropylamino)acridine · 2HCl; IPMS, isopropylmethanesulfonate; MNNG, 1-methyl-3-nitro-1-nitrosoguanidine; 5-MT, 5-methyl-DL-tryptophan; TCPP, 1,2-epoxy-3,3,3-trichloropropane; TEM, triethylenemelamine.

* Present address: Department of Radiation Genetics, State University of Leiden, The Netherlands.

This publication is dedicated to Prof. Dr. R.W. Kaplan on the occasion of his 65th birthday.

and because of its non-pathogenic, symbiotic adaptation to mammalian species. In order to avoid or minimize the selective effects of given mutagenic chemicals to certain genes, systems which detect mutations in several different loci of the *E. coli* chromosome were assayed.

In its present state of development, the strain *E. coli* 343/113(λ) allows the simultaneous detection of several types of DNA alterations, including base-pair changes, reading frameshifts, and deletions; agents which induce vegetative growth of prophage λ can also be detected. This is achieved by plating aliquots of treated bacterial populations on different selective mutation media. The currently used mutation systems in this strain are: (i) mutations from *gal*R^s to *gal*$^+$; (ii) mutations from 5-methyltryptophan sensitivity (MTS) to resistance (MTR); (iii) mutations from arg^-_{56} to *arg*$^+$; and (iv) mutations from nad^-_{113} to *nad*$^+$. The former two systems detect mainly forward mutations (including deletions and frameshifts) in genes *gal*R and *gal*O (i) and, potentially, in five genes belonging mainly to the tryptophan biosynthetic pathway (ii). In the arg_{56} system (iii) most of the *arg*$^+$ revertant colonies appear to arise as a consequence of base-pair changes rather than frameshifts. The nad_{113} system (iv) is sensitive to the mutagenic action of frameshift-inducing agents.

This paper describes procedures for performing various mutagenicity tests with *E. coli* 343/113(λ). Although the strain may be used in spot tests, it was developed primarily for liquid experiments: bacterial suspensions are treated with known concentrations of the chemical under test for given periods of time (up to 240 min), then washed, diluted, and plated on selective mutation media. Protocols for using both stationary and growing cells are presented. Experiments with 9-aminoacridine, proflavine, acriflavine, and acridine yellow confirm that growing cells are very sensitive to the mutagenic action of frameshift-inducing agents, provided that suitable mutation systems are employed. This is the case in the two forward mutation detecting systems *gal*$^+$ and MTR, and in the *nad*$^+$ back mutation system. In the latter, growing cells are even a prerequisite for successful mutagenesis with DNA-intercalating agents. Metal salts such as sodium chromate are also more mutagenic in growing cells (especially in the *arg*$^+$ system), whereas the mutagenic potential of alkylating agents appears not to be strongly dependent on cell growth.

Procedures for testing the bacterial cells together with mammalian organ homogenates are described; they do not differ markedly from other standardized techniques. Experimental protocols employing chemicals which are mutagenic per se are presented: in confirmation of results using other microbes, they demonstrate that in *E. coli* the mutagenic activity of ethyl methanesulfonate is greatly enhanced by rodent liver homogenates, whereas the mutagenic action of 1-methyl-3-nitro-1-nitrosoguanidine is almost abolished after reaction with even small quantities of the same liver fraction. Recommendations for routine performance of intraperitoneal and intrasanguineous host-mediated assays are further given. Finally, diagrams are shown which allow the rapid statistical evaluation of the significance of the differences in mutation frequencies between control and test experiments.

Introduction

Short-term tests for mutagenicity using microbial indicators are presently regarded as necessary because of the high and even increasing number and amount of natural and synthetic chemicals in man's environment, some of which have been shown to possess mutagenic activity [5]. These tests also appear to be of value in assessing the potential carcinogenic activity of environmental chemicals [44]. After the recognition that the mutagenic activity of chemicals may be profoundly altered by mammalian metabolism before they reach the nuclei or the extrachromosomal genetic material of germinal or somatic cells, techniques which take into account the biotransformation of foreign compounds were developed [21,41].

Several strains of different microbial species can be used as genetic indicators in such studies. They include bacteria and lower fungi and some have demonstrated their usefulness [2,26,62]. In 1970 we started the development of a multi-purpose indicator strain of *Escherichia coli* because of two main reasons. (i) Being non-pathogenic and adapted through symbiosis to mammalian species, *E. coli* cells seemed to be very appropriate for host-mediated assays. In fact, preliminary experiments demonstrated the suitability of *E. coli* to be injected into the blood stream of rodents and to be subsequently recovered out of the livers of the animals (intrasanguineous host-mediated assay [49]). (ii) Due to the several different types of DNA alterations leading to mutations, and due to the often encountered selective action of mutagenic chemicals on certain genes or gene loci (especially in back mutation assays), it was necessary to utilize several forward mutation systems, where ideally all known types of mutations would be detected. In order to depict the kind of mutation spectrum induced by chemicals, it was also necessary to include systems (detecting back mutations) which respond to only one or a few types of DNA alterations.

The original strain was *E. coli* K12, *gal*R^s_{18}, isolated by Saedler et al. [56]. The derivative *E. coli* 343/113(λ) allows the simultaneous detection of several types of DNA alterations, including base-pair changes, reading frameshifts, and deletions. Agents which induce vegetative growth of prophage λ can also be detected. The various mutation systems have been calibrated with several mutagenic agents of different modes of action, such as a variety of alkylating agents [14,22,27,48,50,51]; acridine derivatives (see ref. 47 and this paper); nitroheterocyclic compounds including hycanthone, 4-nitroquinoline-1-oxide, nitrofuran, and nitroimidazoles [47,49—52]; hydroxylamine [47]; UV radiation [47]; captan and folpet [47], and aflatoxins B_1 and G_1 [10]. This paper describes procedures for performing mutagenicity experiments with stationary and growing cells of *E. coli* 343/113(λ) in liquid culture, with mammalian organ homogenates, and in host-mediated assays.

Characteristics of genetic markers and mutation systems

A summary of the systems involved in the current mutagenicity tests with strain *E. coli* 343/113(λ) is given in Table I. They include detection of mutations in structural as well as in regulatory genes. Additional mutation systems may also be tested, such as resistance to colicins [8,36], to valine [17,23,59], to phenethyl alcohol [61], to nalidixic acid [6,12,28,30], and to penicillins and

TABLE I

CHARACTERISTICS OF GENETIC MARKERS OF STRAIN *E. coli* 343/113(λ)

(1) Development of strain

Strain	Genotype	Isolation procedure
343	*E. coli* K12, λ^-, *gal*R^s_{18}	*gal*R^s_{18} isolated after UV irradiation
343/56	*E. coli* K12, λ^-, *gal*R^s_{18}, arg_{56}	arg_{56} isolated after MNNG treatment
343/113	*E. coli* K12, λ^-, *gal*R^s_{18}, arg_{56}, nad_{113}	nad_{113} isolated after ICR-191 treatment
343/113(λ)	*E. coli* K12(λ) gal^s_{18}, arg_{56}, nad_{113}	lysogenic derivative isolated after infection with wild type λ bacteriophage

(2) Genetic systems currently used

Marker	Phenotypic trait involved	Type of DNA alteration involved
galR^s_{18}	galactose non-fermentation (*gal*$^-$) to galactose fermentation (*gal*$^+$)	base-pair substitutions, frameshifts, and deletions
MTS	5-methyltryptophan sensitivity (MTS) to resistance (MTR)	base-pair substitutions, frameshifts, and deletions
arg_{56}	arginine auxotrophy (*arg*$^-$) to prototrophy (*arg*$^+$)	mostly base-pair substitutions
nad_{113}	nicotinic acid auxotrophy (*nad*$^-$) to prototrophy (*nad*$^+$)	frameshifts
(λ)	lysogenic state to lytic development of prophage λ	alteration leading to de-repression of prophage

streptomycin [7,16], but until now these additional systems have not provided more information on the mutagenicity of chemicals than the actual battery of systems (unpublished results).

gal$^+$ *system*

The super-repressed mutation in the regulatory gene of the galactose operon, *gal*R^s_{18}, was isolated after UV irradiation [56]. This mutation leads to a galactose non-fermenting phenotype because of the very low levels (less than 5% of wild type) at which the three enzymes of galactose fermentation are (constitutively) produced. Thus, when plated on minimal medium containing galactose as the sole carbon source, *gal*R^s_{18} bacteria grow at a very limited rate. Mutations to a galactose-fermenting phenotype can occur in three ways: (i) by true back mutation to *gal*R^+; (ii) by forward mutation in gene *gal*O, leading to constitutive *gal*O^c type mutants [18]; and (iii) by forward mutation in gene *gal*R, leading to mutants of the constitutive type *gal*R^- [18]. Potentially, all types of mutations leading to an inactive *gal*R gene product (the repressor protein) will produce viable galactose-fermenting cells. In fact, it was shown that the majority of spontaneous and induced galactose-fermenting mutants do arise in gene *gal*R as *gal*R^- mutations [18] and that even deletions and frameshifts extending into the adjacent *lys*A gene (*gal*R^s, *lys*A^+ → *gal*R^-, *lys*A^-; concomitantly leading to lysine auxotrophy) lead to viable detectable colonies [50,56]. Galactose-fermenting mutants are induced by a variety of different mutagenic compounds including both alkylating agents and, especially in growing cells, frameshift-inducing agents.

MTR system

The growth of wild type *E. coli* K12 is inhibited by 5-methyltryptophan. This amino acid analogue interferes with the tryptophan biosynthetic pathway by false feedback inhibition. Forward mutations which lead to 5-MT resistance (MTR) have been shown in *E. coli* K12 to occur in each of the following genes: *trp*R, as constitutive *trp*R$^-$ mutants [11,53]; *trp*O, as constitutive *trp*O^c mutants [32,33]; *trp*E, as mutants with a feedback-resistant anthranilate synthetase [54,58]; *mtr*, as mutants with unknown mechanism of 5-MT resistance [34]; and *aro*P, as mutants lacking the general aromatic amino acid transport system [9]. Mutations in these genes lead to different levels of 5-MT resistance, which also depends on the constitution of the plating medium. Thus the frequency of MTR mutants decreases with increasing 5-MT concentrations [48]. For an optimal detection of mutants, concentrations of 5-MT between 1 and 3 μg/ml of medium are appropriate in strain *E. coli* 343/113. Lower concentrations lead to heavy background growth of the non-mutated residual population, and at higher concentrations, only a few mutants arise as colonies. We have not yet characterized genetically the MTR mutants arising in the *E. coli* K12 strain described here, but examination of the MTR colonies indicates at least three different types which are phenotypically distinguishable by colony size, colony morphology, and by excretion of tryptophan. Being a system detecting forward mutations, a wide variety of differently acting mutagens are active here. Its sensitivity approaches that of the *gal*$^+$ system but, again, there are sometimes substantial quantitative differences between both systems [49,50]. In routine testing we suggest, therefore, that both systems be used to minimize the risk of missing potential mutagenic compounds.

arg$^+$ *system*

The arginine auxotrophy *arg*$_{56}$ was selected among several hundreds of other auxotrophs after treatment of stationary cells of *E. coli* 343 with MNNG (10 μg/ml for 105 min at 37°C in the dark). It was chosen as the most sensitive strain to the mutagenic action of a variety of alkylating agents. A further criterion was its slight leakyness which allows the treated bacteria to perform a few residual cell divisions on the mutation media, and thus to achieve perfection of pre-mutational lesions. The molecular nature of the *arg*$_{56}$ mutation is not yet known. In mutation experiments, two types of revertant colonies can be seen: (i) large colonies which possess wild type-like growth and may be true revertants; they occur at a rather low frequency (about 0.2 per 10^6 plated bacteria) which is not enhanced by mutagenic agents at measurable amounts; (ii) the vast majority of revertants are smaller colonies which arise at frequencies of about 30 to 50 per 10^6 plated bacteria. Although they are clearly distinguishable from the residual background, they do not possess full wild type growth in minimal medium. They arise possibly as forward mutations in (a) gene(s) suppressing the arginine auxotrophy (t_{RNA} gene ?). This would explain the high sensitivity of the *arg*$^+$ system to a wide variety of mostly base-pair changes-inducing chemicals [49—51].

nad$^+$ *system*

In a search for systems detecting specifically frameshift mutations, growing cells of *E. coli* 343/56 (the predecessor of *E. coli* 343/113) were treated with

ICR-191 (at a concentration of 10 μg/ml for 12 h at 37°C in the dark), a known frameshift-inducing agent. Out of the several hundreds of auxotrophs so obtained, the nicotinic acid auxotrophy nad_{113} was retained because of its high sensitivity in back mutation assays to ICR-191 itself and to further acridines (proflavine, acriflavine, quinacrine mustard, acridine yellow). Because of the absolute character of the auxotrophy in mutation experiments we supply the medium with some growth factor (usually in the form of traces of nutrient broth) to ensure some residual growth of the treated bacteria on the mutation plates. Again the molecular nature of the nad_{113} mutation is not yet known; since the revertants are induced by frameshift agents (no alkylating agent tested so far is active here) it seems likely that nad_{113} arose through a frameshift mutation.

Induction of prophage λ

Lysogenic induction was shown relatively early to be achieved with mutagenic agents like UV irradiation and chemicals [40]. Many reports have shown the high correlation between mutagenic, carcinogenic (and carcinostatic), and prophage-inducing activities of chemicals [15,19,29,35,39], and several strains of *E. coli* K12 have already been used as indicators for prophage λ induction [24,29,43]. To allow the concomitant investigation of mutagenic and prophage-inducing effects in *E. coli* 343/113, a lysogenic substrain was isolated after infection with wild type phage gained from spontaneous lysis of strain *E. coli* K12(λ)met_1 (cf. ref. 46). Until now the induction of lytic growth of prophage λ in this strain has been assayed with UV irradiation (unpublished results) and with TCPP (this paper).

Materials and media

Mutagenicity testing with bacteria requires no more than the usual apparatus and equipment needed in bacteriological work. In routine experiments, it may be of advantage to use incubation rooms (37°C) with facilities for adjusting to more than 50% relative air humidity. We further recommend performing experiments under dim lights or under fluorescent lamps emitting light of wavelength above 500 nm (e.g. Philips TL 40W/16, about 580 nm) to avoid photochemical effects.

Biological material

The bacterial strains described in this paper can be obtained from (i) Zentrallaboratorium für Mutagenitätsprüfung, Breisacher Strasse 33, 7800 Freiburg i.Br., G.F.R. and (ii) Environmental Mutagenesis Branch, National Institute of Environmental Health Sciences, Research Triangle Park, N.C. 27709, U.S.A. Long-term cultures (stationary cells) can either be lyophilized (and then stored for years at +4°C) or kept frozen (at −30°C) in suspensions containing 10—15% glycerol in saline buffer. For daily experiments the strains are kept on nutrient agar plates (at +4°C) which are restreaked approximately every four weeks. Until now we have no indication that the genetic markers of *E. coli* 343/113 do change when the strain is serially cultivated for years on nutrient agar plates, but — as will be described later in the text — some occasional, temporary changes in the mutation frequencies may occur. These seem not to be due to hereditary changes in the genetic markers of *E. coli* 343/113, but rather to experimental conditions.

Stock solutions

Solution A. Dissolve 50 g NH_4Cl, 10 g NH_4NO_3, 20 g Na_2SO_4, 39 g $K_2HPO_4 \cdot 3H_2O$, and 10 g KH_2PO_4 in 1000 ml distilled water. Autoclave for 20 min at 120°C.

Glucose solution (50%). Mix 500 g D-glucose with 1000 ml distilled water. Autoclave for 20 min at 110°C.

Galactose solution (20%). Mix 200 g D-galactose (glucose free) with 1000 ml distilled water. Autoclave for 20 min at 110°C.

Trace solution. Solution 1: dissolve 100 mg $CaCl_2 \cdot 2H_2O$ in 10 ml distilled water; solution 2: dissolve 10 g $MgSO_4 \cdot 7H_2O$, 25 mg $FeSO_4 \cdot 7H_2O$, and 10 mg thiamine (vitamin B1 hydrochloride) in 90 ml distilled water. Mix solutions 1 and 2 and sterilize by filtration through a 0.22 μm Millipore filter.

Arginine solution. 20 mg L-arginine monochloride per ml distilled water. Sterilize by filtration.

Lysine solution. 10 mg L-lysine monochloride per ml distilled water. Sterilize by filtration.

Nicotinic acid solution. 1 mg nicotinic acid per ml distilled water. Sterilize by filtration.

Methyltryptophan solution. 1 mg 5-methyl-DL-tryptophan per ml distilled water. Dissolve with 10 N NaOH.

Growth and mutation media

The semi-solid mutation media should always be prepared and stored in essentially the same manner since age, dryness, and medium content of the plates may have some influence on spontaneous (and induced) mutant colony numbers. The growth ingredients are mixed after the autoclaved agar solution has cooled down to approximately 50°C. Then 25 ml are poured into petri dishes (diameter 8—9 cm) and air bubbles are eventually eliminated from the surface with an open flame. The dishes are left at room temperature for 1—2 days and then stored in the cold (+4°C) not more than 14 days before use for mutagenicity tests.

Nutrient broth. Mix 8 g bacto-nutrient broth (Difco) and 5 g NaCl with 1000 ml distilled water. Autoclave for 20 min at 120°C.

Complete agar. Mix 15 g bacto-agar (Difco), 8 g bacto-nutrient broth, and 5 g NaCl with 1000 ml distilled water. Autoclave for 20 min at 120°C. Pour plates as described above.

Complete soft agar. Mix 7 g bacto-agar, 8 g bacto-nutrient broth, and 5 g NaCl with 1000 ml distilled water. Autoclave for 20 min at 120°C.

Saline buffer, pH 7.0. Dissolve 4 g NaCl, 7 g $Na_2HPO_4 \cdot 2H_2O$, and 3 g KH_2PO_4 in 1000 ml distilled water. Autoclave for 20 min at 120°C.

Physiological saline. Dissolve 9 g NaCl in 1000 ml distilled water. Autoclave for 20 min at 120°C.

Medium for detection of gal^+ mutants. Autoclave 10 g bacto-agar in 900 ml distilled water, then add 100 ml solution A, 5 ml galactose 20%, 1 ml trace solution, 0.5 ml arginine solution, 1 ml lysine solution, and 1 ml nicotinic acid solution. Pour plates as described above.

Medium for detection of MTR *mutants.* Autoclave 15 g bacto-agar in 900 ml distilled water, then add 100 ml solution A, 10 ml glucose 50%, 1 ml trace solution, 1 ml arginine solution, 1 ml nicotinic acid solution, and 1 ml methyltryptophan solution. Pour plates as described above.

Medium for detection of arg⁺ mutants. Autoclave 15 g bacto-agar in 900 ml distilled water, then add 100 ml solution A, 10 ml glucose 50%, 1 ml trace solution, and 1 ml nicotinic acid solution. Pour plates as described above.

Medium for detection of nad⁺ mutants. Autoclave 15 g bacto-agar in 900 ml distilled water, then add 100 ml solution A, 2 ml nutrient broth, 10 ml glucose 50%, 1 ml trace solution, and 1 ml arginine solution. Pour plates as described above.

Media for use with organ homogenates

Phosphate buffer, pH 7.4. Dissolve 35.6 g $Na_2HPO_4 \cdot 2H_2O$ in 1000 ml distilled water. Autoclave for 20 min at 120° C. Titrate to pH 7.4 with HCl.

0.15 M KCl. Dissolve 11.2 g KCl in 1000 ml distilled water. Autoclave for 20 min at 120° C.

Cofactors. Shortly before use dissolve 75 mg KCl, 72 mg glucose-6-phosphate · Na_2 salt, 37 mg NADP · Na_2 salt, and 51 mg $MgCl_2 \cdot 6H_2O$ in 15 ml phosphate buffer pH 7.4.

Preparation of liver homogenate (S-9 fraction)

Wash 5 g of fresh liver of mice or rats with ice-cold 0.15 M KCl, then homogenize in 3-fold volume (15 ml) of ice-cold 0.15 M KCl using a 30 ml Potter-Elvehjem-type homogenizer with teflon pestle. Centrifuge the homogenate for 10 min (+4° C) at 9000 × *g* and use the supernatant which contains the microsomal and the soluble enzyme fractions shortly thereafter.

Procedures for direct mutagenicity testing

The strain *E. coli* 343/113 was developed specifically to be used in suspension experiments. It may also be used in a spot test [47] but its sensitivity towards several types of chemical mutagens is then greatly reduced [50,51]. This is valid also with other bacterial indicators [25]. We therefore recommend liquid testing when using *E. coli* 343/113 as indicator. In principle all mutagenicity experiments with this strain are performed in the same way: suspensions of stationary or of growing cells are exposed to different concentrations of the substances under test at 37° C for different periods of time. The cells are then washed and/or diluted, and about 1×10^6 cells are inoculated onto the surface of the different mutation media. The plates are then incubated (2—3 days) until the adequate amount of spontaneous mutant colonies is reached in each mutation system. The mutant colony numbers in all dishes are then determined either manually or with the aid of an electronic colony counter.

1. Stationary cells

(a) Preparation of cell suspension

From the masterplate a small loopful of cells is inoculated into 10 ml of nutrient broth. The suspension is then incubated in a rotary shaker at 37° C for 15 h in the dark (overnight culture). Bubbling of sterile air into the tube is also possible, in any case it is important that the culture receives enough oxygen to reach a final titer of about 2×10^9 viable cells per ml. At this stage, the cells are in the stationary growth phase. The culture is centrifuged two times and resuspended in saline buffer to reach a titer of about 2×10^{10} cells per ml (which means a 10-fold concentration of the original culture). When necessary, several 10 ml cultures can be prepared in parallel and then mixed together. They can be stored at +4°C for at least five days without loss of viability or alteration of mutation frequency.

(b) Treatment of bacteria

Usually six or ten incubation mixtures are prepared containing aliquots of the bacterial population and increasing concentrations of the substance under test.

Incubation mixture	Control	Test
Saline buffer	2.8 ml	2.5 ml
Bacteria	0.2 ml	0.2 ml
Test substance	—	0.3 ml

The final titer of bacteria per ml of incubation mixture is usually 1 to 2×10^9 viable cells. In experiments using stationary cells, DMSO can be used as a solvent of the substance under test up to final concentrations of 20%. In experiments using growing cells, in lysogenic induction experiments, and in microsomal assays the DMSO concentration should not exceed 5%.

The incubation mixtures are incubated in a rotary shaker at 37°C in the dark for different periods of time. Usually 120—180 min are adequate, but longer incubations may also be performed. The pH value, which is usually 7.0, may be adjusted to values between 5.5 and 8.5. Rotating the mixtures is necessary to avoid clumping of the bacteria at the bottom of the tubes. To terminate the treatment, the cells are diluted in ice-cold saline buffer and plated promptly on survival and mutation media.

(c) Plating of the treated cells

Both inactivation of colony forming ability and induction of mutations are scored. To determine the mutation frequency, 0.1 ml aliquots of the 1×10^{-2} diluted incubation mixtures are spread with a sterile glass spatula over the surface of the different mutation media for the detection of *gal*$^+$, MTR, *arg*$^+$, and *nad*$^+$ mutants. For the determination of survival, the mixtures are diluted further down to 1×10^{-6} in saline buffer, and 0.1 ml aliquots are spread with a sterile glass spatula over the surface of complete agar or of fully supplemented minimal agar. Usually three or four plates are spread in parallel. After spreading, the plates are kept for a few minutes at room temperature to allow drying of the inoculated cell suspension, they are then incubated upside down (and, if possible, in one layer) at 37°C in the dark. Survival plates are incubated 20 h. The length of incubation of the mutation plates is important and we have experienced that it may vary from laboratory to laboratory. However, once the adequate incubation conditions have been determined, usually no subsequent changes are necessary. Because of their different characters, the plates of the different mutation systems here described require different incubation times. As a general rule, medium plates for detection of *gal*$^+$ and *nad*$^+$ mutants should be incubated for about 36—40 h, whereas incubation of medium plates for detection of MTR and *arg*$^+$ mutants should be between 60 and 64 h. Shorter incubation times will reduce the spontaneous and induced mutation frequencies, whereas longer times will enhance background growth and mutation frequencies (especially in the *gal*$^+$, MTR, and *arg*$^+$ systems) to numbers too high to be counted easily. The best way to adjust for optimal incubation is to check in initial experiments the spontaneous mutation frequency: the right incubation time is given when plates for *gal*$^+$ mutants show about 40 mutant colonies, plates for MTR mutants 15—20 colonies, plates for *arg*$^+$ mutants 40—50 colonies, and plates for *nad*$^+$ mutants less than 10 colonies. Under these conditions the plates for MTR mutants show practically no background growth, the plates for *gal*$^+$ and *arg*$^+$ mutants show a moderate background, and the plates for *nad*$^+$ mutants a heavy residual growth. In the latter case (*nad*$^+$) even the heavy growth of background cells does not prevent accurate counting of mutant colonies, because these are all large and wild type-like. It must also be stressed that the inoculum of cells per plate should, in these experiments, not be higher than 3×10^6 cells per plate, otherwise the number of mutants per plate becomes too high. During routine serial mutagenicity testing, some batches of plates occasionally show (approximately every 50 experiments) a sudden but temporary increase in background growth and spontaneous mutation frequency affecting independently the *gal*$^+$, MTR, or *arg*$^+$ mutation systems. These changes seem not to be due to changes in the genetic background of *E. coli* 343/113, but rather to unexpected changes in composition of media ingredients or due to natural fluctuation of the spontaneous mutation frequencies in the overnight cultures.

(d) Calibration with triethylenemelamine

In principle, any of the various agents already shown to be mutagenic in *E. coli* 343/113 can be used for calibration. We have selected TEM for the purpose of this paper because its mutagenicity is well documented and because it is easily available from different sources. Finally, this substance is relatively stable and soluble in water. The results of a representative experiment are shown in Fig. 1. The absolute numbers of induced mutants per plate rather than per survivors are plotted on the ordinate. This may lead to a slight underestimate of the actual mutation frequency at higher TEM doses, but a plot per survivors would lead to an overestimation of mutant frequency. This is due to the fact that the actual number of mutant colonies on the plates is not directly proportional to the cell inoculum, and is a common phenomenon of different mutation tests [14,26]. As an example, Fig. 2 demonstrates that isopropylmethanesulfonate-induced *gal*$^+$ colony numbers do not decrease proportionally with decreasing cell inoculum, but that mutation plates contain relatively more mutant colonies at lower inocula than expected from the high inoculum value. In conclusion, to avoid false positive results, it seems safer to consider the absolute mutant numbers at concentrations which do not inactivate or inactivate only moderately the colony forming ability of cells.

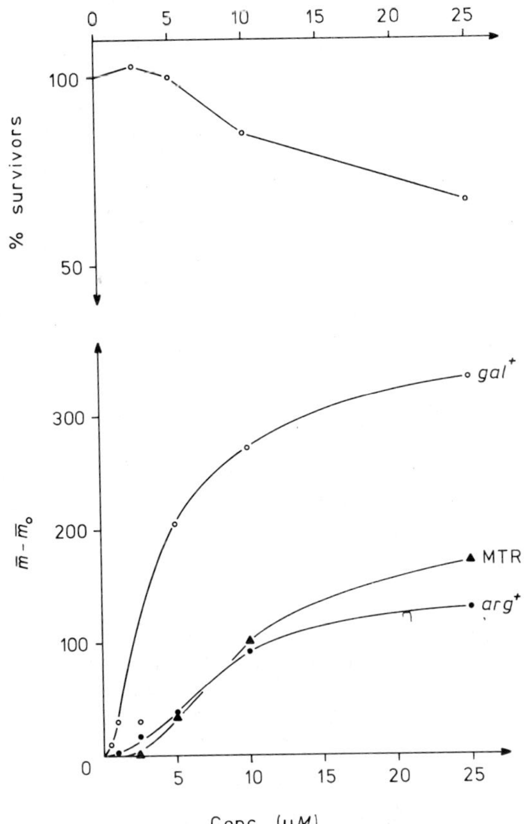

Fig. 1. Inactivation and mutation induction to *gal*$^+$, *arg*$^+$, and MTR by triethylenemelamine in *E. coli* 343/113. Stationary cells were incubated 120 min at 37°C in the dark. m—m_0 = number of induced mutants per plate. Each point is the mean value of three plates of one representative experiment.

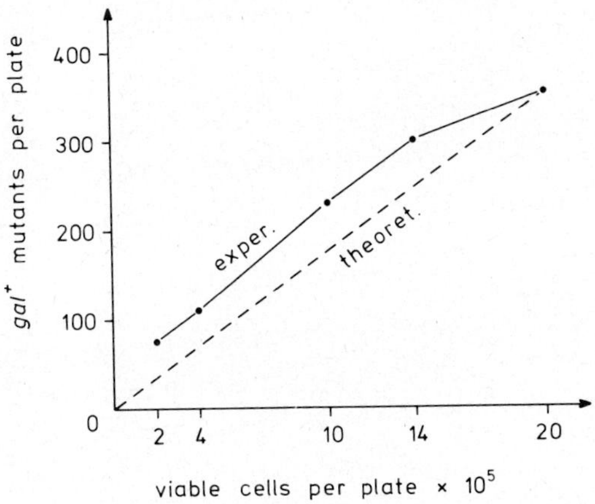

Fig. 2. Non-proportional correlation between inoculum of isopropylmethanesulfonate (IPMS)-treated cells and *gal*$^+$ mutant frequency in *E. coli* 343/113. Stationary cells were treated for 120 min at 37°C in the dark with 0.1% (vol./vol.) IPMS. After treatment, serial dilutions were made to give the desired number of cells on mutation medium plates.

2. *Growing cells*

The procedure is essentially the same as with stationary cells, except that the bacteria are allowed to grow in complete medium before and during chemical treatment: stationary cells of a 10 ml overnight culture (prepared as described in section 1a above) are washed and resuspended in 4 ml saline buffer (which means a 2.5-fold concentration of the original culture). Tubes inoculated with 0.2 ml of this bacterial suspension and with 2.5 ml of nutrient broth are incubated for 60 min at 37°C in a rotary shaker. Then 0.3 ml of the test substance are added to reach a final volume of 3.0 ml. One experiment usually consists of six to ten incubation tubes containing, besides a control tube, different concentrations of the substance under test. After addition of the test compound the tubes are incubated at 37°C in the rotary shaker for an additional period of 120 min. During this 3 h of incubation the viable cell titer rises (at least in the control tube) from about 4×10^8 cells per ml to about $1{-}2 \times 10^9$ cells per ml. To terminate the treatment, the suspensions are diluted 1×10^{-2} in ice-cold saline buffer; the determination of survival and of mutation frequency is then done in the same way as described for stationary cells (see 1c above).

To demonstrate that, for some substances, growing cells are much more sensitive than stationary cells, experiments with sodium chromate are shown in Fig. 3. The results show that both inactivation of colony forming ability and *arg*$^+$ reversion frequency are enhanced in growing cells. The prerequisite of growing cells for successful and effective mutagenesis in the mutation systems *gal*$^+$, MTR, and *nad*$^+$ is further demonstrated in Fig. 4 (mutation induction by 9-aminoacridine) and Table II (induction of *nad*$^+$ mutations by several

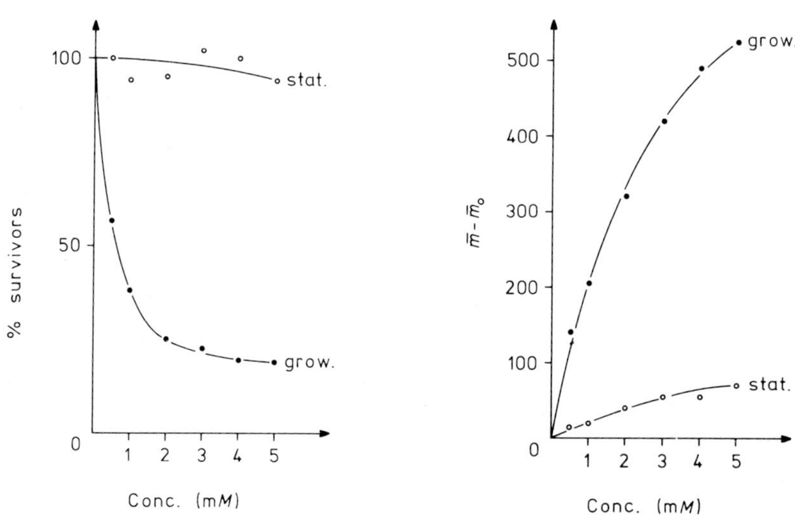

Fig. 3. Survival and induction of *arg*$^+$ mutations in stationary and growing cells of *E. coli* 343/113 after treatment with sodium chromate. a: survival of colony forming ability. b: *arg*$^+$ mutation induction. Stationary cells were treated for 180 min at 37°C, growing cells were treated for 120 min at 37°C. $m - m_0$ = number of induced mutants per plate. The points represent the mean values of two separate experiments.

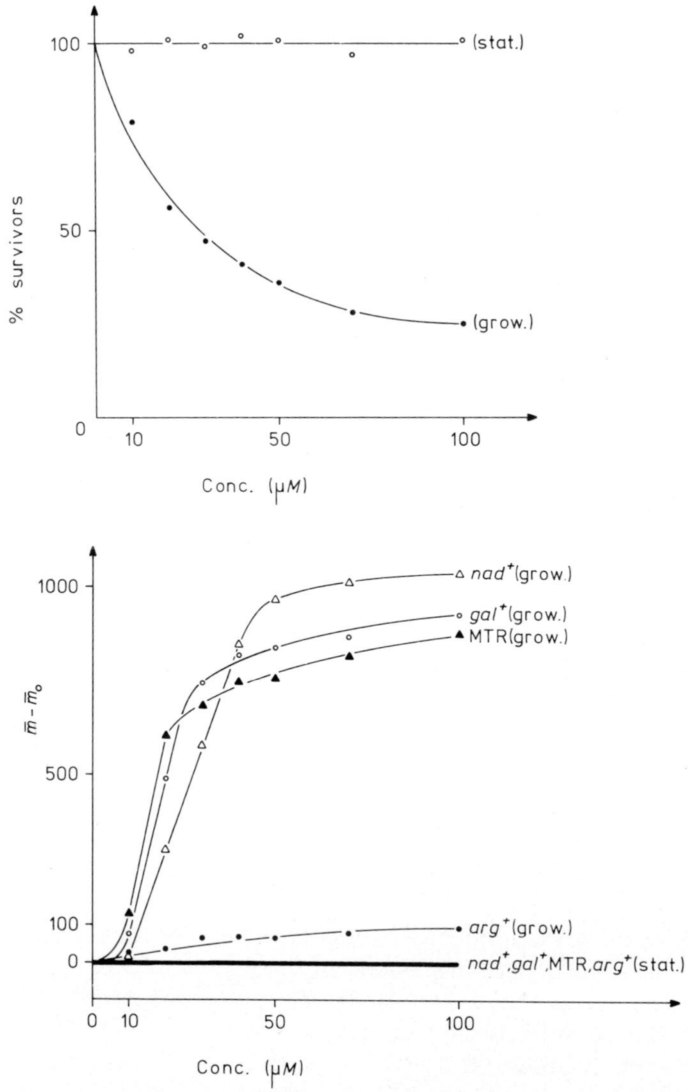

Fig. 4. Survival and induction of *gal*$^+$, MTR, *arg*$^+$, and *nad*$^+$ mutations in stationary and growing cells of *E. coli* 343/113 after treatment with 9-aminoacridine. a: survival of colony forming ability. b: mutation induction. Stationary and growing cells were treated for 120 min at 37°C in the dark. $m - m_0$ = number of induced mutants per plate. The points represent the mean values of two separate experiments.

acridines). These results indicate a sensitive and specific affinity of the *nad*$^+$ system to frameshift-inducing agents, whereas the two forward mutation detecting systems *gal*$^+$ and MTR appear to be sensitive to several types of DNA alterations. The *arg*$^+$ system, which is especially sensitive to alkylating agents [14,49—51], seems to respond only moderately to frameshifts.

In the light of previous findings on the interaction between intercalating agents and DNA [55,57] it is not surprising that growing cells are more sensitive than stationary cells. De-repression of gene activity and increased active

TABLE II

SENSITIVITY OF nad_{113} TOWARDS SEVERAL ACRIDINES IN STATIONARY AND GROWING CELLS

The data represent the mean value of two experiments (3 parallel plates per experiment).

Substance	Concentration [a] (μM)	Stationary cells		Growing cells	
		% Survivors	Number of revertants per plate	% Survivors	Number of revertants per plate
Control		100	2	100	2 (3) [b]
Proflavine	30	86	2	38	27
	50	79	3	25	129
	70	82	2	27	330
Acriflavine	30	88	3	37	22
	50	86	2	26	55
	70	74	2	26	159
Acridine yellow	30	99	2	54	10 (30)
	50	93	1	42	21 (37)
	100	76	4	38	26 (53)

[a] Cells were treated for 120 min at 37°C in the dark in a rotary shaker. To avoid photodynamic effects, preparation of incubation mixtures, dilution of treated bacteria, and plating on the media were carried out under yellow light.

[b] The data in parentheses were obtained after plating twice as much cells on selective medium plates, under otherwise identical experimental conditions.

uptake of foreign molecules during growth may also account for this phenomenon [31,38].

3. *Lysogenic induction*

To perform lysogenic induction several procedures have been proposed [19, 24,29,35]. The protocol described here has essentially the same procedure as for the detection of mutations in growing cells. In fact, both genetic effects can be measured within the same bacterial population. Growing cells are prepared as described in section 2 above: 0.2 ml of stationary cells (titer about 5×10^9 per ml) are incubated together with 2.5 ml nutrient broth for 60 min at 37°C. The cells are then incubated with different concentrations of the substance under test (e.g. for 60 min at 37°C) or they are incubated for different periods of time (0—180 min) with a given concentration of the test compound. The suspensions are then diluted in ice-cold saline buffer down to 1×10^{-5} and 1×10^{-6}. To determine survival, 0.1 ml aliquots of the dilutions are spread over the surface of complete growth medium. To determine lysogenic induction, 0.1 ml aliquots of the dilutions are mixed with 0.1 ml of a 1×10^{-1} diluted overnight culture of the non-lysogenic strain *E. coli* 343/113 in 3 ml of complete soft agar (which is kept molten at 48°C). The mixtures are mixed thoroughly and poured immediately over complete growth agar. Both survival and lysogenic induction plates are then incubated for 24 h at 37°C. The surviving colonies and lytic centers (plaques) are then counted and the titer of viable cells and of induced lysogens are determined. Exemplary results with

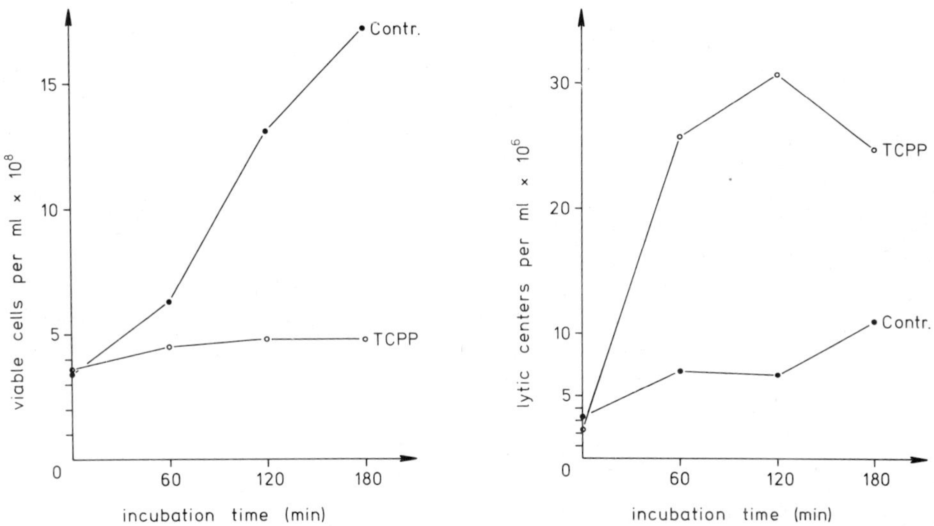

Fig. 5. Survival (a) and lysogenic induction (b) of prophage λ in *E. coli* 343/113(λ) after treatment with 1,2-epoxy-3,3,3-trichloropropane (TCPP) and in untreated cells (Contr.). Growing cells were treated for the periods of time indicated with 1 mM of TCPP at 37°C. Each point is the mean value of three plates of one representative experiment.

the compound 1,2-epoxy-3,3,3-trichloropropane, an alkylating epoxide, are shown in Fig. 5. In experiments not reported here, UV irradiation of wavelength 256 nm was also shown to induce vegetative growth of prophage λ.

Procedures for testing with organ homogenates

Experimental procedures for microsomal mutagenicity tests have already been described [2,20,26]. As already stated, the strain *E. coli* 343/113 was developed especially to be used in quantitative microsomal tests rather than in plate tests. In principle, the procedure is similar to that described in the previous section 1 except that during incubation with the substance under test, homogenates or organ fractions are added together with an NADPH-generating system.

Bacterial cells are harvested as usual by centrifugation from an overnight culture (15 h incubation at 37°C in 10 ml of nutrient broth) and resuspended in saline buffer to reach a final titer of about 2×10^{10} viable cells per ml. S-9 liver fractions and cofactor solutions are prepared as described under Materials and media, shortly before beginning the experiment. Usually six to ten incubation tubes (final volume 3.0 ml) containing different concentrations of the substance under test are then prepared as follows.

Incubation mixture	Control	Test
Phosphate buffer, pH 7.4	0.3 ml	—
S-9 liver fraction	1.0 ml	1.0 ml
Cofactors	1.5 ml	1.5 ml
Bacteria	0.2 ml	0.2 ml
Test substance	—	0.3 ml

The mixtures are incubated for various periods of time at 37°C in a rotary shaker in the dark. The usual time is 180 min, but periods between 30 and 240 min may be chosen. To stop treatment, 10 ml of ice-cold saline buffer are poured into the tubes which are then centrifuged for 10 min at 7000 $\times g$ and +4°C. The pellet containing the bacteria and some residual fraction of the homogenate is resuspended in 3 ml of saline buffer. These suspensions are then diluted 1×10^{-2} into saline buffer; to determine mutation frequencies 0.1 ml aliquots are spread over the surface of the different mutation media. To determine survival, the suspensions are then diluted down further to 1×10^{-6} and 0.1 ml aliquots are spread over the surface of complete medium. Incubation of the plates occurs as described earlier, i.e. suitable incubation times are first determined in preliminary experiments by checking daily the spontaneous mutant colony frequency of the different systems.

The method has been applied successfully for the detection of those mutagens which require metabolic activation through liver enzymes, for example with dimethylnitrosamine, diethylnitrosamine, cyclophosphamide, ifosfamide, trofosfamide, and aflatoxins B_1 and G_1 [10,14,51]. To demonstrate that some information on the fate of directly acting mutagens in the animal body can also be obtained in microsomal assays, the results of experiments using ethyl methanesulfonate and methyl-nitro-nitrosoguanidine are reported in Figs. 6 and 7, respectively. In confirmation of results employing other indicator microbes

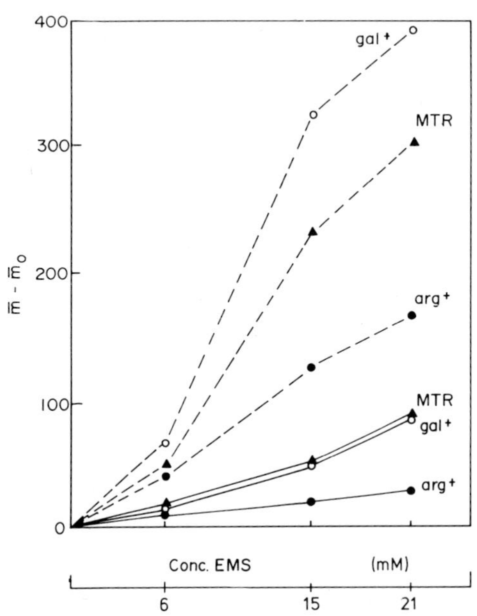

Fig. 6. Influence of microsomal preparations of mouse liver on the mutagenic activity of ethyl methanesulfonate (EMS) towards *E. coli* 343/113. Stationary cells were incubated for 120 min at 37°C with the indicated concentrations of EMS in tubes with (dotted lines) or without (solid lines) mouse liver S-9 fraction, then washed and plated on selective media to determine mutation frequencies to *gal*$^+$, MTR, and *arg*$^+$. $m - m_0$ = number of induced mutants per plate. Each point is the mean value of four plates of one representative experiment.

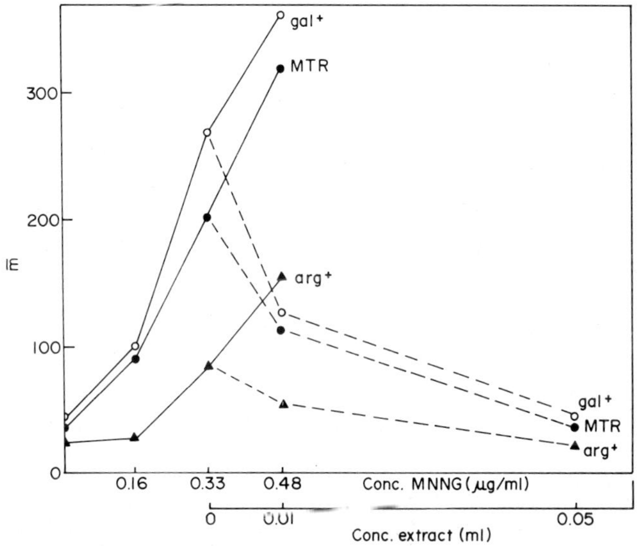

Fig. 7. Influence of increasing amounts of mouse liver microsomal preparations on the mutagenic activity of 1-methyl-3-nitro-1-nitrosoguanidine (MNNG) towards *E. coli* 343/113. Stationary cells were incubated without microsomal preparations for 120 min at 37°C with increasing amounts of MNNG (0—0.48 μg/ml). The cells were then washed and plated on selective media to determine mutation frequencies to gal^+, MTR, and arg^+ (solid lines). In tubes containing 0.33 μg of MNNG per ml, increasing amounts of S-9 mouse liver fraction were added before incubation for 120 min at 37°C and the mutation frequencies to gal^+, MTR, and arg^+ determined as above (dotted lines). m = number of mutants per plate. Each point is the mean value of three plates of one representative experiment.

[4,45], they demonstrate that the direct mutagenic activity of EMS towards *E. coli* is greatly enhanced by the addition of mouse liver homogenates. On the contrary, the high mutagenic action of MNNG is annihilated even with very small amounts of mouse liver homogenate (Fig. 7); whereas usually 1.0 ml of S-9 fraction per incubation tube is used to demonstrate activation of a compound to a mutagen, 0.05 ml of mouse liver fraction abolish and 0.01 ml reduce markedly the MNNG-induced mutation frequencies of all three mutation systems tested, namely gal^+, MTR, and arg^+. These findings explain why in host-mediated assays the mutagenic activity of EMS is easily detected, whereas that of MNNG is much lower than expected from direct, non-microsomal tests [42].

There are several variations of the microsomal assay procedure possible. In experiments not reported here, we have shown that isolated microsomes (100 000 × *g* centrifugation, resuspended in Tris buffer, pH 7.4) from rat liver are able to activate cyclophosphamide to mutagenic product(s) for *E. coli* 343/113. In this case the incubation mixture was 0.2 ml of isolated microsomes instead of 1.0 ml of S-9 fraction. It is also likely that homogenates or fractions of other mammalian organs [60] can be used for activation of foreign chemicals. All these experiments can also be performed with growing cells of *E. coli* 343/113.

Procedures for host-mediated assays

1. Intraperitoneal test

Usually three animals are tested in parallel in experiments involving one control series (3 animals) and five different test doses (15 animals). The animals can be mice or rats. Of course pretreatment with enzyme inducers (PCBs, Aroclor) is possible. 1.0 ml of washed stationary cells (titer about 4×10^9 per ml) are inoculated into the peritoneal cavity of mice (rats received 5.0 ml). The test chemical is then administered via another route. After time periods varying from 15 to 180 min, 1.0 ml of saline buffer is injected again into the peritoneum (rats are injected 5.0 ml). Five minutes later the animals are killed and 0.5 ml of peritoneal fluid is removed and resuspended in 4.5 ml of saline buffer. The suspensions are then diluted 1×10^{-1} in saline buffer. 0.1 ml aliquots are spread over the surface of media for the determination of mutant frequencies. To determine viable cell titer, the suspensions are diluted further down to 1×10^{-5} and 0.1 ml aliquots are spread over the surface of complete agar medium. The plates are incubated at 37°C under the same conditions as described before (see previous section 1c). As reference compound we recommend dimethylnitrosamine at concentrations up to 2% in aqueous solutions, 0.1 ml thereof being applied subcutaneously in mice.

2. Intrasanguineous test

In routine experiments it is appropriate to treat another animal every 4—5 min and to perform experiments with 8 animals or with multiple of 8. This allows the treatment of four animals in parallel and will permit an optimal use of the centrifuge, the rotor of which usually has a capacity for 8 tubes. The total number of animals per experiment should thus be 16 (one control group and three different doses) or 24 (one control group and five different doses of the substance under test). 0.20 to 0.25 ml of washed stationary cells suspended in physiological saline (titer about 4×10^{10} per ml) are inoculated into a lateral tail vein of mice (1.0 ml is injected if rats are used as host animals). The animals are then treated as usual with the test chemical (chemical treatment can, of course, be done before or some time after injection of the bacteria). After time periods of usually 30, 60, 120, or 180 min the animals are killed by cervical dislocation, the livers are removed and resuspended in 10 ml saline buffer. Rat livers are resuspended in 40 ml of saline buffer. These are then homogenized for 60 sec in a Potter-Elvehjem-type tube. The homogenates are then centrifuged for 10 min at $200 \times g$. The supernatant is then taken out of the centrifuge tubes with a pipette, centrifuged two times for 10 min at $7000 \times g$ and +4°C and the pellet resuspended in 10 ml of ice-cold saline buffer. Using rats, the pellets are resuspended in 20 ml of saline buffer. The suspensions are diluted 2×10^{-2} in saline buffer for determination of mutation frequencies and diluted 1×10^{-5} for determination of colony forming ability as described above.

The detailed results of one typical experiment with diethylnitrosamine as reference compound are listed in Table III, showing the fluctuations of parallel

TABLE III

MTR AND *arg*$^+$ MUTATION FREQUENCIES IN *E. coli* 343/113 AFTER TREATMENT OF NMRI MICE WITH DIETHYLNITROSAMINE (DENA) IN THE INTRASANGUINEOUS MUTAGENICITY TEST

DENA dose (mg/kg)	Mouse no.	Recovered viable bacteria per liver			Mutant colonies per plate			
					MTR		*arg*$^+$	
		K	$\overline{K}$	N	m	$\overline{m}$	m	$\overline{m}$
0	1	226	230	2.3×10^9	11	9.3	37	36.3
		250			6		27	
		214			9		46	
					11		35	
	2	277	279	2.8×10^9	9	7.8	33	44.5
		286			3		45	
		275			5		47	
					14		53	
3	3	241	241	2.4×10^9	4	5.8	56	45.5
		234			6		38	
		248			6		43	
					7		45	
	4	252	250	2.5×10^9	6	7.0	53	46.3
		249			6		39	
		250			9		53	
					7		40	
6	5	219	231	2.3×10^9	9	9.5	68	73.0
		infection			10		70	
		242			9		69	
					10		85	
	6	244	244	2.4×10^9	11	14.0	93	77.3
		228			15		69	
		261			15		75	
					15		72	
60	7	210	199	2.0×10^9	10	12.5	155	163.0
		188			15		144	
		infection			8		182	
					17		171	
	8	242	211	2.1×10^9	8	12.3	117	115.5
		208			14		130	
		184			14		107	
					13		108	

K = absolute number of *E. coli* 343/113 colonies per plate; $\overline{K}$ = mean number of K colonies of 3 parallel plates; N = total number of viable *E. coli* 343/113 cells recovered from the liver of inoculated animals; m = absolute number of mutant colonies per plate; $\overline{m}$ = mean number of m colonies of 4 parallel plates. 0.2 ml of a bacterial suspension containing 4×10^{10} cells per ml were injected in the lateral tail vein of female NMRI mice weight about 30 g. Immediately thereafter 0.1 ml of aqueous DENA solutions were applied subcutaneously to give the final concentrations indicated in the table. After 60 min the mice were killed, the livers removed and handled as described in the text.

animals and plates. The experiment was performed with a total of eight animals, two animals being controls and the remainder being treated with different concentrations of DENA (3, 6, and 60 mg/kg applied subcutaneously immediately after injection of the bacteria). There is detectable mutagenicity in the *arg*$^+$ system but not in the MTR system; typical fluctuation between animals is demonstrated in mice nos. 7 and 8.

Statistical evaluation

To estimate the probability by which mutant frequencies from control and from test series differ from each other, the tables calculated by Kastenbaum and Bowman [37] are useful in the mutation systems described here, with some restrictions in those cases where mutant frequencies are not proportional

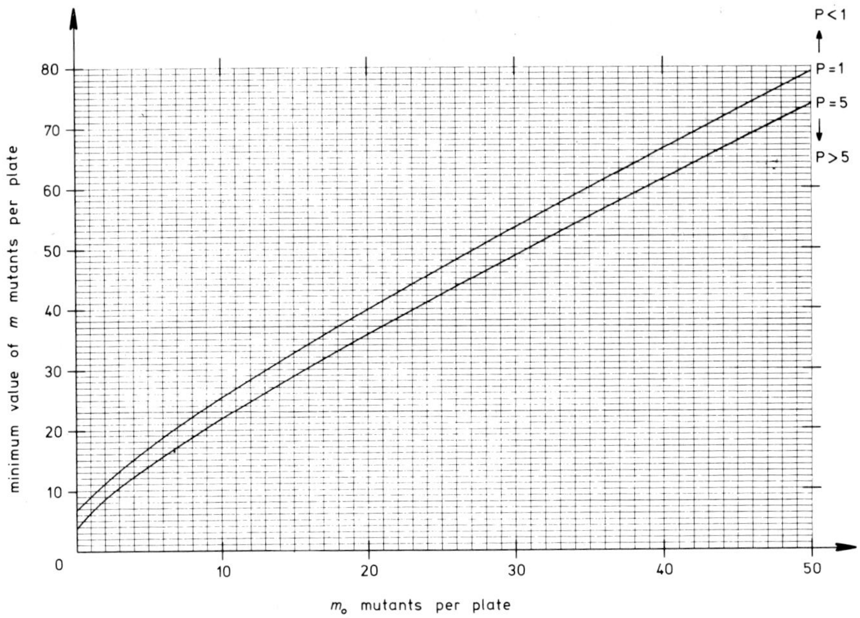

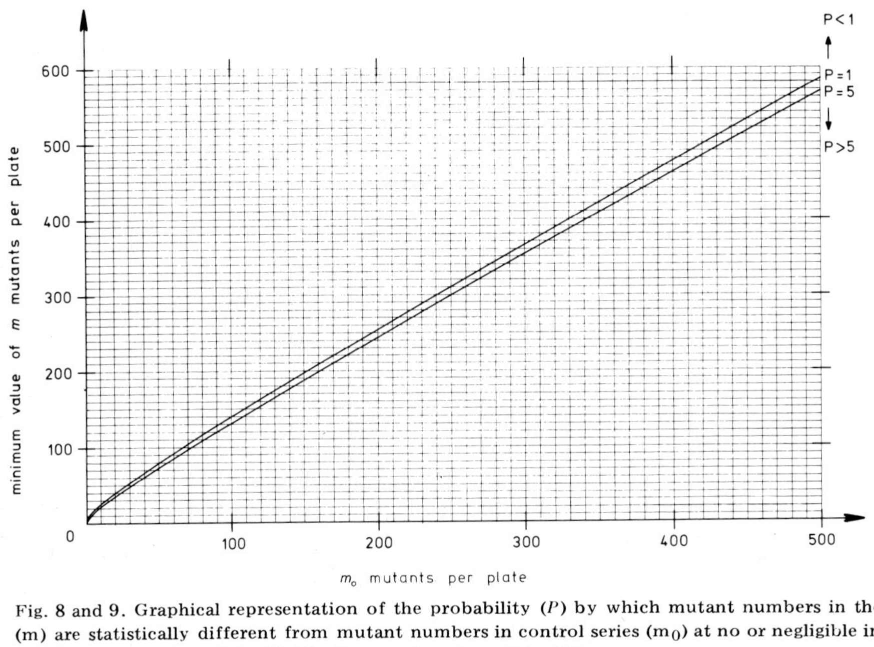

Fig. 8 and 9. Graphical representation of the probability (*P*) by which mutant numbers in the test series (m) are statistically different from mutant numbers in control series (m_0) at no or negligible inactivation. Fig. 8. m_0 values from 0 to 50. Fig. 9: m_0 values from 0 to 500.

to the number of cells per plate and where populations of different size are compared [3].

When the compared populations are of equal size (i.e. at doses which do not inactivate or inactivate moderately the colony forming ability), and provided that the mutant colony numbers are distributed homogeneously within parallel series (which is mostly the case), then it seems plausible to evaluate the significance of the differences between the absolute mutant frequencies per plate in the control and in the test series by means of a χ^2-test [13].

The difference, Δ, between the control mutant colony number (m_0) and the test colony number (m) is significant with the probability P

$$\text{when } \Delta = C + \sqrt{C^2 + 4Cm_0} \text{ where } C = \frac{\chi_1^2}{2}.$$

According to statistical tables, C equals 1.9 at a P of 5%, and C equals 3.3 at a P of 1%. It is thus possible to calculate the smallest value of m which has the probability P to be different from m_0. To serve as a quick way of evaluation, some values have been calculated and drawn in Figs. 8 and 9 for m_0 values up to 50 and 500, respectively.

Conclusions and outlook

In conclusion, we have shown that the strain *E. coli* 343/113 is sensitive to practically all known types of chemical mutagens (and to UV irradiation, see ref. 47) and can be used as a multi-purpose bacterial indicator in various mutagenicity test procedures. Through the introduction of prophage λ, one is able to study concomitantly the mutagenic and the prophage-inducing activity of chemicals.

Even though the mutagen specificity is greatly reduced in the two forward mutation systems gal^+ and MTR, there are still quantitative differences between both which vary from mutagen to mutagen. In routine testing it thus seems of advantage to employ both systems, as well as the two back mutation systems which allow the distinction at least between base-pair changes and frameshifts. At present it is not necessary to add more mutation systems; however, the addition of one (or more) permeability mutation(s) would probably enhance the sensitivity to a number of chemicals of large molecular composition (for instance polycyclic compounds) as shown in other strains [1]. Experiments to reach this are under way.

The strain *E. coli* 343/113(λ) can potentially be used as an indicator in more mutagenicity test procedures than those described here. At present, we are assaying the use of growing cells in microsomal tests for detecting both mutation and lysogenic induction; experiments of this type using other indicators have already been reported [24,26]. It also appears possible to use *E. coli* 343/113(λ) in testing body fluids like blood and urine. Finally, the further development of host-mediated assays to bring *E. coli* 343/113(λ) in contact with blood and lungs of treated animals to gain information on the organ specific activity of chemicals is in progress.

Acknowledgements

The experimental part of this work was supported by the Deutsche Forschungsgemeinschaft. The skilful technical assistance of Mrs. G. Brendel is greatly appreciated.

References

1 Aline, R.F., Jr. and W.S. Reznikoff, Bacteriophage Mu-1-induced permeability mutants in *Escherichia coli* K-12, J. Bacteriol., 124 (1975) 578—581.

2 Ames, B.N., J. McCann and E. Yamasaki, Methods for detecting carcinogens and mutagens with the Salmonella/mammalian-microsome mutagenicity test, Mutation Res., 31 (1975) 347—364.

3 Auerbach, C., Remark on the "Tables for determining the statistical significance of mutation frequencies", Mutation Res., 10 (1970) 256.

4 Barale, R., M. Ducci, S. Baroncelli, C. Bauer, G. Bronzetti, A. Cammellini, R. Guglielminetti, C. Leporini, N. Loprieno, R. Nieri and M. Nozzolini, Liver microsomal assays for evaluating the metabolism of environmental mutagens, Mutation Res., 21 (1973) 277.

5 Barthelmess, A., Mutagenic substances in the human environment, in F. Vogel and G. Röhrborn (eds.), Chemical Mutagenesis in Mammals and Man, Springer, Berlin, 1970, pp. 69—147.

6 Bourguignon, G.J., M. Levitt and R. Sternglanz, Studies on the mechanism of action of nalidixic acid, Antimicrob. Agents Chemother., 4 (1973) 479—486.

7 Breckenridge, L. and L. Gorini, Genetic analysis of streptomycin resistance in *Escherichia coli*, Genetics, 65 (1970) 9—25.

8 Bridges, B.A., R.P. Mottershead and C. Colella, Induction of forward mutations to colicin E2 resistance in repair deficient strains of *Escherichia coli:* experiments with ultraviolet light and captan, Mutation Res., 21 (1973) 303—313.

9 Brown, K.D., Formation of aromatic amino acid pools in *Escherichia coli* K-12, J. Bacteriol., 104 (1970) 177—188.

10 Callen, D.F. and G.R. Mohn, Comparison of the genetic activity of aflatoxins B_1 and G_1 in *Escherichia coli* and *Saccharomyces cerevisiae*, Mutation Res. (1977) in press.

11 Cohen, G. et F. Jacob, Sur la répression de la synthèse des enzymes intervenant dans la formation du tryptophane chez *Escherichia coli*, Comptes Rendus, 248 (1959) 3490—3492.

12 Cook, T.M., W.H. Deitz and W.A. Goss, Mechanism of action of nalidixic acid on *Escherichia coli*. IV. Effects on the stability of cellular constituents, J. Bacteriol., 91 (1966) 774—779.

13 Croxton, F.E., Elementary Statistics, Dover Publ., New York, 1953, pp. 266—283.

14 Ellenberger, J. and G. Mohn, Mutagenic activity of cyclophosphamide, ifosfamide, and trofosfamide in different genes of *Escherichia coli* and *Salmonella typhimurium* after biotransformation through extracts of rodent liver, Arch. Toxicol., 33 (1975) 225—240.

15 Epstein, S.S. and I.B. Saporoschetz, On the association between lysogeny and carcinogenicity in nitroquinolines and related compounds, Experientia, 24 (1968) 1245—1248.

16 Eriksson-Grennberg, K.G., K. Nordström and P. Englund, Resistance of *Escherichia coli* to penicillins. IX. Genetics and physiology of class II ampicillin-resistant mutants that are galactose negative or sensitive to bacteriophage C21, or both, J. Bacteriol., 108 (1971) 1210—1223.

17 Favre, R., A. Wiater, S. Puppo and M. Iaccarino, Expression of a valine-resistant acetolactate synthase activity mediated by the *ilv*O and *ilv*G genes of *Escherichia coli* K-12, Mol. Gen. Genet., 143 (1976) 243—252.

18 Fiethen, L. and P. Starlinger, Mutations in the galactose-operator, Mol. Gen. Genet., 108 (1970) 322—330.

19 Fleck, W., Eine neue mikrobiologische screening-Methode für die Suche nach potentiellen Carcinostatica und Virostatica mit Wirkung im Nucleinsäure-Stoffwechsel, Z. Allgem. Mikrobiol., 8 (1968) 139—144.

20 Frantz, C.N. and H.V. Malling, The quantitative microsomal mutagenesis assay method, Mutation Res., 31 (1975) 365—380.

21 Gabridge, M.G. and M.S. Legator, A host-mediated microbial assay for the detection of mutagenic compounds, Proc. Soc. Exptl. Biol. Med., 130 (1969) 831—834.

22 Gilbert, R.M., S. Rowland, C.L. Dawison and B. Papirmeister, Involvement of separate pathways in the repair of mutational and lethal lesions induced by a monofunctional sulfur mustard, Mutation Res., 28 (1975) 257—275.

23 Glover, S.W., Valine-resistant mutants of *Escherichia coli* K-12, Genetical Res. Camb., 3 (1962) 448—460.

24 Goze, A., A. Sarasin, Y. Moule and R. Devoret, Induction and mutagenesis of prophage λ in *Escherichia coli* K12 by metabolites of aflatoxin B_1, Mutation Res., 28 (1975) 1—7.

25 Green, M.H.L. and W.J. Muriel, Use of repair-deficient *E. coli* strains and liver microsomes to characterize mutagenesis by dimethylnitrosamine, Chem.-Biol. Interact., 11 (1975) 63—65.
26 Green, M.H.L. and W.J. Muriel, Mutagen testing using Trp^+ reversion in *Escherichia coli*, Mutation Res., 38 (1976) 3—32.
27 Greim, H., G. Bonse, Z. Radwan, D. Reichert and D. Henschler, Mutagenicity in vitro and potential carcinogenicity of chlorinated ethylenes as a function of metabolic oxirane formation, Biochem. Pharmacol., 24 (1975) 2013—2017.
28 Hane, M.W. and T.H. Wood, *Escherichia coli* K-12 mutants resistant to nalidixic acid: genetic mapping and dominance studies, J. Bacteriol., 99 (1969) 238—241.
29 Heinemann, B., Prophage induction in lysogenic bacteria as a method of detecting potential mutagenic, carcinogenic, carcinostatic, and teratogenic agents, in A. Hollaender (ed.), Chemical Mutagens. Principles and Methods for their Detection, Vol. 1, Plenum Press, New York, 1971, pp. 235—266.
30 Helling, R.B. and B.S. Adams, Nalidixic acid-resistant auxotrophs of *Escherichia coli*, J. Bacteriol., 104 (1970) 1027—1029.
31 Herman, R.K. and N.B. Dworkin, Effect of gene induction on the rate of mutagenesis by ICR-191 in *Escherichia coli*, J. Bacteriol., 106 (1971) 543—550.
32 Hiraga, S., Operator mutants of the tryptophan operon in *Escherichia coli*, J. Mol. Biol., 39 (1969) 159—179.
33 Hiraga, S., K. Ito, K. Hamada and T. Yura, A new regulatory gene for the tryptophan operon of *Escherichia coli*, Biochem. Biophys. Res. Commun., 26 (1967) 522—527.
34 Hiraga, S., K. Ito, T. Matsuyama, H. Ozaki and T. Yura, 5-Methyltryptophan-resistant mutations linked with the arginine G marker in *Escherichia coli*, J. Bacteriol., 96 (1968) 1880—1881.
35 Ikeda, Y. and T. Iijima, Simplified methods for selecting mutagenic and phage inducing agents from microbial products, J. Gen. Appl. Microbiol., 11 (1965) 129—135.
36 Ishii, Y. and S. Kondo, Spontaneous and radiation-induced deletion mutations in *Escherichia coli* strains with different DNA repair capacities, Mutation Res., 16 (1972) 13—25.
37 Kastenbaum, M.A. and K.O. Bowman, Tables for determining the statistical significance of mutation frequencies, Mutation Res., 9 (1970) 527—549.
38 Kohno, T. and J.R. Roth, Proflavin mutagenesis of bacteria, J. Mol. Biol., 89 (1974) 17—32.
39 Kondo, S., H. Ichikawa, K. Iwo and T. Kato, Base-change mutagenesis and prophage induction in strains of *Escherichia coli* with different DNA repair capacities, Genetics, 66 (1970) 187—217.
40 Lwoff, A., Lysogeny, Bacteriol. Rev., 17 (1953) 269—337.
41 Malling, H.V., Dimethylnitrosamine: formation of mutagenic compounds by interaction with mouse liver microsomes, Mutation Res., 13 (1971) 425—429.
42 Malling, H.V., Host-mediated assay, pro and con, in L. Prakash (ed.), Molecular and Environmental Aspects of Mutagenesis, Charles C. Thomas, Springfield, 1974, pp. 159—175.
43 McCalla, D.R. and D. Voutsinos, On the mutagenicity of nitrofurans, Mutation Res., 26 (1974) 3—16.
44 McCann, J., E. Choi, E. Yamasaki and B.N. Ames, Detection of carcinogens as mutagens in the Salmonella/microsome test: assay of 300 chemicals, Proc. Natl. Acad. Sci. U.S., 72 (1975) 5135—5139.
45 McGregor, D.B., Enhancement by rat liver preparations of the gene conversion frequency induced by ethyl methanesulfonate in *Saccharomyces cerevisiae*, Mutation Res., 23 (1974) 395—398.
46 Mohn, G., Korrelation zwischen verminderter Reparaturfähigkeit für UV-Läsionen und hoher Spontanmutabilität eines Mutatorstammes von *E. coli* K-12, Mol. Gen. Genet., 101 (1968) 43—50.
47 Mohn, G., Microorganisms as test system for mutagenicity, Arch. Toxikol., 28 (1971) 93—104.
48 Mohn, G., 5-Methyltryptophan resistance mutations in *Escherichia coli* K-12. Mutagenic activity of monofunctional alkylating agents including organophosphorus insecticides, Mutation Res., 20 (1973) 7—15.
49 Mohn, G. and J. Ellenberger, Mammalian blood-mediated mutagenicity tests using a multipurpose strain of *Escherichia coli* K-12, Mutation Res., 19 (1973) 257—260.
50 Mohn, G., J. Ellenberger and D. McGregor, Development of mutagenicity tests using *Escherichia coli* K-12 as indicator organism, Mutation Res., 25 (1974) 187—196.
51 Mohn, G., J. Ellenberger, D. McGregor, and H.-J. Merker, Mutagenicity studies in microorganisms in vitro, with extracts of mammalian organs, and with the host-mediated assay, Mutation Res., 29 (1975) 221—233.
52 Mohn, G.R., T.M. Ong and D.F. Callen, Comparison of the mutagenic activity of nitroimidazole derivatives in *Escherichia coli*, *Neurosporg crassa*, and *Saccharomyces cerevisiae*, in preparation.
53 Morse, D.E. and C. Yanofsky, Amber mutants of the *trp*R regulatory gene, J. Mol. Biol., 44 (1969) 185—193.
54 Moyed, H.S., False feedback inhibition: inhibition of tryptophan biosynthesis by 5-methyltryptophan, J. Biol. Chem., 235 (1960) 1098—1102.
55 Newton, A., D. Masys, E. Leonardi and D. Wygal, Association of induced frameshift mutagenesis and DNA replication in *Escherichia coli*, Nature New Biol., 236 (1972) 19—22.

56 Saedler, H., A. Gullon, L. Fiethen and P. Starlinger, Negative control of the galactose operon in *E. coli*, Mol. Gen. Genet., 102 (1968) 79—88.

57 Seeman, N.C., R.O. Day and A. Rich, Nucleic acid-mutagen interactions: crystal structure of adenylyl-3′,5′-uridine plus 9-aminoacridine, Nature, 253 (1975) 324—326.

58 Somerville, R.L. and C. Yanofsky, Studies on the regulation of tryptophan biosynthesis in *Escherichia coli*, J. Mol. Biol., 11 (1965) 747—759.

59 Walker, J.R., *Escherichia coli ras* locus: its involvement in radiation repair, J. Bacteriol., 99 (1969) 713—719.

60 Weekes, U. and D. Brusick, In vitro metabolic activation of chemical mutagens. II. The relationship among mutagen formation, metabolism and carcinogenicity for dimethylnitrosamine and diethylnitrosamine in the livers, kidneys and lungs of BALB/cJ, C57BL/6J and RF/J mice, Mutation Res., 31 (1975) 175—183.

61 Yura, T. and C. Wada, Phenethyl alcohol resistance in Escherichia coli. I. Resistance of strain C600 and its relation to azide resistance, Genetics, 59 (1968) 177—190.

62 Zimmermann, F.K., Detection of genetically active chemicals using various yeast systems, in A. Hollaender (ed.), Chemical Mutagens. Principles and Methods for their Detection, Vol. 3, Plenum Press, New York, 1973, pp. 209—239.

PROCEDURES USED IN THE INDUCTION OF MITOTIC RECOMBINATION AND MUTATION IN THE YEAST *SACCHAROMYCES CEREVISIAE*

F. K. ZIMMERMANN

Mykologie/Genetik, Technische Hochschule, 61 Darmstadt, Schnittspahnstrasse 10 (W. Germany)

SUMMARY

Techniques are described for the use of various yeast strains to detect the induction of *(1)* mitotic crossing-over, *(2)* mitotic gene conversion, *(3)* forward mutation and *(4)* reverse mutation. The technique for the detection of mitotic crossing over is based on a diploid that carries two different alleles of the gene locus *ade2*. These alleles differ in their extent of colony pigmentation engendered on low-adenine media, and they complement each other to the effect that the diploid is white. Mitotic crossing over results in the formation of twin-sectored colonies with a red and a pink sector. The technique for the detection of mitotic gene conversion is based on the use of a heteroallelic diploid carrying two non-complementing alleles that cause a nutritional requirement. Mitotic gene conversion leads to the restoration of intact and dominant wild-type alleles that alleviate the nutritional requirement so that convertant cells can be selected on a minimal medium. The forward mutation technique is based on the use of a haploid strain with a defect in the *ade2*-gene locus which causes the formation of red colonies. Induction of forward mutation in a number of other loci prevents the accumulation of this red pigment so that induction of mutation can be detected by the formation of pink and white colonies. The reverse mutation technique is based on the restoration or compensation of a mutational defect causing a growth requirement. Mutants can be selected for on a minimal medium.

INTRODUCTION

The yeast *Saccharomyces cerevisiae* is a unicellular and uninuclear organism. Strains commonly used by yeast geneticists are either haploid or diploid. The haploid phase is stable, and diploidization can be achieved only by mating two haploid cells of opposite mating types. On the other hand, a diploid cell can become haploid only through meiosis. For a general description of yeast genetics see review by MORTIMER AND HAWTHORNE[4].

Yeast has become popular in environmental mutagenesis research. This is mostly because it is a convenient organism for studying the induction of mitotic recombination of the reciprocal type, mitotic crossing-over, and the non-reciprocal type, mitotic gene conversion (see MAYER[3] and ZIMMERMANN[9]). The use of yeast in environmental mutagenesis work has been reviewed by MORTIMER AND MANNEY[5] and by ZIMMERMANN[11].

In this article the technical procedures are described for testing chemicals for their ability to induce *(1)* mitotic crossing-over, *(2)* mitotic gene conversion, *(3)* forward mutation and *(4)* reverse mutation. In addition, general procedures used in yeast genetics, equipment and media recipes are described.

MITOTIC CROSSING-OVER

Mitotic crossing-over can be shown to occur in vegetative cells without further genetic analysis provided there are appropriate marker combinations. A strain, D5, has been described recently[10] which allows for a simple visual screening for mitotic recombinants. Strain D5 is diploid and combines two different alleles of the gene locus *ade2*. Defective mutants of *ade2* cause a requirement for adenine and the accumulation of a red pigment. D5 carries one allele, *ade2–40*, which causes an absolute growth requirement for adenine and the formation, on low-adenine media, of deep red colonies. The other allele, *ade2–119*, causes only a "leaky" requirement for adenine, *i.e.* cells grow at a reduced rate in the absence of adenine, and the colonies formed on low-adenine medium are only pink instead of red. Moreover, the two alleles complement each other so that the diploid cells carrying this heteroallelic combination form white colonies and do not exhibit an adenine requirement at all. Mitotic crossing-over between the centromere and the *ade2* locus will lead, in 50% of the cases, to the segregation of daughter nuclei, one of which is homoallelic *ade2–40*, the other being homoallelic *ade2–119*. This causes the formation of colonies that have a red and a pink sector twin-spotted colonies. Ever since STERN (cited in ref. 10) demonstrated the occurrence of twin spots in Drosophila in 1936 and explained such twin spots as the result of mitotic crossing-over, many authors have reported not only the occurrence of mitotic crossing-over but also that the cells in such twin spots were actually homozygous as predicted. Consequently, the occurrence of twin spots after mutagenic treatment can be taken as proof of the induction of mitotic crossing-over.

Mutagenic treatment of D5 cells causes not only the formation of cells with red and pink sectors but also colonies with white and pink, white and red doubly sectored, white-pink and red triply sectored colonies besides other colonies that are entirely

TABLE I

INDUCTION OF MITOTIC CROSSING-OVER IN *Saccharomyces cerevisiae* STRAIN D5 WITH NITROUS ACID

	Colonies scored	*Red-pink*	*Red-pink-white*	*Red*	*Pink*	*Red-white*	*Pink-white*	*Hair-line*	*Total aberrant*
Control	2904	0	0	0	0	0	1	3	4
Nitrous acid	3739	55	18	33	60	25	35	15	241

Treatment conditions: 0.2 *M* sodium citrate–HCl (pH 4.5) for control and experimental, $NaNO_2$ (5 mg/ml) in buffer. After 25 min at 21°, cells were diluted 1:100 into a 0.2 *M* potassium phosphate buffer (pH 7.0) to terminate the treatment. Plating: 0.1 ml per petri dish (9 mm inner diameter) with 25 ml of a synthetic complete medium with reduced adenine (5 mg/l). Scoring after 6 days at 28°. Survival: 47.1%. Mitotic recombinants, red-pink and red-pink-white colonies: less than 0.034% in control; 1.95% in experimental. Total of genetically altered colonies: 0.138% in control; 6.45% in experimental.

red or entirely pink. These additional aberrant types are relatively frequent (see Table I). They can be due to mitotic gene conversion, point mutations, chromosomal deletions and aneuploidy induced by the treatment. (See ZIMMERMANN[10] and ZIMMERMANN *et al.*[12].)

It must be emphasized that D5 monitors not only the induction of mitotic crossing-over but all other types of genetic alteration. In this respect, D5 is a multipurpose strain for environmental mutagenesis work. However, only mitotic crossing-over can be unequivocally demonstrated without further genetic analysis.

The design of an experiment with D5 is described in Flowsheet 1.

Special precautions have to be taken to prevent induction of meiosis during treatment. Acetate buffers should be avoided because sporulation is induced in such media. Storage of cells in pain buffers for more than 6 h can have the same effect. Old cultures always contain some sporulated cells, asci, which upon growth may give rise to sectored colonies.

MITOTIC GENE CONVERSION

Mitotic gene conversion is a process of unilateral transfer of small lengths of DNA, up to about 1000 nucleotides long, between homologous regions of chromatids of homologous chromosomes. This process can easily be followed in heteroallelic diploids which carry two defective alleles of the same gene locus. Such a heteroallelic diploid expresses the growth requirement typical of the defect caused by mutation in haploid or homoallelic diploid cells. Conversional transfer of the intact region of one mutant allele to replace the defective nucleotide sequence in the other mutant allele will restore a true wild-type genotype[8]. Wild-type alleles are usually dominant over defective mutant alleles; consequently, the convertant cell will grow in the absence of the originally required growth factor, and can be selected on a synthetic medium. Mitotic gene conversion occurs, on the average, at a rate of one event per 100 000 cells and generation. This incidence can be dramatically increased by mutagenic treatments (see ZIMMERMANN[9]).

Many different heteroallelic diploids have been used over the years by various authors[3,5]. The most commonly used strain is D4 introduced by ZIMMERMANN AND SCHWAIER[14]. It is heteroallelic at two loci: *ade2–1/ade2–2* and *trp5–12/trp5–27*. It has responded to all chemicals mutagenic in yeast.

Induction of mitotic gene conversion can be followed by plating D4 cells on two different selective media. Plating on tryptophan-free medium selects convertants at the *trp5* locus, and plating on adenine-free medium selects convertants at the *ade2* locus. Details of the procedure are given in Flowsheet 2.

Mitotic gene conversion at the *trp5* locus, but not at the *ade2* locus, can be followed by simply plating cells on a tryptophan-free medium and introducing crystals or solutions of the chemical to be tested at the center of the plate. Usually, but for unknown reasons not always, a mutagen will induce conversion which results in a ring of convertants surrounding the site of application (Fig. 1). For details of procedure see Flowsheet 1. It has to be emphasized that such a spot test procedure does not necessarily reflect the quantitative mutagenicity of an agent.

Special precautions have to be taken to prevent sporulation, as discussed above in connection with mitotic crossing-over.

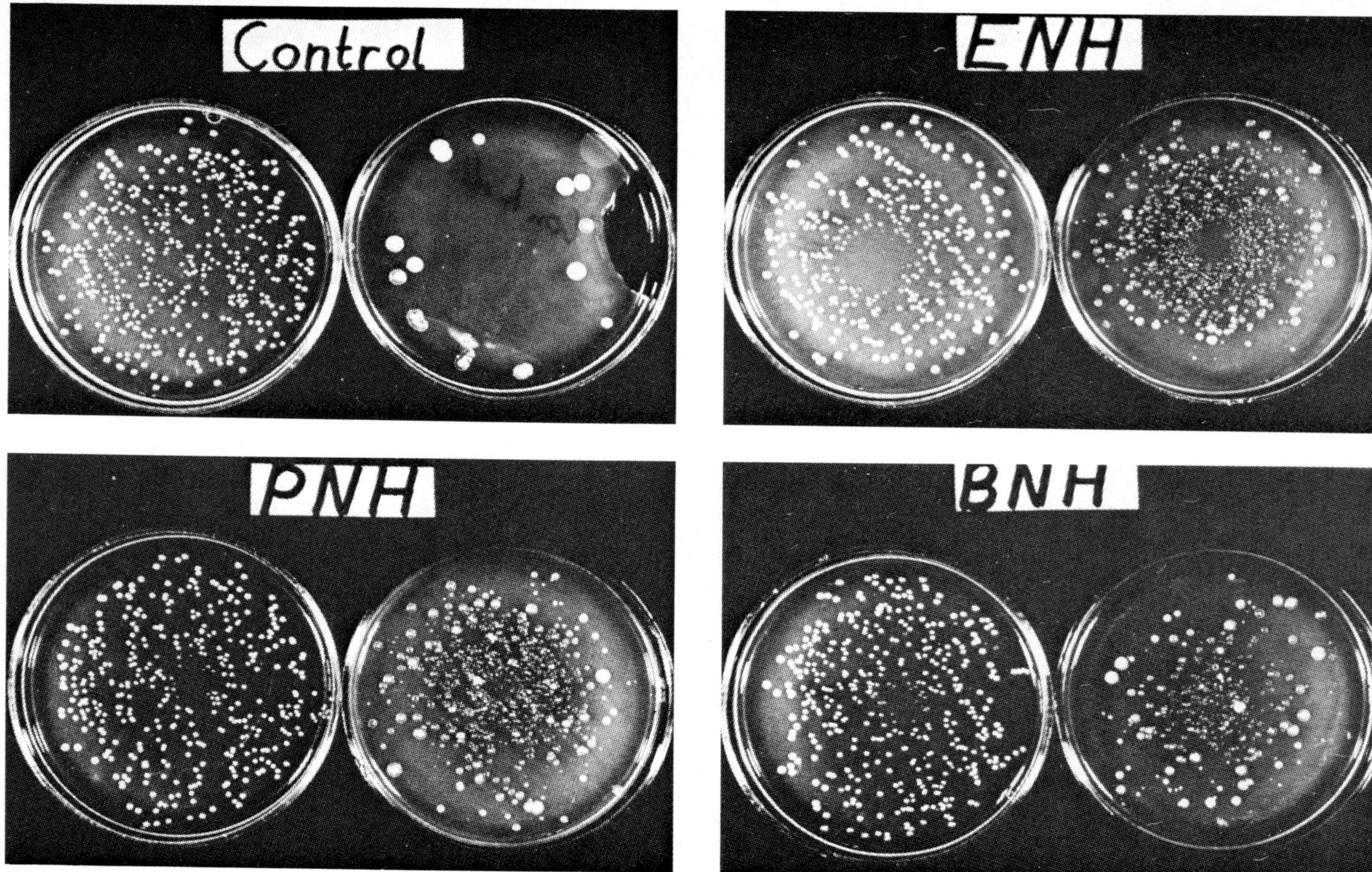

Fig. 1. Induction of mitotic gene conversion on media plates. Strain D4 was grown in liquid YEP glucose medium and manipulated as described in Flowsheet 2. Cells were spread on a synthetic complete medium (250 cells per plate) and on a synthetic medium without tryptophan ($2.5 \cdot 10^5$ cells). Chemicals to be tested were *N*-ethylnitrosourea (ENH), *N*-propylnitrosourea (PNH) and *N*-butylnitrosourea (BNH). A few crystals of a chemical were placed in the center of a plate. Photographed after seven days. Plates to the left of each photograph are complete, those to the right are tryptophan-free media.

MUTATION

Mutation is a rare event, so the determination of the mutagenicity of an agent requires a large number of cells. There are two methods for assessing induction of mutation in yeast: forward mutation as detected by the loss of a function and reverse mutation by the restoration of a function previously abolished by a forward mutation.

Reverse mutation experiments are easy because it is possible to select reverse mutants on a synthetic medium lacking the growth factor required by the original mutants. The drawback with reverse mutation is that it usually requires specific genetic alterations for restoration of function. Not all mutagens induce all possible types of alteration; consequently some mutagens would be classified as non-mutagenic in a reverse mutation system (see MARQUARDT *et al.*[2]). Of course, a large battery of mutant strains could be used to detect all mutagenic agents, but work would then become too much, so it is more convenient to study forward mutation.

Forward mutation

Forward mutation can be detected by the loss of function. Certain forward mutations cause resistance to certain anti-metabolites which allow for selective procedures. However, such resistance mutation systems are not very accurate for various reasons[11]. A relatively convenient system was described by ROMAN[6]. Mutants blocked in *ade1* or *ade2* form a red pigment. Pigment formation in such mutants can be blocked by mutation in several other genes most of which are involved in purine biosynthesis before the blocks caused by *ade1* and *ade2*. Mutagenized cells are plated on a low adenine medium, and mutagenesis produces colonies with white and pink sectors or completely white or pink colonies. A procedure for this type of experiment is given in Flowsheet 3. It is important to use the synthetic medium recommended, to prevent petite induction from mimicking induction of gene mutations. Special precautions have to be taken with *ade1* and *ade2* strains. An *ade2* or *ade1* mutant in combination with a mutation in another *ade* gene has a selective advantage over the single *ade1* or *ade2* mutant. Cultures should always be checked for purity.

Reverse mutation

Reverse mutation is the most convenient test because mutants can be selected. The advantage over mitotic gene conversion which is equally convenient technically is that the events studied are known to be mutations and therefore bear a known relationship to what we are interested in, whereas conversion events are distinct from mutation and may not always obey the same rules in induction.

The procedure for a reverse mutation experiment is given in Flowsheet 4.

It is also possible to use the spot test procedure with reverse mutation, but a quantitative comparison of different mutagens is not possible on this basis. The techniques are similar to those already described for mitotic gene conversion.

Special precautions are required to maintain mutant stocks because most revertants have a selective advantage over non-revertant cells. Regular checking of cultures is helpful (see Flowsheet 4).

EQUIPMENT AND CONSUMABLE MATERIALS

Typical microbiological laboratory equipment is needed for the use of yeast systems in environmental mutagenesis work, including the following items.

(1) An autoclave with the temperature set to 121° and a pressure of 1 kg per cm² above ambient pressure (autoclave 15 min).

(2) An oven for temperatures of 160° to sterilize glassware and other dry goods (3 h at 160°).

(3) A set-up for filter sterilization is useful but not essential.

(4) A water-bath to cool solid media to 45° after they are autoclaved and before they are poured into plates.

(5) A roller drum or a reciprocal shaker for growing cultures in test-tubes in liquid medium.

(6) A gyratory shaker for growing cultures in liquid medium in flasks.

(7) A microscope with objective lenses 5–10×, 25× and 40×; eyepieces 10×.

(8) Three hemocytometers, 0.1 mm deep, for determining cell titers.

(9) Two tally counters or their equivalent.

(10) Two shaking devices to mix liquids in test-tubes (*e.g.* Vortex or Whirlmix mixers).

(11) A small centrifuge, refrigerated if possible, to take test-tubes (very convenient) and centrifuge tubes of volumes 25, 50 and 100 ml.

(12) An analytical balance for weighing mg portions.

(13) A balance for quantities of 1–2 kg for preparation of media and buffers.

(14) A pH meter for adjusting buffers.

(15) Three good quality timers for treatment control.

(16) One or two good quality shaking water-baths with reliable temperature control.

(17) One or two large refrigerators for storing cultures and solutions of media ingredients, buffers, *etc.* Alternatively, a cold room will do.

(18) Freezer or freezing compartment in a refrigerator.

(19) One dozen test-tube racks, preferably stainless steel, with 4×12 or 4×10 holes to store cultures and set up tubes for dilution series.

(20) Glassware:

2000 test-tubes with metal caps.

20 centrifuge tubes 25-ml, 50-ml and 100-ml.

Erlenmeyer flasks, 30×1000-ml (for preparing media), 30×500-ml, 30×250-ml and 30×100-ml.

Beakers, 10×100-ml, 10×500-ml and 10×1000-ml.

Measuring cylinders, 5×25-ml, 5×50-ml, 5×100 ml, 5×250 ml, 5×1000-ml and one or two 2000-ml.

Pipettes, all with blow-out calibration, 50×10-ml, 50×5-ml, 50×2-ml, 200×1-ml (for dilution and plating, 20×0.5-ml, 20×0.2-ml and 100×0.1-ml (for dilution).

100 autoclavable screw-cap bottles for solutions of media ingredients, buffers, sterile water, *etc.*

One accurate automatic pipetting device for volumes between 1 and 10 ml.

Petri dishes, glass, 5000 unless it is planned to use disposable plastic dishes. In either case the inner diameter should be 90–100 mm.

(21) Pipette containers to hold the above-listed pipettes.

(22) Pipette washers and containers to hold used pipettes in water or a sterilizing solution.

(23) Four magnetic stirrers for preparing media, buffers and mutagen solutions *plus* ten sets of teflon-coated spinning bars 10, 15 and 20 mm long.

(24) For the study of mitotic crossing-over and forward mutation using *ade2* mutants, it is necessary to have a dissecting microscope for screening red or white sectors.

(25) A dozen loops for inoculating and streaking cultures.

(26) Half a dozen forceps 10–20 cm long.

(27) Several stainless steel spatulae.

(28) Microscopic slides and cover slips (500 of each).

CHEMICALS AND MEDIA

Potassium phosphate buffers are required for the pH range 5.5–8. They can be prepared by mixing 0.1 *M* solutions of KH_2PO_4 and K_2HPO_4 and can be sterilized by autoclaving. For lower pHs, potassium or sodium citrate buffers can be prepared, the pH of a 0.1 *M* solution being adjusted to the desired value by adding conc. HCl. This buffer cannot be autoclaved because autoclaving renders it toxic. Filter sterilization is best, but a sterile citrate solution can be adjusted with a pH meter in a sterile flask by adding conc. HCl which should be sterile enough for all practical purposes. Acetate buffers should be avoided because sporulation is induced in plain acetate solutions.

Several mutagens are not very soluble. A good solvent is dimethylsulfoxide; alternatively, ethanol or acetone are adequate depending on the chemical. For treatments of up to four hours, 5 to 10% solutions of these solvents are tolerated by yeast.

Most alkylating agents are rapidly inactivated by diluting mixtures of cells and agent into 5% $Na_2S_2O_3$ in water. This can be autoclaved or, better, filter sterilized.

Ethanol, denaturated or not, is useful for sterilizing tools such as spreaders, bent glass rods used to spread cells on media plates, forceps, loops and the like. Adhering alcohol is flamed off before use.

Potassium hydroxide is used to dissolve those amino acids and nucleobases that do not dissolve readily in water.

For plugging flasks, non-absorbent cotton wool is a useful material. It does not require wrapping in gauze. A moderately firm plug remains in the mouth of a 1000-ml Erlenmeyer flask during autoclaving.

Media

There are two basically different media used by yeast geneticists. The first is a rich complete medium called YEP consisting of 1% yeast extract, 2% peptone supplemented with 2% glucose as the usual carbon source. The other is a *synthetic medium* based on Difco yeast nitrogen base without amino acids and supplemented with various amino acids, uracil and adenine as potential growth factors for mutants. Again, the standard carbon source is 2% glucose. For solid media, 1.5% of agar is added. It is not really good microbiological technique to autoclave everything together, but it works reasonably well for these media. After being autoclaved, agar media have to be thoroughly mixed, cooled to 45° and then poured into petri dishes. They are ready for use after two days in a dry room.

Cultures for experiments can be grown in liquid medium, 5 ml YEP glucose, in test-tubes or on the surface of YEP glucose agar plates. The latter procedure gives a better separation of cells which is important with haploid strains.

Synthetic complete medium contains the following special ingredients, at concentrations shown:

Difco yeast nitrogen base without amino acids	6.7 g/l
Adenine sulfate	10 mg/l
or, for better pigmentation of *ade 2* mutants,	5 mg/l
L-Arginine–HCl	10 mg/l
L-Histidine–HCl	10 mg/l
L-Isoleucine	60 mg/l
L-Leucine	60 mg/l
L-Lysine–HCl	10 mg/l
L-Methionine	10 mg/l
L-Tryptophan	10 mg/l
L-Valine	30 mg/l
Uracil	10 mg/l

Stock solutions of these chemicals are prepared to give 1% solutions. In cases of poor solubility, addition of KOH will help. After addition of a few drops of chloroform for sterility, these solutions are best stored in a refrigerator.

Convertants and revertants are detected on selective media. These media are basically complete but lack one or more of the standard ingredients.

GENERAL PROCEDURES

The best growth medium is YEP glucose. Cells needed for experiments on induction of mitotic gene conversion and reverse mutation are best grown in liquid YEP glucose. It is advisable to enrich these media with the required growth factors to prevent accumulation of prototrophic convertants or revertants. Also, it is helpful to grow parallel cultures which are started with about 200 cells per 5 ml YEP glucose liquid in test-tubes. These cultures can be checked, by sample plating, for purity from an excess of prototrophic cells after they reach the early stationary phase (see Flowsheets 2 and 4).

Depending on the strain, yeast cells tend to separate poorly and form clusters. Such clusters do not separate during dilution and spreading; they stick together, and form a single colony. This is disadvantageous in mitotic crossing-over and forward mutation because the altered cells form small sectors in colonies that are hard to detect. In reverse mutation and gene conversion, clusters do not interfere with the expression of the altered genotype. The method of removing such clusters is given in Flowsheet 1.

Preparation of mutagen solutions

Many mutagens and many of the chemicals to be tested are poorly soluble in water. A quick way to achieve solution of a chemical mutagen is to dissolve it first in a small volume of some non-toxic organic solvent and then dilute it in an aqueous buffer. The final concentration of the organic solvent should not exceed 5–10% depending on its effect on cells.

Treatment conditions

Treatments should be carried out under constant temperature conditions and shaking. A temperature-controlled water-bath is indispensable. Treatments should be performed in buffered solutions. The strength of the buffer has to be so adjusted as to compensate possible influences of the dissolved chemical on the overall pH. Genetic activities of various mutagens are strongly dependent on pH. The most obvious example is nitrous acid which is active only in its undissociated form at pH values lower than 5.5 (see SCHWAIER *et al.*[7]). A previously untested chemical should always be used around neutrality and at a pH around 4.5.

Dosage of a chemical should be varied by changing concentrations rather than time of treatment because with some labile chemicals the concentration can decrease sharply during treatment. Mutagenic agents can have a selective affinity for the genetic material which might be expressed less at high than at low concentrations because of competing reactions with cellular components which lead to killing but not to genetic effects. A good span of treatment is about 30 min with reactive chemicals. Longer exposures might lead to the induction of meiosis in diploid cells and thus interfere with mitotic recombination (gene conversion and crossing-over).

Termination of treatments

The accurate termination of treatments is important for quantitative work, but often difficult to achieve. There are, for some reactive alkylating agents, scavengers such as $Na_2S_2O_3$, but usually one can terminate treatments only by separating cells from the chemical. This can be done by diluting into an ice-cold solution and centrifuging in a refrigerated centrifuge. Such a procedure takes only about 5 min from diluting to the addition of washing fluid. Reduction in temperature and dilution together cut down the reaction rate sufficiently.

Plating of cells

It is hard to count more than 400 cells on a petri plate. Consequently, complete plates used to determine survival should not be seeded with more than 400 cells. For mitotic crossing-over, it is better to have an upper limit of 200 cells per plate, because one has to detect twin-spotted colonies. These are the more evident the larger the colony; on more crowded plates with small colonies it becomes difficult to distinguish the hue of the aberrant sectors.

There is also a limit to the number of cells that can be plated on selective media. An auxotrophic cell does not grow on a medium lacking a growth factor, but it is still metabolically active. Heavy lawns of auxotrophic cells, more than $5 \cdot 10^7$ cells, might exhaust the medium of other growth factors, and thus prevent prototrophic revertants or convertants from growing. Another important point is that cells killed by mutagens may lyse and release potential growth factors into the medium which stimulate growth of non-convertant or non-revertant, surviving cells. This situation precludes the use of doses giving high killing and the plating of heavy suspensions on selective media.

Cells are plated on to the surface of solid media by dropping out 0.1 ml aliquots of a cell suspension. These drops are evenly distributed with a bent glass rod. This rod can be sterilized by immersion in ethanol. Before use, the rod is removed from ethanol, adhering alcohol is flamed off, and the glass is cooled by touching the agar surface.

Incubating plates

Normally, yeast grows best between 25 and 35°. However, there can be a very strong effect of post-treatment growth temperature on cell survival and also on mutation yield to the effect that a post-treatment temperature of 35° almost abolishes the mutagenic effect of some treatments (ZIMMERMANN *et al.*[13]). The lower part of the temperature range is preferable.

Scoring plates complete

Colonies on solid media become visible after one to three days. Diploid cells grow faster than haploids. Mutagenic treatment causes growth retardation. Convertant colonies on selective media take one day longer. Revertant colonies can grow very slowly and take up to ten days to develop.

There is no fixed time at which colonies can be counted. This has to be decided for each strain and also for each treatment. As long as there are colonies from normal size down to almost invisible, one has to wait for the final count until all colonies are well visible. In doubtful cases, one has to count plates several days apart to see whether the number of visible colonies remains constant. This is mostly a problem with doses giving high killing. Small colonies are a problem with reverse mutation experiments, because not all revertants restore a wild-type phenotype but are due to suppressor mutations. In contrast with this, gene conversion restores a perfect wild-type condition, and causes less problems with counting colonies.

EVALUATION OF RESULTS

It is important to determine both the number of genetically altered types and the percent survivors. Any mutagen kills cells: there is no mutagen that induces genetic alterations but no killing. For quantitative purposes, frequencies are given as mutants per survivors expressed on a per 10^5, 10^6 and so on basis. The basis is so chosen that the average spontaneous frequency has one digit before the point: control $5.3 \cdot 10^{-6}$, dose $520 \cdot 10^{-6}$. However, an increase in this relative frequency computed on a per survivor basis is not a definite proof of genetic activity of the agent tested. Besides induction, there is selective survival which can engender precisely the same result unless it is shown that the increase above the control is also realized when the frequency is given per treated cells. Therefore, besides giving the frequency per survivors, survival should be given too.

Negative results require two aspects to be considered. First, an agent can be completely inactive in yeast. This is so when neither genetic activity nor cell killing can be detected. Second, an agent might kill cells—it is biologically active—but has no genetic effect. Often, such an agent will give a precipitous killing curve. Within a concentration range having a factor 2, survival can decline from 100 to less than 1%. A highly efficient mutagen will give a flat curve with the genetic effects rising before killing becomes apparent. It is not useful to rely heavily on doses giving high killing. If there is no genetic effect to be detected with a dose giving 90% killing, then the conclusion is that there is no genetic effect at all.

A negative effect can be claimed if no statistically significant increase is observed in a population that is 20–100 times the size of one that would yield an average of one spontaneous mutant or recombinant. This conclusion should be based on results from a dose range running from no killing to about 90% killing, and this in an experimental set-up where the killing *versus* dosage is kept as flat as possible.

Statistical treatment of data should rely on the evaluation of repeat experiments and not so much on the number of mutants or recombinants counted in a single experiment. An appropriate test is the *t*-test. For very small numbers in individual experiments there is KASTENBAUM's test[1]. The advantage of microbial systems is, however, that the populations tested can be quite astronomical without too much cost or time.

FLOWSHEETS

Flowsheet 1—Mitotic crossing-over

(1) Spread about 200 cells onto each of five plates of a synthetic complete medium with adenine at 5 mg/l, and incubate for four days.

(2) Check plates and select completely white and round colonies of normal size. Pick up cells from such colonies and suspend them in the desired buffer until a white suspension is formed. Check under the microscope to see whether there are many cell clusters.

(3) If there are too many cell clusters (more than 15%), centrifuge slowly to sediment clusters (about 5–10 min). The speed of the centrifuge should be so set that it takes about 20 min to sediment all cells. Check the supernatant for clusters. Repeat procedure if necessary.

(4) Adjust cell titer to about $2 \cdot 10^7$ cells/ml. This is the *stock suspension*.

(5) Prepare a solution or suspension of the chemical to be tested in the desired buffer. Some basically water-soluble chemicals dissolve very slowly. They can be dissolved first in a small volume of an organic solvent (acetone, ethanol, dimethylsulfoxide), and this solution can be diluted in buffer. The final concentration of organic solvent should not exceed 10%, and a solvent control should be included along with a buffer control and the treated samples.

(6) Prepare a dilution series of the chemical to be tested in steps of 1:3, 1:10, 1:30, 1:100 and so on. Genetic activity of chemicals has been found in the range from 1 *M* down to 10^{-8} *M*.

(7) Pipette 4.5 ml aliquots of dilutions into test-tubes and place the tubes in the water-bath. Allow time for temperature equilibration.

(8) Start treatment by pipetting 0.5-ml aliquots from the stock suspension into tubes with chemicals. The same pipette can be used for all tubes if the first tube is the buffer control followed by solvent control and then the various dilutions in increasing concentrations.

(9) Treatments should be performed with shaking if the time of treatment exceeds 30 min because yeast cells sediment.

(10) Stopping the treatment. Dilute 1:100 into ice-cold water or some stopping reagent (*e.g.* 5% $Na_2S_2O_3$). Dilute once more 1:10, and plate on to a synthetic complete medium with 5 mg adenine per l. In a first pilot experiment (for determination of effective concentration range), use five plates per concentration. In a final experi-

ment, use at least 20 plates per dose of active agent, and up to 100 with agents that are genetically inactive.

(11) After three to four days, count colonies on a representative sample of five plates per concentration.

(12) After six to eight days, score for aberrant colonies when pigmentation has become intense enough to allow distinction between white, pink and red colonies or sectors. List aberrant colonies in the following categories: *(i)* pink-red; *(ii)* pink-red-white; *(iii)* pink; *(iv)* red; *(v)* white-pink; *(vi)* white-red; *(vii)* hairline sectors, very tiny red or pink sectors where red and pink cannot be distinguished.

(13) Compute survival by dividing the number of colonies on experimental plates by the number of colonies on the same number of control plates.

(14) Compute frequencies of mitotic recombinants by dividing the number of red-pink and red-pink-white colonies by the total number of colonies. Do the same for total frequencies of all aberrant colonies.

Time requirements. Step *(1)*, 20 min; steps *(2)–(10)*, two persons working, 2–4 h depending on size of experiment and experience; step *(11)*, 1–2 h; step *(12)*, 1–4 h, rather variable depending on size of experiment and genetic effects of treatment; steps *(13)–(14)*, 1 h.

Flowsheet 2—Mitotic gene conversion

(1) Inoculate about 200 cells into each of ten test-tubes with 5 ml YEP glucose (use 0.1 ml of a cell suspension with 2000 cells/ml). Place the tubes on the roller drum or reciprocal shaker.

(2) When cultures have reached the early stationary phase, after about 2 days at a temperature between 25 and 30°, remove aliquots of 0.1-ml from each tube, and spread on plates selective for convertants. Incubate the plates, place the tubes in a refrigerator or cold room.

(3) After three to four days, check plates for spontaneous convertants. Discard cultures with high spontaneous convertent frequencies.

(4) Discard the supernatant medium from good cultures, resuspend in distilled water, centrifuge, discard the supernatant, resuspend, centrifuge, and discard supernatant once more. Resuspend the cell pellet in the desired buffer.

(5) Dilute the cells 1:10, count them in a hemocytometer and calculate the cell titer.

(6) Prepare *stock suspension* with $3 \cdot 10^7$ cells/ml.

(7) Prepare a solution or suspension of the chemical to be tested in the desired buffer. Some basically water-soluble chemicals dissolve very slowly. They can be dissolved first in a small volume of an organic solvent (acetone, ethanol, dimethylsulfoxide), and this solution can be diluted in buffer. The final concentration of organic solvent should not exceed 10% and a solvent control should be set up along with a buffer control and the treated samples.

(8) Prepare a dilution series of the chemical to be tested in steps of 1:3, 1:10, 1:30, 1:100 and so on. Genetic activity of chemicals has been found in the range of 1 *M* down to 10^{-8} *M*.

(9) Pipette 4.5-ml aliquots of dilutions into test-tubes and place the tubes in a water-bath. Allow for temperature equilibration.

(10) Start treatment by pipetting 0.5-ml aliquots from the stock cell suspen-

sion into tubes with the chemical. The same pipette can be used for all tubes if the first tube is the buffer control followed by solvent control and then the various dilutions according to increasing concentration.

(11) Treatments should be performed with shaking if the time of treatment exceeds 30 min because yeast cells sediment.

(12) Stopping treatment. Add 10 ml of ice-cold water or some stopping reagent (*e.g.* 5% $Na_2S_2O_3$) and chill the mixture in an ice-bath. Spin down in a refrigerated centrifuge. Resuspend pellet in ice-cold washing fluid, centrifuge, discard the supernatant, repeat once. Finally, resuspend in 5 ml distilled water.

(13) Plate 0.1 ml aliquots of this suspension undiluted on to medium selective for convertants (five plates per concentration).

(14) Dilute 1:1000 in steps of 1:10 or one step 1:100 and another 1:10. Plate 0.1 ml aliquots of the diluted suspension in a complete medium (five plates per concentration).

(15) After four days, score colonies on complete medium; and, after four to six days, on medium selective for convertants.

(16) Compute survival. Divide the number of colonies found on complete plates from treated samples by the colonies counted on the same number of control plates.

(17) Compute the conversion frequency. Calculate the number of viable cells on the plates selective for convertants by multiplying the number of colonies on complete plates by the dilution factor. Divide the number of convertant colonies by the number of viable cells. Express frequencies as *e.g.* $3.5 \cdot 10^{-6}$ survivors.

Time requirements. Step *(1)*, preparing cell suspension with 2000 cells/ml and inoculating ten tubes, 20 min; step *(2)*, 15 min; steps *(3)–(14)*, two persons working, 2–4 h depending on size of experiment and experience; step *(15)*, 1–3 man-hours depending on size of experiment and number of colonies per plate; steps *(16)–(17)*, 1 h.

Spot test procedure: steps (1)–(6). Plate cells from $3 \cdot 10^7$ cells/ml suspension on drop-out medium selective for convertants. Prepare another diluted cell suspension with $3 \cdot 10^4$ cells/ml, plate on complete medium (0.1 ml per plate). Let cell suspension soak in (with normally dry plates, 1 h). Administer chemical directly as crystals or liquid or solution, or place sterile filter disc (5 mm diameter) in center of plate and pipette 0.02 ml or less on disc. Scoring, killing: measure zone of inhibition around site of application. Induction of conversion: count colonies (if possible) outside zone of inhibition and compare with an appropriate area on control plates. Time requirements: spot test, 30–60 min; scoring, 15–30 min; all depending on size of experiment.

Flowsheet 3—Forward mutation with a haploid ade2 strain

(1) Spread about 200 cells on to each of five plates with a complete medium with adenine at 5 mg/l, and incubate for four to six days until red pigmentation of colonies is clearly visible.

(2) Check plates for well-grown, homogeneously red colonies. Pick up cells from such colonies and suspend in the desired buffer until a heavy suspension of pink color is reached. Check under the microscope to see whether there are many cell clusters.

(3) If there are too many cell clusters (more than 15%), centrifuge slowly to sediment the clusters. The speed of the centrifuge should be so set that it takes about

20 min to get a sedimentation of all cells. Spin for about 5–10 min. Check the supernatant for clusters. Repeat if necessary.

(4) Adjust the cell titer to about $5 \cdot 10^7$ cells/ml *stock suspension.*

(5) Prepare a solution or suspension of the chemical to be tested in the desired buffer. Some basically water-soluble chemicals dissolve very slowly. They can be dissolved first in a small volume of an organic solvent (acetone, ethanol, dimethylsulfoxide), and this solution can be diluted into buffer. The final concentration should be included as a solvent control along with a buffer control and the treated sample.

(6) Prepare a dilution series of the chemical to be tested in steps of 1:3, 1:10, 1:30, 1:100 and so on. Genetic activity of chemicals has been found in the range from 1 *M* down to 10^{-8} *M*.

(7) Pipette 4.5 ml aliquots of dilutions into test-tubes and place the tubes in a water-bath. Allow time for temperature equilibration.

(8) Start treatment by pipetting 0.5-ml aliquots from the stock suspension into tubes with the chemical. The same pipette can be used for all tubes if the first tube is the buffer control followed by the solvent control and then the various dilutions according to increasing concentration.

(9) Treatments should be performed with shaking if the time of treatment exceeds 30 min because yeast cells sediment.

(10) Stopping treatment. Add 10 ml of ice-cold water or some stopping reagent (*e.g.* 5% $Na_2S_2O_3$) and chill the mixtures in an ice-bath. Spin down in a refrigerated centrifuge. Resuspend the pellet in 5 ml distilled water and dilute 1:100.

(11) Plate on to a synthetic complete medium with 5 mg adenine per l. In a first pilot experiment (for the determination of effective concentration range), use five plates per concentration. In a final experiment, use at least 10 to 20 plates per dose with active agents, and 50 to 100 with agents that are genetically inactive.

(12) Dilute further by a factor of 20 and plate again on the same medium to count colonies formed by surviving cells, 5 plates per concentration.

(13) After 4–6 days, count colonies on survival plates.

(14) After 6–8 days, when solid red pigment has been formed, score plates with higher cell numbers. With a dissecting microscope, look for pink and white colonies.

(15) Compute survival by dividing the number of colonies on experimental survivor plates by the number of colonies on the same number of control plates.

(16) Multiply the colony number on survival plates by 20 to get the number of surviving cells on plates used to screen for white and pink mutants. Divide the number of mutants by that number to obtain the mutation frequency. Express the mutation frequency on a basis of mutants per 10^3 or 10^4 surviving cells.

Time requirements. Step *(1)*, 20 min; steps *(2)–(12)*, two persons working, 2-4 h depending on size of experiment and experience; step *(13)*, 1–2 h; step *(14)*, 2–6 h, rather variable depending on size of experiment, genetic effects of treatment and experience; steps *(15–(16)*, 1 h.

Flowsheet 4—Reverse mutation with a haploid, auxotrophic mutant

(1) Inoculate about 200 cells into each of ten test-tubes with 5 ml YEP glucose (use 0.1 ml of a cell suspension with 2000 cells per ml). Plate the tubes on a roller drum or reciprocal shaker.

(2) When the cultures have reached the early stationary phase (after about

three days at a temperature between 25 and 30°), remove aliquots of 0.1 ml from each tube and spread on medium selective for revertants. Incubate the plates, place the tubes in a refrigerator or cold room.

(3) After four to six days, check the plates and discard the cultures with high spontaneous revertant frequencies.

(4) Discard supernatant medium from good cultures, resuspend in distilled water, centrifuge, discard the supernatant, resuspend, centrifuge and discard the supernatant once more. Resuspend the cell pellet in the desired buffer.

(5) Dilute the cells 1:10, count them in a hemocytometer and calculate the titer.

(6) Prepare *stock suspension* with $1 \cdot 10^8$ cells/ml.

(7) Prepare a solution or suspension of the chemical to be tested in the desired buffer. Some basically water-soluble chemicals dissolve very slowly. They can be dissolved first in a small volume of an organic solvent (acetone, ethanol, dimethylsulfoxide), and this solution can be diluted in buffer. The final concentration of organic solvent should not exceed 10%, and a solvent control should be set up along with a buffer control and the treated samples.

(8) Prepare a dilution series of the chemical to be tested in steps of 1:3, 1:10, 1:30, 1:100 and so on. Genetic activity of chemicals has been found in the range from 1 *M* down to 10^{-8} *M*.

(9) Pipette 2-ml aliquots of cell stock suspension into centrifuge tubes with 10 ml distilled water, centrifuge, discard and carefully drain off the supernatant.

(10) Start the reaction by pipetting 5 ml aliquots of each dilution of chemical on to cell pellets in centrifuge tubes. Place the tubes in a shaking water-bath. Reagent solution and cell pellets should be equilibrated to bath temperature before the reaction is started.

(11) Treatments should be performed with shaking if time of treatment exceeds 30 min because yeast cells sediment.

(12) Stopping treatment. Add 10 ml of ice-cold water or some stopping reagent (*e.g.* 5% $Na_2S_2O_3$), and chill the mixture in an ice-bath. Spin down in a refrigerated centrifuge. Discard the supernatant, resuspend the pellet in ice-cold washing fluid, cetrifuge, discard the supernatant, repeat once. Finally, resuspend in 2 ml distilled water.

(13) Plate directly on to a synthetic medium selective for revertants. Use five plates per dose.

(14) Dilute further 1:100, 1:100 and 1:3 and plate on a synthetic complete medium for scoring survivors. Use five plates per dose.

(15) After four to six days, score the plates on complete medium, and on media selective for convertants between six and ten days after plating.

(16) Compute survival. Divide the number of colonies found on complete plates of experimentals by the number of colonies on the same number of control plates.

(17) Compute reverse mutation frequencies. Calculate the numbers of viable cells on selection plates by multiplying colony numbers on complete plates with dilution factor, $3 \cdot 10^4$. Express reverse mutation frequencies on the basis of number per 10^7 survivors.

Time requirements. Step *(1)*, 20 min; step *(2)*, 15 min; steps *(3)–(14)*, two persons working, 2–4 h depending on size of experiment and experience; step *(15)*, 1–3 man-hours depending on size of experiment and number of colonies per plate; steps *(16)–(17)*, 1 h.

Spot test procedure: *steps (1)–(5)*. Adjust cell titer to $5 \cdot 10^8$ cells/ml and plate 0.1 ml on medium selective for revertants. Prepare a dilution of about $5 \cdot 10^4$ cells/ml and plate on complete medium. Let the cell suspension soak in (with normally dry plates, 1 h). Administer chemical directly as crystals or liquid or solution, or place sterile filter disc (5 mm diameter) in center of plate and pipette 0.02 ml or less on disc. Scoring, killing: measure zone of inhibition around site of application. Induction or reverse mutation: count colonies (if possible) outside zone of inhibition and compare with an appropriate area on control plates. Time requirements: spot test: 30–60 min; scoring 15–60 min, depending on size of experiment and mutagenic effect of treatment. Remark: revertibility of mutant alleles varies considerably. A highly revertible mutant strain might give good results when only 10^6 cells are plated per selection medium plate instead of the 10^7 indicated above.

REFERENCES

1 KASTENBAUM, M. A., AND K. O. BOWMAN, Tables for determining the statistical significance of mutation frequencies, *Mutation Res.*, 9 (1970) 527–549.

2 MARQUARDT, H., R. SCHWAIER AND F. K. ZIMMERMANN, Die Wirkung von Methyl- und Äthylnitrosourethan, Diäthylsulfat und Methylmethansulfonat auf zwei Genorte von Neurospora sowie einen Mutationsort von Saccharomyces, *Mol. Gen. Genet.*, 99 (1967) 1–4.

3 MAYER, V. W., Induction of mitotic crossing-over in *Saccharomyces cerevisiae* by breakdown products of dimethylnitrosamine, 1-naphthylamine and 2-naphthylamine formed by an *in vitro* hydroxylation system, *Genetics*, 74 (1973) 433–443.

4 MORTIMER, R. K., AND D. C. HAWTHORNE, Yeast genetics, in A. H. ROSE AND J. S. HARRISON (Eds.), *The Yeasts, Vol. 1*, Academic Press, New York, 1969, pp. 385–460.

5 MORTIMER, R. K., AND T. R. MANNEY, Mutation induction in yeast, in A. HOLLAENDER (Ed.), *Chemical Mutagens, Vol. 1*, Plenum, New York, 1971, pp. 289–310.

6 ROMAN, H., A system selective for mutations affecting the synthesis of adenine in yeast, *Compt. Rend. Trav. Lab. Carlsberg*, 26, (1956) 299–314.

7 SCHWAIER, R., F. K. ZIMMERMANN AND U. VON LAER, The effect of temperature on the mutation induction in yeast by *N*-alkylnitrosamides and nitrous acid, *Z. Vererbungslehre*, 97 (1965) 72–74.

8 ZIMMERMANN, F. K., Enzyme studies on the products of mitotic gene conversion in *Sacccharomyces cerevisiae*, *Mol. Gen. Genet.*, 101 (1968) 171–184.

9 ZIMMERMANN, F. K., Induction of mitotic gene conversion by mutagens, *Mutation Res.*, 11 (1971) 327–337.

10 ZIMMERMANN, F. K., A yeast strain for visual screening for the two reciprocal products of mitotic crossing-over, *Mutation Res.*, 21 (1973) 263–269.

11 ZIMMERMANN, F. K., Detection of genetically active chemicals using various yeast systems, in A. HOLLAENDER (Ed.), *Chemical Mutagens, Vol. 3*, Plenum, New York, 1973, pp. 209–239.

12 ZIMMERMANN, F. K., R. SCHWAIER AND U. VON LAER, Mitotic recombination induced in *Saccharomyces cerevisae* with nitrous acid, diethylsulfate and carcinogenic, alkylating nitrosamides, *Z. Vererbungslehre*, 98 (1966) 230–246.

13 ZIMMERMANN, F. K., R. SCHWAIER AND U. VON LAER, Nitrous acid and alkylating nitrosamides: mutation fixation in *Saccharomyces cerevisiae*, *Z. Vererbungslehre*, 98 (1966) 152–166.

14 ZIMMERMANN, F. K., AND R. SCHWAIER, Induction of mitotic gene conversion with nitrous acid, 1-methyl-3-nitro-1-nitrosoguanidine and other alkylating agents in *Saccharomyces cerevisiae*, *Mol. Gen. Genet.*, 100 (1967) 63–76.

HOST-MEDIATED MUTAGENICITY TESTS—YEAST SYSTEMS.

RECOVERY OF YEAST CELLS OUT OF TESTES, LIVER, LUNG AND PERITONEUM OF RATS

R. FAHRIG

Zentrallaboratorium für Mutagenitätsprüfung der Deutschen Forschungsgemeinschaft, D-78 Freiburg (W. Germany)

All studies in the host-mediated assay using yeast cells have been performed with the diploid yeast strains D-3 (ref. 1) and D-4 (refs. 3, 4, 6) of *Saccharomyces cerevisiae* which are suited for the observation of mitotic recombination and gene conversion, respectively. With these strains the recovery of yeast cells from the host animals is good (see later), and a nearly 100% viability among recovered cells is attained. These cells seem to remain in the stationary phase, because no budding can be observed. When treating non-growing cells with mutagens, an increase in the absolute number of mutants after treatment based on the number of yeast cells injected, can be used as evidence against selection of preexisting spontaneous mutants.

Yeast cells incubated in testes of rats may be treated for a period of 30 h. A 24-h treatment time of yeast cells incubated in peritoneum of mice and rats, and 8 h for yeast cells incubated in liver and lung of rats is possible. Compared to the treatment of bacterial strains normally used in the host-mediated assay, these treatment times are relatively long. They may perhaps be the result of the low pathogenic effect of yeast cells on the host animal. The defense mechanism of the host animal against yeast cells, therefore, seems to be quite weak. When incubating yeast cells in untreated animals for a period of 30 h, no enhancement of the frequency of mitotic gene conversion could be observed[3,4] over *in vitro* results (25° or 37°).

The use of the D-3 and D-4 strains in the investigation of either mitotic recombination or gene conversion has encouraged the development of new test strains to detect the induction of gene mutations as well. ZIMMERMANN* has developed the diploid strains D-5, D-6 and D-7, and in our laboratory the test strain MP-1 for the simultaneous detection of recombination, gene conversion, and forward mutation has been constructed.

In the following paragraphs some yeast strains, recommended for use in the host-mediated assay, will be described.

Abbreviations: AO, acridine orange; DENA, diethylnitrosamine; DMF, dimethylformamide; DMNA, dimethylnitrosamine; DMSO, dimethylsulphoxide; EMS, ethyl methanesulphonate; Endoxan, cyclophosphamide; MMS, methyl methanesulphonate; MNNG, *N*-methyl-*N*'-nitro-*N*-nitroso-guanidine; 4-NQO, 4-nitroquinoline 1-oxide; PDT, 3,3-dimethyl-1-phenyltriazene; PyDT, 3,3-dimethyl-1-(3-pyridyl)-triazene; TEM, triethylenemelamine.

*Fachbereich Biologie (10), Technische Hochschule, 61 Darmstadt/Germany.

D-3. This strain offers a relatively simple system for the detection of mitotic crossing over. This event, whereby the heterozygous recessive *ade2* marker is converted to homozygosity, can be detected as a red sector in a normally white colony. Since two reciprocal products must be detectable in order to prove the induction of mitotic crossing over, the D-3 strain was developed with two flanking heterozygous markers *Ac4* and *his8* around the *ade2* gene. These can be checked for simultaneous induction of homozygosity. The D-3 strain has mainly been used by BRUSICK AND MAYER[1]. They have tested 6 mutagens in the host-mediated assay: EMS, MMS, DMNA, DENA, Endoxan, and hycanthone. All substances except hycanthone (100 mg/kg) gave a positive response in the host-mediated assay (mouse peritoneum).

D-4. With this strain which is suitable for detection of gene conversion at the *ade2* and *trp5* loci, most experience in mutagenicity testing *in vitro* and in the host-mediated assay has been gained. In contrast to most reverse mutation systems, mitotic gene conversion could never be specifically caused by only one type of mutagen. A large variety of mutagens, such as ionizing radiation, UV irradiation, alkylating and arylating agents, acridines, nitrous acid and hydroxyurea have been shown to induce mitotic gene conversion *in vitro*. In the traditional host-mediated assay and modifications of this method[3,4,6], 20 different mutagens such as 4-NQO, 2-methoxy-6-chloro-9-(2-chloroethylaminopropylamino)acridine dihydrochloride (ICR-191), hycanthone, MNNG, MMS, EMS, Endoxan, DMNA, TEM and different aryldialkyltriazenes have shown genetic activity.

D-5. The D-5 strain of *Saccharomyces cerevisiae* allows for rapid visual screening of mitotic crossing over in colonies on solid media. D-5 carries two phenotypically distinguishable and recessive alleles of the gene *ade2*. Due to one allele, *ade2-119*, a slight amount of adenine is necessary and the colonies are pink. Due to the other allele, *ade2-40*, a complete requirement for adenine is present and the colonies are deeply red. The two alleles complement each other to the effect that D-5 forms white colonies and does not require adenine for growth. Mitotic crossing over generates two reciprocal products which show up together in colonies as pink and red sectors. Such twin-spotted colonies have been obtained after treatment with ultraviolet light[8], γ-rays[8], EMS[8], ICR-170[2] and DMNA[2].

D-6. This is a test strain for chromosome losses.

D-7. With this test strain mitotic recombination (*ade2-40, ade2-119*) mitotic gene conversion (*trp5-12, trp5-27*), and reverse mutation (*is1-92, is1-92*) may be studied. Treatment *in vitro* with EMS, HNO_2 and UV irradiation has given positive results (ZIMMERMANN, personal communication).

MP-1. This multipurpose strain allows for screening of mitotic crossing over, mitotic gene conversion and forward mutation. The strain has the following constitution:

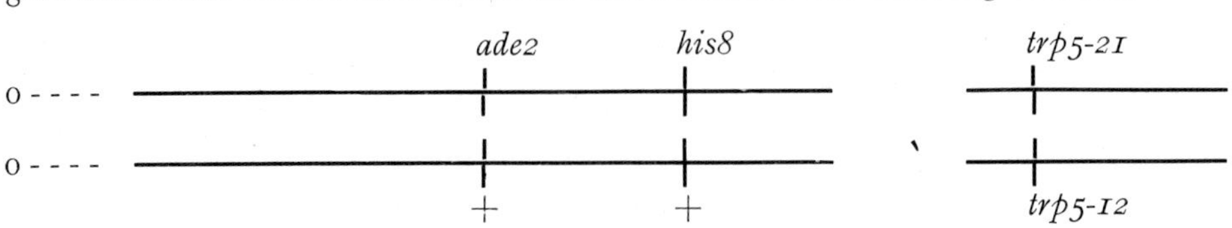

The *ade2* and *his8* markers have been derived from the D-3 strain, the *trp5-21* marker from the haploid strain MB-46, and the *trp5-12* marker from the haploid strain MA-20.

Induction of mitotic crossing over results in homozygosity of the *ade2* marker. In order to exclude mutation and gene conversion, simultaneous induction of homozygosity of the flanking marker *his8* must be demonstrated. Similar to the D-4 strain this system of mitotic gene conversion is based on the induction of *trp*-prototrophic cells in an *trp*-auxotrophic population. Mutation system: since actidione resistance in yeast can be induced by mutations in at least eight different genes, its spontaneous and induced mutation frequency should be much higher than that of only one locus. Actidione resistance is due to an alteration of ribosomal protein. A nonsense or frame-shift mutation in the respective structural gene(s) will certainly lead to the formation of a completely nonfunctional protein which will be lethal. Consequently, frame-shift inducing agents like ICR-compounds and aminoacridines are not able to induce actidione resistance. (For detailed discussion and literature references see ZIMMERMANN[8]).

With this strain different mutagens have been tested *in vitro*: AO, hycanthone, MNNG, EMS, MMS, 4-NQO, TEM, and HNO_2. These agents induced mitotic recombination as well as gene conversion and resistance to actidione. Only the frame-shift inducing mutagens AO and hycanthone were not able to induce actidione resistance (FAHRIG, unpublished results).

Media

Synthetic complete medium. This medium is prepared by adding to Bacto yeast nitrogen base without amino acids (Difco Laboratories, Detroit, Mich., No. 0919-15) the following ingredients per liter of medium:

L-Arginine–HCl	10 mg
L-Aspartic acid	10 mg
L-Glutamic acid	100 mg
L-Leucine	60 mg
L-Lysine–HCl	10 mg
L-Methionine	10 mg
L-Phenylalanine	50 mg
L-Serine	20 mg
L-Tyrosine	30 mg
L-Valine	30 mg
Uracil	10 mg

According to the strain used, the following ingredients are added or not added per liter of medium:

Adenine sulfate	10 mg
L-Histidine–HCl	10 mg
L-Isoleucine	60 mg
L-Threonine	200 mg
L-Tryptophan	10 mg
Actidione	1 mg

The recipe for a medium of equivalent quality without the Bacto nitrogen base has been described previously by ZIMMERMANN[8].

Liquid complete medium. The liquid medium YEP is composed of the following ingredients: 2% Bacto peptone (Difco No. 0118-01), 1% yeast extract (Difco No. 0127-01), 2% glucose.

Reference mutagens

DMNA: Schuchardt, D-8 München, No. ME440. Endoxan: Asta, D-4812 Brackwede. Hycanthone methanesulphonate: Sterling-Winthrop, Rensselaer, N.Y., No. R-015 FF. MMS: Schuchardt, D-8 München, No. N1 240. 4-NQO: Fluka AG, CH-9470 Buchs, No. 73265.

Apparatus needed

Implement. 1 autoclave; 1 refrigerated centrifuge to take centrifuge tubes of volume 50 ml; 3–6 minicentrifuges to take convenient test tubes; 1 shaking apparatus for test tube racks; 1 incubator 25°; 1 analytical balance; 1 electronic balance; 1 microscope; 3–6 whirl-mixers; 1–2 ultra-turrax (TP 18/2 N); 1 vacuum pump.

Glassware. 10 ml and 100 ml beaker glasses; Erlenmeyer flasks of different volume; test tubes with cottonwool plugs; 1–3 hemocytometers, 0.1 mm deep for determining cell titers; glass spatula; 0.1 ml, 1 ml, and 10 ml graduated pipettes; 1 ml pipettes; 2–4 alcohol burners.

Set of surgical instruments. scalpels; shears; pincers; operation table for mice and rats; 1 ml and 5 ml disposable syringes; 5 ml record syringes; tubules No. 12 and 16; silicone hose 0.6 mm.

Culturing of yeast cells

About 10^3 yeast cells are inoculated in a 300 ml Erlenmeyer flask containing 100 ml YEP, set on a shaker and allowed to grow for 3 days at 25° into the stationary phase. 1 ml YEP normally contains $5–10 \cdot 10^7$ cells, depending on the yeast strain used. The cultures of strains for the detection of gene conversion are stored at 4° for 3–4 days. During this time their spontaneous conversion frequency is determined by spreading 0.1 ml from the YEP/yeast-suspension on solid media selective for convertants. Only cultures with a low spontaneous frequency are used for the experiments. The cultures needed for one experiment with 10–20 animals are mixed together in order to obtain a similar spontaneous frequency of convertants in each of the animals.

Description of methods

To determine the effect of a possible metabolic activation or inactivation, all test substances must be tested in both *in vitro* and *in vivo* systems.

In vitro-test (Fig. 1).

The cell cultures are washed twice with distilled water and the cell titers are adjusted to $5 \cdot 10^7$ cells per ml of 0.1 *M* phosphate buffer, pH 7.

These cell suspensions are incubated in a test tube on a shaker at 25° with different concentrations of the test substances. At defined time-intervals treatments are stopped, and 4×0.1 ml of the suspensions containing $5 \cdot 10^6$ cells are spread on four plates of a solid nutrient-deficient media, selective for convertants (mutants). Similarly, after 4 dilutions 1:10 in distilled water four suspensions of $5 \cdot 10^2$ cells are plated out on four plates of complete medium to attain the number of survivors. 10 dishes of complete medium are used when testing the induction of mitotic recombi-

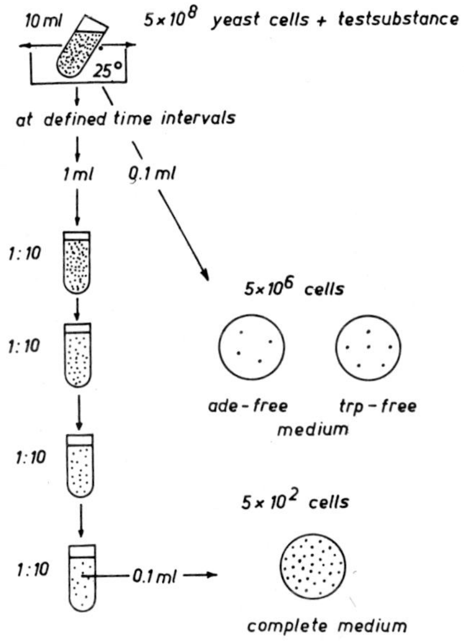

Fig. 1. *In vitro* test—yeast system (D-4-strain).

nation. The cultures are incubated at 25° and both the survivor and the convertant (mutant, recombinant) colonies are counted after 3 and 4 days, respectively. Actidione-resistant colonies are incubated 8 days.

TABLE I

INDUCTION OF MITOTIC GENE CONVERSION AT THE ADE_2 LOCUS OF *S. cerevisiae* D-4 IN HTE HOST-MEDIATED ASSAY

Substance	*Recovery of yeast cells from*	*Concentration (mM)*	*Time of treatment (h)*	*ade Convertants per 5 × 10⁶ recovered cells*				*Significance (P)*
				Experiment		*Control*		
				Number of animals	*Mean and standard deviation*	*Number of animals*	*Mean and standard deviation*	
4-NQO	testes	0.25	1	9	7.5 ± 1.2	5	8.0 ± 4.2	<0.8
	liver			8	10.1 ± 3.3			<0.4
	lung			9	46.8 ± 24.9			<0.01
4-NQO	testes	0.25	4	7	8.0 ± 1.1	5	8.0 ± 1.9	>1
	liver			7	19.5 ± 5.2			<0.005
	lung			5	115.6 ± 33.8			<0.001
Hycanthone	testes	1	1	5	13.5 ± 4.3	7	8.3 ± 1.9	<0.02
	liver			8	10.0 ± 2.6			<0.2
	lung			10	17.2 ± 7.3			<0.01
Endoxan	testes	2	4	8	12.3 ± 3.3	10	8.6 ± 1.9	<0.01
	liver			8	22.2 ± 9.8			<0.001
	lung			8	18.8 ± 6.0			<0.001
MMS	testes	4	1	9	16.2 ± 3.5	6	10.3 ± 2.4	<0.005
	liver			6	32.1 ± 12.1			<0.005
	lung			5	28.2 ± 10.6			<0.005
DMNA	testes	10	4	8	13.1 ± 2.7	10	10.2 ± 1.9	<0.02
	liver			12	76.2 ± 24.0			<0.001
	lung			7	42.2 ± 9.0			<0.001

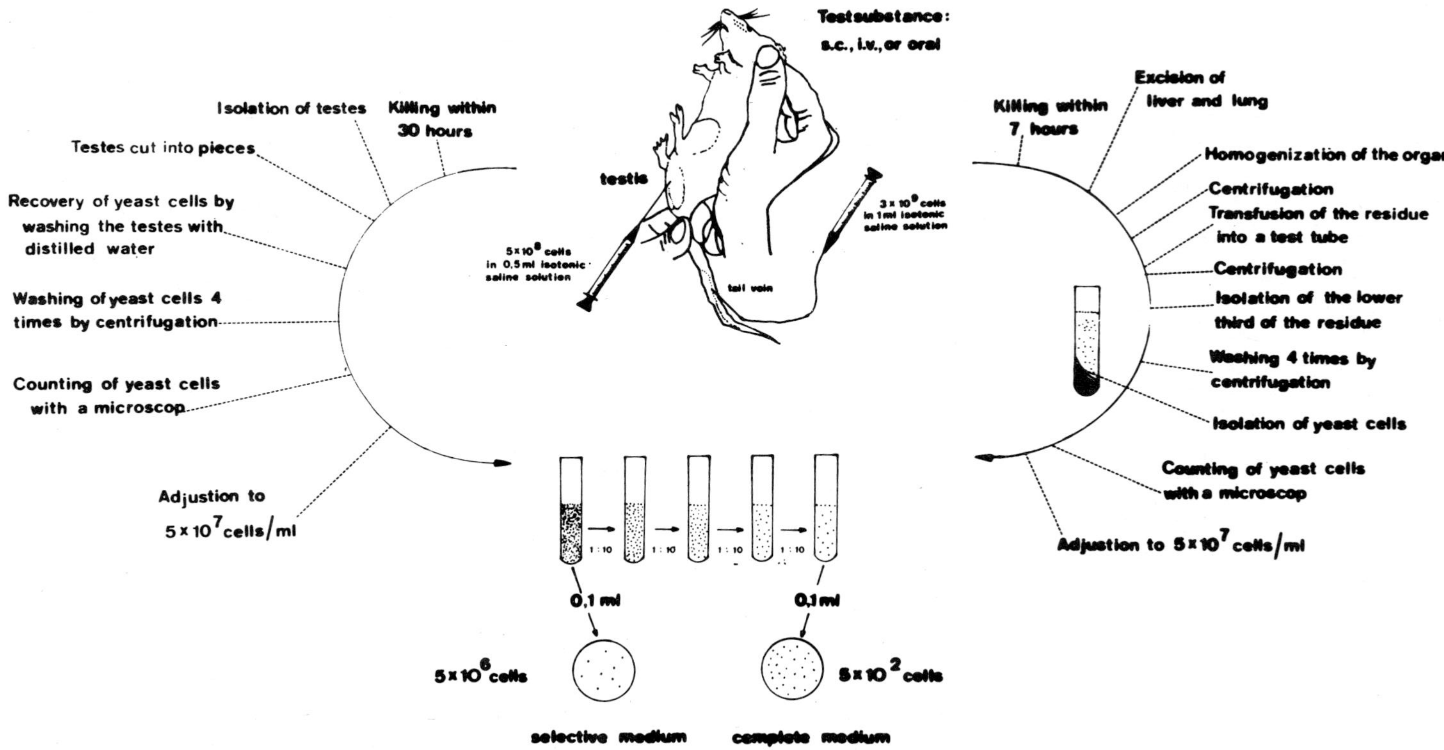

Fig. 2. Host-mediated assay—yeast system.

In vivo-test (Fig. 2)

Host animals. It is possible to use different rodents, but until now mainly mice and rats have been used. For the investigations summarized in Tables I and II male "Wistar"-rats (Jautz/Kisslegg), weight 180–200 g, have been used.

TABLE II

INDUCTION OF MITOTIC GENE CONVERSION AT THE *trp*$_5$ LOCUS OF *S. cerevisiae* D-4 IN THE HOST-MEDIATED ASSAY

Substances	*Recovery of yeast cells from*	*Concentration (mM)*	*Time of treatment (h)*	*trp Convertants per 5 × 10⁶ recovered cells*				*Significance (P)*
				Experiment		*Control*		
				Number of animals	*Mean and standard deviation*	*Number of animals*	*Mean and standard deviation*	
4-NQO	testes	0.25	1	9	7.1 ± 1.4	5	6.7 ± 2.2	<0.8
	liver			8	12.2 ± 3.6			<0.02
	lung			9	55.7 ± 31.5			<0.01
4-NQO	testes	0.25	4	7	8.0 ± 1.6	5	6.7 ± 1.0	<0.25
	liver			7	20.1 ± 4.2			<0.001
	lung			5	112.0 ± 24.2			<0.001
Hycanthone	testes	1	1	5	14.4 ± 6.3	7	7.2 ± 1.8	<0.02
	liver			6	12.9 ± 4.7			<0.02
	lung			10	17.6 ± 5.4			<0.001
Endoxan	testes	2	4	8	12.8 ± 3.9	12	7.6 ± 1.7	<0.001
	liver			8	35.9 ± 24.7			<0.001
	lung			8	20.8 ± 9.6			<0.001
MMS	testes	4	1	9	21.1 ± 2.5	6	12.5 ± 1.5	<0.001
	liver			6	46.5 ± 26.7			<0.02
	lung			5	37.1 ± 18.2			<0.01
DMNA	testes	10	4	7	13.1 ± 2.0	12	8.9 ± 2.7	<0.005
	liver			12	122.8 ± 45.5			<0.001
	lung			7	45.0 ± 13.9			<0.001

Administration of test substances. The test substance can be administered either by subcutaneous injection into the back, intramuscularly, orally, or by intravenous injection into the tail vein. For subcutaneous injection of difficultly soluble substances it is profitable to use solutions in 10–100% DMSO, for the intravenous injection of these substances it is better to use solutions in 10–25% DMF, which is less toxic than DMSO.

Treatment time. For the indirect mutagens a treatment time of 4 h and for the direct mutagens a treatment time of 1 h should be chosen. These treatment times have proved to be favourable because the genetical effect of indirect mutagens so far tested positively correlates with longer incubation times after an initial incubation time of 1 h; on the other hand the genetical activity of direct mutagens so far tested decreases under these test conditions after having reached a maximal effect after 1 h. For unknown test substances a treatment time of both 1 h and 4 h should be chosen. It is possible to lenghten the treatment up to 30 h. The recovery of yeast cells can be summarized as follows:

Injection of yeast cells into	*Recovery of yeast cells from*	*Treatment time (h)*	*Recovery of yeast cells (%)*
Peritoneum	Peritoneum mice/rats	1	80–90
		4	70–80
		16	40–50
		24	30–40
Testes	Testes/rats	1	80–90
		4	40–50
		24	5–10
Tail vein	Liver/rats	1	15–20
		4	8–12
		7	5–7
Tail vein	Lung/rats	1	10–15
		4	7–10
		7	3–4

Compared to *in vitro* results at 25° or 37°, no enhancement in the frequency of genetic alterations could be found, when incubating the yeast cells for (up to) 30 h in otherwise untreated mice or rats.

Intraperitoneal incubation of yeast cells. A 1.5-ml aliquot of a *S. cerevisiae* culture containing 10^9 cells in YEP, is injected into the peritoneal cavity of a mouse, and an 8-ml culture aliquot containing $2 \cdot 10^9$ cells is injected into the peritoneal cavity of rats. At this concentration no visible decrease in vitality of the host animals can be observed.

From 1 to 30 h after administration of the test substances the animal is killed and after retracting the skin an incision is made through the peritoneum. The injected yeast cells are recovered firstly by withdrawing the peritoneal fluid using a syringe without a needle and secondly by washing the peritoneal cavity 2–4 times under sterile conditions with distilled water or phosphate buffer, pH 7. The viability of recovered cells is almost 100%.

The peritoneal fluid samples of one animal are poured together, centrifuged at about 2000 rpm for 10 min in a minicentrifuge taking convenient test tubes, and the supernatant is decanted. The yeast cells in the residue are washed three times (suspension) in 10 ml distilled water (10 min, 2000 rpm). The cell titers, determined photometrically or by microscopic counts, are adjusted to $5 \cdot 10^7$ cells per ml. Samples of 0.1 ml from these suspensions containing $5 \cdot 10^6$ cells are spread on each of four plates of solid nutrient-deficient media selective for the detection of convertants or mutants. Similarly, 4 suspensions containing $5 \cdot 10^2$ cells are plated out on four dishes of complete medium to determine the percentage of recovery. Ten dishes of complete medium are used when testing the induction of mitotic recombination.

The cultures are incubated at 25° and the colonies are counted after 3 days to determine the percentage of recovery. 4–5 days of incubation are necessary to determine the frequency of conversion and recombination, respectively. Actidione-resistant colonies are allowed to grow for 8 days.

Incubation of yeast cells in the testes. $3–5 \cdot 10^8$ yeast cells, suspended in 0.5 ml isotonic saline solution are injected into each testis of a rat. At this concentration no visible decrease in vitality of the host animals can be observed. During injection the testes are kneaded to improve the distribution of the yeast cells. In the light microscope histological preparations of testes show that the yeast cells are concentrated in the interstitital tissue as well as in the lumen of the seminiferous tubules. That means, they are not only in contact with the testicular lymph but also with the rete testis fluid.

1 to 30 h after administration of the test substance the rat is killed and the testes are isolated under sterile conditions. After cutting the testes up, the yeast cells are recovered by washing the pieces with distilled water. The viability of recovered cells is about 100%.

The fluid samples are centrifuged in a minicentrifuge taking convenient test tubes at 2000 rpm for 10 min at room temperature. The supernatant is decanted and the yeast cells in the residue are washed 4 times in 10 ml streptomycin solution (1 mg/ml) by centrifugation (10 min, 2000 rpm). The cell titers are determined by microscopic counts and adjusted to $5 \cdot 10^7$ cells per ml. These suspensions which still contain some cellular tissues of the testes are treated in the same way as samples from the peritoneum.

Intravenous injection[5] *of yeast cells and recovery from liver and lung*. $3 \cdot 10^9$ yeast cells, suspended in 1 ml isotonic saline solution are injected into the tail vein of a rat using a tubule No. 15 connected by a 0.6 mm silicone hose with a 1 ml disposable syringe. Blood in the silicone hose shows that the tubule is properly injected into the tail vein. Indirect mutagens can be injected intravenously together with the yeast cells. 1 to 7 h after the administration of the test substance the animal is killed and its liver and lung are excised. Each organ is supended in 20 ml phosphate buffer, pH 7 and homogenized 15–30 sec at room temperature in a 50-ml centrifuge cup using an ultra-turrax (TP 18/2 N).

Following homogenization, the material is centrifuged at about 2500 rpm for 15 min in a refrigerated centrifuge taking the 50-ml centrifuge cups at 0–4°. The supernatant is decanted and the residue transfused into a test tube. The material now is centrifuged at about 2000 rpm for 10 min in a minicentrifuge taking convenient test tubes at room temperature. The supernatant and the upper 2/3 of the residue are decanted and drawn off with a vacuum pump, respectively.

The lower third of the residue is washed 4 times by centrifugation (10 min, 2000 rpm) in 10 ml streptomycin solution (1 mg/ml). The yeast cells can normally be seen as a white film on the residue. In this case a considerable part of the residue can be drawn off with a vacuum pump. If the yeast cells don't separate clearly from the residue, the whole residue has to be used as yeast suspension. In each case the yeast suspension will contain a considerable amount of cellular tissue. But presence of cellular tissue in yeast cell suspensions has no detectable influence on the frequency of convertants. The yeast cell titers of these suspensions are determined by microscopic counts and adjusted to $5 \cdot 10^7$ cells per ml. These suspensions are treated in the same way as samples from the peritoneum.

Registration and statistical treatment of experiments. In the following an example of how to register an experiment will be given. All rats of the experiment are treated in the same way: yeast cells are injected into the tail vein and into the testes, and with the exception of the control animals 4-NQO is applied subcutaneously into the back. In our example only the data of yeast cells recovered from testes will be recorded. The data of yeast cells recovered from liver and lung are given in Tables I and II. A discrepancy in the number of animals for the different organs indicates that the experiments have either failed or that yeast cells have been recovered from only one or two organs. For the statistical treatment of the experiments we have made use of the t-distribution, described in detail in all books dealing with elementary statistical methods. Using the t-distribution, only the data of one experiment can be compared. The protocol of an experiment should be as follows:

Experiment No. **HMA 112**.

Rat No. **1**. Weight **200** g.

Yeast cells injected into *tail vein, testes,* peritoneum. i.v., oral, *s.c.* injected with

0.25 ml 4-NQO (100 mg 4-NQO/5 ml DMSO).

Rat killed after **1** h.

Concentration of recovered yeast cells, diluted in 10 ml H_2O = **$1.8 \cdot 10^7$ cells/ml**.

Centrifugation and dilution to **3.6 ml** = $5 \cdot 10^7$ cells/ml.

Convertant colonies per plate		Colonies per plate
ade-free medium	trp-free medium	complete medium
33	37	486
36	32	443
39	40	459
41	38	472
149	147	1860
$X_{1-ade} = 8.01$	$X_{1-trp} = 7.9$	

Similar results are gained with the other 8 rats of the experiment, treated in the same way as rat No. 1 with 4-NQO:

$X_{2-ade} = 7.94$; $X_{3-ade} = 8.84$; $X_{4-ade} = 8.34$; $X_{5-ade} = 6.01$;
$X_{6-ade} = 6.44$; $X_{7-ade} = 5.84$; $X_{8-ade} = 8.75$; $X_{9-ade} = 7.71$.

$X_{2-trp} = 6.82$; $X_{3-trp} = 7.58$; $X_{4-trp} = 8.97$; $X_{5-trp} = 5.78$;
$X_{6-trp} = 8.08$; $X_{7-trp} = 4.04$; $X_{8-trp} = 7.06$; $X_{9-trp} = 7.33$.

$n_{X-ade/trp} = 9$; $\bar{X}_{ade} = 7.54 \pm 1.15$; $\bar{X}_{trp} = 7.06 \pm 1.44$.

The data of the 4-NQO-untreated control animals are:

$n_{Y-ade/trp} = r$; $\bar{Y}_{ade} = 8.05 \pm 4.17$; $\bar{Y}_{trp} = 6.88 \pm 2.22$.

There is no statistically significant difference between 4-NQO-treated and control animals since $t_{ade} = 0.35$; $0.6 < P < 0.8$, and $t_{trp} = 0.39$; $0.7 < P < 0.8$. But it is not possible to prove or disprove a mutagenic effect statistically: we merely doubt it more or less. Experimentally, a mutagenic effect might be considered proved or disproved after carrying out many consistent repetitions. The minimum permissible size of one experimental group is five treated and five untreated animals because the standard deviation of the experiments is relatively high. After at least two consistent repetitions a mutagenic or nonmutagenic effect might be considered proved. Although in Tables I and II the data of only one experiment per substance are shown, at least two consistent repetitions of each experiment exist.

Comparative studies with reference mutagens

A comparison of the genetical effects of different mutagens on the D4-strain incubated in testes and peritoneum of rats and an investigation of their genetical effects in rats and mice, respectively, has been previously[4] published.

Six mutagens were tested: Three were mutagenic *per se*, and the other three had indirect effect, *i.e.* showed strong genetic activity only after metabolic activation by a mammal. The indirect mutagens were PyDT, 2-bis-(chloroethyl-)amino-tetrahydro-2H-1,3,2-oxazaphosphorine-2-oxide (Endoxan) and DMNA. The direct mutagens were TEM, PDT, and MMS. The induction of mitotic gene conversion was the indicator of genetic activity. With the exception of DMNA, all substances were genetically more active in the testes than in the peritoneum of rats, and all substances except MMS were genetically more active in rats than in mice. The sensitivity of testes and peritoneum to the different mutagens can be summarized as follows: TEM, PyDT, PDT, Endoxan—rat testes > rat peritoneum > mouse peritoneum, MMS—mouse peritoneum > rat testes > rat peritoneum; DMNA—rat peritoneum > mouse peritoneum > rat testes.

A comparison of the genetical effects of different mutagens on the D-4-strain incubated in testes, liver, and lung of rats shall be described in the following. Five mutagens were tested, three being mutagenic *per se* and two after metabolic activation. These mutagens are recommended as reference compounds.

DMNA, a monofunctional alkylating indirect mutagen. Endoxan, a bifunctional alkylating indirect mutagen. Hycanthone methanesulfonate, a partial frameshift-inducing direct mutagen. MMS, a monofunctional alkylating direct mutagen. 4-NQO, a partial frameshift-inducing mutagen, known to be metabolically activated in microorganisms as well as in mammals (for literature see refs. 1, 4, 8).

With the exception of endoxan administered as solution in 25% DMF by intravenous injection, all other substances were administered as solutions in absolute DMSO by subcutaneous injection into the back of the animal. As mentioned before, a treatment time of 1 h was chosen for the direct mutagens, and 4 h for the indirect ones. These treatment times have proved to be favourable.

All five mutagens tested were able to induce *ade*- and *trp*-convertants. In Tables I and II the substances are arranged according to their activity, *i.e.* according to the concentration administered: 4-NQO > Hycanthone > Endoxan > MMS > DMNA. A comparison of the genetical activity of the substances in testes, lung and liver is shown in fig. 3. Here the number of convertants which were induced in that incubation site where each substance had its highest level of effectiveness are set to equal 100. The convertants induced at the other two incubation sites were set into relation to this standard. It is apparent that all five mutagens show a characteristic organ-specific genetic activity. While 4-NQO and hycanthone have their highest level of effectiveness in the lung, for endoxan, MMS and DMNA the liver is the organ which is effected. With the exception of hycanthone all substances have their lowest level of effective-

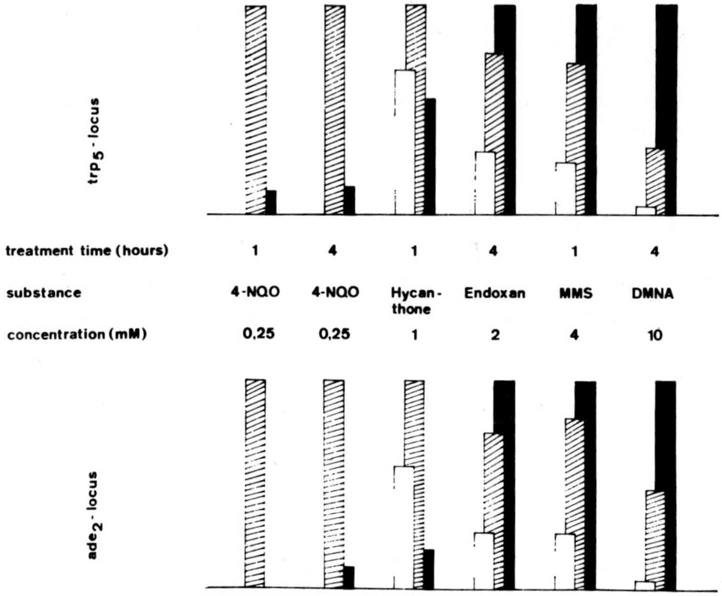

Fig. 3. The effectiveness of mutagens in testes, lung and liver of rats. The number of convertants which were induced in that incubation site where each substance had its highest level of effectiveness are set to equal 100. □, testes; ▨, lung; ■, liver.

ness in the testes. The genetical effectiveness of the different mutagens in testes, lung and liver can be ranked as follows: 4-NQO—lung > liver > testes (no effect), hycanthone—lung > testes > liver, Endoxan, MMS, DMNA—liver > lung > testes.

The most remarkable result is the exact correlation between the organ-specific mutagenic activity of substances described here and the organ-specific DNA-repair synthesis and tumor formation following injection of mutagens. Two examples for this correlation shall be taken of a paper of STICH AND KIESER[7] describing the use of DNA repair synthesis in detecting organotropic actions of chemical carcinogens:

"Following injection, 4-NQO accumulates in the lung where it is rapidly metabolized into the proximate carcinogen 4-HAQO. It induces a high frequency of lung adenomas and adenocarcinomas, and causes a DNA damage in this organ as revealed by a DNA repair synthesis. The liver of these animals contained only traces of 4-NQO and 4-HAQO, no hepatomas were seen in 4-NQO-injected animals, and there was no detectable level of DNA repair synthesis in hepatic cells. In contrast, DMNA appears to be mainly activated in the liver of rodents, although other organs including kidney and respiratory tracts also appear capable of a low but definite metabolism of nitrosamines. DNA repair synthesis occurred in those tissues which give rise to neoplasms: mainly in cells of the liver and lung."

These findings are in full agreement with the findings in the host-mediated assay described in this paper.

This result underlines the importance of the problem of organ-specific activity of mutagens for the host-mediated assay in particular and mutagenicity testing in general. The point to be made is that, if the host-mediated assay is a relevant mutagenicity test it is not enough to use only one incubation site. False negative results would be unavoidable.

The minimal permissible size of experiment should, therefore, be the recovery of yeast cells out of liver and lung.

ACKNOWLEDGEMENT

The expert technical assistance of Miss H. Drescher, Mrs. B. Fahrig, Mrs. A. Forstmeier, and Mrs. B. Velden is gratefully acknowledged. I would like to thank Dr. G. F. Kolar (Heidelberg) for reviewing the manuscript.

REFERENCES

1 Brusick, D. J., and V. W. Mayer, New developments in mutagenicity. Screening techniques with yeast, *Environ. Health Perspect.*, 6 (1973) 83–96.
2 Brusick, D., and H. Andrews, Comparison of the genetic activity of dimethylnitrosamine, ethyl methanesulfonate, 2-acetylaminofluorene and ICR-170 in *Saccharomyces cerevisiae* strains D3, D4 and D5 using *in vitro* assays with and without metabolic activation, *Mutation Res.*, 26 (1974) 491–500.
3 Fahrig, R., Metabolic activation of mutagens in mammals. Host-mediated assay utilizing the induction of mitotic gene conversion in *Saccharomyces cerevisiae*, *Agents and Actions*, 3 (1973) 99–110.
4 Fahrig, R., Development of host-mediated mutagenicity tests. I. Differential response of yeast cells injected into testes of rats and peritoneum of mice and rats to mutagens, *Mutation Res.*, 25 (1974) 29–36.
5 Mohn, G., and J. Ellenberger, Mammalian blood-mediated mutagenicity tests using a multipurpose strain of *Escherichia coli* K12, *Mutation Res.*, 19 (1973) 257–260.
6 Siebert, D., A new method for testing genetically active metabolites. Urinary assay with cyclophosphamide (endoxan, cytoxan) and *Saccharomyces cerevisiae*, *Mutation Res.*, 17 (1973) 307–314.
7 Stich, H. F., and D. Kieser, Use of DNA repair synthesis in detecting organotropic actions of chemical carcinogens, *Proc. Sci. Exp. Biol. Med.*, 145 (1974) 1339–1342.
8 Zimmermann, F. K., Detection of genetically active chemicals using various yeast systems, in A. Hollaender (Ed.), *Chemical Mutagens*, Vol 3, Plenum, New York, 1973, pp. 209–239.

THE ISOLATION AND DETECTION OF MUTAGENIC SUBSTANCES IN BODY FLUID AND TISSUES OF ANIMALS AND BODY FLUID OF HUMAN SUBJECTS

M.S. LEGATOR, T.G. PULLIN and T.H. CONNOR

Division of Environmental Toxicology and Epidemiology,
Department of Preventive Medicine and Community Health,
University of Texas Medical Branch, Galveston, Tex. 77550 (U.S.A.)

Background

The analysis of blood and urine of treated animals for mutagenic activity was first described as a modification of the host-mediated assay procedure in the initial publication on the assay [5]. This analysis of body fluids for either direct acting mutagens or active metabolites is a versatile procedure that can be used with experimental animals, but, of special importance, this technique can be used to detect mutagenic substances in man. As pointed out in another publication [12], this analysis combined with cytogenetic and genetic repair studies greatly extends the ability to detect mutagenic agents in the human population [9].

Since the introduction of the body fluid analysis, several modifications to this analysis have occurred attempting to extend the utility of the procedure. Some modifications, such as lyophilization of the body fluids, the use of chromatography and solvent extraction of active compounds are employed to increase the sensitivity and aid in isolating and detecting active metabolites or metabolic intermediates [2,11]. In urine analysis β-glucuronidase has been used to disassociate various conjugates since it has been shown that many metabolites occur in the urine as glucuronides [3].

Protocol

A. Selection of indicator organisms

A major advantage of this procedure is our ability, at least in principle, to utilize a wide variety of indicator organisms. The restriction inherent in the host-mediated assay, where the optimum indicator organism is one that can grow at a near normal rate in the treated animal, is not a limiting factor with this technique. Microbial systems, fungi, insects, and mammalian cells theoretically could be or are being used. The only restriction in the selection of an appropriate indicator is the sensitivity of the system and the type of genetic information desired. Although a variety of indicator organisms can be utilized, the histidine auxotroph of *Salmonella typhimurium* has been the most extensively used because of its obvious technical advantages. Studies have also been carried out with yeast [13], and the utility of Drosophila in this regard is under current investigation [15].

In the case of the histidine auxotroph of *S. typhimurium* one should be aware of the possible increase in revertants due to histidine normally present in urine, especially in human urine. However, from studies in our laboratory, where control urine from normal individuals was followed over a two-year period, it was found, using *S. typhimurium* tester strains TA1535 and TA1538, that the mutation frequency is within the range normally encountered with these strains (Fig. 1).

The histidine auxotroph of *S. typhimurium* is a convenient indicator organism, and the same culture conditions, media and tester strains as described by Ames et al. [1] can be used by this procedure. For each sample of blood, urine or tissue, TA1535, TA1537, TA1538, TA100 or TA98 should be used.

B. Treatment of animals or man

In animal studies, chemicals can be administered by different routes, formulations, and for varying lengths of time. These tests can be performed utilizing any experimental animal where body fluids or tissues can be obtained for analysis. Obviously in human subjects, agents that are administered by appropriate route for the usual lengths of time (therapeutic regimen of a given drug) will have to be utilized.

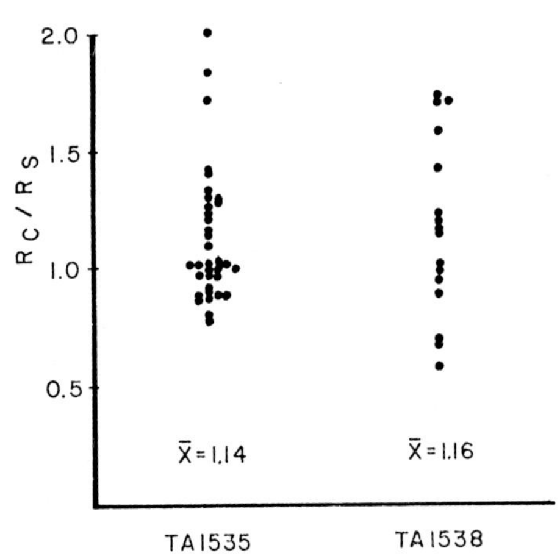

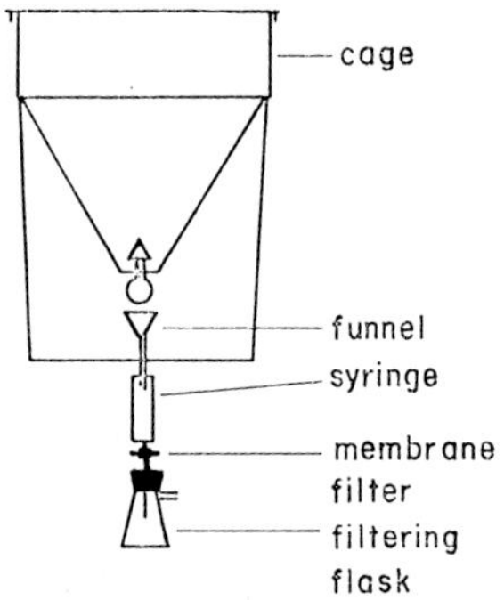

Fig. 1. The ratio of the number of histidine revertants of control urine sample (R_C) to the number of spontaneous histidine revertants (R_S) using *S. typhimurium* TA1535 and TA1538. Urine samples were sterilized by passage through a membrane filter and 0.2 ml was added to each plate.

Fig. 2. Metabolism unit modified for collection of urine. Urine is filtered through a membrane filter by means of a vacuum attached to filtering flask. Filtered urine is kept at 0—4°C and is protected from the light.

C. Collection of samples

With rodents, the collection of blood from the retroorbital sinus has proven to be a valuable technique. Blood samples are withdrawn with a Pasteur pipette and added to a tube containing heparin. As much as 0.2 ml can be taken from a single adult mouse without sacrificing the animal, and several samples can be taken periodically from the same animal. Samples from 8—10 animals are usually pooled for a single determination. In human subjects peripheral blood samples of 10—30 ml can be conveniently collected in heparinized tubes.

In the collection of urine from rodents, special metabolism cages are used, as illustrated in Fig. 2. Ten animals are placed in the cage, and the urine is drawn through a Millipore filter; 5 ml of urine is usually obtained in 4—5 h from 10 animals.

The following procedure can be used for evaluating blood, urine, or tissue samples.

(a) Incorporation of urine, blood, or tissue homogenates directly into the media to which the appropriate indicator organism is added;
(b) lyophilization of blood, urine, or tissue prior to addition to the detection system;
(c) general extraction procedure such as the use of solvent of different polarity or the use of Amberlite columns [11];
(d) specific extraction procedures when prior information is available as to the chemical properties of the substances under investigation [2].

D. Example of a specific procedure

The following is a general extraction procedure that we have found useful. This procedure is a modification of the method used for identifying toxic substances proposed by Goldbaum and Dominguez [6]. It can be applied to either blood, urine, or tissue samples. The following illustrates the procedure after collecting a 40 ml human urine sample which has then been passed through a Millipore filter.

Materials and methods

Reagents

(1) Sodium borate solutions.
  (a) Concentrated sodium borate: add 20 g of sodium borate to 100 ml of distilled water, heat to dissolve and keep warm to prevent precipitation.
  (b) Dilute sodium borate: dilute the warmed concentrated sodium borate 1 : 100.
(2) Isobutanol in chloroform (USP), 5% solution.
(3) 0.1 N H_2SO_4.
(4) 0.75 N KOH.
(5) Concentrated HCl.

Isolation procedure

A small portion (approximately 1 ml) of the urine specimen is removed and

25 and 100 μl samples are tested directly on *S. typhimurium* tester strains. The remaining sample (approximately 40 ml) is divided into two equal aliquots. β-Glucuronidase (1.5 mg/ml urine, 100,000 units/g) is added to one aliquot and incubated at 37°C for 16—18 h. After this incubation period, approximately 1 ml is removed and 25 and 100 μl samples are tested. The remainder of this aliquot and the one containing no β-glucuronidase are extracted by the following procedure:

Twenty milliliters of urine are placed into a 250 ml separatory funnel containing 60 ml 5% isobutanol in chloroform. The urine solution is adjusted to pH ± 0.2 by the addition of 4 ml of warm concentrated sodium borate solution. The contents of the funnel are then manually mixed for 30 sec. The upper aqueous layer is removed and 25 and 100 μl samples are tested with the various tester strains.

Ten milliliters of dilute sodium borate is then added to the solvent phase and mixed manually for 15 sec. The aqueous layer is removed and discarded. Two milliliters of 0.1 N H_2SO_4 is added to the solvent remaining in the separatory funnel and mixed manually for 30 sec. The upper aqueous layer is then removed and transferred to a 5 ml narrow-tip centrifuge tube. This tube contains the basic constituents.

The solvent remaining is used for the extraction of acid constituents. Two milliliters of 0.75 N KOH is added to the solvent in the separatory funnel and mixed manually for 30 sec. The upper aqueous layer is removed and transferred to a 5 ml narrow-tip centrifuge tube.

The remaining solvent is used for extraction for the presence of neutral drugs. Twenty milliliters of this solvent are evaporated to dryness by means of a gentle air stream. Approximately 0.2 ml of dimethylsulfoxide (DMSO) is added to the residue and 25 μl is tested by the bacterial mutagenic assay system.

The 2 ml of 0.1 N H_2SO_4 extract is made alkaline by the addition of 0.75 N KOH. Approximately 0.2 ml of 5% isobutanol in chloroform is added and mixed on a Vortex mixer for 30 sec. The upper aqueous layer is removed and discarded. The solvent layer is then evaporated to dryness by means of a gentle air stream. Approximately 0.2 ml DMSO is added to the residue, and 25 μl is tested by the bacterial mutagenic assay system.

The 2 ml of 0.75 N KOH extract is acidified with concentrated HCL. Approximately 0.2 ml of 5% isobutanol in chloroform is added and mixed on a Vortex mixer for 30 sec. The upper aqueous layer is then removed and discarded. The solvent layer is then evaporated to dryness by means of a gentle air stream. Approximately 0.2 ml DMSO is added to the residue and 25 μl is tested by the bacterial mutagenic assay.

Test plates for the fractions from the aliquot containing no β-glucuronidase are made in duplicate and β-glucuronidase (200 units/plate) is added to one-half of these plates. In this way, the fractions were tested directly on the plates with and without β-glucuronidase (Fig. 3).

Results

A. Recording of data

The most convenient manner of presenting the data using the *S. typhimu-*

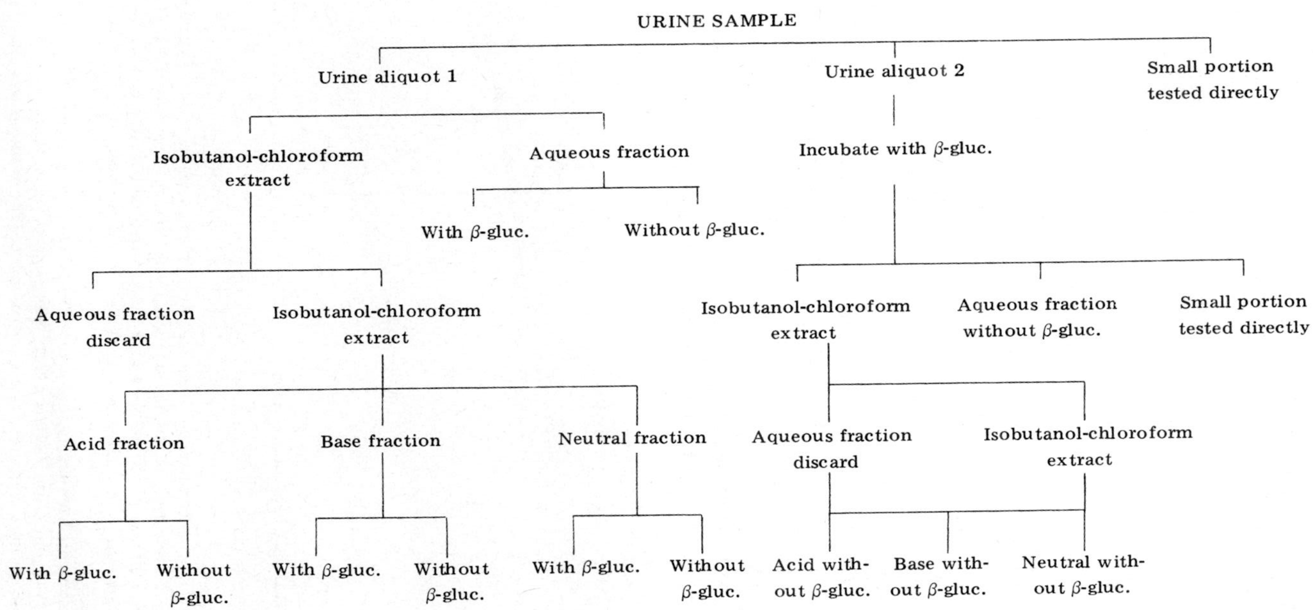

Fig. 3. Extraction method for detection of mutagenic constituents in urine.

rium assay system is to record the data as the ratio of number of survivors in the treated over the number of survivors in the control. It is impossible to rule out the role of selection in studies based on mutation fluctuation without reconstruction experiments. A rough approximation of the role of selection can be obtained by comparing antibacterial effect, if any, of the chemical or metabolites on the tester strain and also revertants.

B. Example of extraction procedure

The specific extraction procedure described in this report was used with hycanthone methyl sulfonate, after the compound was administered to mice orally at a concentration of 50 mg/kg. In initial studies, it was determined that the chemical occurs as a water soluble urinary glucuronide. To determine if the active compound would appear in either the neutral, basic or acid fraction, the urine of treated animals was incubated at 37°C for 16—18 h with β-glucuronidase prior to use of the extraction procedure. Fig. 4 records the results when urine and liver were analyzed. Mutagenic activity was found in the acid fraction of both liver and urine, with the urine showing about twice as much activity as the liver. Mutagenic activity was also demonstrated in the neutral fraction of the urine, but not the liver, suggesting the possibility of a second active metabolite after hycanthone administration (Fig. 4).

C. Detection of mutagenic activity in two drugs given to human subjects

With two therapeutic drugs: the antischistosomal compound niridazole and the widely used antitrichomonal agent metronidazole, active mutagenic substances were detected when this procedure was used to analyze body fluids from both animals and human subjects. Fig. 5 records the mutagenic activity detected in the blood of mice treated with niridazole and Fig. 6 illustrates the

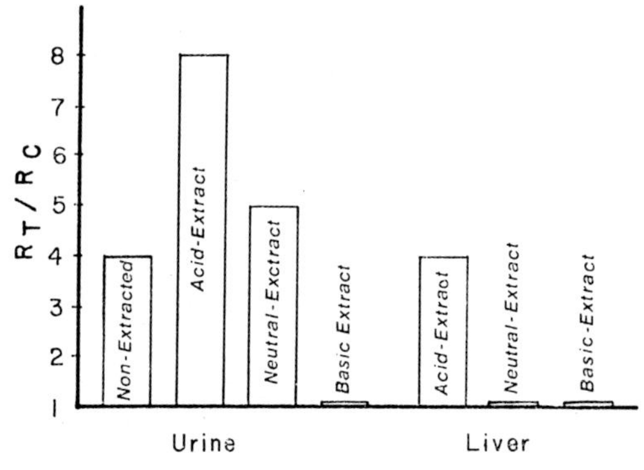

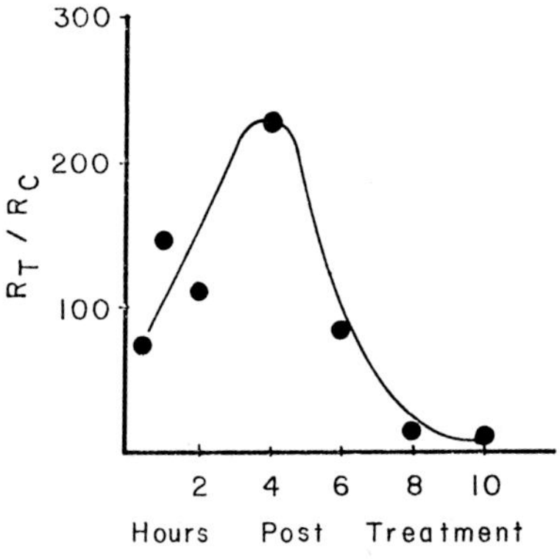

Fig. 4. Detection of mutagenic substances in urine and liver of mice treated with hycanthone, using *S. typhimurium*, TA1538. Mice received 50 mg/kg and the urine was collected for 24 h after treatment. Livers were removed from a similar group of mice at 2 h after treatment. All plates contained 200 units β-glucuronidase. R_T/R_C = revertants in treated/revertants in control samples.

Fig. 5. Mutagenic activity detected with *S. typhimurium* (TA1538) in blood samples of mice treated with niridazole. The mice received one dose (200 mg/kg), and samples were collected at 2-h intervals from five mice and pooled. Blood samples were not concentrated, and 0.2 ml of blood was added per plate.

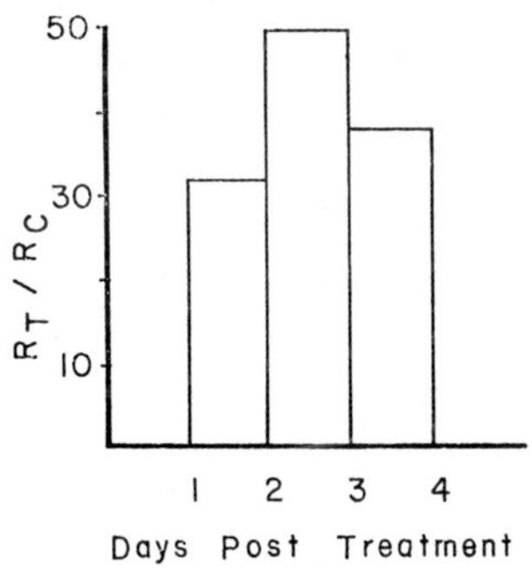

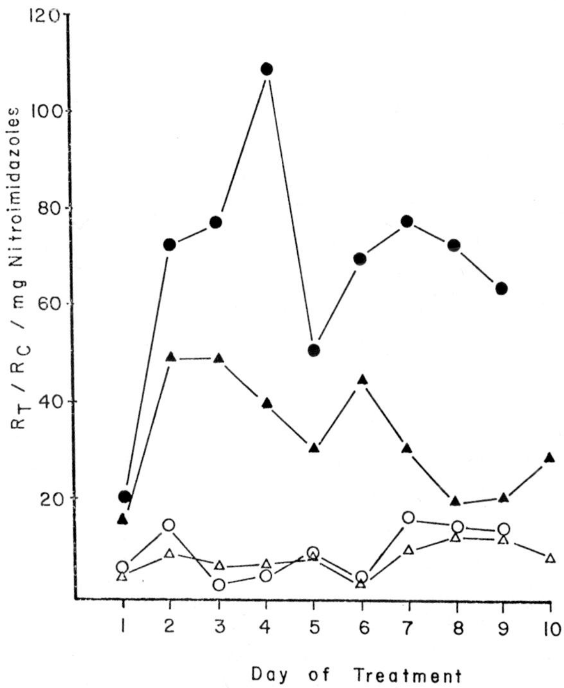

Fig. 6. Mutagenic activity detected with *S. typhimurium* (TA1538) in urine samples of a single patient treated with niridazole. The patient received one dose (25 mg/kg), and three 24-h samples were collected. Urine samples were not concentrated and 0.1 ml of urine was added per plate.

Fig. 7. Mutagenic activity detected with *S. typhimurium* (TA1535) in urine samples of two patients treated with metronidazole. Patients received 750 mg/day, and the urine was collected 1 h after the morning dose on the day indicated. Circles, patient 1; triangles, patient 2. Solid figures show actual detected activity and open figures show expected activity based on equivalent amount of metronidazole.

mutagenic activity in the urine of a single subject treated with niridazole. Fig. 7 depicts the mutagenic activity in two patients treated with metronidazole (solid figures) as compared to the expected activity (open figures) based on the presence of metronidazole, thus indicating that the active compound is a metabolite, not detectable in *S. typhimurium* assay without metabolic activation by the intact host.

Statistical analysis

Several statistical methods have been suggested as appropriate for the various mutation tests now available. Frantz and Malling stated that the basic statistical approach is a chi-square comparing number of bacteria and number of revertants in experimentals and controls [4]. Tables were constructed by Kastenbaum and Bowman for determining the statistical significance between mutation frequencies from control and test series [7].

We have found that the Wilcoxon Rank Sum Test is a very useful statistical method for the bacterial mutagenic assay [14]. This is a non-parametric statistical method for determining the significance of differences between two groups or treatments. In the bacterial system we have tested the difference in means in unpaired replicates, where the experimenter wishes to decide if the two groups have been drawn from the same population. The number of replicate observations may or may not be the same for the two groups.

In this procedure let M equal the number of observations in the smaller group and N equal the number of observations in the larger group. Rank numbers from one to M + N are assigned to all the observations, giving the smallest observation a rank of one and the largest a rank of M + N. Tied values are assigned the average rank. The M rank numbers are added and the sum called T. This value, T, is compared with tables that contain, for specified M and N, the critical upper and lower rank sums, T_U and T_L respectively, associated with specific probabilities. If $T \geqslant T_U$ then the mean of the smaller group is significantly larger than the mean of the other group at the specified probability level. If $T \leqslant T_L$, then the mean of the smaller group is significantly smaller than the mean of the other group at the specified probability level. If $T_L < T < T_U$ then there is not sufficient evidence at the specified probability level to say that the means of the two groups differ.

The following example illustrates this statistical method: numbers of mutant bacteria in urine from a group of patients receiving therapeutic doses of metronidazole (Flagyl) for one day and from a group receiving no Flagyl are given. The investigator wishes to know if the group receiving Flagyl differs significantly from the group receiving no Flagyl.

Flagyl patients (M)		Non-Flagyl patients (N)	
Mutant bacteria	Rank	Mutant bacteria	Rank
156	10	65	9
281	12	28	3
266	11	61	7
62	8	39	4
286	13	11	1
774	*14*	12	2
T=	68	59	6
		58	5

In this example M = 6, the number of observations in the Flagyl patients and N = 8, the number of observation in the non-Flagyl patients. Rank numbers are assigned to the data from 1 to 14.

The sum of the ranks, T, for the smaller group, M, is 68. In the table [14] for M = 6, N = 8 and P = 0.01, $T_L = 25$ and $T_U = 65$. Since $T > T_U$ the investigator may conclude that the average number of mutant bacteria in the urine of Flagyl patients is significantly higher than that of the non-Flagyl patients at the P = 0.01 level.

Discussion

1. Advantages

A. The measurements of genetic lesions by indicator organisms

There are few procedures that indicate a variety of chemically induced genetic lesions in experimental animals and man. Indeed, the majority of tests carried out in the intact mammal directly detect only chromosomal damage. The analysis of body fluids in man and animals as well as body fluid and tissue in experimental animals greatly expands our capability of measuring gene mutation as well as cytogenetic abnormalities. Although, up to this point, a great deal of our experimental results have been derived from the Salmonella histidine auxotroph system, it is anticipated that future developments will utilize mammalian cells where a variety of induced genetic lesions can be measured by analysis of tissue or body fluid of the intact host. The importance of this can be appreciated when one realizes that a major class of genetic abnormalities in man that can be induced by chemicals, non-disjunction, cannot suitably be detected in prokaryotic systems. It is well within our capability to utilize mammalian cells to detect non-disjunctional errors, gene mutations, cytogenetic effects and induction of repair in body fluid and tissues of experimental animals and body fluid of man. Of special interest are the ongoing studies that indicate the suitability of Drosophila as an indicator organism.

B. The analysis of metabolites produced in the intact host

One of the major and perhaps insurmountable disadvantages of in vitro tests is the inability of any standard in vitro system to measure exactly the dynamic metabolic reaction that occurs in the intact host; additionally, metabolites that are formed by means other than liver microsomes are missed by the standard in vitro activation procedures. Illustratively, the metabolite of metronidazole found in urine of treated human subjects is not detected by in vitro procedures with or without activation. Compounds such as cycasin and numerous steroids are metabolized by intestinal flora and illustrate the need to analyze for metabolites that are found in the intact animal.

C. Industrial monitoring

At the present time we have instituted a program of periodically examining workers who are exposed to potentially mutagenic substances *. The selection of the industrial population for monitoring is based on (a) prior identification of compounds in animal systems that are mutagenic or carcinogenic, (b) chemicals that are suspected of being mutagenic by our knowledge of structure-activity relationship and (c) compounds to which workers are exposed in large quantities. Just as we have demonstrated our capability to detect therapeutic drugs by this method, so also should we be able to detect mutagens to which workers are exposed during their working day. The analysis of body fluid should always be carried out following an industrial or laboratory accident.

---

* Dow Chemical Co., Freeport Texas Division

2. Disadvantages of the procedure

The analysis of body fluids and tissues like any indirect measurement for mutagenicity has decided limitations. Host defenses are of course not taken into account and consequently the effect of genetic repair cannot be determined. The measurement being confined to the genetic effect on the indicator organisms will obviously not tell us of the ultimate genetic lesion which is induced in the host. The use of a variety of indicator organisms would minimize this problem.

Overall assessment of the procedure

The analysis of body fluids in animal and man is an important procedure for determining a varient of genetic lesions. Its flexibility as to the choice of indicator and the ability to carry out the analysis on a variety of species, including man, make it a unique test for detecting chemically induced mutations. This procedure greatly extends our ability to evaluate chemically induced mutations in animals and is the only practical procedure that will allow us to determine gene mutation of the parent compound or its metabolites, regardless of where the metabolites are produced. Chemicals that cannot be detected by a standardized in vitro activation procedure can be found by this method. Of special importance is our ability now to evaluate the mutagenic effect of chemicals after human exposure, which, along with cytogenetic analysis, greatly extends our capabilities for screening human populations. This, like any other toxicological procedure, has definite limitations. In this procedure we are measuring an indirect response and this procedure is a compromise measure that attests to our inability to carry on a direct study with the intact mammal.

The cost of this procedure will depend to some extent upon the indicator organism chosen, but it is one of the least expensive procedures that can be carried out. If the Salmonella tester strains are used as the indicator organisms the cost will be slightly greater than in vitro testing.

The validation of this procedure, like all mutagenic tests, and indeed the great majority of routinely used non-genetic toxicological methods, has yet to be accomplished. Cytogenetic studies, where coded slides are read by several laboratories, is probably the only mutagenicity study where we have documentation as to variation and reproducibility [8,10]. The nature of this procedure where Salmonella tester strains are the indicator organism, from our experience to date, is a reproducible, simple economic procedure. The general application of this test, in conjunction with other procedures should detect the majority of mutagenic agents.

Acknowledgement

This work was supported by Research Grant ES01111-01 from the National Institute of Environmental Health Sciences, National Institutes of Health.

References

1 Ames, B.N., J. McCann and E. Yamasaki, Methods for detecting carcinogens and mutagens with the Salmonella/mammalian microsome mutagenicity test, Mutation Res., 31 (1975) 347—364.

2 Commoner, B., A.J. Vithayathil and J.I. Henry, Detection of metabolic carcinogen intermediates in urine of carcinogen-fed rats by means of bacterial mutagenesis, Nature, 249 (1974) 850—852.

3 Durston, W. and B.N. Ames, A simple method for detection of mutagens in urine. Studies with the carcinogen, 2-acetylamino-fluorine, Proc. Natl. Acad. Sci. (U.S.), 71 (1974) 737—741.

4 Franz, C.N. and H.U. Malling, The quantitative microsomal mutagenesis assay method, Mutation Res., 31 (1975) 365—380.

5 Gabridge, M.G., A Denunzio and M.S. Legator, Microbial mutagenicity in animal-mediated assay, Nature, 68 (1969) 221.

6 Goldbaum, L.R. and A.M. Dominguez, A system for the toxicological analysis of drugs in biological specimens, Prog. Chem. Toxicol., 5 (1974) 101—149.

7 Kastenbaum, M.A. and K.O. Bowman, Tables for determining the statistical significance of mutation frequencies, Mutation Res., 9 (1970) 527—549.

8 Kilian, D.J., F. Moreland, M. Benge, M.S. Legator and E. Whorton, A collaborative study to measure inter-laboratory variation with the in vivo bone marrow metaphase procedure. Submitted for publication.

9 Legator, M.S., T. Connor and M. Stoeckel, Detection of mutagenic substances in the urine and blood of man in occupational monitoring for genetic hazards, Ann. N.Y. Acad. Sci., 269 (1975) 1—45.

10 Legator, M.S., K.A. Palmer and I.-D. Adler, A collaborative study of in vivo cytogenetic analysis. Toxicol. Appl. Pharmacol., 24 (1973) 337—350.

11 Legator, M.S., M. Stoeckel and T.H. Connor, Techniques for isolating mutagenic substances from urine and blood of treated mammals using histidine auxotroph as the indicator organism, Mutation Res., 26 (1974) 456.

12 Legator, M.S., S. Zimmering and T.H. Connor, The use of indirect indicator systems to detect mutagenic activity in human subjects and experimental animals, in A. Hollaender (ed.), Chemical Mutagens, Principles and Method for their Detection, Vol. 4, Plenum, New York, 1976, pp. 174—191.

13 Siebert, D., Induction of mitotic conversion in *saccharomyces cerevisiae* by lymph fluid and urine of cyclophosphamide treated rats and human patients, Mutation Res., 21 (1973) 202.

14 Wilcoxon, F. and R.A. Wilcox, Some Rapid Approximate Statistical Procedures, Lederle Laboratories Publication, New York, 1964.

15 Zimmering, S., Utility of Drosophila for detection of potential environmental chemical mutagens. Ann. N.Y. Acad. Sci., 269 (1975) 26—33.

LABORATORY PROCEDURE FOR ASSESSING SPECIFIC LOCUS MUTATIONS AT THE TK LOCUS IN CULTURED L5178Y MOUSE LYMPHOMA CELLS

D. CLIVE AND JO-ANNA F. S. SPECTOR

Environmental Mutagenesis Branch, National Institute of Environmental Health Sciences, P.O. Box 12233, Research Triangle Park, N.C. 27709 (U.S.A.)

CONTENTS

Abbreviations: BUdR, 5-bromodeoxyuridine; CM, cloning medium; DMSO, dimethyl sulfoxide; F_0, Fischer's medium *plus* antibiotics; F_{10}, F_0 *plus* 10% horse serum; F_{10P}, F_{10} *plus* sodium pyruvate; FM, "freezing" medium; G, glycine; H, hypoxanthine; M, methotrexate; PPLO, pleuro-pneumonia-like organisms; T, thymidine; THG, thymidine, hypoxanthine, glycine; THMG, thymidine, hypoxanthine, methotrexate, glycine; TK, thymidine kinase; TMP, thymidylate; TS, thymidylate synthetase; VC, viable counts.

I INTRODUCTION

The development of techniques for measuring mutant frequencies at specific loci of cultured mammalian cells offers several advantages for the study of mammalian cell genetics and testing potential environmental mutagens[1]. Among these are simplicity, low cost, sensitivity and biological relevance to mammalian systems.

Our studies of this type have employed L5178Y mouse lymphoma cells. We initially became attracted to these cells because of a number of their characteristics which seemed to favor their use in an *in vitro* mammalian cell mutational assay system. These include growth in suspension culture, a short generation time of about 11 h, a stable, near-diploid chromosome number, and a cloning efficiency of up to 100% in a simple soft agar cloning medium (CM).

The specific locus we have chosen to work with is one which specifies the enzyme thymidine kinase (TK). We have isolated two homozygous and the expected single heterozygous genotypes (denoted as: $TK^{+/+}$, $TK^{-/-}$ and $TK^{+/-}$), and developed an assay which permits quantitation of forward and reverse mutational events at this autosomal TK locus.

In this paper we report the specific methodologies currently in use in our laboratory along with materials, instrumentation and formulary.

II METHODOLOGY

In this section, the basic forward and reverse mutation assays will be outlined and the individual techniques involved in these assays described.

(A) Forward mutation system

1 Treatment day

a. Glassware. The following glassware is prepared (Section IIIE): cell count tubes; 125-ml screw-capped boiling flasks; tubes for mutagens; centrifuge tubes; a 1000-ml Erlenmeyer flask.

b. Solutions and media. The following solutions are needed: 0.1% trypsin in saline; solvents for mutagens; and F_{10P} (non-selective growth medium).

c. Cells. A recently cleansed (Section IIIF) cell culture is diluted to a concentration (Section III A, B) of $0.3 \cdot 10^6$ particles/ml, and thoroughly mixed. Then 50-ml aliquots are dispensed into prepared 125-ml flasks, and placed into a shaker incubator.

d. Mutagenizing (use light sparingly). The desired concentration range of putative mutagen to be covered is determined and appropriate stock solutions prepared (Section III G). The mutagen is then added to the 50-ml aliquots of cells with swirling and then the culture flasks are replaced in the shaker incubator for 2 h.

e. Mutagen removal. The cells are collected by centrifugation ($500 \times g$, 10 min) and the supernatants containing the non-incorporated mutagen discarded.

f. Washing. The cells are resuspended in 25 ml of fresh F_{10P} with gentle vortex mixing and pelletted as before. The supernatant is discarded and the washing step is repeated.

g. Preparation for expression. The washed cells are quantitatively transferred into appropriately labeled 125-ml flasks that contain 50 ml of F_{10P} and allowed to express their genetic lesions in the shaker incubator.

2 Expression days

The cell cultures are counted daily and diluted to $0.3 \cdot 10^6$ particles/ml on each day of expression and on the cloning day.

3 Cloning day (see Fig. 1)

a. Preparation. (i) The 5-bromodeoxyuridine (BUdR) stock solution is thawed and melted agar is added to the previously prepared CM (Section III (*I*)). *(ii)* Six Petri plates per culture are prepared and labeled (viable count and BUdR).

b. Cloning. 10 ml of each culture at $0.3 \cdot 10^6$ particles/ml (or a total of $3.0 \cdot 10^6$ particles) are centrifuged ($500 \times g$ for 10 min) and the cells collected and resuspended in approximately 1.0 ml of retained supernatant. The cell suspensions are transferred

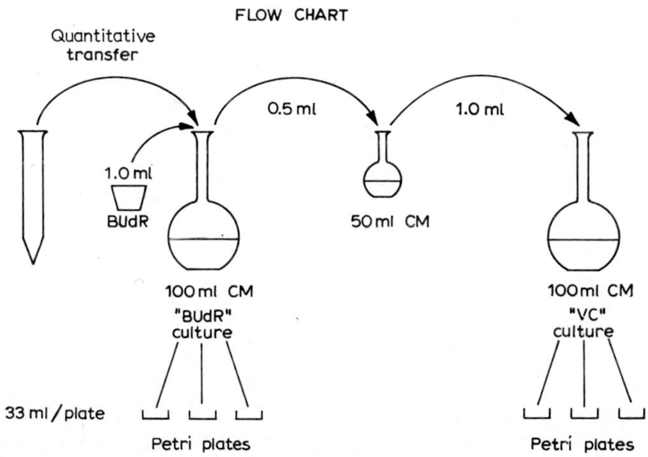

Fig. 1. Dilution sequence for cloning. See text for details.

to 300-ml flasks containing 100 ml CM (labeled "BUdR") and mixed in the shaker incubator for 15 min (the concentration is now $3 \cdot 10^4$ particles/ml). 0.5-ml aliquots of the "BUdR" suspensions are then diluted 1:100 (0.5 ml to 50 ml CM in 125-ml flasks; $3 \cdot 10^2$ particles/ml) and 1.0 ml of BUdR stock is added to the remainder in the "BUdR" flasks. The diluted aliquots and BUdR containing cultures are placed in the shaker incubator. After 15 min, the diluted aliquots (300 particles/ml) are further diluted 1:100 (1 ml to 100 ml CM in 300-ml flasks, 3 particles/ml) and labeled "VC", viable counts. (When low plating efficiencies are expected, this cell concentration can be increased. Calculations are adjusted accordingly.) The "VC" cultures are added to the shaker incubator. After 15 min "VC" and "BUdR" flasks are removed and 33-ml portions poured into labeled petri plates. The plates are gelled at $-20°$ for 12 min, warmed at room temperature for 5 min, equilibrated with 5% CO_2-in-air in a sealed lucite box for 1 h, and incubated at 37° for 10 days. The colonies are then counted and results calculated (Section III J, K).

(B) Reverse mutation system

The only differences between the forward and reverse mutational assay system

stem from: *(1)* the 300-fold difference in the rates of the $TK^{+/-} \rightarrow TK^{-/-}$ and the $TK^{-/-} \rightarrow TK^{+/-}$ mutations, and *(2)* the nature of the mutant and wild-type cells involved. Therefore, instead of plating $3 \cdot 10^6$ $TK^{+/-}$ cells in 100 ml of BUdR-supplemented CM (33 ml/petri plate), approximately $2 \cdot 10^8$ of $TK^{-/-}$ cells are cloned in 2000 ml of CM supplemented with the selective agent thymidine, hypoxanthine, methotrexate, glycine (THMG). (10^5 cells per ml; 1000 ml per 2800-ml low form culture flask.) Calculations are adjusted to allow for this difference in the total plated cell number.

III INDIVIDUAL TECHNIQUES

(A) Procedure for cell counting

The L5178Y cells tend to grow in small clumps in growth medium. To ensure accurate particle counts, the clumps are dissociated into single cells (without harm) by incubating them for a short period of time in trypsin. This is accomplished by adding 9.0 ml sterile 0.1% trypsin in saline and 1.0 ml of the appropriate cell cultures to each of a number of cell count tubes, and mixing each tube for 10 sec on a Vortex stirrer (setting No. 3). The tubes are incubated at 37° for 10 min, remixed for 10 sec, and counted using a Coulter or comparable particle counter.

(B) Diluting cultures

The generation time of the mouse lymphoma cells is roughly 10–11 h. This means that in 24 h, the number of cells will have increased nearly 5-fold, and in 48 h, there will be nearly 25 times as many. Therefore, for daily maintenance, one volume of culture is diluted with 4 volumes of fresh F_{10P}. Likewise, if the culture is to grow unattended for the entire weekend, one volume of the culture is diluted with about 100 volumes of fresh F_{10P}.

(C) Freezing cells

The cells to be frozen are collected by centrifugation (500 × *g* for 10 min) and resuspended in "freezing" medium (FM) to a final concentration of about $5 \cdot 10^6$ cells/ml. 1.0-ml aliquots are dispensed into freezing ampoules which are then tightly sealed and placed in the ring of a biological freezer (8 ampoules/container). The gasket ring is set to position "A", and the freezer is placed in a liquid nitrogen tank for 2 h. The ampoules are then transferred to appropriately labelled cardboard canes which are placed in aluminum canisters and gently lowered into the liquid nitrogen tank for long term storage.

Protective face shield, rubber gloves and lab. coat should be worn during the thawing procedure. The appropriate ampoule is quickly removed and stored in a padded, lidded styrofoam ice bucket while the cane and canister are replaced in the liquid nitrogen tank. The frozen cell sample is then thawed without unnecessary warming by dipping the ampoule into a 37° water bath. Next, the ampoule is swabbed with alcohol, opened in the hood and its contents gently transferred to a 125-ml flask containing F_{10P}. The fragile cells are kept in the upright incubator overnight without agitation to allow them to adapt to growth medium before placing the flask in the shaker incubator.

(D) Stock solutions

The mutagen solutions are prepared as either 100 × or 1000 × stocks depending on the solvent to be used. Mutagens soluble in saline or 10% dimethyl sulfoxide (DMSO) in saline are prepared as 100 × stocks and 0.5 ml is added to each 50-ml culture. Those soluble only in organic solvents such as ethanol, ether, or 100% DMSO, are prepared as 1000 × stocks and 0.05 ml is added to each culture. The organic solvents can be toxic to the cells; therefore it is best to keep their final concentration at or below 0.1%.

(E) Glassware

All glassware used in the laboratory is sterilized, screw capped, gassed with 5% CO_2 in air, and warmed to 37° before use. Each aliquot of cells during the course of an experiment will require:

3 125-ml screw capped boiling flasks (2 for mutagenesis, 1 for cloning);
2 300-ml screw capped boiling flasks (both for cloning);
2 centrifuge tubes (1 for mutagenesis, 1 for cloning);
2 or more cell count tubes (1 per day);
assorted pipettes and graduated cylinders.

(F) Cleansing cultures

Before cells are used in a mutagenesis experiment, it is desirable to have the frequency of spontaneously accrued mutants as low as possible. The techniques used to accomplish this follow together with their rationales.

TK-competent cell line (e.g. 3.7.2.)

TK-competent heterozygote ($TK^{+/-}$) cells grown in suspension culture spontaneously mutate to TK-deficiency ($TK^{-/-}$) at the rate of $2 \cdot 10^{-6}$ mutations/locus/generation. To ensure a low background frequency of mutants in a mutagenesis experiment, the culture is first grown in medium supplemented with THMG. This selective medium permits the normal growth of TK-competent cells while killing TK-deficient mutants. It works as follows: *(1)* methotrexate (M) blocks folate metabolism. As a result, thymidylate synthetase (TS)-mediated production of thymidylate (TMP) is arrested. *(2)* TMP production, however, can be maintained by TK phosphorylation of exogenously supplied thymidine (T). *(3)* A similar situation exists with respect to purine metabolism; this block is overcome by the addition of hypoxanthine (H). *(4)* Glycine (G) is present as a single-carbon source to replace the blocked folate metabolic pathway. In this manner, TK-competent cells can survive the methotrexate present in THMG. However, TK-deficient cells are totally devoid of pathways leading to TMP production after TS is blocked and hence they die.

The cleansing technique currently employed follows: on Thursday, 0.5 ml THMG stock is added to a 50-ml aliquot of cells. The next day, Friday, between 5 and

10 ml of this culture is pelletted and resuspended in 300 ml F_{10P} supplemented with 3.0 ml thymidine, hypoxanthine, glycine (THG)*. The cleansed culture will be ready to use on Monday.

$TK^{-/-}$ cell lines (e.g. 3.7)

The rate at which cells spontaneously revert from a TK-deficient to a TK-competent phenotype is about 300 times lower ($6 \cdot 10^{-9}$ mutations/locus/generation) than is a forward mutational event. The induced reversion rate for these cells is also lower. For this reason, TK-deficient cell lines are rarely used in safety testing suspect mutagens, but are used when hunting for new TK-heterozygote cell lines. In such a hunt it is important to use cleansed cultures if we wish to know with a high degree of confidence that any selected revertant arose from the mutagen treatment in question rather than through some ill-defined "spontaneous" mechanism.

We accomplish this by prolonged growth in BUdR. BUdR is an alternate substrate (with T) of TK. The phosphorylation product of BUdR (BUdR monophosphate) in TK-competent cells is lethal to cells following incorporation into DNA. TK-deficient mutant cells lack this enzyme activity and suffer no ill effects from this treatment. Therefore, BUdR can select for TK-deficient cells and against TK-competent ones.

(G) Concentration range of mutagen

A typical forward mutation experiment involves 24 culture aliquots distributed as 2 untreated controls, 21 different concentrations of the mutagen being tested, and 1 positive control. The concentrations are geometrically centered about a "guesstimated" LD_{50} dose, with each pair of adjacent concentrations in the ratio $1:\sqrt[8]{10}$ (*i.e* , 1:1.334; in this manner 8 concentrations cover a 10-fold range in concentration and the total experiment spans a 316-fold concentration range—*e.g.*, 1 μg/ml to 316 μg/ml). To produce a dilution series, 20 tubes, each containing 0.667 ml of sterile solvent are prepared. 2.0 ml of sterile stock solution are transferred to the first of these tubes which is thoroughly mixed. Serial transfers of 2.0-ml samples effects the dilution series.

(H) Expression time

After cells have been exposed to a mutagen, they need time to divide, fix the induced genetic lesions and produce cells in which these errors are phenotypically expressed. During this period the mutated cells dilute out or otherwise lose their previous stores of TK and become BUdR-resistant (in the forward mutation assay; TK must be generated in the reverse system). This two-or-more-day period is called the expression time.

(I) Cloning medium (CM)

CM (see MEDIA section also) used in this laboratory is of soft agar type. It is

* M binds very tenaciously to the enzyme dihydrofolate reductase which is not totally released after washing the cells. Relief from this temporary residual methotrexate inhibition is provided by the addition of THG following washing.

usually made 1 or 2 days in advance and stored at 4°. The evening before cloning, the CM (*minus* agar) is put into the 37° upright incubator and the agar is put into a waterbath that is connected to an automatic timer. The timer will switch the waterbath on at 6 *a.m.* and have the agar melted and ready to use as early as 9 *a.m.* It is important to swirl the medium vigorously when adding the agar so that it will not gel immediately. After mixing the complete CM thoroughly, it may be aliquoted into the appropriate flasks.

(J) Counting petri plates

The individual cells in soft agar grow into countable colonies after about 10 days of incubation. Instead of counting the number of colonies on each plate by eye, which would be both tedious and time consuming, we use an automatic colony counter. This instrument is particularly worthwhile as it essentially replaces a technician and ensures long term consistency in counting results.

(K) Calculations

The results of part of a mutagenesis experiment are calculated in Table I to illustrate the typical way our data is treated for comparative analysis. These data are plotted in full in Fig. 2.

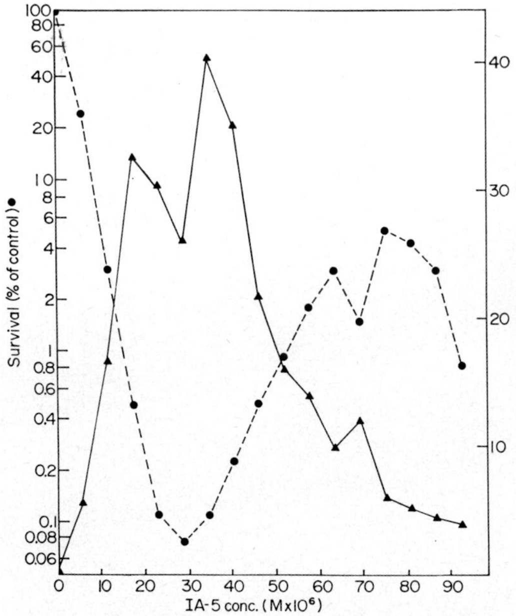

Fig. 2. Mutagenicity of IA-5 in the $TK^{+/-} \rightarrow TK^{-/-}$ mutational assay system. Induced mutant frequencies (number of BUdR-resistant ($TK^{-/-}$) mutant colonies per 10^4 viable cells) and cell survival (% of control) following 2-h exposure to the indicated mutagen concentrations are compared.

TABLE I

SAMPLE DATA AND CALCULATIONS

1	2	3	4	5	6	7	8	9	10	11	12	13
Mutagen conc. (μM)	*"Cell" counts (·10⁻⁶)*			*Suspension growth*		*Cloning data*		*Cloning*		*% Growth*[f]	*Mutant frequency*[g]	*Induced mutant frequency*[h]
	Day 1	*Day 2*	*Day 3*	*Total*[a]	*%*[b]	*VC*	*BUdR*	*VC*[d]	*%*[e]			
0	1.378	1.576	1.296	104.2	100	98, 117, 106 (3/ml)[c]	76, 85, 73	107	100	100	$0.73 \cdot 10^{-4}$	—
34.6	0.596	0.492	0.312	3.36	3.20	34, 56, 36 (30/ml) 352, 358, 372 (300/ml)	115, 122, 109	3.90	3.65	0.11	$29.57 \cdot 10^{-4}$	$28.84 \cdot 10^{-4}$ ($40.38 \cdot 10^{-4}$)
Hycanthone												
55.3	1.016	1.344	1.230	62.2	60.0	106, 70, 76 (3/ml)	160, 179, 188	84.0	78.5	45.3	$2.09 \cdot 10^{-4}$	$1.36 \cdot 10^{-4}$ ($1.90 \cdot 10^{-4}$)

[a] $\frac{\text{Day 1 cell count}}{0.3 \cdot 10^6} \times \frac{\text{Day 2 cell count*}}{0.3 \cdot 10^6} \times \frac{\text{Day 3 cell count*}}{0.3 \cdot 10^6}$. (* Previous day's cell count if culture was not diluted.)

[b] $\frac{\text{Suspension growth of treated culture}}{\text{Suspension growth of control culture}} \times 100.$

[c] Concentration of cells cloned.

[d] Average from column 7 and adjusted to a plated cell concentration of 3/ml.

[e] $\frac{\text{VC of treated culture}}{\text{VC of control culture}} \times 100.$

[f] $\frac{(\text{Column 6}) \times (\text{Column 10})}{100}$

[g] $\frac{\text{Column 8 (average)}}{\text{Column 9 (VC)}}$

[h] Mutant frequency of reference hycanthone control at 45% growth is $1.90 \cdot 10^{-4}$. All data are normalized by the factor 1.90/1.36 for comparison with other experiments. Control frequency subtracted.

(L) *Mycoplasma and cultured cells*

A major concern in tissue culture lies in maintaining contamination-free cell lines. Sterile glassware and materials, careful techniques and the filtration, heat-treatment and addition of antibiotics to the media all contribute to this goal. However, in spite of these elaborate precautions, there is always the threat that Mycoplasma (or pleuro-pneumonia-like organisms (PPLO)) may appear in a culture.

PPLO have the insidious characteristics of being so small as to be able to pass through filter pores; of being present (in concentrations as high as 10^7 organisms/ml) yet still undetectable with regard to culture growth, morphology and other parameters; and of being of unknown origin and highly infectious with L5178Y mouse lymphoma cell lines[4].

Our cultures acquire a number of unusual characteristics when infected with PPLO, which have been previously summarized[2]. Most importantly, infected $TK^{+/-}$ cultures cannot be "mutagenized"; *i.e.*, no BUdR-resistant colonies can be recovered from such cultures either spontaneously or following mutagen treatment.

Attempts have been made to eliminate PPLO from cultures by using a high level of kanamycin (400 μg/ml), heat treatment, or high levels of BUdR. Kanamycin suppresses the growth of PPLO and yields cell cultures which temporarily behave as PPLO-free. However, all such cultures develop the symptoms of PPLO-infection soon after removal from kanamycin. The use of heat (41° overnight) to kill PPLO has not been very successful either. Our L5178Y cells have about the same temperature sensitivity as the Mycoplasma making the heat treatment equally lethal to both organisms. (Our hope with this method is to develop a heat stable cell line.) To date, only high levels (50 μg/ml) of BUdR have routinely succeeded in maintaining $TK^{-/-}$ cultures free of PPLO. Contaminated heterozygote cultures must be discarded. Aliquots of all PPLO-free $TK^{+/-}$ cultures are preserved in liquid nitrogen, and serve as back-up cultures.

IV MEDIA

(A) *Procedures*

The media used are principally Gibco products which have been supplemented in order to optimize growing conditions for the L5178Y mouse lymphoma cells. Unless indicated otherwise, they are prepared within the Institute by trained personnel under aseptic conditions.

The cells grow in suspension culture in Fischer's medium supplemented with penicillin and streptomycin (F_0) and 10% horse serum (to give F_{10}), under a 5% CO_2-in-air atmosphere, with a generation time of 10.6–11.0 h at 37°. The addition of sodium pyruvate (100 μg/ml to F_{10}) produces a medium (F_{10P}) that will support cell growth from a single cell inoculum and is the growth medium most commonly used at present.

Cells grown in suspension can be cloned in soft agar CM employing a modification of the soft agar cloning technique of CHU AND FISHER[3] with a cloning efficiency of 80–100% relative to Coulter particle counts. Nonselective CM consists of horse serum (20%), sodium pyruvate (220 μg/ml) and Noble agar (0.37%) in F_0. The addition of either BUdR or THMG to the above makes the CM selective for one or the other phenotype.

Long term storage of the cells grown in suspension can be accomplished with minimal losses by replacing the suspension growth medium with FM and maintaining the sealed ampoules in liquid nitrogen. FM consists of F_0 (60% v/v), horse serum (20%), glycerol (12%), and DMSO (8%). The F_0 and F_{10P} are usually prepared in 5-l lots. The solutions are filter sterilized and aliquoted into 500-ml sterile bottles in a sterile laminar flow hood. The bottles are then heated to 57° in a water bath for 45 min to kill PPLO which might have escaped exclusion during filtration, and stored at 4°. A portion of these media-containing bottles (*ca.* 12) are maintained at 37° for daily use.

Horse serum is purchased fully prepared and frozen. Serum used in CM is filter-sterilized and heated to 57° for 30 min beforehand, while that used in F_{10P} is sterilized after addition to that medium.

(B) Formulary

1 F_0 ($93/100 l)
Source:

Fischer's Medium for Leukemic Cells of Mice, Cat. No. H-11.
Instant Tissue Culture Powder Medium Packets.
1 packet/5 l final volume.
Company: Grand Island Biological Co. (Gibco), Grand Island, New York.

Supplements to 1 packet:

Penicillin "G" sodium (1650 units/mg)	155 mg
Streptomycin sulfate	250 mg
$NaHCO_3$	5.625 g
Distilled H_2O	5000 ml

Adjust pH (before filtering) to 6.6. After filtering, working pH will be 6.8.

2 F_{10P}
Source:

Fischer's Medium for Leukemic Cells of Mice, Cat. No. H-11.
Instant Tissue Culture Powder Medium Packets.
1 packet/5 l final volume.
Company: Grand Island Biological Co. (Gibco), Grand Island, New York.

Supplement to 1 packet:

Penicillin "G" sodium (1650 units/mg),	155 mg;
Streptomycin sulfate	250 mg;
Pyruvic acid (sodium salt),	500 mg;
$NaHCO_3$	5.625 g;
Horse serum	500 ml;
5% Pluronic solution	100 mg; (Pluronic F 68 BASF, Wyandotte Corp., Wyandotte, Michigan)
Distilled H_2O	4400 ml
Total	5000 ml

Adjust pH (before filtering) to 6.8. After filtering, working pH will be 7.0.

3 Horse serum ($20/0.5 l)
Source:

Frozen Horse Serum, Cat. No. 605 (500-ml bottles)
Company. Grand Island Biological Co. (Gibco), Grand Island, New York.

Supplements:

None

4 Cloning medium (CM) (see Table below)

Final volume (ml)	*Volume of components (ml)*			
	F_0	*Serum*	*100 × Pyruvate*	*4% Agar*
100	69.75	20	1.0	9.25
250	174.4	50	2.5	23.1
500	348.8	100	5.0	46.2
1000	697.5	200	10.0	92.5
1500	1046.3	300	15.0	168.7
2000	1395	400	20.0	185
2500	1744	500	25.0	231
3000	2092	600	30.0	278
4000	2790	800	40.0	370
5000	3488	1000	50.0	462
6000	4185	1200	60.0	555

5 Freezing medium (FM) (100 ml total)

DMSO	8 ml
Glycerol	12 ml
Horse serum	20 ml
F_0	60 ml

6 Selective media

Selective media are made by the addition of a selective agent to F_{10P} or CM. Specifically, BUdR-supplemented F_{10P} or CM selects for TK-deficient cells (against TK-competent cells), and THMG-supplemented F_{10P} or CM selects for TK-competent cells (against TK-deficient cells).

V EQUIPMENT AND MATERIALS

The equipment and materials that are essential for the efficient functioning of our tissue culture lab. and the sources from which they are obtained are presented below:

(A) Equipment

1 Laminar flow hood	$2500
2 Upright incubators (2)	$3000 each
Model 1651 (Window covered to ensure absence of light within)	Hotpack Corp., Philadelphia, Pa.

3	Shaker incubators (2) G25 Controlled Environment Incubator Shaker (Window covered to ensure absence of light within)	\$1500 each New Brunswick Scientific, New Brunswick, N.J.
4	Coulter Counter Model ZB	\$6000 Coulter Electronics, Hialeah, Fla.
5	Automatic Bacterial Colony Counter Model 870	\$5000 Artek Systems Corp., Farmingdale, N.Y.
6	Centrifuge Size 2 Model K	\$1500 International Equipment Co., Needham Hts., Mass.
7	Water Baths Thelco Models 83, 84	\$300 each Precision Scientific, Chicago, Ill.
8	Microscope (Inverted) Leitz	\$1000 Wetzlar, Germany
9	Gas Apparatus 5% CO_2-in-air cylinders Solenoid regulator Gas outlet in sterile hood	
10	Vortex-Genie Mixer Model K-550-G	\$65 Scientific Industries, Inc., Springfield, Mass.
11	Liquid Nitrogen Storage and Filling Tanks	\$600 (total) Union Carbide Corporation
12	Biological Freezer Type BF-5	Union Carbide Corporation

(B) Materials

1	Filter sterilizers 0.20 micron Plain Membrane Filter Unit	NALGE Sybron Corp., Rochester, N.Y.
2	Freezing ampoules Cat No. 1076-01 Screw cap sterile tubes	\$55/case of 500 Inhalation Therapy – Anesthesia Sales and Services, Medical Products Division of: National Welders Supply Co., Inc. 5303 Old Dowd Road, P.O. Box 8436 Charlotte, N.C. 28208

(C) Glassware

All glassware is screw-capped, sterilized, glassed with 5% CO_2 in air, and warmed to 37° prior to use.

1. 125-ml, 300-ml and 500-ml flat bottom boiling flasks.
2. 50-ml centrifuge tubes.

3. 200-ml centrifuge bottles.
4. 25 × 150 mm culture tubes (used for cell-counts and mutagen stocks).
5. Assorted Erlenmeyers, cylinders and pipettes.

(D) Solutions

All the solutions listed below are 100 × stocks which are filter sterilized and stored in the freezer unless indicated otherwise.

1 BUdR

0.5 g/100 ml F_0

2 THG

30 mg	Thymidine
50 mg	Hypoxanthine
75 mg	Glycine
100 ml	F_0

3 THMG

100 ml	THG (as above)
0.2 ml	Methotrexate Stock

4 Methotrexate Stock (5000 μg/ml)

One 50-mg vial Methotrexate in 10 ml F_0

5 Sodium pyruvate

2.2 g	Sodium pyruvate
100 ml	F_0

6 Agar	Autoclaved and stored at 4°
8 g	Noble agar/200 ml distilled water/flask

VI REFERENCES

1 Clive, D., W. G. Flamm, M. R. Machesko and N. J. Bernheim, A mutational assay system using the thymidine kinase locus in mouse lymphoma cells, *Mutation Res.*, 16 (1972) 77–87.

2 Clive, D., W. G. Flamm and J. B. Patterson, Specific locus mutational assay systems for mouse lymphoma cells, in A. Hollaender (Ed.), *Chemical Mutagens: Principles and Methods for their Detection*, Vol. 3, Plenum, New York, 1973, pp. 79–103.

3 Chu, M.-Y., and G. A. Fischer, The incorporation of [^{3}H]cytosine arabinoside and its effect on murine leukemic cells (L5178Y), *Biochem. Pharmacol.*, 17 (1968) 753–767.

4 Stanbridge, E., Mycoplasmas and cell cultures, *Bacteriol. Rev.*, 35 (1971) 206–227.

MUTAGENICITY TESTING WITH V79 CHINESE HAMSTER CELLS

C.F. ARLETT

MRC Cell Mutation Unit, University of Sussex, Falmer, Brighton BN1 9QG (Great Britain)

Contents

Introduction

The forward mutation assay system based upon 8-azaguanine resistance (AG) using V79 Chinese hamster cells has been in widespread use for almost a decade [6,7]. Interest has been focused at a number of levels: as a mutagen assay system [2]; the genetical validity of resistant clones [18], and at the level of cell biology [15]. While there will not be an opportunity to discuss the significance of all these observations the existence of these various levels of interest indicates that the system is capable of generating information at all these levels. A consequence of the widespread use of the V79 cells is that there are almost as many ways of performing experiments as operators. The aim here is to give the procedures which work in our laboratory with supplementary comments. More recently ouabain resistance (OUA) has been introduced as an alternative or complementary assay system [4,5] and there are indications that the use of 6-thioguanine (TG) is replacing AG in many laboratories [29,30,33].

The descriptions here refer to a work load of approximately two full-scale experiments per week, which is appropriate to one full-time tissue culture technician with independence from washing up and with half-time support for the setting up and treatment phases of an experiment.

Tissue culture

The biological material

The original isolation of a fibroblast cell type from a male Chinese hamster (*Cricetulus griseus*) embryo lung was made by Ford and Yerganian [13]. Since that time the material has been circulated through a large number of laboratories and might be said to be as ubiquitous as HeLa. During the passage of time many changes both obvious and hidden have taken place. We are in possession of cells labelled V79 with changes in cellular or clonal morphology [21], the ability to grow in suspension or monolayer culture with equal facility, the ability to grow with different serum concentrations, and with enhanced sensitivity to some DNA damaging agents (V7979) [31]. As many of these properties may influence the response of the cells to mutagens [21] care must be exercised to determine the pedigree of any material.

Two cultures are in regular use in our laboratory. The first, V79 I, was obtained through Professor B.A. Bridges from E.H.Y. Chu in 1969 and has always been maintained in 10% foetal calf serum (FCS). The second, V79 II, was obtained as a recloned isolate from B. Dean (Sittingbourne) who had received a culture from D. Wild (Freiberg) who had, in turn, obtained the cells from Dr. Chu. This second culture has always been maintained in 15% FCS and produces clones quite distinct in appearance from V79 I. We would recommend V79 II for general use; it produces more robust clones than V79 I and has, despite the additional cost of using 15% FCS, the advantage of exhibiting a low ($\sim 10^{-7}$) spontaneous frequency of mutation to OUA resistance [4].

Media

The enormous cost of manufacture of tissue culture media from the simple components is largely related to labour charges. Few, if any, laboratories do not now take advantage of the large batch and sterility testing facilities of the commercial laboratories (Flow Laboratories or Gibco-Biocult). Eagle's minimal essential medium (MEM) [12] with Earle's salts is the most commonly used medium although some workers prefer Ham's F10 [17]. The bicarbonate buffering system is most commonly employed. MEM is normally supplied as a 10× concentrate, although if you wish to purchase sterile water 1× strength is available. This medium may also be available in a powdered form suitable for filter sterilization. The manufacturers are normally happy to modify the formulation of media to customers' own requirements, thus hypoxanthine as an alternative substrate to AG or TG may be excluded from F10.

Sterile solutions of sodium bicarbonate (7.5%), glutamine (200 mM) and penicillin (5000 I.U./ml) plus streptomycin (5000 μg/ml) are also obtained from the commercial houses on the basis of cost effectiveness. All the V79 cells which we have received have already been cultured in media containing penicillin and streptomycin and we have chosen to maintain these antibiotic supplements.

Foetal calf serum is ordered in 50 litre reserve batches from the supplier and is pre-tested for its effectiveness in promoting cloning ability (we have only ever recorded one instance where a reduced cloning efficiency was observed), and as a component of medium with selective agent, AG or OUA, since serum components influence the effectiveness of the selective system [5,21,33,35]. For the OUA system it may be considered worthwhile to record K^+ levels by flame photometry. Calf serum which is used in conjunction with F10 in some laboratories [33] is subjected to the same evaluation procedure, it has been our experience when working with human cell strains that there is a larger batch-to-batch variability than with FCS.

We have found the 500 ml medium/serum bottles re-usable many times and containing a convenient volume, the volume markings are also very helpful. The composition of our complete medium is:

340 ml	sterile double-distilled water (DDW)
45 or 68 ml	FCS (aliquots are dispensed into 100 ml medium bottles from 500 ml stock)
12.8 ml	sodium bicarbonate
4.0 ml	glutamine
4.5 ml	penicillin/streptomycin
40 ml	10× stock MEM with Earle's salts and without glutamine or sodium bicarbonate

Normally medium is made up on a day-to-day basis to reduce the need to store large quantities of complete medium and so avoid any loss of growth promoting activity. We discard any complete medium which is not used up within 14 days from production.

Storage (Table I) of medium and components requires a deep freeze, fridges or access to cold or freezer room facilities. Modern domestic fridges may be of

TABLE I

STORAGE CONDITIONS AND ACCEPTABLE LIFE OF MEDIUM COMPONENTS

Material	Storage temperature (°C)	Storage life
Double distilled water	RT [a]	1 month
Medium 1 × and 10 × without glutamine and sodium bicarbonate	4	1 year
Foetal calf serum	−20	1 year
L-glutamine (200 mM)	−20	1 year
Antibiotics	−20	1 year
Sodium bicarbonate	RT	1 year

[a] RT = room temperature.

a flimsy plastic construction, commercial catering models can be more robust with shelving strong enough to support the weight of large quantities of liquid. Upright freezers are more economical of floor space and are certainly easier to manage than the less expensive chest type. For materials stored at room temperature cupboards with glass doors to exclude dust are recommended.

Special solutions

(1) AG stock. 8-Azaguanine (Sigma) is stored at room temperature, 30 mg is added to 10 ml 0.5% sodium carbonate, autoclaved at 5 p.s.i. for 5 min and stored in the incubator for the duration of an experiment, it may take up to 6 h to dissolve completely. The stock solution is added at 0.1 ml to 10 ml to achieve 30 μg/ml.

(2) TG stock. 6-Thioguanine (Sigma) is stored at 4° C, 30 mg is added to 10 ml 0.5% sodium carbonate (dissolves rapidly although we have noted a loss of solubility in aged batches), filter sterilized (Millipore filter 0.45 μm pore size) and stored in the incubator for the duration of an experiment. The stock solution is added at 0.1 ml to 10 ml to achieve 30 μg/ml.

(3) OUA stock. Ouabain octahydrate (Sigma) is stored at room temperature, 0.735 g is added to complete medium and dissolved in a water bath at 50° C — this may take up to 4 h — filter sterilized and stored in the incubator for the duration of an experiment. The stock is added at 0.5 ml to each 10 ml medium to achieve 1 mM.

(4) Gentamycin reagent (Schering) sterile solution supplied at 50 mg/ml and stored at 4° C. Used at final concentration of 50 μg/ml.

(5) Phosphate-buffered saline (Dulbecco "A") tablets supplied by Oxoid, stored at room temperature, each tablet to make 100 ml is dissolved in double-distilled water (DDW) and autoclaved at 120° C for 20 min. The buffer is stored at room temperature for up to one month.

(6) Sodium carbonate solution is prepared as 0.5 g Analar grade reagent in 99.5 ml DDW and sterilized in the autoclave at 120° C for 20 min. The solution is stored at room temperature indefinitely.

(7) Trypsin solution. One ampoule of Bacto desiccated powder is dissolved in 200 ml buffer and filter sterilized and stored at —20° C for up to one month. This working stock is diluted 10-fold in buffer for use, the supply for one week being stored at 4° C.

(8) Dimethyl sulphoxide (DMSO). Laboratory reagent grade (BDH) material is autoclaved at 120°C for 20 min as 50 ml aliquots and stored indefinitely at room temperature.

(9) Methylene blue. The stain (East Anglia Chemicals) is weighed out at 5 g/ 500 ml DDW (1%) in a medium bottle and we have found solubility is enhanced by autoclaving at 120° C (about 20 min).

Washing up; glassware and plastics

Axiom I. The success of tissue culture is dependent upon the quality of the washing up.

It seems inevitable that operators are likely to require a mixture of glass and plastic materials. We find it impossible to function with plastics alone since the cost and particularly the storage requirements of such an operating mode are inhibiting.

Newly delivered glassware is soaked for two days in a bath containing 2% hydrochloric acid and then subjected to the standard washing up routine. At the end of the working day excess medium buffer etc. is poured off, the vessel rinsed in cold tap water and totally immersed in a large bath of 0.5% Decon 75 (Decon Labs.) in single-distilled water. This bath is changed once weekly using spent distilled rinsing water. The next day the soaked items are removed from the bath, brushed if required and immersed in a sink of running cold tap water (RTW) for 2 h, transferred to a second sink with RTW for a further 2 h followed by immersion in SDW for 0.5 h, twice. Final rinsing is with DDW for 0.5 h, the items are then warm air dried overnight.

Assembly of apparatus, caps to bottles, plugging pipettes, manufacture of simple solutions or aliquots of water to be sterilized and the packaging of these items can take place during the rinsing periods. After packaging in sterilization packs (John Dickinson and Co.) in various sizes and sealing with autoclave tape, the items are dry heat or autoclave-sterilized as appropriate. Sterility testing tubes (Albert Browne Ltd.) are used in both cases as a check on proper operation.

Glass pipettes reach the washing up area in rubber pipette buckets (P.B. Cow Ltd.) containing 3—5% hypochlorite solution (Deosan Hypochlorite-Diversey Ltd.) in SDW. The pipettes are unplugged, rinsed in pipette washer in RTW for 2 h, soaked in SDW for 0.5 h, twice, and DDW for 1 h. Accelerated drying in a drier (Shandon Southern) permits the processing of the pipettes within one day.

Working area

We have found a 12 × 8 ft. module to be a sufficient area for a culture room where two people may conveniently cooperate in an experiment. The furniture constists of:

(i) 7 ft. laminar flow cabinet (Envair Ltd.) a vertical flow model with a filtered exhaust which may be vented from the room should satisfy all requirements;
(ii) 5 ft. bench with sink;
(iii) 5% CO_2/air gas cylinder;
(iv) bench centrifuge (MSE minor);
(v) incubators (pattern to be discussed below);
(vi) under-bench units for storage;
(vii) two swivel stools;
(viii) inverted microscope.

The room might well be air-conditioned and be capable of being blacked out when using photo-sensitive materials. Overnight air and surface sterilization is provided by germicidal ultra-violet lamps.

Culturing cells

Axiom II. The success of mammalian cell mutation experiments is entirely dependent upon the quality of the tissue culture.

TABLE II

SUB-CULTURE PEDIGREE FOR MAINTENANCE AND EXPERIMENTATION

M [a]	F	M	F	M
I ↘	I	I → bench	——————→	discard
	↘ II ↘	II	II ——→	bench
	↓	↘ III ↘	III	III
	↓		↘ IV ↘	IV
	2×10^5 cells in 75 cm^2 culture vessel		↓	↘ V
	↓		2×10^5 cells in 75 cm^2 culture vessel	
	Monday trypsinize count set up two 75 cm^2 culture vessels at 2×10^6 cells per bottle			
	Tuesday experiment A reinoculate at 2×10^6 per bottle or set up fresh bottles			
	Wednesday experiment B			

[a] M = monday, F = friday.

We insist on a total embargo on mouth pipetting on all occasions to protect both ourselves and the cells. We have found the Pi Pump device (Schuco Scientific Ltd.) easy to manipulate; it does, however, require the use of plugged pipettes.

Cultures are normally maintained in 125 ml medical flat bottles, capped with disposable (but re-usable) plastic caps (Sterilin Ltd.) or with metal caps with silicon rubber liners. Alternatively, plastic tissue culture flasks (Falcon Plastics) with 75 cm^2 growth area may be used.

A sub-culture routine is illustrated in Table II; a culture inoculated with 2—4 $\times 10^4$ cells on Monday is sub-cultured on Friday. Examine the culture by eye and microscope; is the medium cloudy, are the cells detaching, do they look healthy? Suck off the medium, rinse once with about 2 ml 0.025% trypsin, add 5 ml trypsin and return vessel to the incubator for 10—15 min. Tap the vessel two or three times to detach cells, check at this stage at the microscope that the cells have detached because the trypsin sometimes becomes inactivated on storage, then pipette vigorously with a 5 ml pipette. The shear forces help to generate a single cell suspension without reducing cloning ability.

The suspension of cells in trypsin is transferred with the same pipette to a plastic disposable centrifuge tube (25 ml Universal container, Sterilin Ltd.) and centrifuged at a setting of 2—3 for 2 min on the M.S.E. minor. These centrifuge tubes are also useful as 25 ml bottles and are very convenient for use in dilution series. The supernatant is discarded and the pellet broken up by tapping the end of the tube with the index finger. The cells are resuspended in 5—10 ml of complete medium and counted either by Coulter counter or by using a Haemocytometer (Hawksley) with an improved Neubauer grid BS748 (depth 0.1 mm, 1/400 mm^2). This has the virtue of being cheap, easy to use and forces the experimenter to look at the cells. It is also possible to determine the frequency of cells which are not single. No mathematics are required to calculate the concentration of cells on selection of the appropriate grid, thus a mean count of 20 = 20×10^4 cells/ml. The number of cells inoculated and harvested is noted in the sub-culture log and may be used to monitor the vigour of the culture. It may thus be possible to recognize the introduction of some toxic substance to the system. It should also be remembered that having resuspended the cells in medium to stop the action of trypsin, the cells will now commence to stick to the walls of any vessel unless agitated at frequent intervals — how to lose your cells.

A new bottle is inoculated with 2—4×10^4 cells in 10 ml complete medium, fresh medium is added to the harvested bottle which retains enough cells for adequate growth. The new bottle is flushed with a 5% CO_2/air mixture for 15 sec, the gas mixture being filtered through the sterile cotton wool plug in a Pasteur pipette, before incubation. The bottle which had fresh medium on the preceding sub-culture day is taken out of the incubator and kept on the bench until the next round of sub-culture is seen to be successful. If it is incubated for a longer period the medium becomes exhausted and the cells detach and die. We have thus generated a pedigree of sub-cultures with at least two old bottles in reserve as a precautionary measure.

Since most of the cells are discarded at each passage there are normally sufficient to set up extra or larger bottles for experimental requirements. We normally separate material for sub-culture and experiments as in Table II. It should be clear that the number of culture vessels used is determined by the size of the experiment. Two treatment modes are common: trypsinize, plate, treat or treat, trypsinize, plate.

In addition to regular sub-culture adequate stocks of cells must be held in liquid nitrogen storage to replenish or recommence sub-cultures. For preservation in liquid nitrogen a large culture vessel is inoculated and the culture grown until virtually in stationary growth phase. The culture is then trypsinized, resuspended in complete medium, counted, centrifuged and resuspended in complete medium supplemented with 10% DMSO to achieve a final concentration of about 3×10^6 cells/ml. One ml aliquots are then dispensed into sterile plastic screw cap ampoules (Nunc) and after a slow freeze period using a Union Carbide BF-6 Biological Freezer set to freeze at a rate of 2° C/min these are stored either in the liquid or gas phase. Resurrection is by rapid thaw, the ampoule is removed from the storage vessel, placed in a disposable tube which is then placed in a box containing water at about 37° C. After cleaning the outside of the ampoule with alcohol the suspension is diluted 10 : 1 with complete medium, centrifuged and resuspended for a second time in complete medium and then inoculated into a 75 cm^2 culture vessel. Good growth is normally achieved in 2 days and the cells are then subjected to a normal sub-culture routine.

One other aspect of routine sub-culture should be mentioned at this point; it has been our practice to expose cells in the first three to five passages after thawing to culture in the presence of gentamycin as an anti-mycoplasma therapy. Routine testing for mycoplasma may take the form of sending out the material for contract assay or the use of agar plate tests (Flow Labs.), biochemical [22] or staining assays [23]. We have found the techniques to demonstrate non-nuclear thymidine uptake [22] to be a good indication of infection. Positive or suspect cultures are immediately discarded and new material set up from liquid N_2. Since mouth pipetting is completely banned we can only assume that any such infections are introduced with medium components.

Platings of cells are made into 5 or 9 cm vented plastic petri dishes (Nunclon T.C. grade) and these may be incubated under a variety of conditions. It is our normal practice to store the dishes in plastic lunch or cake boxes (Payne and Co.). The CO_2 enrichment for the gas atmosphere is provided by the appropriate weight of a gassing mixture in sterile water in a scintillation vial within the box which is then sealed using a gas-tight vinyl tape (Intech. Tapes). The gassing powder is compounded from sodium bicarbonate and tartaric acid in the ratio 50 : 14 and stored under desiccated conditions. Newly delivered boxes are heated at 50°C overnight to drive off any monomer. For routine use they are washed in the Decon bath followed by tap water and finally rinsed with alcohol, drained dry and stored with the lid on. These boxes may be stacked in an anhydric incubator or any suitable temperature controlled area. They have the advantage of allowing tissue culture material to be grown in an incubator which is used for any purpose. Alternatively, the dishes can be incubated in a CO_2 enriched incubator; we favour the LEEC injection model where a graduated pulse of CO_2 is introduced into the chamber on shutting the door. This model has the advantage of simplicity and, therefore, cost over the continuous flow models.

The incubation period may vary depending upon treatment, as a general rule plates for survival responses are stained after 7 days and for mutant selection 10 days after the selective agent is added. One ml of 1% methylene blue is added to the medium on the plate for at least 30 min, the stain is then poured off and the plate allowed to drain and air dry (overnight) before being given a light rinse in tap water which increases the intensity of the stain. Care should be exercised in not leaving the stain on too long or "throwing" the stain off since the clones sometimes detach, particularly if overgrown. The middles of large clones sometimes fall out since these cells may have died.

Experiments

Toxicity tests

The size of the actual mutation experiment is inhibiting and it would, therefore, be foolish not to have some very good indications of the lethal effect of a potential mutagen before embarking on such experiments. While it is anticipated that some prior knowledge of toxicity is available, the first determination would be based upon "guestimates". Since the results are known within

TABLE III

MODEL TOXIC RESPONSE EXPERIMENT

Concentration (μg/ml)	0	0.1	1.0	10	100
	100 [a]	100	100	100	100
		500	500	500	500
			1000	1000	1000
				2000	2000
				5000	5000

[a] cells per 5 cm dish.
Three petri dishes are used for each point, five for control.
Requirements 50 × 5 cm dishes, 300 ml complete medium (excess) treatment time 2 h.

7 days an unlucky guestimate is not usually disastrous.

A simple toxic response test using a chemical agent might take the form outlined in Table III. It would normally be our practice to trypsinize and plate the appropriate cell numbers, allow 4 h for the cells to attach and recover from the trauma of trypsinization, then either remove the medium aseptically for re-use at the end of the treatment period or add the agent to the medium. An alternative is to apply the treatment in pre-warmed medium without serum or in Dulbecco A buffer. These two treatments reduce the possibilities of interactions between the agent and components of the medium. At the end of the treatment time (2 h) the medium is removed and the plates rinsed, once with buffer and once with complete medium (3 ml/dish) and then reincubated with either fresh or the initial medium at 5 ml/dish. If the initial medium is used it will require supplementation with 10% fresh medium to make up for losses on removal.

Proper attention must be paid to hazards when handling dangerous agents and we recommend applying the treatments in a fume cupboard. At the end of the treatment period all media and washings should be disposed of in the appropriate manner.

After incubation for 7 days the plates are stained and the colonies counted using a simple binocular microscope. We adopt the criterion of a colony containing 50 or more cells qualifying as a survivor. It is, therefore, useful to have a zoom facility for those instances where it is necessary to check the number of cells in a colony. As a general rule, we find that colonies which are visible to the naked eye will, with very rare exceptions, qualify as survivors. No difficulties should be experienced with the computation of the results, except to remember to correct for dilution, which are expressed as per cent of control.

Mutation experiments

(i) Action of the selective agents

Dose-response curves for the selective agents are published [4,30]. It is clearly worthwhile when setting up a series of experiments to confirm that the cells are responding in the appropriate fashion and that the selective agent is unchanged. A recent experience may illustrate this point. The "average" spontaneous frequency for V79 II cells for AG resistance is of the order of 1×10^{-5};

a population which we had been using for OUA resistance was exposed to AG and found to have a level of spontaneous resistance of about 2×10^{-3}, it was also seen to be high for resistance to 30 μg/ml thioguanine (8×10^{-4}). The cause of this increase was almost certainly cell selection during an episode when it was necessary to rescue cells from a grossly overgrown old sub-culture bottle. Reduction to the "average" spontaneous mutation frequency was achieved by setting up a number of cultures from a low level of inoculum (2×10^1/ml) which were tested for the frequency of mutants when the appropriate population size was achieved. Satisfactory populations were then multiplied and frozen down for future use. It should be noted, however, that experience over some years suggests to us that resistant cells accumulate in in populations stored under liquid N_2.

The sub-culture protocol outlined above appears to be adjusted to select the average spontaneous AG resistance frequency and rigid attention to such a fixed programme is recommended on these grounds.

(ii) In situ tests

This term refers to the fact that the mutagenized population is undisturbed after treatment except for the addition of the selective agent. This protocol must mean that at late expression times [3] colonies rather than cells are exposed to the agent, it also means that there is no opportunity for segregation of damage except within a clone and thus any processes such as metabolic cooperation [27] which act against the phenotypic expression of the mutants will reduce the efficiency of recovery.

An example of this design in an experiment which attempted to obtain a dose-response effect for OUA resistance with UV as a mutagen is shown in Table IV. This particular experiment had the treatment mode of trypsinize, plate and treat which represents the most cumbersome situation but is given to indicate the size of an experiment which can be performed by two people. The alternative, where a large population is treated on a 15 cm diameter petri dish followed by trypsinization and plating is simpler since the problems of removing medium and the irradiation of multiple replicates is avoided.

Day 1: set up 3×75 cm^2 bottles, 2×10^6 cells/bottle. Label plates, prepare medium, estimated requirement = 1200 ml, 300 ml spare.

Day 2: Trypsinize, spin, count all three in 130 ml = 1.5×10^5/ml.
Dilute 82 + 41 = 1×10^5 cells/ml;
dilute 122 + 1098 = 1×10^4 cells/ml;
dispense 10 ml per 9 cm dish with Cornwall pipette;
dilute 1 + 9, 7 + 63, = 1×10^2/ml, D1 for survival series, dispensed at 5 ml per 5 cm dish after dilution sequence completed (Table IV).

After all plates are dispensed they are boxed, gassed and placed in the incubator, these manipulations take approximately 1.5 h. The cells are treated after a 4 h incubation period; the medium is removed aseptically, the cells irradiated and the medium is replaced using a Cornwall pipette. The plates are then boxed up in batches according to the time at which the selective agent is to be added, flushed with CO_2 and returned to the incubator. Ouabain is added to the zero time plates and at the beginning and end of each working day for the further expression times until the experiment is complete. It is clear that it is

TABLE IV

THE INDUCTION OF OUABAIN-RESISTANT MUTANTS BY ULTRAVIOLET LIGHT

Dose (J/m^2)	Expression time (h)						5 cm Survival plates		
	16	24	40	48	64	72	Cells plated	Colonies	Survival %
0	0 0	0 0	0 0	0 0	0 0	0 0	100	81 100 86	100
	0 0	0 0	0 0	0 0	0 0	0 0		89 95	
2.6	1 1	1 1	3 5	0 3	0 0	0 0	100 } D_3	92 80 80	96.5
	1 2	1 2	6 5	0 1	0 0	0 0		91 92	
5.2	0 0	0 0	4 14	7 13	0 0	0 0	100	67 58 79	75.6
	0 0	1 1	9 11	15 13	0 0	0 0		67 70	
7.8	0 0	1 1	1 2	10 7	14 11	5 11	200 = D_2	49 57 41	27.6
	0 0	0 1	1 1	2 6	9 15	5 6		52 —	
10.4	0 0	0 0	0 0	0 1	1 1	6 5	500 = D_1	26 29 30	6.7
	0 0	0 0	0 0	1 1	1 4	4 —		34 31	

Four 9 cm plates for each dose/expression time, at 1×10^5 cells per dish.

Calculation of mutation frequencies

Dose (J/m^2)	Cells surviving ($\times 10^4$ per dish) [a]	Mean number of mutant colonies/dish [b]	Mutants per survivor [c]
0	9.02	0	—
2.6	8.70	4.75	5.46×10^{-5}
5.2	6.80	12.0	1.76×10^{-4}
7.8	2.49	12.25	4.92×10^{-4}
10.4	0.6	5.0 [d]	8.83×10^{-4}

[a] Although the target was 1×10^5 cells/dish the survival plates, which are an actual measure of the cells plated, indicates that 9.02×10^4 were recovered. The number for each dose is corrected using the appropriate survival level.

b Mean number of resistant clones at the time for optimum expression of mutants [3].

[c] Computed as $\frac{b}{a}$, with a zero frequency in the control series there is no correction for the spontaneous level and these levels represent the induced mutation frequencies.

[d] It is possible that this value is not from the optimal expression time.

wasteful to fill in the expression times for each dose of mutagen. The "shape" of the experiment evolves with the knowledge of when the optima appear, but in the absence of this information all expression times ought to be tested for each dose. When it is also clear that the control response level is very low (usually only with OUA) the number of control treatment plates may be reduced to a level compatible with a check on the stability of the system (one per expression time?).

It should also be clear that we cannot conclude that a given treatment is *not* mutagenic unless a number of experiments are performed and evaluated against a positive control.

(iii) Respreading tests

In this design, unlike the in situ tests, cells are dispersed and replated after treatment and before addition of the selective agent. Two variations of this design are available. In the first (Fig. 1) a large population of cells is mutagenized and maintained in exponential growth phase such that the population is subcultured and sampled for survival and mutation at daily intervals. This proto-

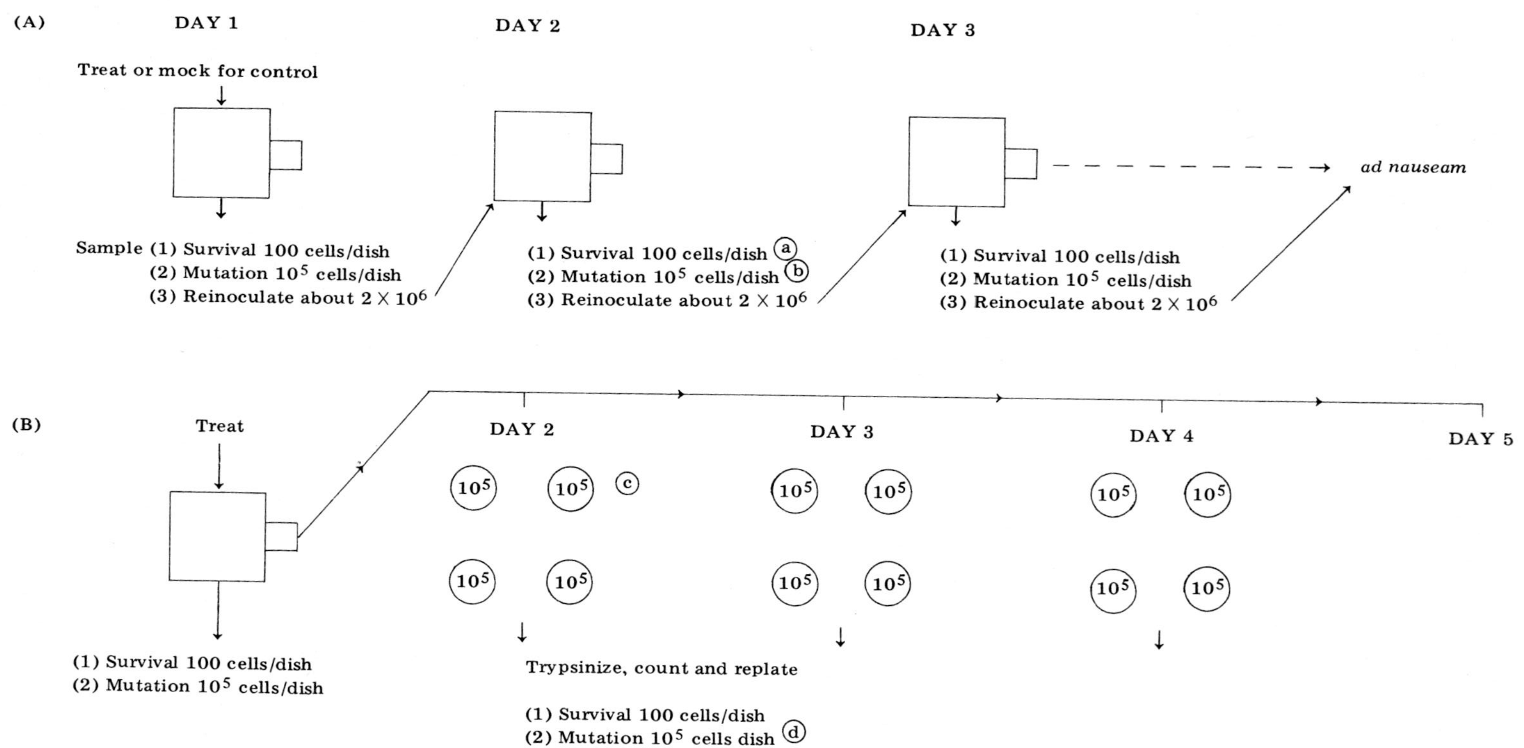

Fig. 1. Respread protocols for mutation assays with Chinese hamster cells. Protocol A: respread and reinoculation of flasks. Protocol B: respreading of plates
a = number adjusted for anticipated survival level. Normally recovers to control, untreated, level after 3—4 days;
b and d = mutation frequency calculated using: Mean number mutants on day n/No. cells plated as estimated from survival on day n;
c = sets of 4—6 dishes set up at the end of the treatment are respread at daily intervals.

col is analogous with the standard experimental design for L5178Y mouse lymphoma cells [10] except for the necessity for dispersal of the monolayer cultures with trypsin. In this way mutant and sensitive cells are segregated before selection. Regrettably, proper corrections for the recovery of mutant cells cannot always be made without prolonged study of the recovery characteristics of resistant mutants induced by the mutagen of interest [26]. In the case of OUA we will show an extreme instance where a substantial loss of mutants follows from the action of trypsin on the mutagenized population.

The second variation of this design (Fig. 1) takes the form of setting up groups of petri dishes as in the in situ protocol but sampling them (respreading) at intervals.

An example of the first of these variations together with an in situ experiment performed with the same material and at the same time for comparative purposes and using AG, TG and OUA as selective agents, is shown in an abbreviated form below and the results are shown in Fig. 2.

Day 1: set up $2 \times 75\ cm^2$ bottles, 1.2×10^6 cells/bottle

Day 3: trypsinize, spin, count 1 bottle = 9.6×10^6 cells, plate into 15 cm diameter petri dish (Nunc) in 50 ml medium, incubate for 5 h

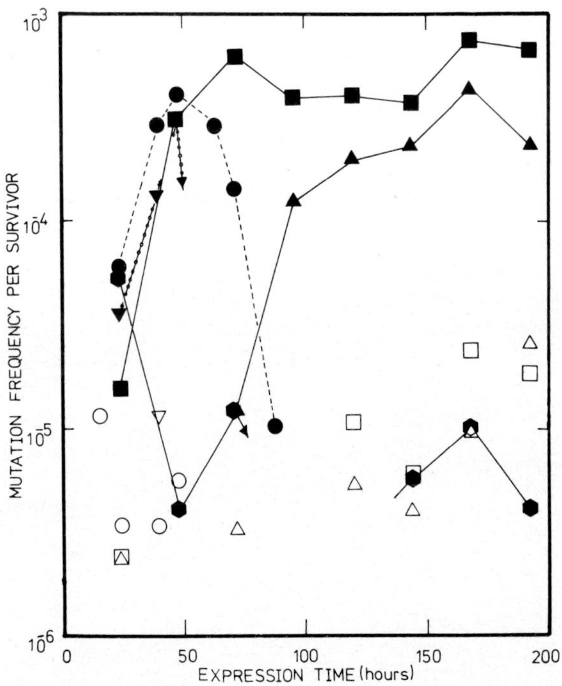

Fig. 2. A comparison of the in situ and respread mutagenicity protocols after 6.1 J/m² UV treatment and using the selective agents AG, TG (both at 30 μg/ml) and OUA (at 1 mM). ●- - - - - -●, in situ AG, UV-treated; ○, in situ AG, control; ■———■, respread AG, UV-treated; □, respread AG, control; ▼-○-○-○▼, in situ OUA, UV-treated; ▽, in situ OUA, control (1 colony only observed); ♦———♦, respread OUA, UV-treated; ▲———▲, respread TG, UV-treated; △, respread TG, control. In the OUA series *no* colonies were observed on the respread plates for control and for a number of expression times in the treated samples. In the AG and TG respread series at some expression times for the control samples no colonies were observed.

remove medium and irradiate with 6.8 J/m^2 fluence ultra-violet light,
trypsinize, spin down, count = 8.7×10^6 cells;
set up (i) series of in situ plates at 1×10^5 cells/dish
(ii) five 5 cm dishes at 100 cells/dish for survival estimates
(iii) a 75 cm^3 bottle containing 15 ml medium and 2.6×10^6 cells
add selective agent to the zero expression time plates
A "dummy" irradiation series is produced for control with the second bottle

Day 4: (= day 1 for respread series)
trypsinize count = 3.8×10^6 cells (for treated)
dilute, plate (i) at 1×10^5 cells/dish for mutation assay, add selective agents
(ii) at 100 cells/dish for survival assay return 2×10^6 cells to bottle and incubate
a similar series is handled for control

Day 5: (day 2 for respread series)
trypsinize count = 1×10^7 cells
dilute plate as in Day 4
return cells to bottle

Continue for at least a further five days.

Results and interpretation

Properly conducted mutation experiments should provide convincing evidence of the existence of a dose-response curve. The dose-response curves for cell killing and mutation induction in the mutation experiment described in the previous "In situ tests" section is shown in Fig. 3.

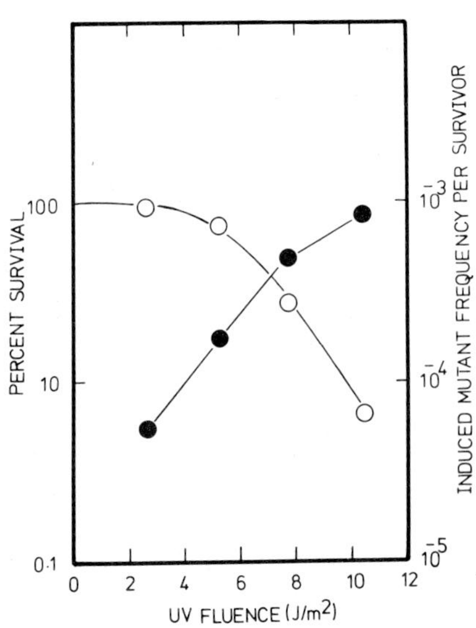

Fig. 3. Cell survival and induced mutation to OUA resistance following treatment with UV light. ○, survival, ●, mutation. The data for this figure are taken from Table IV.

With regard to cell killing the data display is as a plot of the logarithm of surviving fraction (treatment/control) against dose on a linear scale. Such a plot may not always be appropriate with mutation experiments where cell killing tends not to exceed two log scales and which, therefore, may be reasonably encompassed on a linear-linear plot. The dose scale for chemical agents will frequently show such units as mM, M, μM per unit time (i.e. dose × time). An alternative is to use a fixed dose with variable time, e.g. 1 μg/ml for 1 - - - - n min. We take the view that the use of molarity units implies knowledge of the purity of the material under test and that it is preferable to employ a weight unit, e.g. μg/ml. Proper attention to dosimetry is required and is absolutely necessary for comparisons between experiments or laboratories.

For mutation frequencies the plot of the logarithm of induced mutation frequency against linear dose is most often encountered. The major justification for this plot is convenience of scale. Linear-linear plots provided a ready visual indication of changes in shape of the dose-response curve and the slope of log-log plots allows a formal interpretation of the single or multi-hit components of the system. Difficulties are presented when a comparison of more than one mutagen is required, there seems some justification here for a plot of induced mutation frequency against survival [4].

The data for the expression time responses may also be conviently displayed in a graphical form. This information may be important in showing how a particular mutagen is acting. We have observed that with OUA as a selective agent ethyl methane sulphonate (EMS), in contrast to other mutagens, induces mutants at zero expression [4].

With regard to the respread experimental protocol again the object is to produce a dose-response curve. Survival curves and mutation induction curves do not have the same relationship as with in situ experiments since the former is determined at zero time and the mutation frequencies per survivor are calculated from the survival response of the particular respread population.

The information from the comparison of the respread and in situ protocols (Fig. 2) using different selective agents produces some interesting observations. The first is the dramatic loss of OUA-resistant clones on respreading. We have obtained supporting evidence [32] that newly arising OUA-resistant clones are less able to withstand the action of tryspin than wild-type or stabilized OUA-resistant cells. Thus the respreading protocol is inappropriate when using the OUA selective system. This observation should act as a reminder that mutant clones within any system need not necessarily have the same recovery potential as wild-type cells. Some workers have countered this problem with assays of the cloning efficiency [26,29] of mutants by mixing mutant and wild-type cells together under the assay conditions. The appropriate correction factor may then be applied to the data.

The second observation is that for AG resistance with this particular isolate and mutagen virtually identical induced mutation frequencies are obtained using the two protocols. We have been able to confirm this result with EMS and γ-irradiation and with at least one other V79 isolate. In addition, the time course for the expression of mutants may be very different for any two selective agents. Thus TG-resistant mutants arise significantly later than AG-resistant mutants. This situation would suggest that in those experiments where TG is used as a selective agent few, if any, mutants migth be anticipated in the in situ experiments and the respreading protocol might well be required to continue for many days. These observations force us to conclude that the in situ technique represents the most efficient experimental protocol.

A further conclusion from the in situ results in this UV experiment is that the frequency of mutation to OUA or AG-resistance is approximately equal. We have shown that this is not always the case for all mutagens [4]. With γ-irradiation we were unable to recover any OUA-resistant mutants with both Chinese hamster and L5178Y mouse lymphoma cells. With methyl methane sulphonate (MMS) fewer OUA- than AG-resistant mutants were observed in both species [4,9]. The apparent "false negatives" with γ-irradiation is of interest as a possible diagnostic whatever its explanation [4,11]. Thus, if any suspected mutagen induces AG- but not OUA-resistant mutants we may be in a position to suggest that its mutagenic action is analogous to ionizing radiation, the low frequency with MMS is an example. The use of OUA is also indicated by the low spontaneous mutation frequencies, providing the potential for the study of weak or low levels of mutagens. We would like, therefore, to recommend the use of at least two selective agents: AG and OUA, since clearly there is the potential for the recovery of more than two times the information produced from the use of any one selective agent.

Finally the use of a positive control mutagen is worthwhile in order to monitor the reliability of the system and to act as a reference. Our experience would indicate the choice of EMS as a most efficient chemical mutagen which is relatively easy to handle; UV is an alternative which is available in most laboratories.

Additional techniques and developments

Many mutagens/carcinogens require metabolic activation to an active form and such mutagens would, because of the low level of metabolic activity shown by V79 cells, probably prove non-mutagenic in experiments of the type already described. Two techniques are available to help to resolve this situation. Experiments have been described where V79 cells are co-cultivated on a feeder layer of metabolically active embryonic Syrian hamster cells [19] and which show feeder layer-mediated mutagenesis in the target cells. Metabolic activation might also be achieved by the use of rat liver microsome fractions similar to those employed in bacterial mutation studies [1]. We have recently found that whole liver cells, freshly prepared from rat, can be used in this capacity with L5178Y mouse lymphoma cells and anticipate that they may be successfully combined with V79 cells.

Green et al. [16] have shown that the modification devised by Ryan [24] of the classic Luria-Delbruck fluctuation test gives an increase in sensitivity with bacterial tests and permits the detection of low levels of mutagens or weak mutagens. A similar improvement in sensitivity of the test system has been achieved with L5178Y mouse lymphoma cells by the use of this technique [11] which, again, may be applicable to V79 Chinese hamster cells, particularly those isolates which have been selected for growth as suspension cultures.

The experiments described herein have concentrated on the AG, TG and

OUA resistance systems. Back mutation assays are possible with the cells selected by TG and AG [8]. A second round of selection of hypoxanthine-guanine phosphoribosyl transferase-negative cells selected initially by TG can be performed using azaserine medium [30]. A whole array of mutant phenotypes have been achieved in cultured mammalian cells ranging from glutamine or proline auxotrophs [20] through glucose independent mutants [25] to the metabolic inhibitor resistant phenotypes. It is this latter class, because of the availability of a ready selective technique for mutant enrichment, which will continue to be the most useful in the assays for potential mutagens. Toxic analogues of the pyrimidine series, such as cytosine arabinoside, excess thymidine, bromo- and iododeoxyuridine and methotrexate, which have been used successfully as selective agents with mouse cells [10,14], may also find a use with Chinese hamster cells. Toxic amino acid analogues such as azetidine [34] may also have a use. There is thus much scope for the development of new, additional, assay systems.

Therefore two hopes for the future; one being the development of single cell mutation systems, i.e. mutants detected as single cells by the use of a cytochemical staining procedure [28] and the second being the development of techniques for the measurement of chronic in addition to acute effects. It might be suggested that experiments taking the form of treating the whole animal (and thereby allowing for the action of metabolism) followed by assaying for mutants in primary cell cultures may yet be possible. The ease of culture of Chinese hamster cells and the substantial body of information available on cellular mutagenesis make this species a prime candidate for study in this field.

Acknowledgements

I am grateful to S.A. Harcourt and B.C. Broughton for technical studies reported in this contribution. I have attempted to incorporate as much mammalian cell mutagenesis "lore" as possible and thereby am indebted to that large group of scientists who have interests in the field. Dr. D.R. McCalla (McMaster, Ontario, Canada) provided many useful comments.

References

1 Ames, B.N., W.E. Durston, E. Yamaski and F.D. Lee, Carcinogens are mutagens: a simple test system combining liver homogenates for activation and bacteria for detection, Proc. Natl. Acad. Sci. (U.S.), 70 (1973) 2281—2285.

2 Arlett, C.F., Mutation testing with cultured mammalian cells, Lab. Practice, 21 (1972) 420—423.

3 Arlett, C.F. and S.A. Harcourt, Expression time and spontaneous mutability in the estimation of induced mutation frequency following treatment of Chinese hamster cells by ultraviolet light, Mutation Res., 16 (1972) 301—306.

4 Arlett, C.F., D. Turnbull, S.A. Harcourt, A.R. Lehmann and C.M. Colella, A comparison of the 8-azaguanine and ouabain resistance systems for the selection of induced mutant Chinese hamster cells, Mutation Res., 33 (1975) 261—278.

5 Baker, R.M., D.M. Brunette, R. Mankovitz, L.H. Thompson, G.F. Whitmore, L. Siminovitch and J.E. Till, Ouabain resistant mutants of mouse and hamster cells in culture, Cell, 1 (1974) 9—21.

6 Bridges, B.A. and J. Huckle, Mutagenesis of cultured mammalian cells by X-rays and ultraviolet light, Mutation Res., 10 (1970) 141—151.

7 Chu, E.H.Y. and H.V. Malling, Mammalian cell genetics II. Chemical induction of specific locus mutations in Chinese hamster cells in vitro, Proc. Natl. Acad. Sci. (U.S.), 61 (1968) 1306—1312.

8 Chu, E.H.Y., P. Brimer and H.V. Malling, Reversion studies of chemically induced 8-azaguanine resistant mutants in Chinese hamster cells in culture, Genetics, 61 (1969) S10—S11.
9 Cole, J., unpublished results.
10 Cole, J. and C.F. Arlett, Ethyl methanesulphonate mutagenesis with L5178Y mouse lymphoma cells: a comparison of ouabain thioguanine and excess thymidine resistance, Mutation Res., 34 (1976) 507—526.
11 Cole, J., C.F. Arlett and M.H.L. Green, The fluctuation test as a more sensitive system for determining induced mutation in L5178Y mouse lymphoma cells, Mutation Res., 41 (1976) 377—386.
12 Flow Manual, 1974.
13 Ford, D.K. and G. Yerganian, Observations on the chromosomes of Chinese hamster cells in tissue culture, J. Natl. Cancer Inst., 21 (1958) 393—425.
14 Fox, M. and D. Anderson, Characteristics of spontaneous and induced thymidine and 5-iodo-2-deoxyuridine resistant clones of mouse lymphoma cells, Mutation Res., 25 (1974) 89—105.
15 Fox, M., J.M. Boyle and B.W. Fox, Biological and biochemical characterization of purine analogue resistant clones of V79 Chinese hamster cells, Mutation Res., 35 (1976) 289—310.
16 Green, M.H.L., W.J. Muriel and B.A. Bridges, Use of a simplified fluctuation test to detect low levels of mutagens, Mutation Res., 38 (1976) 33—42.
17 Ham, R.G., An improved nutrient solution for diploid Chinese hamster and human cell lines, Exptl. Cell Res., 29 (1963) 515—526.
18 Harris, M., Mutation rates in cells at different ploidy levels, J. Cell Physiol., 78 (1971) 177—184.
19 Huberman, E., Mammalian cell transformation and cell-mediated mutagenesis by carcinogenic polycyclic hydrocarbons, Mutation Res., 29 (1975) 285—291.
20 Kao Fa-Ten and T.T. Puck, Genetics of somatic mammalian cells. IV. Properties of Chinese hamster cell mutants with respect to the requirement for proline, Genetics, 55 (1967) 513—524.
21 Newbold, R.F., P. Brookes, C.F. Arlett, B.A. Bridges and B. Dean, The effect of variable serum factors and clonal morphology on the ability to detect hypoxanthine guanine phosphoribosyl transferase (HGPRT) deficient variants in cultured Chinese hamster cells, Mutation Res., 30 (1975) 143—148.
22 Perez, A.G., J.H. Kim, A.S. Gelbard and B.D. Djardjevic, Altered incorporation of nucleic acid precursors by mycoplasma-infected mammalian cells in culture, Exptl. Cell Res., 70 (1972) 301—310.
23 Russell, W.C., C. Newman and D.H. Williamson, A simple cytochemical technique for demonstration of DNA in cells infected with mycoplasmas and viruses, Nature, 253 (1975) 461—462.
24 Ryan, F.J., Spontaneous mutation in non-dividing bacteria, Genetics, 40 (1955) 726—738.
25 Shapiro, N.I., A.E. Khalizev, E.V. Luss, E.S. Manuilova, O.N. Petrova and N.B. Varschaver, Mutagenesis in cultured mammalian cells II. Induction of gene mutations in Chinese hamster cells, Mutation Res., 16 (1972) 89—101.
26 Simons, J.W.I.M., Dose-response relationships for mutants in mammalian somatic cells in vitro, Mutation Res., 25 (1974) 219—227.
27 Subak-Sharpe, J.H., R.R. Burk and J.D. Pitts, Metabolic cooperation by cell-cell transfer between genetically marked mammalian cells in tissue culture, Heredity, 21 (1966) 342—343.
28 Sutton, H.E. and G.W. Karp, Jr., Somatic variation in human 6-phosphate dehydrogenase, Genetics, 64 (1970) S63.
29 Thacker, J. and R. Cox, Mutation induction and inactivation in mammalian cells exposed to ionizing radiation, Nature, 258 (1975) 429—431.
30 Thacker, J., M.A. Stephens and A. Stretch, Factors affecting the efficiency of purine analogues as selective agents for mutants of mammalian cells induced by ionizing radiation, Mutation Res., 35 (1976) 465—478.
31 Todd, P.W. and A.B. Hellewell, Phenotype and karyotype of a cultured Chinese hamster cell line defective for ultraviolet light resistance, Mutation Res., 7 (1969) 129—132.
32 Turnbull, D., Factors Affecting the Response of Cultured Chinese Hamster Cells to Mutagenic Alkylating Agents, D. Phil Thesis, University of Sussex, 1975.
33 Van Zeeland, A.A. and J.W.I.M. Simons, The effect of calf serum on the toxicity of 8-azaguanine, Mutation Res., 27 (1975) 135—138.
34 Wasmuth, J.J. and C.T. Caskey, Biochemical characterization of azetidine carboxylic acid resistant Chinese hamster cells, Cell, 8 (1976) 71—77.
35 Wild, D., Serum effect on the yield of chemically induced 8-azaguanine resistant mutants in Chinese hamster cell cultures, Mutation Res., 25 (1974) 229—234.

CHEMICAL MUTAGENESIS WITH DIPLOID HUMAN FIBROBLASTS

LOIS JACOBS and ROBERT DeMARS

Laboratory of Genetics, University of Wisconsin, Madison, Wisc. 53706 (U.S.A.)

Contents

Introduction

The development of methods for the rapid and accurate determination of the mutagenic effects of chemicals on human cells in vivo would permit the identification of compounds that may cause unnecessary genetic damage to the human population. Although it has not been possible to evaluate the mutagenic effects of compounds by directly testing human subjects, the effects of chemicals on intact test organisms or on cells in culture have been studied in vitro. The difficulty with these approaches is that of extrapolating the results obtained to humans. Studying mutagenesis with normal human cells in vitro, however, should reduce the degree of extrapolation necessary to provide information about the in vivo effects of compounds on the human population.

Cultures of normal skin fibroblasts can be obtained from virtually any human. The comparison of cultures from different humans will give the opportunity to relate genetic behavior in vitro to the inheritance in the human population of genetic factors that affect mutability. In addition, cultures from mutant persons in known pedigrees may provide evidence that somatic cells with altered phenotypes in vitro are true genetic variants.

Low cloning efficiencies and a limited number of cell doublings are still mistakenly cited as two overriding technical limitations of diploid human fibroblasts. However, we find (a) that we achieve cloning efficiencies of 20—50%, (b) that most induced and spontaneous variants have sufficient doubling potential to permit karyotype analyses, measurement of enzyme activities, determination of phenotypic stability and sensitivity to mutagens, and evaluation of enzyme-dependent growth properties in various media, (c) that many variants also can be subcloned with cloning efficiencies of

20—50%, and (d) that identical aliquots of variant populations can be stored for years in liquid nitrogen, allowing continuous experimentation with the same cell population.

Mutagenesis in normal human fibroblasts in vitro cannot be considered equivalent to the mutagenic process in an intact human, however, since many compounds must undergo metabolic activation in order to produce mutagenic effects. If the enzymes needed for activation are not present in the human fibroblasts, such compounds will not be detected as mutagens. This problem may be partly alleviated in the future by including a liver microsomal fraction that will provide the necessary activation. Even so, the detailed application of results obtained with human fibroblasts and other in vitro model detection systems, to mutagenesis in vivo will be difficult.

A suitable method of detecting mutational events in human diploid fibroblasts is the selection for 8-azaguanine-resistant cells. Apparently, hereditary cell variants that are resistant to 8-azaguanine (AG) can be detected in fibroblast cultures derived from any human, male or female. A simple principle is applied in detecting the mutants. AG, as such, is innocuous to cells at 10^{-6} M to 10^{-4} M concentrations, but it is converted to a toxic ribonucleotide by hypoxanthine-guanine-phosphoribosyltransferase (HPRT, EC 2.4.2.8) [13] in normal cells that have the enzyme [12]. HPRT is a dispensible enzyme under ordinary culture conditions, allowing HPRT-deficient mutants to grow out as resistant colonies in the presence of AG.

The mutational study of AG resistance has genetic advantages. Forward mutations that cause HPRT deficiency exist in humans; the cultures established from these mutant persons are prototypes to which variants arising in vitro can be compared [2,9,27]. The in vivo prototype mutant alleles are X-linked [20,21,32]. This means that single mutational changes on an X chromosome will be phenotypically expressed since cells from males have only one X chromosome and cells from females express genes on just one of the two Xs that are present [14—16,25,31]. Since HPRT is non-essential under ordinary culture conditions, gene mutations that cause any degree of HPRT deficiency resulting in resistance to azaguanine will not reduce the viability of the mutant cells. In contrast, HPRT deficiencies resulting from deletions or losses of the X chromosome are likely to occur in inviable cells that will not be detected as AG^r mutant clones. Finally, fluctuation tests provide evidence that most, if not all, in vitro AG^r variants result from changes in cell heredity before analog selection begins [8,9].

The detailed natures of the changes causing HPRT deficiency in AG^r fibroblasts are still unknown in most cases. However, all of the many proliferating AG variants we have isolated have stable hereditary changes and we will refer to them as mutants, using a broad meaning of the word [29]. The rationale for using AG resistance for detecting mutations in normal human diploid fibroblasts has been discussed in more detail elsewhere [6,7].

This paper describes the methods we use (a) to start and to propagate strains of normal human fibroblasts, (b) to detect and quantitate the in vitro induction of human diploid AG^r fibroblasts with *N*-methyl-*N'*-nitro-*N*-nitrosoguanidine (MNNG) and 2-nitrosofluorene (2-NOF), and (c) to isolate, propagate, and characterize the AG^r mutants that are observed after mutagenesis. In addition, we briefly discuss the properties of the AG^r mutants that we have isolated. Data from experiments involved in the development of the mutagenesis methods will be published elsewhere. These are exemplary methods and we emphasize that each mutagen and each fibroblast strain used may require some modification of these procedures.

Starting strains of fibroblasts

Details of the procedures that we use to start suitable human diploid fibroblast strains are included here, since our experience contradicts the claims that low cloning efficiency and limited propagability restrict their use for mutagenesis detection. Most of these procedures have been reported previously by us and others.

Medium

We find medium F10 [17] to be superior to Eagles's medium [11] for culturing human diploid fibroblasts. Commercially available * hypoxanthine-free powdered F10 is suitable for general cell culture and for selection with purine analogs. Powdered F10 is dissolved in double distilled water, sterilized with a Millipore filter apparatus, and stored at 4° C.

For routine cell growth, 85 volumes of F10 containing 100 units/ml penicillin and 100 μg/ml streptomycin are combined with 15 volumes of fetal calf serum to constitute FCS-F10. The antibiotics are used in all procedures described here because the risk of contamination is increased in mutagenesis experiments that require repeated handling of hundreds of cultures over long periods of time. The antibiotics are stored frozen as a mixed 100-fold concentrated stock solution that is added to bottles of filtered F10 when they are first used. The concentrations of penicillin and streptomycin that we use do not reduce the growth rate or the cloning efficiency of the fibroblasts and do not mask the presence of mycoplasma. Fungal contaminations are controlled with Fungizone (Squibb) at 1 μg/ml, but it is not routinely included in the medium.

Serum and F10 are freshly combined just prior to use because the growth promoting ability of the FCS-F10 is diminished when it is stored pre-mixed at 4° C. It is possible that the decline in growth promoting ability can be avoided by freezing the premixed medium, but the large volumes required for mutagenesis experiments make this impractical.

Trypsin

We use 0.25% trypsin for dispersing fibroblasts. The solutions are obtained frozen and are stored at −20° C until used. Thawed solutions that are stored at 4° C retain adequate tryptic activity for at least four weeks provided that aliquots are removed and warmed for use rather than repeatedly warming the entire solution.

---

* The sources of supplies and equipment are given in an Appendix.

Incubators

Optimum cell growth is achieved when cell cultures are kept in a 37° C incubator with a humidified atmosphere containing about 5% CO_2 and 95% air. The pH of FCS-F10 after equilibration in the incubator should not exceed 7.4. Gauges on incubators are often unreliable, so temperature and pH should be determined with a suitable thermometer and pH-meter. The culture-holding capacities of some models of incubators can be more than tripled by putting in additional shelf brackets and trays. Cell growth is not reduced in incubators filled to capacity.

Skin biopsies

Almost all of our work is performed with fibroblasts derived from the inner surface of the forearm or from foreskins (see the following paragraph), since the in vitro doubling potential of skin fibroblasts varies with the part of the body from which the biopsy is taken [24]. The forearm skin is cleansed with 70% ethanol, allowed to dry and then biopsied by elevating a papilla of skin with a sterile 27-gauge needle and removing the biopsy with sterile iris scissors or with a scalpel. Biopsies are at least 2—3 mm in diameter and deep enough to cause slight bleeding. Samples are immersed immediately in FCS-F10. It is preferable to cultivate biopsies as soon as possible, but they may be refrigerated at 4° C for several days before the preparation of primary cultures, if necessary. We have also derived cultures from skin fragments that we stored for prolonged periods at liquid nitrogen temperature.

Foreskins obtained from routine hospital circumcisions generally produce very good fibroblast strains with greater in vitro doubling potential than smaller biopsies taken from older donors. For this reason, we use foreskin-derived cultures for most mutagenesis experiments. Each foreskin is immersed immediately in FCS-F10. Foreskins may also be stored for several days at 4° C before primary cultures are prepared.

Preparation of primary cultures

Primary cultures are prepared by transferring the tissue with sterile forceps to a 35 mm tissue culture dish to which a few drops of FCS-F10 have been added. Using a *sharp* sterile iris scissors, the tissue is divided into several pieces, usually one or two for small skin biopsies and five or six for foreskins. Each of these fragments is then transferred to a separate 35 mm glass culture dish with a few drops of medium where it is minced as finely as possible with the scissors. The fine pieces are immobilized under a sterile 25 mm round glass coverslip that is pressed down firmly. The volume of medium is brought to 2 ml, making sure that the coverslip remains immobilized after the medium is added, and the dishes are incubated undisturbed for one week.

The medium is replaced at the end of the first week and changed twice the second and subsequent weeks. Biopsies vary, but fibroblastic outgrowth is usually apparent in 7—10 days.

Primary subculture

When the fibroblastic outgrowths have formed a monolayer that covers most of the dish bottom and that contains 10^5 to 5×10^5 cells/dish (within 2—4 weeks), they are subcultured. The medium is removed and the coverslip is placed in a second culture dish because cells grow out from the tissue fragments onto the coverslip surface as well as onto the culture dish bottom. Both the original dish and the coverslip are rinsed with 1 ml of trypsin and incubated in 0.5 ml of trypsin until the cells begin to loosen (about 5—10 min). An aliquot of 1.5 ml of FCS-F10 is then added and the cells are pipetted loose from both sides of the coverslip and from the dish bottom using a Pasteur pipette. Fragments of tissue and clumps of cells are usually present at this stage, but they disappear after the second subculture. For each original 35 mm dish, the total volume of suspension is brought to 10 ml with FCS-F10 and is distributed into one 60 mm dish or three 35 mm dishes. After the cells have attached (4—6 h), the medium is replaced with fresh FCS-F10 (4 ml/60 mm dish and 1.5 ml/35 mm dish). This removes debris and reduces the residual trypsin concentration. The FCS-F10 of these cultures is replaced every second day, and the second subculture usually is made within one week.

Propagation of fibroblasts

Routine subculture

The medium is aspirated from freshly confluent cultures of $1—2 \times 10^6$ cells per 60 mm dish and each monolayer is rinsed with 1 ml of trypsin. The cultures are incubated at 37°C with 0.5 ml of trypsin until a quick shaking motion causes many of the cells to detach from the dish (3—10 min). It is *very* important that trypsinization be thorough enough to allow the cells to be dispersed readily into a suspension of single cells by gentle pipetting; it is equally important to avoid damaging the cells by overlong trypsinization or excessive pipetting. The length of time for appropriate trypsinization varies with the age and quality of the trypsin, among cell strains and with population density. Therefore, it is necessary to check cultures during trypsinization to determine when they are ready for pipetting. If all of the cells come off the dish in clumps without shaking or pipetting, the culture may have been trypsinized too long. Conversely, if some cells remain firmly attached to the dish after pipetting, the culture may be inadequately trypsinized.

When the cells are detached, 5 ml of FCS-F10 are pipetted into the dish and the majority of loosely attached cells can be seen coming off the surface under the force of the liquid. Firmly pipetting 10—15 times with a 5 ml pipette gives a suspension of single cells. If trypsinization has been appropriate, most cells still appear elongated. Overlong trypsinization results in a spherical appearance. The suspension is then diluted to 25 ml with FCS-F10 and is distributed in aliquots of 4 ml/60 mm dish or 1.5 ml/35 mm dish. The dishes are immediately put into the incubator. It is unnecessary to remove trypsin by centrifugation or other means provided it is diluted at least 10-fold in culture medium. Whenever subcultures are made from several dishes of one strain, all cultures are pooled before plating.

Mitotic activity is apparent 24 h after subculture. Thereafter, mass populations of a good strain can be expected to double at least once every 24 h in FCS-F10 until they approach confluence. Certain kinds of cell strains (e.g. from patients with Bloom's syndrome or xeroderma pigmentosum) usually perform less well. Excessive trypsinization may result in cells that are slow to attach and spread out on the dish, or cells that are slow to resume division. Fewer than 1—5% of all cells remain unattached 24 h after subculture when the procedures described here are used.

Convenient and apparently optimum culture conditions are maintained for most strains by subculturing and diluting the populations 1 : 6 once per week, with two medium changes equally spaced between subcultures. Usually the cells are confluent, but not densely packed, after one week. Dense cultures are difficult to trypsinize into suspensions of single cells and the resumption of cell division is delayed. It is interesting to note that human fibroblasts are often difficult to trypsinize free from the dish during the first 24—48 h after a previous trypsinization.

Cloning fibroblasts

$$\text{Cloning efficiency} = \frac{\text{Clones formed}}{\text{Cells plated}} \times 100 \, .$$

Cloning efficiency is independent of the number of cells in the inoculum under the conditions for cell growth described above. Freshly confluent cultures are trypsinized and diluted in FCS-F10 to about $0.5\text{—}1.0 \times 10^5$ cells/ml. The cells are counted with a hemocytometer, which permits observation of the condition of the cells and of the suspension. A good suspension has fewer than 5% of its cells in clumps of two or more cells.

For most strains, final cloning suspensions are adjusted to give an average of 100 cells/60 mm dish, resulting in about 20—50 clones/dish. Larger inocula result in underestimated cloning efficiencies because contiguous clones can appear to be a single colony. When the cloning efficiency is reduced by lethal treatments, the inoculum may be increased to 10^3 cells/dish and more dishes may be used to give enough clones for statistical accuracy. Plates inoculated with more than 10^3 cells can be hard to score because small diffuse clones are difficult to distinguish clearly from the background of inviable, but attached, cells.

To ensure the uniform distribution of the cells during the plating process, the cloning suspension is stirred by an automatic magnetic stirrer at low speed or swirled frequently by hand. Four ml volumes are pipetted into 60 mm dishes and, if necessary, the dishes are tilted gently to distribute the medium over the bottom of the dish. These cultures are immediately placed in the incubator and are left undisturbed for one week. Swirling freshly inoculated dishes or allowing them to sit at room temperature before placing them in the incubator can cause the cells to concentrate near the center of the dish.

Seven and ten days after plating, the medium is changed. Frequent medium changes are avoided because manipulation of the cultures increases the probability that loosely attached dividing cells will detach from developing clones and reattach elsewhere in the dish. Such cells may give rise to satellite colonies that erroneously increase the observed cloning efficiency. Few satellite colonies occur if the fluid is pipetted against the side of the dish when the medium is changed rather than onto the developing clones.

Colonies are stained and counted when they range from 1—7 mm in diameter, usually 12—14 days after inoculation. Smaller clones are difficult to count because they are diffuse and lightly stained, while larger clones may grow together to form a single colony. Both conditions cause underestimation of the cloning efficiency.

For staining, the dishes are emptied, rinsed once with 5 ml of 0.9% saline to remove serum residue, and fixed for 10 min in 95% ethanol. The ethanol is then removed and the dishes are stained for 3 min with 0.2% (w/v) methylene blue in water, rinsed in distilled water and air dried. Magnified images of the colonies are projected onto white paper using an overhead projector. Individual cells can be seen this way and only clones of at least 50 cells are counted. However, the majority of clones counted for determining the cloning efficiencies of 20—50% referred to above contain thousands of cells. We believe that strains with cloning efficiencies of less than 20%, using the large clone standard, are poor choices for routine mutagenicity work.

Storing fibroblasts in liquid nitrogen

A large supply of viable cells from a single biopsy can be stored in liquid nitrogen for many years. Work can be carried out over long periods of time using the same cell population, because aliquots of cells with maximum doubling potential can be thawed for each experiment. As a very conservative example, a foreskin may yield six 60 mm dishes at its first subculture and 36 at its second, or about 7.2×10^7 cells when all 36 are confluent, enough for 72 ampules of 10^6 cells. For freezing, freshly confluent cultures containing 1—2×10^6 cells/60 mm dish are rinsed with 1 ml of trypsin and then incubated at 37°C with 0.5 ml of trypsin until the cells are just ready to come off the dish bottom. The trypsin is gently removed from the monolayer and the cells are resuspended in freezing medium (72 volumes F10, 18 volumes FCS, and 10 volumes glycerin that has been sterilized by autoclaving) to give an approximate concentration of 10^6 cells/ml. Although at least 0.5 ml of freezing medium is added immediately to each dish to prevent drying, 3—5 ml volumes can be transferred from dish to dish for easy pipetting. *All dishes are pooled* and thoroughly mixed so that aliquots are as nearly identical as possible. This is essential for reproducible mutagenesis experiments. One ml aliquots of the pooled suspension are then distributed to freezing ampules with a sterile Pasteur pipette. The ampules are sealed, frozen by lowering the temperature 1°C/min to —60°C and then stored in liquid nitrogen. No more than 1 h should elapse from the time the first cells are suspended until the freezing process is started. Cells that are kept longer at room temperature in the freezing medium or that are frozen at a different rate usually have reduced viability when they are thawed.

Often, strains are frozen using the first subculture. If further work confirms that a strain is a good one for mutagenesis experiments, ampules are thawed, grown through two subcultures, pooled and refrozen as the working stock for that strain.

Thawing cells from liquid nitrogen

Ampules are thawed rapidly by removing them from the liquid nitrogen and submerging them in 37° C water. Persons thawing cells should do so in a secluded area and should wear protective face masks and gloves because improperly sealed ampules may explode upon warming.

When thawed, the ampule is sterilized by dipping it in acetone for several minutes. It is then air dried and the top is snapped off. The cells are suspended with a Pasteur pipette and diluted with at least 10 ml of FCS-F10 to reduce the glycerin concentration. Centrifuging the cells to remove the glycerin-containing medium is not necessary. This suspension is inoculated into two 60 mm dishes.

The FCS-F10 is replaced the following day and every second day until the cells are confluent and ready for subculture, usually within 3—7 days. Some debris from non-viable cells may be present in the freshly thawed cultures, but it is eliminated by the first subculture. Of the thawed cells 30—80% should attach to the dish because recovery of fewer cells means that a small and possibly unrepresentative fraction of the original population gives rise to the new culture.

Mutagenic treatment

Preparation of the mutagen

It is important to consider carefully the choice of solvent, treatment medium, dose range, length of exposure, and safety precautions for each mutagen that is used. Control over the *effective* applied dose must be achieved if results are to be compared from experiment to experiment and from laboratory to laboratory. Even if stable preparation and application of the mutagen cannot be achieved, careful definition of the procedures used will reduce variation that is caused by inadvertent procedural differences.

Solvent

The solvent used to dissolve the mutagen should not alter it and should have minimal effect on the cloning ability of the fibroblasts. Solvent concentrations that do not reduce cloning efficiencies are the most desirable, but some reduction may be unavoidable.

The stability of the compound in the solvent is determined by comparing the survivals of cells treated immediately after dissolving the compound with cell survivals after various intervals between the preparation of the solution and the treatment of the cells. Mutagenesis procedures are then carried out within time periods that minimize the loss of mutagen activity. Failure to consider this point results in the application of highly variable *effective* doses of the mutagen.

Specifically, we have found that MNNG dissolves in 10—20 min in double distilled water and is nearly stable for at least an hour at room temperature. We are presently quantitatively evaluating its long-term stability as a frozen concentrated stock solution.

We have found that 2-NOF dissolves in 15—20 min in absolute ethanol and is stable for at least an hour at room temperature. Although we originally dissolved 2-NOF in dimethylsulfoxide (DMSO), DMSO from different sources varies extensively in its effects on the stability of 2-NOF and we no longer use it.

Treatment medium

The mutagens are applied to the cells in serum-free F10 because serum reacts with some mutagens and reduces the amount interacting with the cells. MNNG is more stable in F10 buffered with 0.02 M HEPES to pH 6.8 than in F10 alone. This concentration of HEPES has a negligible effect on cell cloning efficiency when it is used in a 4-h treatment.

Mutagen dose range

The range of mutagen concentrations that gives from 100% to 0.1% cell survival after the mutagen treatment is the most useful for mutagenesis experiments. Concentrations of $0.5\text{—}10 \times 10^{-6}$ M 2-NOF and MNNG (our results) and *N*-acetoxy-2-acetylaminofluorene (Maher and Wessel [23]) result in the desired range of cell survivals when these compounds are applied to the cells in F10 for four hours.

Length of mutagen treatment

Exposure of the cells to the mutagen may be much shorter than the interval of treatment if the compound has a short half-life in the treatment medium. We use a 4-h treatment for MNNG and 2-NOF because it is convenient. In a series of experiments, the consistent use of the same treatment period is important if reproducibility is to be achieved.

Safety precautions

The most convenient way to minimize chemical risk is to prepare and to apply the mutagen using plastic disposable pipettes, test tubes and flasks in an area protected by plastic disposable underpads. Persons handling the mutagens wear disposable plastic gloves, and face masks when appropriate. No mouth pipetting is done. All mutagen-containing fluids and all paper and plastic wastes are disposed of in the same way as radioactive materials. The mutagen-containing medium is removed from cultures with an aspirating system that is used only for that purpose. In the case of long-lived or volatile compounds, aspirating is done by hand to prevent harmful aerosols from entering the air.

Other factors influencing mutagen toxicity

In addition to the factors already discussed, the observed cell survival after mutagen treatment also depends on the cell population density, the possible detachment of cells during treatment and whether or not wild-type and mutant cells are equally sensitive to the mutagen.

Cell density at the time of treatment

The survival of cells as a function of 2-NOF or MNNG dose is the same for cells that are treated and allowed to proliferate in situ as for cells that are treated and then subcultured immediately after treatment. In contrast, the cell population density at the time of the mutagen treatment markedly affects cell survival. Population densities in the range of 10^2 to 10^5 cells/60 mm dish give identical cell survivals. However, at population densities of more than 4×10^5 cells/60 mm dish, the toxic effects of the mutagen are greatly reduced.

Mutagenesis experiments can be done by treating the cells at 10^4/60 mm dish, the density used for AG-selection, with no replating after treatment. However, this requires the treatment of large numbers of dishes and is an inefficient use of materials and labor. We prefer to treat the cells at 10^5/dish, and then to replate them at 10^4/dish just prior to selection (see *Cell population density for selection*, p. 205, and *Expression time—phenotypic lag*, p. 205). The effects of population density should be investigated for each substance being tested.

Cell detachment during exposure

Determinations of whether cells detach and are lost during any of the medium changes involved in the mutagen treatment are made, since the cell survival is finally expressed in relation to the total number of cells treated. If many cells detach, it is possible that the surviving fraction and the incidence of mutants differ in the attached and detached cell populations. A thorough study of a substance may then require that separate determinations of the survival and of the mutant incidence be made in the two subpopulations. Detachment of cells is prominent during treatments of human fibroblasts with hycanthone at 50—100 μg/ml [8], for instance, but is insignificant when MNNG and 2-NOF are used at concentrations of $0.5\text{—}10 \times 10^{-6}$ M. The few cells that do detach at high concentrations are not viable.

Mutant cell sensitivity to the mutagen

Some chemicals being tested for mutagenic activity may select for pre-existing mutants rather than, or in addition to, inducing new ones. This can cause the overestimation of the mutant fraction in treated cell populations, and can result in false or inflated mutagenic effects. A correction for this source of error is made by comparing the survivals of cells from mutant clones and from normal clones derived from the same cell strain after chemical treatments similar to those used in mutagenesis experiments. Different AG^r clones and different normal clones from the same strain have slightly different sensitivities to 2-NOF and to MNNG. However, AG^r mutants do not demonstrate a *pattern* of sensitivity to these mutagens that differs from the pattern of untreated control clones. If mutants are found that are significantly more resistant to the mutagen than are controls, a rigorous quantitation of the resistance will be required if *induced* mutagenic effects are to be estimated from the mutant frequencies observed.

Selection for AG-resistant mutants

AG-selective medium (CS-F10-AG) contains 85 volumes F10, 15 volumes *calf* serum and 1.5×10^{-5} M 8-azaguanine. Calf serum contains less hypoxanthine than fetal calf serum and permits effective selection against wild-type human diploid fibroblasts at low concentrations of AG without serum dialysis. AG is dissolved in a minimal amount of 1 M NaOH, diluted in double distilled water and stored at room temperature as a sterile 100-fold stock solution.

Although we have also observed that AG loses its toxic effects when incubated in the presence of calf serum, as reported by Van Zeeland and Simons [33], none of the quantitative aspects of the work on AG resistance reported from our laboratory have been affected by this potential source of error; CS-F10-AG, as defined above, completely inhibits the growth of normal cells when it is freshly prepared and replaced twice per week. Mitotic activity in the background wild-type population is not detected by microscopic examination after the first 24—48 h in CS-F10-AG. The number of cells does not increase detectably during selection.

Alternatively, the purine analog 6-thioguanine (TG) can be used to select for HPRT-deficient cells. A concentration of 3×10^{-5} M TG in FCS-F10 is as effective in inhibiting wild-type cell growth as the CS-F10-AG just described. TG is advantageous for some purposes, because it does not lose its toxic effect when incubated with serum and it is effective in the presence of fetal calf serum, which is better than calf serum for the growth of human cells.

TG has two disadvantages. (a) Some AG^r mutants have only partial deficiencies of HPRT and some variants among this group are partially sensitive to TG at the concentrations needed for efficient selection. While some partially TG-sensitive mutants of this sort might be undetected after selection with TG, we emphasize that they are a small fraction, perhaps less than 5%, of all mutants so far isolated by selection with AG. The incidences of spontaneous and induced resistant clones are not consistently or significantly different when selection is imposed with AG and TG in paired cell populations. Therefore, TG could be used in place of AG when only the number of resistant mutants is being determined, but it may be unsuitable for the isolation of some mutants with qualitatively altered HPRT. (b) Wild-type cell proliferation is completely inhibited in FCS-F10-TG at 3×10^{-5} M, but few of the cells come off the dish during selection. In contrast, during AG-selection a significant number of wild type cells detach from the dish, decreasing the probability of contaminating the clone with normal cells when it is isolated (see *Characterization of AG^r mutants*, p. 212). Also, mutants cells form more compact and easily identified clones in CS-F10-AG than they do in FCS-F10-TG.

The recovery of AG^r mutants is dependent on the AG concentration used for selection, on the cell population density during selection and on the length of time allowed for induced mutations to become phenotypically expressed.

AG concentration

Different cell strains and different subclones of the same strain express slightly different sensitivities to azaguanine. If the azaguanine concentration used for selection marginally inhibits normal cell proliferation, the most resistant wild-type cells will slowly develop into diffuse "clones". After staining, these "clones" will be counted as mutants when, in fact, their proliferation was within the range of normal cell response to the concentration of azaguanine used.

AG at 1.5×10^{-5} M is 3—5 times the concentration required to completely inhibit our normal strains, if culture medium is replaced twice per week with fresh CS-F10-AG. We have isolated over 500 AG^r clones that continued to proliferate in CS-F10-AG after isolation (see also *The azaguanine resistance phenotype is stable*, p. 216). None of these mutant colonies showed significant or consistent differences in replating efficiencies in CS-F10-AG as compared to CS-F10.

Cell population density for selection

AG^r cells that touch wild-type cells become sensitive to the AG through contact feeding, making it necessary to select for the AG^r mutants in cell populations where cell-cell contacts are minimal [3]. The correct population densities are determined by reconstruction experiments in which selection is applied to populations containing known small numbers of mutant cells and various excess numbers of normal cells. We have found that 10^4 normal cells/ 60 mm dish at the start of selection is the largest number that permits as many AG^r colonies to form as when no normal cells are present. Reconstruction experiments are performed for each mutagen treatment in each experiment by adding known small numbers of mutant cells to a sample of the population of treated normal cells that is plated for selection.

A by-product of our reconstruction experiments is the conclusion that under our experimental conditions a human fibroblast has the same chance of forming a clone in complete isolation as it does in a background of many non-proliferating cells, i.e. no feeder-layer effect is discernible. This means that the cloning efficiencies determined by placing 100 cells in a dish are applicable to the dishes containing 10^4 or more cells, an extrapolation that is necessary for the calculation of mutation rates.

Expression time — phenotypic lag

Segregation of the mutant allele from the normal allele by cell division (segregation lag) and loss of the normal HPRT due to degradation or dilution by cell division (phenotypic lag) are required before a new AG^r mutant can form a clone in CS-F10-AG. Although much of our mutagenesis work has been carried out with 72 h between the treatment with mutagen and the start of AG-selection, recent experiments show that nearly twice as many mutants are recovered after 96—120 h as are recovered after 72 h. Longer times do not increase mutant recovery. The length of time appropriate for *maximum* mutant recovery may need to be determined for each mutagen and each cell strain.

Protocol for quantitating mutagenic effects

The protocol described here overcomes the problems of segregation lag, phenotypic lag and contact feeding simply by non-selectively propagating the control and treated cell populations for enough cell divisions after treatment to ensure the phenotypic expression of all induced mutants. Cells are then dispersed into sparse subpopulations to ensure the maximum recovery of the AG^r mutants after selection. This can be done because genotypically AG^r mutant cells and normal cells proliferate at nearly the same rate in FCS-F10, which keeps the proportion of genotypically mutant cells in the cell population after it has proliferated non-selectively following treatment nearly the same as the proportion of genotypically mutant cells in the population immediately after the induction of the the mutations by the chemical treatment.

However, the proportion of mutants in a proliferating population will not remain constant if there is a significant spontaneous mutation rate. For both treated and control populations, the increase in the proportion of mutants during non-selective proliferation due to spontaneous mutations is $\frac{f\mu}{2 \ln 2}$, where f is the number of population doublings and μ is the spontaneous mutation rate (see ref. 26). The spontaneous mutation rate for AG resistance is of the order of 10^{-6} mutants/cell generation [8,9]. For both the treated and the control populations, the number of viable cells will increase about 10—30 fold, or 3.3—4.9 cell doublings, during the period of non-selective proliferation. For $f = 3.3$ and 4.9 $\mu = 10^{-6}$, $\frac{f\mu}{2 \ln 2} = 2.4 \times 10^{-6}$ to 3.6×10^{-6}.

This increase can be compared to the proportion of pre-existing AG^r mutant cells in normal fibroblast strains, which is of the order of 10^{-5} mutant cells per clone-forming cell tested [9]. Following MNNG treatments that result in cell survivals of 5—80%, of the order of 10—100 mutant cells are detected as AG^r clones in 10^5 progeny tested. However, only about one of these clones would be expected to be the progeny of a new spontaneous or a pre-existing mutant. In any case, the incidence of mutants in the treated population is corrected for new spontaneous and pre-existing mutants by subtracting the control incidence of mutants. A further indication of the validity of the approach is the fact that experiments in which the cells were treated and immediately replated have given induced mutation rate estimates similar to experiments in which the cells were treated at the same dose and allowed to proliferate prior to replating.

This method is not suitable for the isolation of assuredly independent mutant clones for further characterization, because each mutant cell that is initially present produces several mutant colonies. Modifications of the protocol that are suitable for the isolation of independent mutants are described on p. 211.

A. Preparation of cultures

Steps in the following protocol are diagrammed in Fig. 1. Throughout the protocol all of the dishes mentioned are 60 mm plastic tissue culture plates. The medium is FCS-F10 unless otherwise specified.

(1) Thaw one ampule of AG-sensitive (AG^s) cells and one ampule of AG-resistant (AG^r) cells, each into two 60 mm dishes, approximately two weeks before treatment (see *Thawing cells from liquid nitrogen*, p. 201).

(2) When the dishes are confluent (3—7 days), subculture the AG^s cells into a total of 12 dishes. Subculture the AG^r cells 1 : 6, keeping only 2—3

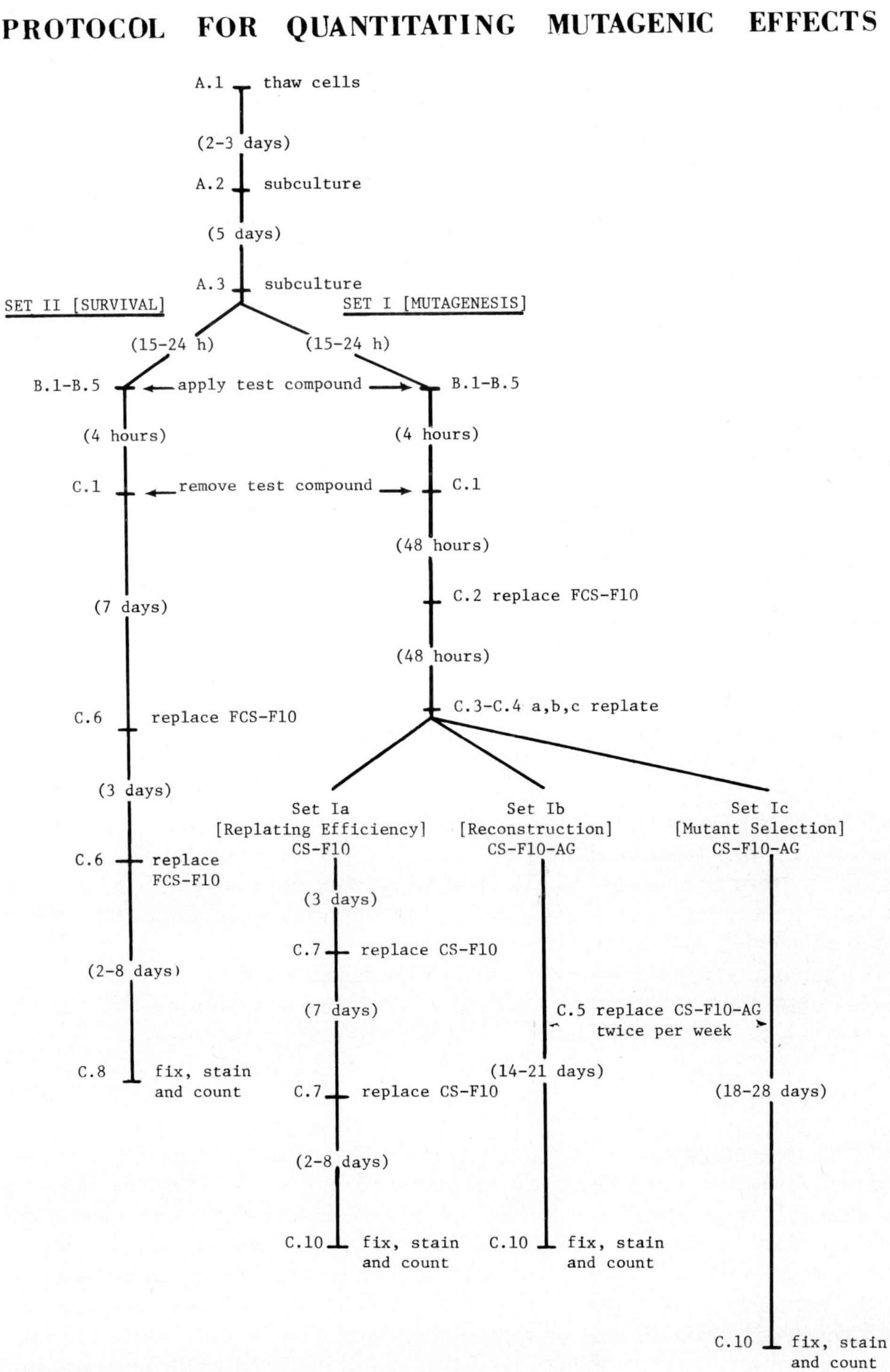

Fig. 1. Protocol for quantitating mutagenic effects in human diploid fibroblasts. The lettered steps refer to the protocol in the text.

dishes. Replace the FCS-F10 every two days.

(3) Five days after subculture, there should be 1.0—1.5 $\times 10^6$ cells/dish resulting from divisions after thawing. At this time, use these cultures to inoculate a set of subcultures (I) with 10^5 cells/dish (usually about 12 dishes per treatment dose), which will be used to quantitate mutagenesis. A second set (II) is inoculated with a number of cells calculated to produce 30—40 clones/dish if the cells are untreated (see *Cloning fibroblasts*, p. 199). Set II dishes will be used to determine the survival of cells after treatment with the mutagen. Enough Set II dishes should be inoculated to permit the counting of at least 50—100 surviving clone-formers for each treatment. The determination of cell survival in this manner proves to be very convenient. We have ascertained that the survival of cells is the same in Set I and Set II dishes that are subjected to the same treatment.

B. Treatment of cells with a mutagen

The treatment procedure works most efficiently if two people work together.

(1) At 15—24 h after plating, set up the work area with disposable pipettes and test tubes, plastic gloves, plastic underpads for the counter, containers for solid and liquid mutagen wastes, Biopette (Schwarz/Mann) and sterile Biotips, and necessary media and solvents.

(2) One person prepares the mutagen dilutions to give the correct final concentrations when 0.05 ml is added to the final 5 ml volume (see *Preparation of the mutagen*, p. 201). Simultaneously, the second person removes the FCS-F10 from the dishes to be treated and replaces it with 5 ml of F10 or F10 buffered to pH 6.8 with 0.02 M HEPES.

(3) Using a Biopette, one person adds 0.05 ml of the mutagen solution to each of the mutagenesis (Set I) or survival (Set II) dishes. The use of a 0.2 ml Biopette permits the distribution of 3—4 drops of the mutagen solution in a circular pattern around the dish. Immediately after the mutagen is added, the second person gently swirls the dish to ensure complete mixing. The control plates are treated with solvent only.

(4) Incubate the cultures at 37°C in a CO_2 incubator for 4 h.

(5) Count the cells in four untreated Set I dishes to determine the number present at the time of treatment.

C. Post-treatment

(1) At the end of the 4-h treatment, remove the mutagen-containing medium from all dishes as completely as possible and add fresh FCS-F10. Rinsing the dishes is unnecessary for MNNG or 2-NOF because both compounds are highly unstable in the presence of the fetal calf serum that is added. The residual activity of more stable compounds may require that the dishes be rinsed.

(2) Forty-eight hours after the mutagen treatment, replace the medium of both the treated and control mutagenesis cultures (Set I) with fresh FCS-F10. The medium on the survival (Set II) dishes is not replaced until one week following treatment.

(3) (Steps 3 and 4 are repeated for each dose.) One hundred twenty hours after

the treatment, trypsinize and pool all Set I cultures in CS-F10. Count and dilute the cells in CS-F10 to a total volume of 500 ml, 2.5×10^3 cells/ml (suspension A); not all Set I cells are replated. Calculate and record the number of cells per dish at the time of replating.

(4) *(a) Replating efficiency of Set I cells.* Using CS-F10, make a 1 : 100 dilution of suspension A and plate 10 dishes at 10^2/dish (Set Ia) to determine the replating efficiency for each dose.

(b) Reconstruction. Combine a 50 ml aliquot of suspension A, AG to a final concentration of 1.5×10^{-5} M, and 625 AG^r cells. Using 4 ml aliquots, plate 10 dishes for each treatment (Set Ib) for determination of the recovery of AG^r cells during selection. Plate 10 dishes with 50 AG^r cells alone (see *Cell population density for selection*, p. 205).

(c) Selection for AG^r mutants. Add AG to a final concentration of 1.5×10^{-5} M to suspension A and plate 100 dishes at 10^4/dish (Set Ic), 4 ml/dish.

(5) Replace the CS-F10-AG in the reconstruction (Set Ib) and mutagenesis (Set Ic) dishes twice per week for 2—3 weeks.

(6) Seven and ten days after treatment, replace the medium on survival dishes (Set II) with FCS-F10.

(7) Seven and ten days after replating (step C.4a), replace the medium on replating efficiency dishes (Set Ia) with CS-F10. If cloning efficiencies are the same in FCS-F10 and CS-F10, replating efficiency determinations may be made earlier if the medium is replaced with FCS-F10 at this time.

(8) Twelve to eighteen days after treatment, fix, stain and count the Set II survival dishes (see *Cloning fibroblasts*, p. 199).

(9) Twelve to sixteen days after the cultures are replated (step C4), start scanning for clones. Living clones for propagation and study are located macroscopically by holding the dishes 2—5 cm above a black background with obliquely incident light provided by a fluorescent lamp (see Appendix). Clones consisting of as few as 10^3 cells appear as whitish colonies against a clear, black background.

(10) When the clones are 3—5 mm in diameter, fix, stain and count the replating efficiency (Set Ia), reconstruction (Set Ib) and mutagenesis (Set Ic) plates. The optimum clone size for counting may be reached on different days for the Set Ia and Set Ib cultures; it is not necessary to fix the two sets on the same day. The Set Ic cultures may be fixed and stained later than the Set Ia for Set Ib cultures in order to permit the mutant colonies to be more easily identified.

Calculation of mutation frequencies (see Table I)

Computation of the number of AG^r colonies

Satellite colonies can erroneously inflate the number of AG^r colonies. The true number of primary AG^r colonies can be estimated by assuming that they are randomly distributed and by using the number of Set Ic dishes lacking AG^r colonies to solve for m in the P_0 term of the Poisson series: $P_0 = e^{-m}$, where m is the average number of primary AG^r colonies per dish. The total number of primary colonies is then estimated by (n)(m) where n is the total number of dishes.

TABLE I

COMPUTATIONS OF INDUCED MUTATION FREQUENCIES FOR ONE MNNG EXPERIMENT

MNNG dose	Control	0.5×10^{-6} M	3.0×10^{-6} M
Survival fraction (Set II dishes) [a]	1.00	1.00	0.05
Dishes scored = n (Set Ic dishes)	216	215	144
Dishes with 0 clones	210(210) [c]	203(203)	48(48)
Dishes with 1 clone	6(5.9)	12(11.6)	43(52.8)
Dishes with 2 clones	0.08	0(0.33)	30(29.0)
Dishes with 3 clones	0	0	16(10.6)
Dishes with 4 clones	0	0	3(2.9)
Dishes with 4+ clones	0	0	4(0.76)
Total clones observed	6	12	183
Mutants per dish [b] = m	0.0281	0.0574	1.10
Total clones computed (n)(m) = C	6.1	12.34	158.4
Recovery of AG^r colonies = R (Set Ib dishes)	1.00	1.00	1.00
Total cells = N (Set Ic dishes)	2.16×10^6	2.15×10^6	1.44×10^6
Replating efficiency = RE (Set Ia dishes)	0.36	0.40	0.08
Total viable cells scored = (RE)(N)	7.69×10^5	8.58×10^5	1.15×10^5
Mutants per viable cell $\frac{(n)(m)/R}{(RE)(N)} = M$	7.93×10^{-6}	1.40×10^{-5}	1.38×10^{-3}
Induced mutants per viable cell [d] $M_I = M_T - M_C$	—	5.60×10^{-6} [e]	1.37×10^{-3} [f]

[a] Sets of dishes refer to the *Protocol for Quantitating Mutagenic Effects* (p. 205).
[b] The average number of mutants per dish (m) calculated from the 0 term of the Poisson series, $P_0 = e^{-m}$, using the observed number of dishes with no clones.
[c] Figures in parenthesis indicate the Poisson expectations.
[d] The treated frequency minus the control frequency.
[e] Not significantly different from the control frequency by the chi-square method using a 2 × 2 table [4] or the Kastenbaum and Bowman method [19].
[f] Highly significant increase over the control frequency, $P << 0.001$.

Correction for incomplete recovery of AG^r colonies because of contact feeding

The Set Ib dishes provide a direct estimate for each treatment dose of the fraction, R, of all AG^r cells present at the beginning of selection that are able to form a countable colony in spite of contact feeding.

$$R = \frac{AG^r \text{ colonies in the presence of } AG^s \text{ cells}}{AG^r \text{ colonies in the absence of } AG^s \text{ cells}}$$

The final corrected number of primary AG^r colonies, $C = (n)(m)/R$, is estimated for each treatment dose.

Correction for replating efficiency

The mutant incidence per *viable* cell is given by $M = \frac{(n)(m)/R}{(RE)\ (N)}$, where N is the number of cells subjected to selection in the Set Ic dishes and RE is the replating efficiency of the population at the time of replating (step C.4a).

Computation of the induced incidence of mutants

The induced mutant incidence is given by $M_I = M_T - M_C$, where M_T is the mutant incidence in the treated population and M_C is the mutant incidence in the untreated population. The significance of differences between M_T and M_C values is calculated using appropriate statistical methods. The chi-square test, using a 2 × 2 table [4], or the method of Kastenbaum and Bowman [19] is suitable.

Isolation and characterization of independent AG^r mutants

In order to avoid studying, unknowingly, multiple isolates derived from the same AG^r mutational event, procedures that provide independent mutants must be used when AG^r mutants are to be characterized. Truly independent *spontaneous* mutants can be studied by using either just one AG^r clone from each of a set of strains derived from different humans, or by using AG^r colonies that must result from different mutations that have occurred in a single strain of cells in a fluctuation test [9]. The isolation of independent *induced* mutants can be accomplished by simply modifying the protocol for quantitating mutagenic effects (p. 205). Each Set I culture that contains an induced mutant after treatment at step C.1 will contain a small clone derived from the mutant at step C.3. If the cells in each Set I dish are then dispersed and subjected to selection separately from the cells in other Set I dishes, any AG^r clones that appear in different sets of Set Ic subcultures will have resulted from mutational events that were induced in different Set I dishes. The probability that the mutant clones that are isolated from different sets of Ic dishes are descended from the same *pre-existing* mutant depends on the magnitude of the mutagenic effect and the incidence of pre-existing mutants.

Alternatively, independent mutants can be obtained by subculturing Set I cultures immediately after treatment (step C.1). The cells are replated as in step C.4., except that all of the replating is done using FCS-F10 with no AG. Allowing for lags, 96—120 h later the medium on Set Ib and Set Ic cultures is replaced with CS-F10-AG. Set Ic cultures that are expected to have high survivals may be plated at less than 10^4 cells/dish to avoid cell densities at the start of selection that significantly reduce the number of AG^r clones that are recovered.

Contact feeding among cells during the segregation of the genetically mutant from the genetically normal cells at the earliest stages of clone formation is a potential problem here because the cells are not dispersed just prior to selection. This problem is extremely difficult to evaluate, but it may be a minor one in our work because human fibroblasts tend to migrate away from each other after a cell division [5]. We have found that the mutant incidences determined after the dispersal of such small clones are similar to the mutant incidences determined without replating.

Two or more clones that occur in the same dish can be due to the development of satellite colonies from a single primary clone. Since it is impossible to distinguish the dishes containing multiple clones due to more than one independent mutational event from the dishes containing satellites, only one clone is isolated from each dish.

Isolation of AG^r colonies

(1) When the clones are about 5—10 mm in diameter, remove the medium and rinse the dish with 1.0 ml of sterile 0.9% NaCl. Rinsing with trypsin is avoided because it loosens parental cells and the risk of contaminating the clone with wild-type AG^s cells is increased.

(2) A sterile forceps is used to manipulate a sterile glass cylinder (10 mm inner diameter by about 10 mm in length). One edge of the cylinder is coated by pressing it into sterile silicone stopcock lubricant. The cylinder is then positioned around the colony and pressed down firmly to ensure a leakproof seal. Leaks increase the probability of contamination by parental cells as well as the loss of cells from the desired clone.

(3) Add 0.1 ml of trypsin to the ring and incubate the clone at 37°C for 5—10 min. The trypsinization may be slow if the clones are dense.

(4) When the monolayer is loosened, add about 0.5 ml of FCS-F10 and pipette the cells 10—15 times with a Pasteur pipette. Be careful not to dislodge the cloning ring.

(5) Transfer the cells to a total volume of 4 ml and inoculate four 13 mm dishes (1 ml/dish) or two 35 mm dishes (2 ml/dish). Most clones do not give single cell suspensions directly, but after growth to confluence, trypsinization will give suspensions of single cells. The more dense the clone, the more clumps in the suspension.

(6) Replace the medium the following day with FCS-F10-TG or CS-F10-AG and every second day until confluent monolayers are formed.

Characterization of AG^r mutants

When undertaking mutagenesis studies using AG resistance, it is important for individual laboratories to clearly characterize a representative sample of the AG^r colonies that are observed. The mutations that affect HPRT activity in humans fall into two broad categories. Type I mutants have greatly reduced or undetectable HPRT activity and the fibroblasts are unable to use hypoxanthine for growth; males with the Lesch-Nyhan syndrome are prototypes for Type I mutations. Type II mutants have altered HPRT but the cells retain the ability to use hypoxanthine for growth; males with less severe clinical consequences of their HPRT deficiency than patients with the Lesch-Nyhan syndrome are prototypes for Type II mutations. The cells from both Type I and Type II humans are resistant to AG [9]. Both types of mutants also arise de novo in cultures of fibroblasts from humans who have normal HPRT [9].

We are usually interested in making four kinds of determinations with the newly isolated AG^r clones: (a) the cloning efficiency, (b) the stability of the AG^r phenotype after proliferation in the absence of AG, to ascertain if the cells have a true hereditary change, (c) the HPRT activity, and (d) the ability to grow in [AS]H medium, to determine if the cells can utilize hypoxanthine (H) when the de novo pathway of purine nucleotide synthesis is blocked with azaserine (AS), a glutamine analog that inhibits phosphoribosyl-formyl-glycinamidine synthetase [22].

In at least 90% of our isolations, the primary subcultures prepared from an AG^r colony (*Isolation of AG^r colonies*, p. 212) are free enough of the parental AG^s cells to permit immediate and reliable execution of all four determinations. However, we have found that AG^s cells can persist and even multiply slowly in the presence of AG, if they occur in dense populations of AG^r cells. AG^s cells that are picked up when an AG^r clone is isolated may result in an erroneous partial HPRT activity. Persistent parental cells can also confuse tests for the ability of the cell population to grow in [AS]H medium, as well as tests for reversion, which detect the ability of individual cells to use hypoxanthine. The first kind of confusion results in the misclassification of a Type I mutant as a Type II mutant and the second kind of confusion results in the attribution of revertability to mutants that are not truly revertable.

If the mutant population is released from selection, as in testing for the stability of the AG^r phenotype, the contaminating parental cells can become a significant fraction of the mutant culture because parental cells usually have more doubling potential than the cells in a mutant clone. The phenotypic shifts that may be observed in the population are not due to epigenetic changes or the dependence of the AG^r phenotype on the presence of AG; they simply indicate impure clonal isolates.

These errors can be completely avoided if the tests are performed with subclonal populations that are derived from single cells grown in isolation, either with or without AG. A convenient and reliable compromise procedure, especially for the mutants with low cloning efficiencies, is to inoculate 60 mm dishes with 10^3 cells from primary subcultures of AG^r clones and to propagate them to confluence (more than 10^6 cells/60 mm dish). If propagated in the presence of selective agent, the resultant cells can be used to determine HPRT activity or to determine the ability of the population to use hypoxanthine for growth. If propagated in the absence of selective agent, the resultant cells can then be used to test for the persistence of AG resistance. However, this procedure is not suitable for obtaining cells to be used in rigorous tests for rare revertants.

The cultures used to make the determinations of the cloning efficiency, stability of the AG^r phenotype, HPRT activity, and the ability to grow in [AS]H medium can be started simultaneously for each AG^r mutant, if a 15 ml suspension containing 10^4 cells per ml in FCS-F10 (suspension A) is prepared. A normal AG^s strain and a prototype AG^r strain are included as controls in each set of determinations.

HPRT assays

(A) The cells are plated in glass cylinders, with about 9 mm inner diameter, that are mounted on microscope slides with a clear silicone adhesive [30]. The cultures are grown to confluence, fixed and assayed in situ. This method requires about 10^5 cells and allows mutants with limited growth potential to be assayed. Single-use slides with growth chambers are now commercially available and they may be satisfactory substitutes for our home-made, reusable, glass cylinder slides.

(1) For each mutant, distribute 0.5 ml volumes (5×10^3 cells) of suspension A to each of 8 assay wells using a Biopette: two wells for enzyme assays, two wells for protein determinations, and four wells as duplicates for reassay, if necessary.

(2) Replace the medium in the wells with 0.5 ml FCS-F10 every second day until 1—2 days after the cells have reached confluence. Microscopic observation should show all eight of the monolayers to be uniform and of similar cell density. Holes in the monolayers are usually the result of damage from the aspirator needle.

(3) When the monolayers are 1—2 days post-confluence, remove the medium.

(4) Rinse each well three times by filling it with 0.9% saline.

(5) Fix the cells by filling the wells with acetone at room temperature for 1 min.

(6) Remove the acetone and air dry the wells. Replace the lids and freeze the slides at —20°C until the cells are assayed. The HPRT activity in normal cells is stable for weeks under these conditions.

(B) The HPRT assay measures the conversion of radioactive hypoxanthine (H) to inosinic acid (IMP). Since human fibroblasts have the enzyme 5′-nucleotidase, which readily converts IMP to inosine (I), the formation of both IMP and I are measured.

(1) To assay 26 wells, prepare the following cocktail on ice:

150 μl $MgSO_4$ (5×10^{-2} M)
150 μl Tris (0.1 M) pH 7.4
150 μl PRPP (0.01 M)
150 μl [^{14}C]hypoxanthine (10^{-4} M, 50—60 mCi/mmole)
150 μl hypoxanthine (1.9×10^{-3} M)
750 μl double distilled water

(2) Remove the slides containing the fixed cells from the freezer and incubate them at 37°C for 20 min.

(3) Remove the slides from the incubator and add 50 μl of the cocktail to each of two wells for each mutant. Tap the edge of the slide on the counter top, if necessary, to completely cover the monolayer with the cocktail.

(4) Replace the lids and incubate the wells with the cocktail at 37°C, 30 min for normal controls or 2 h for mutants.

(5) At the end of the incubation period, stop the reaction with 10 μl of 4 M formic acid per well.

(6) Add 50 μl of marker mix per well. (Marker mix is composed of equal volumes of 10^{-2} M I, 10^{-2} M IMP, and 10^{-2} M H.)

(7) Spot 10 μl samples from each well on Whatman 3MM paper. Air dry.

(8) Chromatograph (ascending) in 5% Na_2HPO_4 for 2.5 h at room temperature.

(9) Locate the marker I, IMP, and H spots with an ultraviolet lamp. The spots appear purple. IMP migrates the fastest, H the slowest. Cut out the I and IMP spots together and the H spot separately.

(10) Count the papers in a scintillation counter using a mixture of PPO (2,5-diphenyloxazole, 22.74 g), POPOP (1,4-bis-2-(5-phenyloxazolyl)-benzene, 0.037 g) and toluene (3.79 liter).

(11) Determine the total amount of protein per well according to the method of Oyama and Eagle [28], using two wells for each mutant. The cultures contain about 10—50 μg of protein per well when assayed. The volumes of reagents and the standard curves used for the protein determination should be adjusted accordingly.

(12) Calculate the picomoles of radioactive hypoxanthine that has been converted to I plus IMP per μg of protein per hour. Caution: the microbial contamination of stock radioactive hypoxanthine can result in its conversion to inosinic acid, making it useless for assay purposes.

Use of hypoxanthine for growth

(1) For each mutant, plant 0.1 ml volumes (10^3 cells) of suspension A in each of six wells in a plastic Micro-Test II tissue culture plate.

(2) Twenty-four hours after plating, replace the medium for each mutant as follows: 2 wells, 0.1 ml CS-F10; 2 wells, 0.1 ml CS-F10 + 10^{-6} M azaserine; and 2 wells, 0.1 ml CS-F10 + 10^{-6} M azaserine + 3.0×10^{-5} M hypoxanthine (this is [AS]H medium). H and [AS] are dissolved *separately* in a minimal amount of 1 M NaOH, diluted in double distilled water and stored at 4°C as separate 200-fold stock solutions.

(3) Replace the media every second day as in step 2 and make microscopic observations at each medium renewal. The cells in the [AS] control wells should be inhibited within 24 h after their first exposure to [AS]-containing medium. The cell growth in [AS]H wells may be slightly slower than it is in CS-F10 control wells, but normal strains have no difficulty in reaching confluence. The cell proliferation should be uniform throughout each well; an isolated area of mitotic activity may result from parental cell contamination, but could also result from some kind of reversion.

Stability of the azaguanine-resistant phenotype

(1) Dilute 1.25 ml of suspension A with 48.75 ml of FCS-F10 to give suspension B (2.5×10^2/ml). Inoculate each of ten 60 mm dishes with 4 ml of suspension B, containing 10^3 cells.

(2) Replace the FCS-F10 twice per week until the plates are confluent.

(3) When the 60 mm dishes are confluent, trypsinize the cultures and replate the cells in FCS-F10 at 2.5×10^3 cells/well in plastic Micro-Test II plates (duplicate wells for each AG concentration). Include an AG^r prototype strain and a normal control.

(4) The following day replace the medium with CS-F10 or CS-F10 plus AG at concentrations from 2.0 to 30×10^{-5} M. Replace the media every second day.

(5) Count the cells with a Coulter counter on day 8 to evaluate the stability of the AG^r phenotype.

Cloning efficiencies

(1) Dilute 5 ml of suspension B with 45 ml of FCS-F10 to give suspension

C (25 cells/ml). Distribute 4 ml volumes of suspension C (10^2 cells) to ten 60 mm dishes for each mutant.

(2) Replace the FCS-F10 after 7 and 10 days.

(3) Fix and stain the clones after 12—14 days.

Properties of mutants we have isolated

Using the protocols described in this paper, we have isolated and characterized numerous spontaneous and induced AG^r colonies in order to determine what cell changes have permitted clonal growth under the conditions of AG selection. It is not surprising that isolated AG^r mutants vary in their doubling potentials and subcloning efficiencies, since individual cells within a population of parental human fibroblasts have widely different doubling potentials [24]. We estimate, however, that about 50% of the primary AG^r colonies that we observe have a size and an appearance indicating that they could be propagated and characterized. Of 492 colonies actually isolated, 380 (or 40% of all AG^r mutants that we have observed) grew well enough to be partly or completely characterized. These characterizations, which will be described in detail elsewhere, are summarized here.

The azaguanine resistance phenotype is stable

Sixty-five independent AG^r clones, spontaneous or induced by X-rays, MNNG or 2-NOF and isolated with 1.5×10^{-5} M AG in CS-F10, were grown in the absence of selection through at least a 1000-fold increase in cell number. Upon rechallenge, all populations so derived were still resistant to AG at concentrations of $2—30 \times 10^{-5}$ M.

Possible origins of stable resistance to AG, other than DNA changes in the X chromosomal locus that determines the structure of HPRT, and some ways of detecting them if they occur, have been described previously [7]. For example, if HPRT deficiency in an AG^r mutant were to result from a change on an autosome or from an extrachromosomal epigenetic change, then segregation of the X chromosome with its normal allele for HPRT from the sites of those changes should result in the reappearance of HPRT activity. Such segregation does occur in hybrids of mouse and human cells.

We have prepared hybrids of four AG^r mutants (two spontaneous, one X-ray-induced and one 2-NOF-induced) from four unrelated fibroblast strains by fusing them with non-reverting HPRT-deficient mouse cells. After their initial isolation, the reappearance of HPRT in hybrid cells was looked for by subjecting large hybrid populations to selection with [AS]H medium. So far, no [AS]H-selectable cells have appeared in 31 independent hybrids in which the presence of the X chromosome from the human AG^r mutant was verified by independent means [10,18]. These results are additional strong evidence that AG resistance in human fibroblasts results from stable hereditary changes that occur on the X chromosome and not from readily reversed, extrachromosomal changes.

Most AG^r mutants are unable to use hypoxanthine for growth

We previously thought that the majority of the AG^r mutants that originated in vitro were Type II [1,9]. We now believe that many of the mutants were misclassified, *although the numbers of mutants reported were accurate.* The misclassifications resulted from an inadequate appreciation, at early stages of our work, of the subcloning ability of AG^r mutants and a failure to take advantage of this in order to completely free them of parental cells. As noted above, inadequately purified populations of mutant cells may be erroneously classified as having HPRT activity, as being Type II mutants or as having reverted to AG-sensitivity. These comments may also be applicable to work with cells other than human fibroblasts.

Our more recent and more extensive work with properly purified mutant isolates indicates that at least 90% of the numerous mutants induced by 2-NOF and MNNG are Type I. An extensive survey of spontaneous and X-ray-induced mutants has not yet been made, but it is likely to have a similar outcome.

Most AG^r mutants lack HPRT activity

More than 90% of the spontaneous and chemically induced AG^r mutants that we have studied have undetectable or greatly reduced HPRT activity (less than 5% of the AG^s parental activity); nearly all Type I mutants so far studied fall into this category. The remaining 10% of all mutants have various subnormal amounts of HPRT activity; most of the Type II mutants that we have studied fall into this category, which may be why we have observed that the Type II mutants are somewhat less resistant to 6-thioguanine than are the Type I mutants.

In summary, the spontaneous and induced AG^r mutants that are isolated by using the protocols described in this paper have stable hereditary changes that are not perceptibly reversed by propagation in the absence of AG. In most cases the changes appear to occur on the X chromosome. The two kinds of mutants that arise in vitro resemble in every respect cells from humans who inherit various mutant alleles for HPRT. We do not know cogent reasons for thinking that the changes to AG resistance that we observe in vitro differ from similar changes that might occur in the somatic or germ cells in vivo.

Appendix

Items listed in this appendix are those used in our laboratory. Similar items may be available from other equally suitable sources.

Filter apparatus for preparing serum and medium

Sanitary sterilizing 142 mm filter, No. YY30 142 05
Dispensing pressure vessel 5 liter ASME, No. XX67 00L 05
(a) F10 filters: 1 prefilter (127 mm) AP25 127 50
1 0.22 μm (142 mm) GSWP 142 50
(b) FCS filters: 1 prefilter (127 mm) AP25 127 50
1 0.22 μm (142 mm) GSWP 142 50
1 0.45 μm (142 mm) HAWP 142 50
1 1.2 μm (142 mm) RAWP 142 50
1 3.0 μm (142 mm) SSWP 142 50

Millipore Corporation
Bedford, Mass. 01730, U.S.A.

Culture dishes

Lux plastic 60 × 15 mm culture dishes, No. 5220
Biolabs, Inc.
2910 MacArthur Boulevard
Northbrook, Ill. 60062, U.S.A.
Falcon plastic 60 × 15 mm culture dishes, No. 3002
Falcon plastic 35 × 10 mm culture dishes, No. 3001
Falcon Micro Test II tissue culture plates No. 3040
Falcon Micro Test II lids No. 3041
Falcon
1950 Williams Drive
Oxnard, Calif. 93030, U.S.A.

Light box for locating clones

A Stir-lite magnetic stirrer equipped with a fluorescent light can easily be modified for use as a light box. A piece of black paper 65 × 65 mm square is placed over the center of the light source, leaving 20—30 mm uncovered on all four sides.
Scientific Products
1210 Waukegan Road
McGaw Park, Ill. 60085, U.S.A.

Serum and medium

Unfiltered fetal calf serum
St. Louis Serum Company
114 St. Clair Avenue
East St. Louis, Ill., U.S.A.
Filtered calf serum, No. 617
Nutrient Mixture F-10 (dry powder medium, with L-glutamine, without $NaHCO_3$ and without hypoxanthine — special order)
Grand Island Biological Company
3175 Staley Road
Grand Island, N.Y. 14072, U.S.A.

Chemicals

Azaserine, A grade, No. 19236
Calbiochem
P.O. Box 17
Elk Grove Village, Ill. 60007, U.S.A.
Azaguanine, No. 90 2205
Schwarz/Mann
Mountain View Avenue
Orangeburg, N.Y. 10962, U.S.A.
Glycerin G-33 CERTIFIED ACS
Fisher Scientific Company

711 Forbes Avenue
Pittsburgh, Penn. 15219, U.S.A.
HEPES (*N*-2-hydroxyethylpiperazine-*N*′-2-ethanesulfonic acid)
Calbiochem
P.O. Box 17
Elk Grove Village, Ill. 60007, U.S.A.
Hypoxanthine (pfs), No. H 9377
Sigma Chemical Company
P.O. Box 14508
St. Louis, Mo. 63178, U.S.A.
Hypoxanthine [8-^{14}C]NEC-149, 40—60 mCi/mmole
New England Nuclear Corporation
575 Albany Street
Boston, Mass. 02118, U.S.A.
Methylene blue NA 0652
Allied Chemical
P.O. Box 431
Morristown, N.J. 17960, U.S.A.
PRPP (5-phosphorylribose 1-pyrophosphate, magnesium), No. 4500
P.L. Biochemicals, Inc.
1037 West McKinley Avenue
Milwaukee, Wisc. 53205, U.S.A.
POPOP (1,4-bis-2-(5-phenyloxazolyl)-benzene) Scintillation Grade, No. 6002030
PPO (2,5-diphenyloxazole) Scintillation Grade, No. 6002022
Packard Instrument Company, Inc.
200 Warrenville Road
Downers Grove, Ill. 60515, U.S.A.
Trypsin (0.25%), No. 505
Grand Island Biological Company
3175 Staley Road
Grand Island, N.Y. 14072, U.S.A.

Acknowledgements

This work was supported by research grants from The National Institutes of Health (GM-06983, GM-15422 and NIEHS-722095). L. Jacobs was supported by N.I.H. training grant GM-00398. This is paper No. 1990 from the Laboratory of Genetics, University of Wisconsin, Madison.

References

1 Albertini, R.J. and R. DeMars, Diploid azaguanine-resistant mutants of cultured human fibroblasts, Science, 169 (1970) 482—485.
2 Benke, P. and N. Herrick, Azaguanine-resistance as a manifestation of a new form of metabolic overproduction of uric acd, Am. J. Med., 52 (1972) 547—555.
3 Cox, R.P., M.R. Krauss, M.E. Balis and J. Dancis, Communication between normal and enzyme deficient cells in tissue culture, Exptl. Cell Res., 74 (1972) 251—268.
4 Croxton, F.E., Elementary Statistics with Applications in Medicine and the Biological Sciences, Dover Publications Inc., New York, 1959.
5 DeMars, R., Investigations in human genetics with cultivated human cells: a summary of present

knowledge, in T.M. Sonneborn (ed.), The Control of Human Heredity and Evolution, MacMillan Co., New York, 1965.

6 DeMars, R., Mutation studies with human fibroblasts, Environm. Health Perspectives, Experimental Issue 6 (1973) 127—136.

7 DeMars, R., Resistance of cultured human fibroblasts and other cells to purine and pyrimidine analogues in relation to mutagenesis detection, Mutation Res., 24 (1974) 335—364.

8 DeMars, R., unpublished data.

9 DeMars, R. and K. Held, The spontaneous azaguanine-resistant mutants of diploid human fibroblasts, Humangenetik, 16 (1972) 87—110.

10 DeMars, R. and J. Jackson, unpublished data.

11 Eagle, H., Amino acid metabolism in mammalian cell cultures, Science, 130 (1959) 432—437.

12 Elion, G.B., Biochemistry and pharmacology of purine analogs, Fed. Proc., 26 (1967) 898—904.

13 Enzyme nomenclature, Recommendations of the International Union of Biochemistry, 1964, Elsevier, Amsterdam, 1965.

14 Felix, J.S. and R. DeMars, Detection of females heterozygous for the Lesch-Nyhan mutation by 8-azaguanine-resistant growth of cultured fibroblasts, J. Lab. Clin. Med., 27 (1971) 596—604.

15 Fujimoto, W.Y., J.H. Subak-Sharpe and J.E. Seegmiller, HG-PRT deficiency: clinical agents selective for mutant or normal cultured fibroblasts in mixed and heterozygous cultures, Proc. Natl. Acad. Sci. (U.S.), 68 (1971) 1516—1519.

16 Goldstein, J.L., J.F. Marks and S.M. Gartler, Expression of two X-linked genes in human hair follicles of double heterozygotes, Proc. Natl. Acad. Sci. (U.S.), 68 (1971) 1425—1427.

17 Ham, R.G., An improved nutrient solution for diploid Chinese hamster and human cell lines, Exptl. Cell Res., 29 (1963) 515—526.

18 Kahan, B. and R. DeMars, Localized derepression in the human inactive X chromosome in mouse-human cell hybrids, Proc. Natl. Acad. Sci. (U.S.), 72 (1975) 1510—1514.

19 Kastenbaum, M.A. and K.O. Bowman, Tables for determining the statistical significance of mutation frequencies, Mutation Res., 9 (1970) 527—549.

20 Kelley, W.N., M.L. Greene, F.M. Rosenbloom, J.E. Henderson and J.E. Miller, Review: HG-PRT-deficiency in gout, Ann. Int. Med., 70 (1969) 155—206.

21 Lesch, M. and W. Nyhan, A familial disorder of uric acid metabolism and central nervous system function, Am. J. Med., 36 (1964) 561—570.

22 Lewenberg, B., I. Melnick and J.M. Buchanan, The effect of aza-1-serine and 6-diazo-5-oxo-norleucine on inosinic acid biosynthesis de novo, J. Biol. Chem., 225 (1957) 163—176.

23 Maher, V.M. and J.E. Wessel, Mutations to azaguanine resistance induced in cultured diploid human fibroblasts by the carcinogen, *N*-acetoxy-2-acetylaminofluorene, Mutation Res., 28 (1975) 277—284.

24 Martin, G.M., C.A. Sprague and C.J. Epstein, Replicative life-span of cultivated human cells, Lab. Invest., 23 (1970) 86—92.

25 Migeon, B.R., V.M. Der Kaloustian, W.L. Nyhan, W.J. Young and B. Childs, Hypoxanthine-guanine phosphoribosyltransferase deficiency: heterozygote has two clonal populations, Science, 16 (1968) 425—427.

26 Newcombe, H.B., Delayed phenotypic expression of spontaneous mutations in *Escherichia coli*, Genetics 33 (1948) 447—476.

27 Nyhan, W.L., Seminars on the Lesch-Nyhan syndrome, Fed. Proc., 27 (1968) 1017—1112.

28 Oyama, V.I. and H. Eagle, Measurement of cell growth in tissue culture with a phenol reagent (Folin-Ciocalteau), Proc. Soc. Exptl. Biol. Med., 91 (1956) 305—307.

29 Rieger, R., A. Michaelis and M.M. Green, A Glossary of Genetics and Cytogenetics, Classical and Molecular, Springer-Verlag, New York, 1968.

30 Russel, J.D. and R. DeMars, USP-glucose: α-D-galactose-1-phosphate uridylyl transferase activity in cultured human fibroblasts, Biochem. Genet., 1 (1967) 11—24.

31 Salzmann, J., R. DeMars and P. Benke, Single allele expression at an X-linked hyperuricemia locus in heterozygous human cells, Proc. Natl. Acad. Sci. (U.S.), 60 (1968) 545—522.

32 Shapiro, S.L., G.L. Sheppard, Jr., F.E. Dreifuss and D.S. Newcombe, X-linked recessive inheritance of a syndrome of mental retardation with hyperuricemia, Proc. Soc. Exptl. Biol. Med., 122 (1966) 609—611.

33 Van Zeeland, A.A. and J.W.I.M. Simons, The effects of calf serum on the toxicity of 8-azaguanine, Mutation Res., 27 (1975) 135—138.

HOST-MEDIATED CYTOGENETIC ASSAY*

J. G. BREWEN

Biology Division, Oak Ridge National Laboratory, Oak Ridge, Tenn. 37830 (U.S.A.)

Chromosome aberrations constitute a significant portion of the genetic damage produced by most mutagenic agents. Since the ultimate practical endpoint of any mutagenicity testing program is to estimate man's hazard, it behooves researchers to perform some tests on human material in order to make direct comparisons to other eukaryotic systems**. The test described here allows for a direct comparison of chromosome aberration production in prokaryotes and eukaryotes, as well as human cells, in the same environment, Host-mediated assays have utilized bacterial systems[7], yeast[5], Neurospora[9], and mammalian cells[6], and also human peripheral leukocytes[1,2], and although direct cytogenetic analysis cannot be done on bacteria, yeast, and Neurospora, certain inferences concerning chromosome breakage can be made.

The mutagenic events analyzed in the present test are structural alterations of the chromosomes. The types and classes of the aberrations expected are described in detail in several review articles (*e.g.* see EVANS[4], and KIHLMAN[8]), as are the times during the proliferative cell cycle when the various types are produced.

The technique for using human peripheral leukocytes is based on the procudere of BUCKTON AND NETTESHEIM[3] and is quite simple. A sample of freshly drawn, heparinized human venous blood is divided into 4-ml aliquots and centrifuged at maximum speed in a clinical centrifuge for at least 5 min. The resulting light buffy coat that is on top of the packed red cells is removed with a pasteur pipette and put into a 1:1 mixture of Hank's balanced salt solution (HBSS) and phytohaemagglutinin (PHA) at pH 6.8 to 7.0 at a ratio of 1 buffy coat per 0.2 ml of the HBSS–PHA mixture. The buffy coats from all the aliquots of the same donor can be pooled at this time. This mixture is then loaded into diffusion chambers prepared in the following way.

Lucite rings[a], with outside diameter 22 $\pm$ 0.5 mm, inside diameter 14.0 $\pm$ 0.2 mm, height 6.0 $\pm$ 0.1 mm, and a 1-mm hole drilled through the wall, have Millipore filter discs (catalogue No. VCWPO2, 2MM) glued onto each side with DEK–Adhese cement[b]. The chambers are then dry-sterilized at 180° for 48 h. Twenty-four to 36 h

---

Abbreviations: CHA, cyclohexylamine; HBSS, Hanks' balanced salt solution; MMS, methyl methanesulfonate; PHA, phytohaemagglutinin.

* Research sponsored by the U.S. Atomic Energy Commission under contract with the Union Carbide Corporation.

** See other articles in this series.

[a] Obtained from Piedmont Plastics Inc., 4927 Selabert Ave., Charlotte, N.C. 28211 (U.S.A.). Specify that soap lubricant must be used in lieu of cutting oil.

[b] Obtained from Donald Tulloch, Jr., P.O. Box 146, Chadds Ford, Pa. 19217 (U.S.A.).

before the chambers are to be used they are filled with a sterile 1% solution of mouse serum in HBSS and stored in a container of this solution at 4°. This procedure removes the majority of toxic substances (*i.e.* detergents) from the filters. At the time the chambers are to be loaded with the human leukocytes, the mouse serum solution is removed with a 22-gauge needle that is inserted through the hole drilled in the Lucite ring (see Fig. 1). The leukocyte mixture is then loaded into the chamber with a 22-gauge needle *via* the hole, and melted paraffin is placed in the hole to completely seal the chamber. The loaded chambers are then covered with HBSS until they are put into the host animal.

Any host animal can be used. The animal is anesthetized, the hair is shaved from the abdomen, a small incision is made to the side of the medial line, and the diffusion chamber is slipped into the peritoneal cavity. The incision is then closed with metal clips.

The host animal can then be treated with the test compound at any stage of the leukocytes' transformation and ultimate cell division. Since all of the leukocytes are in G_0 at the time the chambers are inserted into the host animal, the technique offers the opportunity to treat a relatively homogenous population of cells. At later times, usually 24–36 h after implantations, the great majority of cells that will enter mitosis at 56–62 h are replicating their DNA. The maximum mitotic acticity is reached 56–72 h after implantation.

Mitotic cells are harvested by sacrificing the host animal and removing the chamber. The Millipore filters are cut off the ring, and they along with the contents of the chamber are put into 3 ml of TC 199, or other suitable culture medium, $1 \cdot 10^{-6}$ *M* colchicine and incubated at 36° for 2–4 h. One ml of 1% pronase is then added, and the culture is agitated by magnetic stirring for 15 to 30 min to dissolve the clots that invariably form inside the chamber. After the pronase treatment the cells are centrifuged out of the medium at full speed on a clinical centrifuge. Slide preparation then follows any of several methods, depending on the individual investigator.

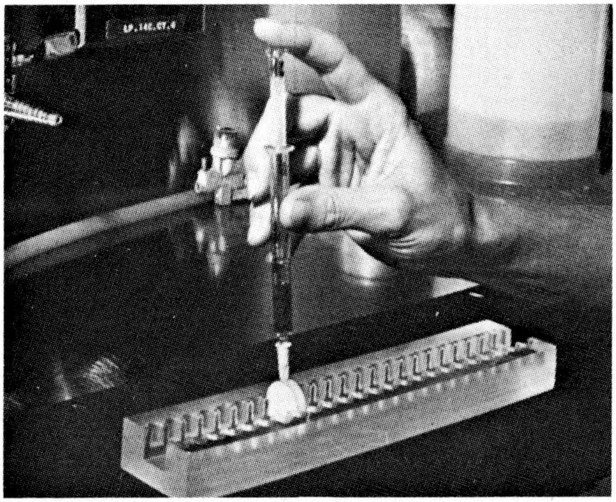

Fig. 1. The loading of a diffusion chamber through the access hole drilled through the lucite ring. This hole is then sealed with melted paraffin.

The technique we employ is to wash the cells once with HBBS, centrifuge, resuspend in quarter-strength HBSS in distilled water, incubate 20 min at 36° centrifuge, and fix the cell pellet with 3:1 methanol:glacial acetic acid. The fixed cells are then washed 2 to 4 times with fixative, and a drop of the final cell suspension is dropped on a clean, wet slide and the slide is air-dried. The stain used is optional; we routinely use 2.5% Giemsa or 1% aceto-orcein.

This technique has been used to test the clastogenic effects of two compounds—methyl melthanesulphonate (MMS)[5] and cyclohexylamine (CHA)[9]. The MMS experiments were conducted with mice as the host animals, and in the CHA experiments Chinese hamsters were used. The MMS experiments consisted of injecting mice with doses of MMS of 65 or 130 mg/kg body weight at 12 or 24 h prior to removal of the chambers. The injections were both intravenous or intraperitoneal. Table I summarizes the data obtained. The CHA experiments varied from the above protocol in that the host hamsters received daily intraperitoneal injections, beginning on the day of implantation, for three consecutive days. The daily doses were 50, 150, or 450 mg/kg body weight. The results of these experiments are presented in Table II.

Although the technique is cumbersome at first attempts, it allows.for the use of human cells as the target cells in an *in vivo* situation where detoxification of a com-

TABLE I

CHROMATID ABERRATIONS 12 AND 24 h AFTER INJECTION OF MMS

Route of injection[a]	*Concentration (mg/kg)*	*Exposure time (h)*	*Number of cells scored*	*Number of iso- and chromatid deletions*	*Number of chromatid exchanges*	*Total*[b] *breaks (% ±S.E.)*
i.v.	65	12	200	7	1	4.5 ± 1.5
i.p.	65	12	400	20	6	8.0 ± 1.4
i.v.	130	12	200	25	3	15.5 ± 2.8
i.p.	130	12	400	17	13	10.8 ± 1.6
i.v.	65	24	300	4	0	1.3 ± 0.7
i.p.	65	24	200	7	5	8.5 ± 2.1
i.v.	130	24	300	19	9	12.3 ± 2.0
i.p.	130	24	400	12	5	5.5 ± 1.2
Control			300	3	0	1.0 ± 0.6

[a] i.v., Intravenously; i.p., intraperitoneally.
[b] Deletions + twice exchanges.

TABLE II

CYTOGENETIC OBSERVATIONS ON HUMAN LEUKOCYTES TREATED *via* A HOST-MEDIATED ASSAY SYSTEM AFTER INTRAPERITONEAL INJECTIONS OF CHA

Concentration (mg/kg/day)	*Number of cells scored*	*Number of rings and dicentrics*	*Chromatid breaks*[a] *(% ±S.E.)*	*Achromatic lesions (% ±S.E.)*
Control	550	0	0.72 ± 0.36	0.72 ± 0.36
50	450	1	1.11 ± 0.50	1.33 ± 0.54
150	550	0	2.18 ± 0.63	2.00 ± 0.60
450	550	0	2.18 ± 0.63	0.72 ± 0.36

[a] All isolocus breaks of both chromatids were considered to be isochromatid deletions and are included under this heading.

pound, or metabolic conversion of a harmless compound into a mutagen, must be considered. The use of human cells offers the advantage that the results obtained can be compared directly to results in the host animal and in leukocytes of other species tested in the same host.

REFERENCES

1 BREWEN, J. G., P. NETTESHEIM AND K. P. JONES, A host-mediated assay for cytogenetic mutagenesis: preliminary data on the effects of methyl methanesulfonate, *Mutation Res.*, 10 (1970) 645–649.

2 BREWEN, J. G., F. G. PEARSON, K. P. JONES AND H. E. LUIPPOLD, Cytogenetic effects of cyclohexylamine and *N*-OH-cyclohexylamine on human leukocytes and Chinese hamster bone marrow, *Nature New Biol.*, 230 (1971) 15–16.

3 BUCKTON, K. E., AND P. NETTESHEIM, *In vitro* and *in vivo* culture of mouse peripheral blood for chromosome preparations, *Proc. Soc. Exptl. Biol. Med.*, 128 (1968) 1106–1110.

4 EVANS, H. J., Chromosome aberrations induced by ionizing radiations. *Intern. Rev. Cytol.*, 12 (1962) 221–321.

5 FAHRIG, R., Metabolic activation of aryldialkyltriazenes in the mouse: Induction of mitotic gene conversion in *Saccharomyces cerevisiae* in the host-mediated assay, *Mutation Res.*, 13 (1971) 436–439.

6 FISCHER, G. A., The host-mediated mammalian cell assay, *Agents and Actions*, 3 (1973) 93–98.

7 GABRIDGE, M. G., AND M. S. LEGATOR, A host-mediated microbial assay for the detection of mutagenic compounds, *Proc. Soc. Exptl. Biol. Med.*, 130 (1969) 831–834.

8 KIHLMAN, B. A., *Actions of Chemicals on Dividing Cells*, Prentice–Hall, Englewood Cliffs, N.J., 1966.

9 LEGATOR, M. S., AND H. V. MALLING, The host-mediated assay: a practical procedure for evaluating potential mutagenic agents in mammals, in A. HOLLAENDER (Ed.), *Chemical Mutagens*, Plenum, New York, 1971.

IN VITRO ANAPHASE AND METAPHASE PREPARATIONS IN MUTATION TESTING

W.W. NICHOLS, R.C. MILLER and C. BRADT

Institute for Medical Research, Camden, N.J. 08103 (U.S.A.)

Cytogenetic techniques have found widespread use in mutagenicity testing. It is generally recognized that in vivo cytogenetic methodologies are preferable to those in vitro for these purposes. This is because of the desirability of metabolic activation and/or detoxification obtained in the intact animal. There are, however, instances in which in vitro methods provide valuable ancillary information. This is especially true with the demonstration of in vitro metabolic action using liver polysomes with cytogenetic methodologies [4].

Some of the situations in which in vitro methods are valuable are when it is desirable to accurately control the concentration and exposure time of cells to the compound under test, when it is desirable to use higher doses than can be tolerated in vivo, when correlative in vivo and in vitro data are sought, and when side-arm bridges or pseudochiasmata are sought which are more easily detected in anaphase preparations. In this latter case, it is possible to prepare in vivo anaphase preparations from bone marrow, but these have been of poor quality and very difficult in most hands. The present paper describes methods used successfully in our laboratory for in vitro anaphase and metaphase preparations.

Anaphase

Anaphase preparations may be made by any of several methodologies that eliminate pretreatment of dividing cells with agents that interfere with the cell spindle. In metaphase preparations it is necessary to interfere with the spindle, and this is usually accomplished with agents as colchicine or Colcemid and hypotonic expansion. It is these steps then, that should be eliminated in making anaphase preparations. One convenient approach has been to grow cells directly on coverslips in petri dishes. When human fibroblasts are the cells to be examined, approximately 400,000 cells are added to a 60 mm diameter petri dish containing a 25 × 50 mm coverslip and grown in Eagle's-Earle's or any one of a number of other media supplemented with 20% fetal calf serum and placed in a 5% CO_2 atmosphere. The coverslips are fixed without pretreatment when adequate numbers of mitoses are visualized by inspection in an inverted microscope. In our hands this is usually 48—92 h after planting. At this time 1 ml of glacial acetic acid is added directly to the approximate 5 ml of medium in the petri dish in an effort to partially fix the mitotic cells that tend to be round and more loosely attached than interphase cells. After 2—3 min the coverslip is transferred to a rack for rectangular coverslips (12 coverslip capacity) and submerged in absolute methanol for 5 min. Next the coverslips are transferred immediately to Giemsa stain for approximately 30 sec, and then into distilled water. At this point the wet coverslip is placed on a slide, cell side down, and examined in the microscope to check for adequacy of staining. If too light they are returned to the stain solution for an additional 10—15 sec and again checked. If they are too dark they can be destained by immersing the coverslip in 100 ml of distilled water to which has been added approximately 1 ml of glacial acetic acid. After destaining the coverslips are washed thoroughly in distilled water and then the staining procedure is repeated. When the coverslips are properly stained they are removed from the water and placed in acetone for 5 min, rinsed in acetone/xylene 50/50 and placed in xylene for at least 10 min. They are then mounted with cells down by putting a drop of coverbond mounting medium on the slide and gently placing the coverslip on the drop. If it is desired to use Feulgen stain Kahle's modified fixative is substituted for the methanol fixative, and if orcein stain is to be used a fixative consisting of 60% glacial acetic acid and 10% 1 N HCl is substituted. While this is the method for anaphase preferred in our laboratory, preparations can also be made from cells grown in suspension or cells grown as a monolayer that have been put into suspension after removing from the glass with trypsin or some other similar enzyme. In this case the procedure is to centrifuge the cells into a button, remove the trypsin and/or medium and add either the fixative described for orcein stain or air dry fixative slowly to the button of cells. In the case of the 60% glacial acetic acid and 10% 1 N HCl fixative, it is permitted to sit on the button for 20 min, at which time the cells are suspended in the fixative, again centrifuged, and the fixative removed. The cell button is then suspended in a few drops of orcein stain and a drop of this suspension is placed on a slide and covered with a coverslip. The excess stain is gently expressed and the slide is examined under the microscope to assess the cell density. If too many cells are present a further dilution is made with stain and the procedure repeated. When expressing the excess stain from under the coverslip this should be done gently without squashing, since anaphases are fragile and squashing introduces artifacts. When air dry fixative is used slides are prepared by dropping a heavy suspension of cells and fixative onto slides and either air dried or put on a slide warmer at 32°C. In some materials in which a very low mitotic index is the general rule, it has been possible to arrest cells at metaphase in order to collect mitotic figures and then release the cells and permit them to progress to anaphase. This is accomplished by adding Colcemid at a final concentration of 0.1 μg/ml for approximately 2 h. After this the Colcemid-containing medium is removed and the cells are re-fed with prewarmed complete medium. After 1—2 h to permit the arrested metaphase cells to recover and progress to anaphase the coverslips are processed as described above. A series of comparisons have been made between the same cells grown on coverslips and fixed in situ without other treatment, coverslips with Colcemid arrest and release, and air dry preparations on a series of human diploid fibroblast cell lines. It was found that when early or midpassage cultures were examined the frequency of abnormal anaphase was approximately the same by all three methods. However, when the last 5—10 passages in the lifespan of human diploid fibroblasts was examined, a time when an increase in chromosome abnormalities is detected, there was a significant difference in the numbers of abnormal anaphase seen between the three methods. Coverslips without treatment had the lowest frequency, coverslips with Colcemid arrest and release an intermediate frequency, and those preparations made by the air dry technique the highest frequency (Table I).

The types of anaphase aberration that can be distinguished include acentric fragments which are single or paired chromosomal segments left at the equator of the cell (Fig. 1E—G), and correspond to chromatid and chromosome breaks in metaphase, an attached fragment that lags behind the main body of anaphase chromosomes but is in the same orientation with the chromosomes and seems to be attached by an attenuated segment (Fig. 1C,D) which corresponds to gaps seen in metaphase. A chromosome bridge is viewed as a single or double unit stretched between the two groups of anaphase chromosomes (Fig. 1H—J). The single unit results from an asymmetrical chromatid intra- or interchange and the double unit from an asymmetrical chromosome interchange. Another defect recognized in anaphase much more readily than in metaphase is the side-arm bridge or pseudochiasma (Fig. 1K,L). This name arose from the resemblance of this structure in mitotic anaphase to a rod bivalent with non-terminalized chiasma seen in meiosis. When side-arm bridges are produced by ionizing radiation they occur only in cells treated during prophase. There has been considerable discussion related to whether side-arm bridges represent subchromatid or chromatid events [2,3,5]. In some instances the appearance of side-arm bridges can be mimicked by induced stickiness and does not represent a true chromosomal defect. Lagging chromosomes and multipolar spindles (Fig. 1M,N) that result in abnormalities of chromosome number are also well visualized in anaphase.

TABLE I

FREQUENCY OF ANAPHASE ABNORMALITIES BY THREE TECHNIQUES

	Coverslip no treatment	Coverslip with Colcemid arr. and release	Air dry
Tissue culture, passage 2—11			
Cells analyzed	2328	2233	2350
% cells with abnormalities	6.2	6.1	6.4
Tissue culture, passage 12—21			
Cells analyzed	1400	1400	1350
% abnormalities, anaphase	7.0	7.0	7.9
Last 10 tissue culture passages			
Cells analyzed	1667	1593	1703
% abnormalities	10.3	13.6	17.6

The data on tissue culture passage 2—11 is cumulative from 6 independent cultures, passage 12—21 is cumulative data from 3 independent cultures, and the last 10 passages from 4 independent cultures.

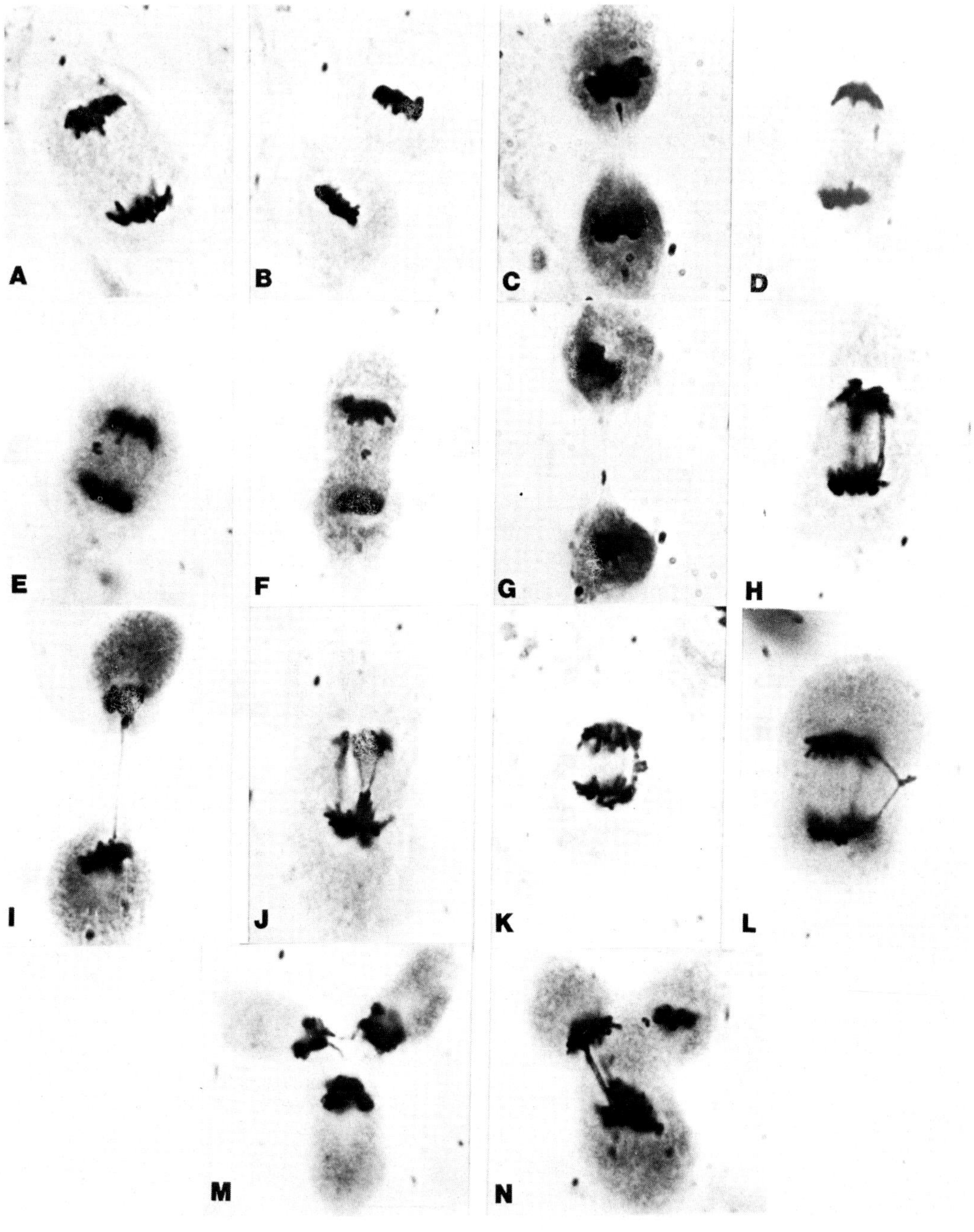

Fig. 1. Examples from anaphase preparations of human diploid fibroblast-like cells. A and B: normal anaphase. C and D: attached fragment. E—G: a variety of acentric fragments. H—J: a variety of anaphase bridges. K and L: side-arm bridges. M and N: multipolar spindles with bridges and fragments.

Metaphase

Conventional metaphase stains such as orcein, Giemsa and Feulgen, are still useful in assessment of acute damage that results in chromosome breakage and unstable rearrangements. However, since much more information, such as precise localization of acute damage and many stable rearrangements not detectable by conventional stains, can be assessed by banding procedures, it is these techniques that are recommended for in vitro metaphase analysis. While any of several banding methods are suitable for these techniques, the Giemsa-trypsin method is the one most commonly employed in our laboratory and it is this protocol, a modification of the Seabright method [6], that will be described here. This method for harvesting cultured cells was designed for monolayer cultures of fibroblast-like or epithelial-like cells. Since proper spreading and state of contraction of chromosomes is essential to optimal banding, attention to the variables of slide preparation will pay large dividends.

In preparation for harvesting, cultures are re-fed with fresh medium 18—24 h before it is desired to make preparations. On the day the cultures are to be harvested they are inspected on the inverted microscope to insure that there is a sufficient number of dividing cells. If there are too few dividing cells and the culures have a light cell density they can be re-fed again and checked 24 h later. If the cell density is heavy it is necessary to pass the culture to induce log phase growth and thereby have sufficient dividing cells for harvest. When the cells are suitable for harvest Colcemid 0.1 μg/ml (Velban 0.1 μg/ml may be substituted) is added to the culture 0.5—1 h prior to harvesting. Next the culture medium is removed and placed in a conical centrifuge tube. The cells are then rinsed briefly with 3 ml of Puck's EDTA at room temperature and this solution is added to the centrifuge tube with medium. The medium and EDTA solution are preserved for subsequent centrifugation in order not to lose the loosely attached mitotic figures that separate from the cell sheet. The cells in the flask are now treated with 4 ml of Puck's EDTA-trypsin solution at 37°C for at least 5 min, but not more than 10 min, and after vigorous shaking to free attached cells the contents are added to the previously preserved medium and EDTA. If necessary the flask can be rinsed with the contents of the centrifuge tube. After centrifugation at 35 $\times g$ for 10 min the supernatant is gently poured off and the cell button is resuspended in the small amount of supernatant remaining. Next, approximately 1 ml of sodium citrate-potassium chloride hypotonic solution prewarmed to 37°C is added with constant agitation and allowed to stand at 37°C for 15 min. The cells are then centrifuged at 500 rpm for 4—5 min, supernatant poured off, and 2 ml of 3 : 1 methanol : acetic acid fixative is added, dropwise with continued mixing, after which the cells are centrifuged at 50 $\times g$ for 2—3 min. Again the supernatant is decanted and the cells resuspended in the remaining few drops. Additional fixative is then added dropwise with agitation to a volume of 2 ml. The suspended cells are left standing for an hour, and after settling out, the fixative is again changed with or without centrifugation. At this point the cell suspensions can be stored in stoppered tubes at 4°C for several days without ill effects. Prior to making slides the fixative is changed again and enough fixative added to make a relatively thin cell suspension. Well-spread metaphase chromosome preparations may be made by applying 3–6 drops of the cell suspension to the end of a clean wet slide held at a slight angle so that the suspension spreads down the length of the slide. Chromosome spreading is enhanced by blowing gently on the slide to assist the flow of the cell suspension down the length of the slide (care should be taken not to inhale the volatile fixative from the slide). The slide is then allowed to air dry in a vertical position and immediately checked for proper spreading with low power phase optics. If the metaphase spreads are not satisfactory several variations can be employed. Slides may be cooled in ice water prior to use. This procedure appears to be most useful if the relative humidity is low (⩽20%). If the chromosomes do not appear to be flat when observed under phase optics, the cell button can be resuspended in 1 : 1 absolute methanol : glacial acetic acid fixative and the procedure repeated. While this alteration in fixative does not impair chromosome banding, it may sometimes interfere with bromodeoxyuridine-induced sister chromatid differentiation. If the relative humidity is high (⩾70%) it may be necessary to speed up the drying process by placing the slides flat on a slide warmer at 56° C instead of allowing them to air dry. Flaming the slides, as is sometimes done for ordinary Giemsa staining, is not recommended, since in our hands this results in metaphase chromosomes that are often not suitable for banding. When the slides are obtained with satisfactory metaphase plates they are then stored for 3–5 days at room temperature prior to banding. After storage the Giemsa-trypsin banding is accomplished by dipping the slide in tap water and transferring immediately to a Coplin jar containing 0.04% EDTA-trypsin at room temperature for 10 sec to 2 min. The slide is then rinsed in running tap water and dipped into a stop bath consisting of 5% serum in Hank's balanced salt solution. The slide is removed immediately from the stop bath, rinsed in tap water, and placed wet into Giemsa stain for 3–10 min. When removed from the stain the slide is rinsed well in deionized water and blotted dry with filter paper. Drying is completed on a slide warmer at 56° C for 1 h and then the slide is mounted in Coverbond or another suitable mounting medium.

The minimal categories that should be scored in evaluating metaphase preparations are chromosome number, chromatid deletions (open chromatid breaks), chromatid exchange, chromosome deletion (open chromosome break), chromosome exchange, premature chromosome condensation (chromosome pulverization) and one category for "other" observations. Gaps are generally not recognizable in banded chromosome preparations. At the present time in our laboratory when unbanded chromosome preparations are made gaps are recorded but not included in abnormalities per cell or total cells with one or more abnormalities.

Conduct of Experiments

While experimental design is frequently modified depending on the circumstances of the investigation, the suggestions made by the Ad Hoc committee of the Environmental Mutagen Society [1] serve as a basis for in vitro cytogenetic testing for mutagens. These suggestions hold for either anaphase or metaphase studies. That report indicated that three time intervals and three dose levels should be used. The time intervals should be selected so that in one set cells are exposed to mutagens only during G2 and mitosis in order to detect side-arm bridges or subchromatid aberrations. In a second set the mitotic cells should have been exposed during the entire cell cycle with a third set exposed for a period of time equivalent to approximately two cell cycles in an attempt to take into account the possibility of a mitotic delay. Thus in human diploid fibroblasts, exposure times of 2—3 h, 24 h and 48 h fulfill these requirements. The three concentrations of test compound suggested are the therapeutic or exposure blood level, 10 times the therapeutic or exposure blood level, and the maximum concentration that permits cells to proceed to mitosis. In addition to these times and concentrations, positive as well as negative controls should be utilized and all scoring should be done on coded slides. When a positive effect is observed a dose-response curve should be constructed to aid in evaluation of potential mutagenic risk.

Discussion

It is emphasized that in vitro cytogenetic studies are an adjunct to other studies rather than a primary test system. This is because of the desirability of in vivo metabolic activation and/or detoxification as mentioned earlier. They do, however, possess considerable utility under specific circumstances, and this utility is increasing with the demonstration of successful in vitro metabolic activation with liver polysomes.

For acute studies to detect unstable chromosomal abnormalities anaphase methods offer rapid and accurate assessment. If localization of defects or chronic and stable abnormalities are sought, then banding of metaphase chromosomes must be utilized.

Materials

Glass and plastic ware

1. Tissue culture flasks 25 cm^2
   Falcon Plastics, Oxnard, Calif. 93030, U.S.A.
2. Rite-on glass micro slides 3 in. × 1 in.
   Clay Adams division of Becton, Dickenson and Co., Parsippany, N.J. 07054, U.S.A.
3. Glass coverslips 24 × 50 mm No. 1
   Clay Adams, Parsippany, N.J. 07054, U.S.A.
4. Coplin staining jars
   Clay Adams, Parsippany, N.J. 07054, U.S.A.
5. Pasteur pipets — 5—3/4 in.
   Dolbey Scientific, Philadelphia, Pa., U.S.A.

Reagents,

1. Absolute methanol
2. Glacial acetic acid
3. Velban lyophilizer

Grand Island Biological Co., Grand Island, N.Y. 14072, U.S.A.

4. Colcemid
   CIBA Pharmaceuticals Co., Summit, N.J. 07901, U.S.A.
5. Hank's balanced salt solution
   GIBCO, Grand Island, N.Y. 14072, U.S.A.
6. Calf serum (for stop bath any available serum will do)
   GIBCO, Grand Island, N.Y. 14072, U.S.A.
7. Culture medium: depending upon cell type
8. Giemsa blood stain, original azure blend type
   Harleco, Philadelphia, Pa., U.S.A.
   (a) Anaphase stain
      5 ml Giemsa
      3 ml absolute methanol
      3 ml 0.1 M citric acid
      3 ml 0.2 M Na_2HPO_4
      100 ml distilled water
   (b) Banding stain
      1% Giemsa in Gurr's buffer — pH 6.8
9. Hypotonic soln: 40 ml deionized water
   (fresh) 5 ml calf serum
   0.1 g KCl
   0.1 g sodium citrate
   1.0 μg/ml Colcemid
   or suitable substitute
10. Gurr's buffer tablets pH 6.8
   Bio/medical Specialties, Box 1687, Santa Monica, Calif. 90406, U.S.A.
11. Coverbond synthetic resin (mounting medium)
   Harleco, Philadelphia, Pa. 19143, U.S.A.
12. Puck's EDTA solution:
   8 g NaCl
   0.4 g KCl
   0.4 g sodium bicarbonate
   0.2 g EDTA
   1.0 g glucose
   1.0 liter distilled water
   Stored at 4° C, used at room temperature
13. Puck's EDTA-trypsin solution:
   Same as above plus 16.7 ml 2.5% trypsin solution.
   Stored at 4° C, used at 37° C.
14. Acetone
15. Xylene

Miscellaneous equipment

1. Constant temperature incubation chamber,
   maintained at 37° C for most cell cultures. Many culture systems require CO_2 incubator — consult appropriate culture methods.
2. Heated water bath 37° C.

Chicago Surgical and Electrical Co., Div. of Lab-line Instruments, Inc., Melrose Park, Ill., U.S.A. or suitable substitute.

3. Lint-free absorbent paper for blotting slides.
4. Slide warmer
   Clinical Scientific Equipment Co., Melrose Park, Ill., U.S.A.

Acknowledgements

This work was supported by research grants from the National Institutes of Health: RO-1-ES01136-02, No. 3-P01-AG-00378-05 and contract N01-AG-4-2865; and American Cancer Society grant VC-178. Dr. Nichols is the S. Emlen Stokes Professor of Genetics at the Institute for Medical Research.

References

1 Chromosome methodologies in mutation testing. Report of the ad hoc Committee of the Environmental Mutagen Society and the Institute for Medical Research, Toxicol. Appl. Pharmacol., 22 (1972) 269—275.

2 Heddle, J.A., The strandedness of chromosomes: evidence from chromosomal aberrations, Canad. J. Genet. Cytol., 11 (1969) 783—793.

3 Kihlman, B.A., Subchromatid exchange in the strandedness of chromosomes, Hereditas, 65 (1970) 171—186.

4 Natarajan, A.T., A.D. Tates, P.O.W. van Buul, M. Meijers and N. de Vogel, Cytogenic effects of mutagens/carcinogens after activation in an in vitro microsomal system. I. Induction of chromosome aberrations and sister chromatid exchanges by diethyl nitrosamine (MN) in CHO cells in the presence of rat liver microsomes, Mutation Res., 37 (1976) 83—90.

5 Nuti-Ronchi, V., The effect of colchicine on subchromatid exchanges in root meristem cells of *Vicia faba* and *Allium cepa*, Mutation Res., 9 (1970) 385—394.

6 Seabright, M., A rapid banding technique for human chromosomes, Lancet, ii (1971) 971—972.

THE MICRONUCLEUS TEST

W. SCHMID

Division of Medical Genetics, Department of Pediatrics, University of Zürich, Zürich (Switzerland)

INTRODUCTION

The micronucleus test (m.t.) *in vivo* is a method devised primarily for screening chemicals for chromosome-breaking effects. The test substances are normally applied sub-acutely to small mammals, and the effect is read in direct smears from bone marrow. This testing procedure, developed by SCHMID and coworkers[1,3,5,6] has a number of important advantages over the analysis of bone-marrow metaphase chromosomes. In technique of preparation as well as the reading of the slides, it is simpler and faster than chromosome analysis in the same material, but not at the expense of accuracy. In the monitoring of chromosome breakage, the test is at least as sensitive as the metaphase method; in addition it includes effects on the spindle apparatus. All these properties render the m.t. highly suitable for routine toxicological screening.

The method is based on the following principles and observations: in anaphase, acentric chromatid and chromosome fragments lag behind when the centric elements move towards the spindle poles. After telophase the undamaged chromosomes, as well as the centric fragments, give rise to regular daughter nuclei. The lagging elements are included in the daughter cells, too, but a considerable proportion is transformed into one or several secondary nuclei which are, as a rule, much smaller than the principal nucleus and are therefore called micronuclei. Similar events occur if the functioning of the spindle apparatus is impaired, *e.g.* under the influence of colchicine; in this event, however, the main nucleus is often replaced by a whole group of small nuclei, which, in general, are considerably larger than typical micronuclei. In bone-marrow smears from mammals treated with chromosome-breaking agents, micronuclei are found in numerous cell types, always provided that these cells have completed—under the influence of the mutagen—one or a few mitoses. Micronuclei can be found *e.g.* in myeloblasts, myelocytes and erythroblasts. In cell types with little cytoplasm they are not always easily distinguishable from normal nuclear lobes or projections (Fig. 1f). An analysis restricted to nucleated bone-marrow cells is rather tedious and unreliable if the effects are small. Fortunately, the great majority of micronuclei present themselves in a cell type much more suitable for analysis, namely the class of the youngest erythrocytes. A few hours after completion of the last mitosis, erythroblasts expel their nucleus; for unknown reasons, micronuclei remain in the cytoplasm of the young erythrocyte and there they are easily recognizable. As an ad-

Abbreviation: m.t., micronucleus test.

ditional advantage for the micronucleus test, young erythrocytes stain differently from older forms. For the duration of their adolescence, lasting approx. 24 h, they stain not red but bluish; to use a technical term: they are polychromatic. If scoring of micronuclei is restricted to this cell type, then it is known that the anomalies were bound to arise mostly, if not exclusively, during the course of the immediately preceding mitosis and under the influence of the temporally limited action of the mutagen. The morphology of all other types of bone marrow cell does not reveal whether they were mitotically active during the test period.

Micronuclei in polychromatic erythrocytes are characteristic and obtrusive elements which can be scored by personnel without special training in cytogenetics. In hematological routine these micronuclei have long been known as Howell-Jolly bodies. Their differentiation from occasional artifacts will be dealt with in a later paragraph. Typically, micronuclei are *round* with a diameter of about 1/20 to 1/5 of an erythrocyte. (Fig. 1a–c) Even after treatment with high doses of chromosome-breaking agents, most of the micronucleated cells contain just *one* micronucleus. There are, however, always some cells with two or more of these bodies as well as some with almond-shaped micronuclei (Fig. 1d). If the findings deviate from the picture described above, the observations must be considered as *atypical* and have to be clarified by additional means.

A particular virtue of the m.t. resides in its permitting a fast comparison of a test result in any number of different mammalian species. Number and morphology of the chromosomes do not pose limitations to this technique[3]. In cases of particular importance, bone marrow from primates or man can be studied as well.

On the sensitivity of the method, compared with the analysis of metaphase chromosomes, the following can be said: experienced investigators agree that the spontaneous aberration rate for chromatid-type aberrations in direct preparations of bone marrow from numerous mammalian species is very low, generally below half a percent. This situation is different from that often found in tissue culture cells where the spontaneous aberration rate is often in the region of several percent and is influenced by numerous uncontrollable factors acting in cell cultures. In practice, however, the reliability of metaphase chromosome analyses in bone marrow all too often proves to be unsatisfactory. Many results suffer from too small numbers of metaphases and those few chromosome spreads that are analyzed often exhibit a technical quality which simply does not permit a satisfactory evaluation of the karyotype. Difficulties stemming from too small numbers are not met in the micronucleus test: the number of polychromatic erythrocytes is almost unlimited and the spontaneous rate of micronucleated cells is low and consistent, in the region of three per thousand. The test reacts positively to very low dosages of standard chromosome-breaking mutagens[4,7]. As mentioned above, the m.t., in contrast with metaphase scoring, is capable of indicating loss of entire chromosomes due to disturbances of the spindle apparatus. One thing the test cannot provide is the characterization of the nature of the nuclear disturbance that has led to a positive score. Therefore, it is clear that any unexpected test result obtained with the m.t. must be clarified with appropriate studies *in vivo* or *in vitro*. This is the place where metaphase chromosome studies come into their right.

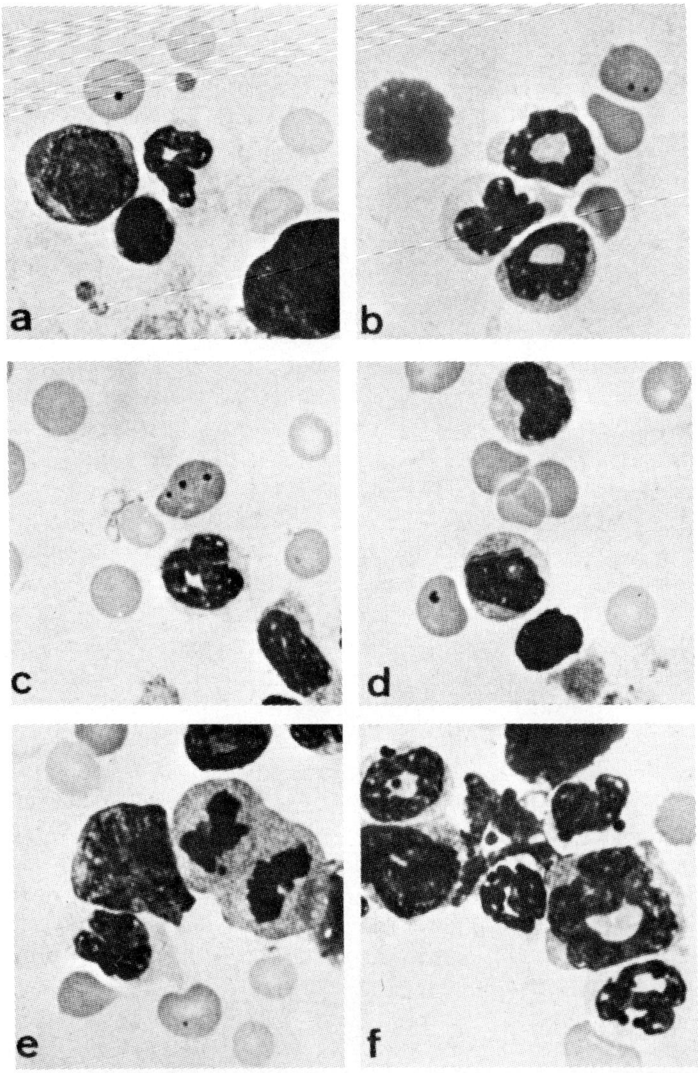

Fig. 1. (a) Typical morphology of micronuclei in polychromatic erythrocytes: a single, round (Feulgen-positive) body with a diameter of 1/20 to 1/5 of an erythrocyte. (b) and (c) Even after treatment with high doses of chromosome-breaking agents, multiple micronuclei are present in a minority only of all micronucleated cells. (d) A small proportion of the micronuclei are almond- or ring-shaped. (e) A small micronucleus in an erythrocyte and a lagging element in an anaphase. (f) Micronuclei in nucleated bone marrow cells are difficult to distinguish from normal nuclear lobes or projections; therefore, they are *not* evaluated quantitatively in this test.

THE TECHNICAL PROCEDURE

Normally, test substances are applied intraperitoneally or *per os*. It should not be overlooked that the i.p. injections, especially in rats, sometimes end up unnoticed in the cecum. The consequence is that the test substance gets into the blood to a very reduced extent, at best. This is one more reason not to work with too small numbers of animals. Experience with known chromosome-breaking substances suggests that, when compounds of unknown mutagenic activity are tested, a subacute treatment over about 30 h should be chosen[5]. The intention is to submit a great proportion of the cell population to the tested compound during two cell cycles[6]. Whether this expectation can be fulfilled with two applications, *e.g.* 30 and 6 h before the animals are killed, depends on the pharmacodynamic properties of the test compound. The dose should always cover a wide range, from the usage level up to the highest tolerable dose. Before a test is done, problems of solubility, mode of application and dosage must be clarified with the aid of experts in toxicology.

(a) Extraction of bone marrow from small animals

Before the animals are killed, a 5-ml centrifuge tube is filled with fetal calf serum for each individual. This admittedly expensive fluid has proved superior to all other rinsing solutions or cheaper sera tried so far. With substitutes the cells were often damaged or the erythrocytes were partially agglutinated resulting in preparations that were unusable or consumed too much time for analysis.

From a single mouse one can easily obtain enough bone marrow cells for several slides. For the m.t. the animals are, of course, *not* pretreated with metaphase-blocking agents. From the freshly killed animal both femora are removed *in toto*, which means that one is cutting through pelvis and tibia. The bones are then freed from muscle by the use of gauze and fingers. By gentle traction the distal epiphyseal portion is torn off together with the rest of the tibia and the surrounding muscle. The proximal end of the femur is carefully shortened with scissors until a small opening to the marrow canal becomes visible. With the needle of appropriate size mounted, about 0.2 ml serum is pulled from the tube into a disposable plastic syringe. Then the needle is inserted a few mm into the proximal part of the marrow canal which is still closed at the distal end. Next, the femur is submerged completely in the serum and squeezed against the tube to prevent the bone from slipping off the needle. Subsequently, the marrow is aspirated; should the needle have become obstructed, the serum in the syringe is first pressed out. After several gentle aspirations and flushings, the process is repeated from the distal end of the femur. The bone marrow cells should get into the serum as a fine suspension and not in the form of gross particles.

(b) Preparation of the smears

The tube is centrifuged at 1000 rev./min for 5 min. The supernatant is removed with a Pasteur pipette. If the sediment is large, half a drop of serum is left; if it is minute, all the supernatant is drawn off. The cells in the sediment are carefully mixed by aspiration into the capillary part of a fresh, siliconized Pasteur pipette. A small drop of the viscous suspension is put on the end of a slide and spread by pulling the material behind a polished cover glass held at an angle of 45 degrees. The size of the droplet is chosen so that *all* material is used up at a distance of 2–3 cm. The preparations are then air dried.

(c) Staining

The best results are obtained if staining takes place the day following preparation. If staining is done immediately, the slides must be flamed shortly. Staining is carried out in ordinary vertical staining jars according to the following procedure:

stain for 3 min in undiluted May-Gruenwald solution; stain for 2 more min in May-Gruenwald diluted with distilled water 1:1; stain for 10 min in Giemsa diluted with distilled water 1:6; rinse in distilled water; blot dry with filter paper; clean back side of slide with methanol; clear in xylene for 5 min, and mount with cover glass. (*Note:* contrary to hematological routine it is necessary to mount these preparations so that they are analyzable at high *and* medium magnifications.)

(d) Analyzing the slides

First the slides are screened, at medium magnification, for regions of suitable technical quality, where the cells are well spread, undamaged and perfectly stained. Such regions are normally located in a zone close to the end of the smear. A perfect morphology of the nucleated cells serves as criterion for good quality, even though the nucleated cells are not evaluated in the test. The erythrocytes must be well spread, neither globular nor having slurred contours. Their staining has to be vigorous, red in mature erythrocytes (anulocytes) and with a strong bluish tint in the immature forms (polychromatic erythrocytes).

In control preparations the proportion of erythrocytes in the regions described is in the order of 8%, and roughly half of these are polychromatic. At medium to high magnification one thousand polychromatic erythrocytes are screened for the presence of micronuclei. The scored elements are the micronucleated cells and not the number of micronuclei. As an important control it is necessary to register, separately, the number of micronucleated mature red erythrocytes. If their incidence exceeds a limit of five in the fields containing one thousand polychromatic erythrocytes the suspicion arises that the preparation contains artifacts resembling micronuclei. Such a suspicion has to be substantiated or repudiated by careful study. Although a single preparation contains thousands of analyzable erythrocytes it is, for many reasons, preferable to increase the number of animals investigated instead of the number of cells studied per animal.

REMARKS CONCERNING SPECIFIC POINTS OF THE METHOD

Test animals

To ensure good quality preparations it is advisable to use adolescent animals, *e.g.* outbred mice 7–12 weeks old. Fat tissue in the bone marrow of older animals has a deleterious effect on the smears. In very young animals the percentage of polychromatic erythrocytes is higher, reflecting a higher turn-over. Real differences in the sensitivity between animals of different age groups have not been demonstrable, nor were differences found between the two sexes (SCHMID AND SUTTER, unpublished results).

In the classical method the entire content of the femora is used for the preparations; therefore bone-marrow cells and peripheral blood are always in a given relation to each other which is influenced only by the treatments the animal received. If, in larger animals or in man, bone marrow is obtained by puncture, then the admixture of peripheral blood varies greatly, owing to unpredictable lesions of blood vessels. Again, large amounts of fat render it difficult to obtain satisfactory slides.

The technique of preparation

For good reasons the technical steps of preparation have been described explicitly. Despite its simplicity the method demands ample practice, not only with controls but especially with animals treated with strong mutagens. In these, the amount of bone-marrow cells can be reduced drastically, even within 30 h, and preparation becomes much more demanding. Trying to analyze preparations of unsatisfactory quality wastes time and is likely to produce unreliable results.

Artifacts

The micronucleus test is a relatively simple method yet it was not devised for evaluation by fools. The test offers numerous possibilities for checking the validity of the results obtained. If, *e.g.*, there should be a mechanism producing inclusions in erythrocytes by acting on the expulsion of the nuclei from erythroblasts instead of by breaking chromosomes or causing non-disjunction, a thinking investigator would immediately recognize this hypothetical mechanism by the mere observation that micronuclei are missing from the nucleated cell types on the same slides.

Artifacts that are sometimes barely distinguishable from micronuclei do occur, although rarely. They can be formed by residues of stain, or, especially in the bone marrow of rats, by basophilic granules of destroyed precursors of basophilic leucocytes. The presence of such artifacts can be recognized by their pattern of distribution in the preparation. In contrast with real micronuclei, which under the conditions of the experiment are bound to be restricted to the polychromatic forms, artifacts of the origins mentioned are distributed indiscriminately over all types of erythrocyte. This is why the analyst has to register "micronuclei" even when they appear in red, mature erythrocytes.

A different kind of observation was reported by JENSSEN[2] and confirmed in our laboratory. After treatment of animals with aminoacridines, a high proportion of polychromatic erythrocytes exhibit multiple dust- or thread-like, basophilic inclusions (Fig. 2). The morphological picture of this phenomenon, however, is completely different from that of micronuclei produced by chromosome-breaking agents or inhibitors of mitosis. Furthermore, they are Feulgen-negative and, being fragments of nucleoli, they stain with RNA-specific reagents (MAIER AND SCHMID, in preparation).

The protocol of analysis

In the section on analysis the routine procedure consisting of evaluation of

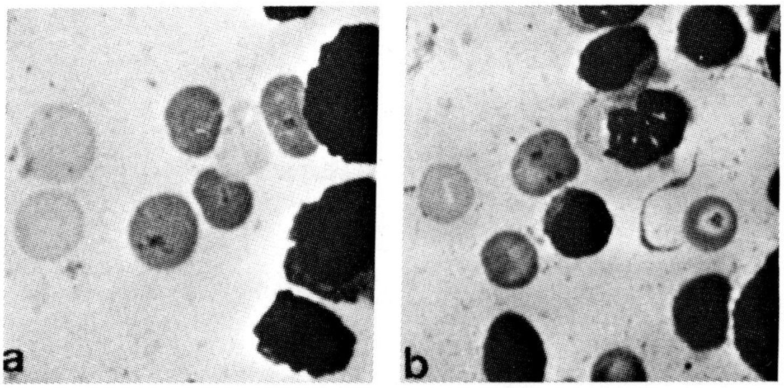

Fig. 2. Multiple, dust- or thread-like, basophilic inclusions in polychromatic erythrocytes after treatment of the animals with aminoacridines. Their morphology is completely different from micronuclei produced by chromosome-breaking agents or inhibitors of mitosis.

1000 polychromatic erythrocytes per animal was described. A second possibility may be considered, depending on the nature of the substances to be tested[7]. This second procedure is based on the following considerations. The transition from polychromatic to mature erythrocytes is a short but nevertheless continuous process. Whether an erythrocyte in transition should be scored as polychromatic or as red will always be a more or less subjective decision made by the investigator. In controls, as well as after treatments not markedly affecting the proliferation of the bone-marrow cells, the relation of polychromatic to mature erythrocytes is fairly consistent, in the region of 1:1. Analysis of 2000 total erythrocytes in place of 1000 polychromatic erythrocytes reduces the number of subjective decisions; furthermore, the results of different investigators are easier to compare. If this protocol is restricted to the use in tests where there is no marked reduction in the proportion of young erythrocytes it is perfectly acceptable in toxicological screening. As soon as dose–effect relationships are studied on chromosome-breaking compounds, or if the normal proliferation of bone marrow cells is disturbed for other reasons, *e.g.* by antimetabolites, scoring has to be restricted to the population of polychromatic erythrocytes. Under such influences the bone marrow regularly becomes flooded with peripheral blood and the mentioned 1:1 relation becomes badly distorted in favor of the irrelevant mature red cells.

If the m.t. is carried out only occasionally, and on a wide spectrum of substances, the laboratory should use only the classical method of analysis; if the test is used frequently and on substances from which it is known in advance that they have no drastic hematological effects, then the second mode of evaluation is preferable.

Time consumption for the m.t.

Apart from the training of the investigator, time consumption for the m.t. depends, to a great extent, on the quality of the preparations. An unproblematical test using four different dosages and five animals per dose keeps the tester busy for about a week. For preparation and staining he needs about a day and half, and the analysis of 1000 erythrocytes per animal takes from 30 to 60 min. The test can take much longer when low positive results ask for repeats, additional dose ranges or other modes of application.

REFERENCES

1 Boller, K., and W. Schmid, Chemische Mutagenese beim Säuger. Das Knochenmark des Chinesischen Hamsters als *in vivo*-Testsystem. Hämatologische Befunde nach Behandlung mit Trenimon, *Humangenetik*, 11 (1970) 35–54.

2 Jenssen, D. S., Chromosomal damage induced by some aminoacridines in mouse bone marrow cells, (Abstract), *4th Annual Meeting of the European Environmental Mutagen Society, Heidelberg, Germany, 1974*.

3 Matter, B., and W. Schmid, Trenimon-induced chromosomal damage in bone marrow cells of six mammalian species, evaluated by the micronucleus test, *Mutation Res.*, 12 (1971) 417–425.

4 Matter, B. E., and J. Grauwiler, Micronuclei in mouse bone marrow cells. A simple *in vivo* model for the evaluation of drug-induced chromosomal aberrations, *Mutation Res.*, 23 (1974) 239–249.

5 Schmid W., D. T. Arakaki, N. A. Breslau and J. C. Culbertson, Chemical mutagenesis. The Chinese hamster bone marrow as an *in vivo* test system, I. Cytogenetic results on basic aspects of the methodology, obtained with alkylating agents, *Humangenetik*, 11 (1971) 103–118.

6 Schmid, W., Chemical mutagen testing on *in vivo* somatic mammalian cells, *Agents and Actions*, 3 (1973) 77–85.

7 Von Ledebur, M., and W. Schmid, The micronucleus test. Methodological aspects, *Mutation Res.*, 19 (1973) 109–117.

ADDENDUM

Results obtained with the micronucleus test have been summarized by Schmid[2] and new data obtained with ten model mutagens: alkylating agents, cytostatic antibiotics, spindle poisons and antimetabolites, were published by Maier and Schmid[1].

1 MAIER, P., AND W. SCHMID, Ten model mutagens evaluated by the micronucleus test, *Mutation Res.*, 40 (1976) 325–338.
2 SCHMID, W., The micronucleus test for cytogenetic analysis, in *Chemical Mutagens, Vol. 4*, Academic Press, New York, 1976, in press.

A COLLABORATIVE STUDY TO MEASURE INTERLABORATORY VARIATION WITH THE IN VIVO BONE MARROW METAPHASE PROCEDURE *

D. JACK KILIAN [1], FRANCES M. MORELAND [2], MARJORIE C. BENGE [1], MARVIN S. LEGATOR [3] and ELBERT B. WHORTON, JR. [4]

[1] *Occupational Health and Medical Research, Dow Chemical U.S.A., Freeport, Tex.,* [2] *Genetic Toxicology Branch, Food and Drug Administration, Department of Health, Education and Welfare, Washington, D.C.,* [3] *Division of Environmental Toxicology and Epidemiology, Department of Preventive Medicine and Community Health, University of Texas Medical Branch, Galveston, Tex. and* [4] *Division of Biometry, Department of Preventive Medicine and Community Health, University of Texas Medical Branch, Galveston, Tex. (U.S.A.)*

Summary

Six cytogenetic laboratories joined in a collaborative study of rat chromosome aberrations to measure interlaboratory variation in results of standardized procedures and to devise methods to minimize interlaboratory differences. A preliminary workshop was held to resolve scoring differences, to develop a joint protocol and common glossary, and to reach agreement on uniform reporting methods. Osborne-Mendel rats from a common source were sent to each laboratory. Triethylenemelamine (TEM) was used at doses of 100, 200, 300 and 400 μg/kg to induce clastogenic effects; results were compared to those of a control group of untreated animals. Femoral bone marrow cells were evaluated with the scorers unaware of the dosage. Final results showed highly significant dose effects with the test compound, and most laboratories showed a similar pattern of dose response. The most significant index dealt with the percentage of abnormal cells. All other variables and related transformations showed greater degrees of insensitivity. This study illustrated that rat cytogenetic analysis is an effective test system for evaluation of a compound for mutagenic potential, particularly for the index reflecting the proportion of abnormal cells.

---

* This paper was presented in part at the Fifteenth Annual Meeting of the Society of Toxicology, March 14—18, 1976, Atlanta, Ga., U.S.A.

Introduction

Cytogenetic analysis is a widely employed indicator system for the evaluation of chemically induced mutations. It allows objective evaluation of insults to the genetic material and is the only method that permits direct visual analysis of damage to chromosomes. Chromosomal abnormalities — numerical and structural — occur in one out of 200 live births.

It is generally agreed that a mutagenic test system should meet certain basic requirements. First, the system must be relevant to human concern and indicative of the type of genetic damage implicated in human disorders. Second, the test should take into account the activation and detoxification features of mammalian metabolism. Finally, the material and time costs should not be prohibitive. (In the view of many investigators, banding techniques are too time consuming and too variable to be considered in routine cytogenetic analysis for clastogens [4].) In vivo cytogenetic evaluation meets these criteria and is one of the basic tests recommended for chemicals, foods, cosmetics, and the like.

Difficulties have been encountered, however, in attempts to make meaningful comparisons of cytogenetic results reported by different laboratories. These difficulties may lie primarily in the lack of agreement on a standardized nomenclature/classification system for scoring and reporting chromosome aberrations. Problems have also been seen, although to a lesser extent, in adoption of uniform laboratory techniques.

Similar difficulties have been found in other areas of chemical toxicology. In what might appear to be the relatively straighforward evaluation of chemically induced animal eye and skin irritation, Weil et al. [6] and Weil and Scala [5] reported considerable variation among 25 laboratories. These authors concluded that the primary reason for interlaboratory differences lay in the reading of reactions, as opposed to variations in the interpretation and performance of procedures. It was pointed out, however, that procedural difference could not be completely excluded as a component of the problem.

A step toward resolution of the problems of variability in cytogenetic analysis was taken in 1973, when a collaborative study of rat chromosome aberrations was performed by four laboratories [1], three of which also participated in the common effort reported here.

In the 1973 project, participants jointly carried out procedures recommended by the Ad Hoc Committee on Chromosome Methodologies in Mutation Testing of the Environmental Mutagen Society [3]. Participants worked together at a central laboratory and subsequently evaluated jointly prepared slides in their own facilities with available equipment. Upon analysis of the results, significant interlaboratory variability was found to exist. The differences appeared to vary with the category of aberration scored.

In an independently sponsored, follow-up attempt to enlarge the area reported in 1973, it was decided to proceed with a further study, utilizing a greater number of laboratories and a revised protocol. The purpose was first, to study and test interlaboratory variation following the application of standardized procedures at independent facilities and second, to devise reporting methods that could minimize interlaboratory differences. Government, academic, industrial, and independent commercial laboratories were invited to take part *.

A cooperative workshop, preceded and followed by rounds of practice scoring, was held initially to isolate areas of potential conflict in scoring and testing procedures and to attempt to reach consensus in problem areas. Following the workshop, animals from a common source were tested with a known mutagen and scored by participants at each of the laboratories according to a definitive, jointly devised protocol. Data were submitted to a coordinator for tabulation and, subsequently, to an independent facility for statistical analysis.

In an effort to test agreement among the participants, it was decided to compare scoring terminology and to confer about the results. All participants were sent identical photocopies of 10 human and 12 rat metaphase spreads (Fig. 1), five of which were marked to show a commonly encountered, but undefined, chromosome aberration. Participants were asked to score the marked abnormalities in their accustomed manner.

Fourteen scorers from eight laboratories responded, and a striking degree of variation was seen. For none of the five abnormalities did all 14 participants give the same definition. In one photograph, ten varying forms of terminology were used to score the single, marked aberration.

Following this initial round of practice scoring, a two day problem-solving workshop was held. Three major areas of consensus evolved as a consequence: general agreement as to interpretation of a common glossary (Fig. 2); formats for a raw data score sheet (Fig. 3); an 80-column computer punch card (Fig. 4), and data-card column design (Fig. 5); and a definitive protocol for use at the separate laboratories. A listing of field and column code definitions is given in the Appendix. Mark-sense score sheets for animal data (Fig. 6) † and human data were also developed as the study progressed.

In contrast to the pre-workshop findings, a post-workshop round of practice scoring showed good agreement on interpretation of the nomenclature and good adaptation to the standardized format of the raw data scoring sheet. The greatest variation was seen for the category of gaps. In view of this, it was decided to proceed with optional scoring of gaps during the independent laboratory phase of the study, but to omit this category from consideration in the

* Laboratories from ten institutions were involved in the initial phase of the project: Division of Genetics, Roger Williams General Hospital, Providence, R. I. (Dr. Legator); Occupational Health and Medical Research, Dow Chemical U.S.A. (Dr. Kilian and Ms. Benge); Toxicology Section, Health Safety Laboratory, Eastman Kodak Company, Rochester, N.Y. (W.J. Krasavage); Biomedical Science, Chronic Toxicology Branch, Consumer Product Safety Commission, Washington, D.C. (R.E. Osterberg and J. Murphy); Genetic Toxicology Branch, Food and Drug Administration (Ms. Moreland); Reproductive Genetics Unit, George Washington University School of Medicine, Washington, D.C. (C.B. Jacobson and J. Varhidy); Cell Biology Section, Hazelton Laboratories, Vienna, Va. (R. Kapp); The Hine Laboratories, Inc., San Francisco, Calif. (J. Nemenzo); Medical Research Laboratory, Pfizer, Inc., Groton, Conn. (H. Holden); and Life Sciences Division, Stanford Research Institute, Menlo Park, Calif. (A. Mitchell). Six Laboratories completed the study. Statistical analyses were performed by Dr. Whorton, University of Texas Medical Branch, and J. Springer, Division of Mathematics, Food and Drug Administration. T.B. Lloyd, Occupational Health and Medical Research, Dow Chemical U.S.A., provided editorial assistance.

† The mark-sense sheet for animal data was not used for final analysis of the results in this study, but it is felt that the format could be ideal for future cytogenetic work requiring rapid analysis of standardized data based on large numbers of observations.

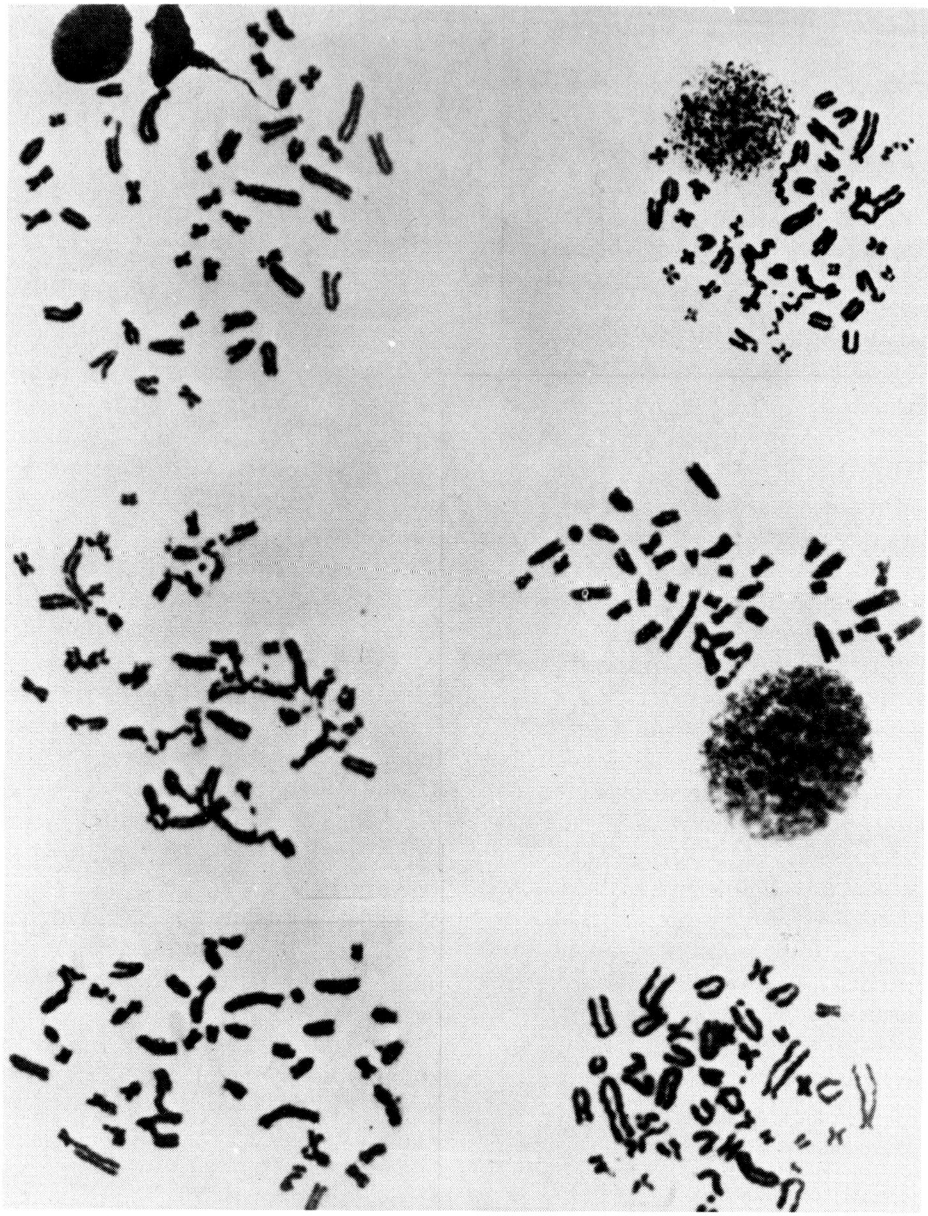

Fig. 1. Composite of rat metaphase spreads.

final statistical analysis. In general, it appeared that agreement among laboratories in the post-workshop practice round was good for cells with relatively few aberrations and best for the categories of dicentrics, rings, and exchanges.

An unexpected problem surfaced during the practice slide evaluation phase of the cooperative workshop when it was found that the quantity and quality of aberrant mitoses were less than expected with use of a known mutagen

GLOSSARY

Chromatid Aberration: change seen on only one arm of chromosome.

Chromosome Aberration: change seen on both arms of chromosome at identical sites.

Gap: aligned, non-staining area of discontinuity that is equal to or less than the width of the chromatid or that contains visible material connecting the proximal and distal portions of the arm; classified as either chromatid or chromosome aberration.

Break with Fragment: any separation that exceeds the width of the arm, or a discontinuity accompanied by disturbance of axial integrity, in association with an identifiable piece of displaced chromatid material; noted for either chromatid or chromosome aberrations.

Break without Fragment: any chromatid separation that is wider than the arm itself, but without identifiable chromatid fragments; can be classified as either chromatid or chromosome aberration.

Fragment: any free, displaced chromatid piece seen either in association or not in association, with a parent chromatid.

Marker: aberrant configuration involving the entire chromosome.

Exchange Figure: chromatid interchange, involving two or more chromosomes, with either symmetrical or asymmetrical distortion of the usual chromatid pattern.

**Translocation:* transfer of chromosomal material involving two or more chromosomes.

**Inversion:* exchange figure seen following two-site break with reverse, 180° turn of intermediate segment; can be paracentric or pericentric; difficult to identify.

Dicentric: chromosome with two centromeres; a type of marker.

Ring: chromosome whose ends have joined to form a double or single circle, with or without a centromere; a type of marker.

Miscellaneous Aberrations: events seen more rarely than gaps, breaks, and markers; includes numerical variations of genome.

Dot Fragments: a minute, usually circular, displacement of chromatid material.

Aneuploidy: irregular variation of the chromosome number; hypodiploid or hyperdiploid.

Polyploidy: increase in chromosome number in excess of the diploid and in multiple of the haploid number.

Endoreduplication: chromosome reduplication occurring in absence of any separation of sister chromatids.

Telomere Bleb: distal gap in non-satellite area.

Severely Damaged Cell: cell with ten or more aberrations of any type or with pulverization.

Pulverization: extreme fragmentation of the chromatid material.

*For tabulation of animal data, these relatively rare figures are included under exchanges.

Fig. 2. Glossary.

RAW DATA CYTOGENETIC SCORE SHEET

Normal Cells w/Gaps ____________________

Normal Cells w/o Gaps ____________________

Date: ________
Investigator: ________
Code #: ________
Slide #: ________
Slide quality: Excellent ______ Good ______ Poor ______ Unscorable ______

VERNIER	CHROMATID			CHROMOSOMAL			MARKERS			MULT AB	MISC AB	OBSERVATION
	GAP	BwF	Bw/oF	GAP	BwF	Bw/oF	EX	DIC	RG			
1.												
2.												
3.												
4.												
5.												
6.												
7.												
8.												
9.												
10.												
11.												
12.												
13.												
//												
50.												

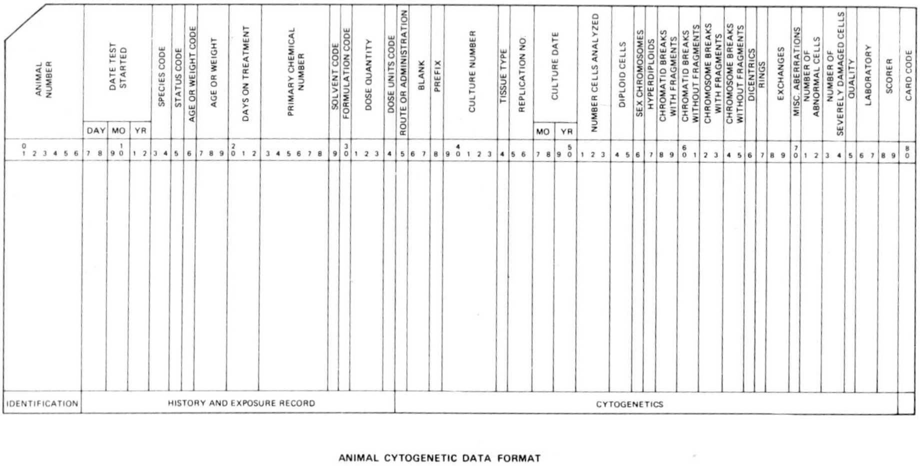

Fig. 4. Animal cytogenetic data card.

ANIMAL CYTOGENETIC DATA CARD FORMAT

	FIELD	*CARD COLUMN(S)*
IDENTIFICATION	Animal Number	1 - 6
HISTORY AND EXPOSURE RECORD	Date test started	7 - 12
	Species code	13 - 14
	Status code	15
	Age or weight code	16
	Age or weight	17 - 19
	Days on treatment	20 - 22
	Primary chemical number	23 - 28
	Solvent code	29
	Formulation code	30
	Dose quantity	31 - 33
	Dose units code	34
	Route of administration	35
	Blank	36 - 37
CYTOGENETIC DATA	Prefix	38
	Culture number	39 - 43
	Tissue type	44
	Replication number	45 - 46
	Culture date	47 - 50
	Number of cells analyzed	51 - 53
	Diploid cells	54 - 55
	Sex chromosomes	56
	Hyperdiploids	57
	Chromatid breaks with fragments	58 - 59
	Chromatid breaks without fragments	60 - 61
	Chromosome breaks with fragments	62 - 63
	Chromosome breaks without fragments	64 - 65
	Dicentrics	66
	Rings	67
	Exchanges	68 - 69
	Miscellaneous aberrations	70
	Number of abnormal cells	71 - 72
	Number of severely damaged cells	73 - 74
	Quality	75
	Laboratory	76 - 77
	Scorer	78 - 79
	Card code	80

Fig. 5. Animal cytogenetic data card format.

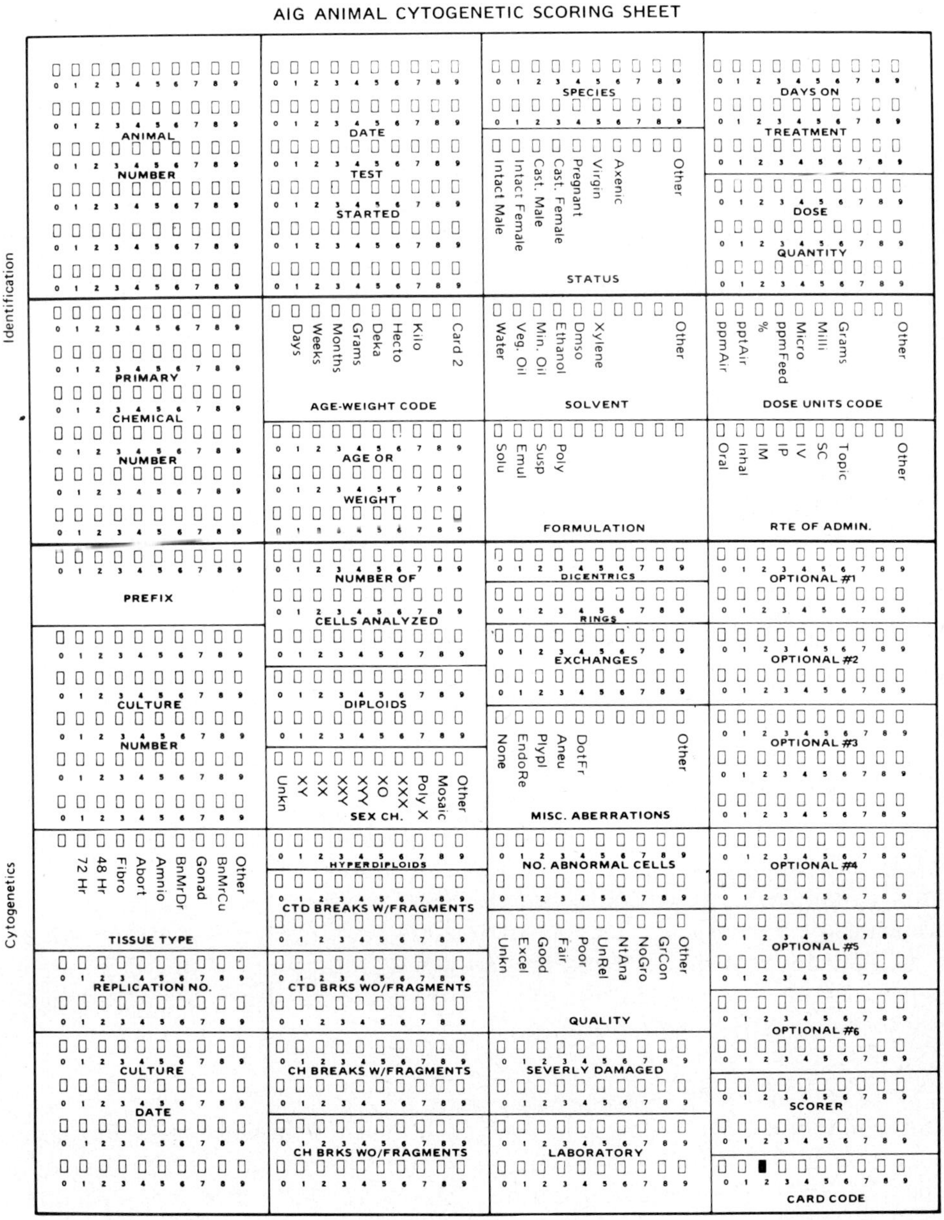

Fig. 6. AIG animal cytogenetic scoring (sensor) sheet.

(triethylenemelamine, TEM). After consultation with the supplier, it was concluded that the TEM had been stored without proper desiccation and, as a result, had polymerized. A new lot was obtained and tested on 12 animals. Results were scored by three of the participants, all of whom agreed that the fresh supply would be satisfactory for use in the common project.

Materials and methods

Treatment of animals

Fifty male Osborne-Mendel rats weighing between 125 and 150 g were sent to each laboratory at the same time from a common source (Rockland Farms, Gilbertsville, Pa.). Ten rats were used for each of four levels of the mutagenic agent, triethylenemelamine (TEM) (Lot 2075-J0239 supplied through the courtesey of E.W. Cantrall, Lederle Laboratories, Pearl River, N.J.), and ten rats were used as controls. The TEM was administered via intraperitoneal (IP) injection at dose levels of 100, 200, 300, and 400 μg/kg of body weight. The TEM was freshly prepared immediately prior to injection, using distilled water as the solvent. Intraperitoneal injections of distilled water, corresponding in amount to the TEM preparation, were given to the control animals.

Mitotic arrest and sacrifice

An IP injection of a standardized preparation of colchicine (Grade A, Lot 101195, Calbiochem, La Jolla, Calif.) was given to inhibit mitosis in each animal 22 h after treatment with the mutagen. The colchicine was prepared in Hanks balanced salt solution (HBSS) at a concentration of 4 mg/ml, pH adjusted to 7.4 with 0.1 N NaOH and injected i.p. at a dose of 4 mg/kg.

The animals were sacrificed by cervical dislocation 2 h after injection of the colchicine and 24 h after administration of the TEM and distilled water injections.

Bone marrow cell recovery

Skin and muscle tissue were removed from both femurs immediately after sacrifice, care being taken to avoid cutting the femoral artery. The femurs were severed distally with bone snips, and the bone marrow was aspirated through an 18-gauge (1.5 in.) needle into a 10-ml syringe containing 5 ml warm HBSS. The HBSS and enough bone marrow to give a total volume of 10 ml were dispensed into a 15-ml centrifuge tube containing 3 ml HBSS and mixed vigorously to assure disassociation of the cells.

Incubation and harvest

The cell-HBSS mixture was centrifuged in an International Centrifuge (Model UV) at 1000 rpm for 10 min, and the supernatant fraction was discarded. An 8-ml, hypotonic solution of 0.075 M KCl was added to the residual cell pellet, and the material was mixed thoroughly with a disposable, siliconized pipette. The cells were incubated in a water bath maintained at 37°C for 20 min and then recentrifuged at 1000 rpm for 10 min. The supernatant was discarded. Freezer-chilled Carnoy's fixative (methanol : acetic acid, 3 : 1) was added to give a total volume of 5 ml/tube. The cells were gently resuspended and then refrigerated at 4°C for 2 h. The cells were again centrifuged for 10 min at 1000 rpm, and the supernatant was discarded. Fresh Carnoy's solution was added to give a volume of 5 ml/tube; the cells were resuspended; and the tubes were refrigerated for a minimum of 12 h at 4°C before slide preparation was started.

Slide preparation

The fixed cells were centrifuged for 10 min at 1000 rpm; the supernatant was discarded; and 0.8—1.0 ml fresh Carnoy's solution was added to each tube. The cells were resuspended and then dropped on labeled slides that had been chilled in ice water or shaved ice. Slides were flame-dried and drained perpendicularly for 1—2 min. Slides were stained in 10% Giemsa; dehydrated with double washes of acetone, acetone : xylol (1 : 1), and pure xylol; and mounted with Permount.

Fifty metaphase spreads per animal were analyzed, when possible. Dose records and slides were coded for blind analysis and to assure anonymity of each laboratory.

Results

Statistical analyses of the data generated by this study were related to eight primary indices of aberration: (1) chromatid breaks, (2) chromosomal breaks, (3) exchange figures, (4) total aberrations, (5) severely damaged cells, (6) abnormal cells, (7) dicentrics, and (8) rings. In all, we investigated 30 variables, including 21 primary variables and 9 variable transformations. Of the 30 variables, 6 were judged to provide the most meaningful interpretations of the overall laboratory differences and are presented here.

The analyses and interpretations centered on questions of consistency among the laboratories across the doses of TEM used. All variables — specific and general — were measured using the metaphase scoring definitions outlined in the glossary. It should be noted that calculations for chromatid breaks, chromosome breaks, and exchange figures were made on the basis of aberrations found per *non-severely damaged cell.* The rationale for our choice of this unit of measurement is presented below in the Discussion. Figs. 7—12 are graphic representations of the mean values by laboratory according to dose levels administered for each variable.

The first variable presented in this report is that of AA CT, defined as the average number of *chromatid breaks* per non-severely damaged cell. Referring to Fig. 7, it is apparent that the average number of chromatid breaks increases with increased dose. There is variation among the laboratories, and this is greater at higher doses.

The second variable, AA CH, is defined as the average number of *chromosomal breaks* per non-severely damaged cell. When the results of calculations involving AA CH are graphed (Fig. 8), it can be seen that there is variation among laboratories with increased doses, despite the fact that the overall dose-response pattern appears to be smoothly increasing across doses.

The third in our series of variables involved AA EX, the graph for which is shown in Fig. 9. AA EX is defined as the average number of exchange figures per non-severely damaged cell. It was found that the general pattern of response across doses was quite similar to that observed for chromatid and chromosome breaks, although there was less variation among the laboratories.

The fourth variable, TR AB CL treats the data for the proportion of *abnormal cells* as

$$\text{TR AB CL} = \arcsin \sqrt{\text{proportion of abnormal cells}},$$

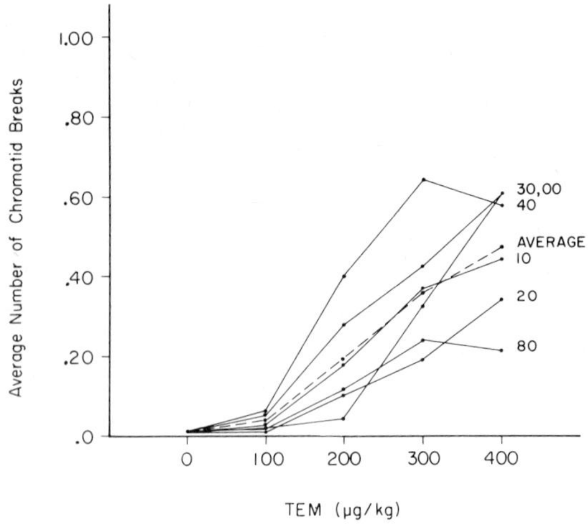

Fig. 7. Graph showing the average number of chromatid breaks per non-severely damaged cell according to laboratory and dose level of TEM.

and is measured in degrees. When *no* abnormal cells were observed, the proportion was defined as

$$\frac{1}{4\ (\text{number of cells scored})}$$

before applying the transformation [2].

Each animal's proportion of abnormal cells (as well as the proportion of severely damaged cells) was transformed in this manner in an attempt to stabilize the animal-to-animal variability prior to using the analysis of variance procedures. A graphic display of the results for TR AB CL is shown in Fig. 10. A

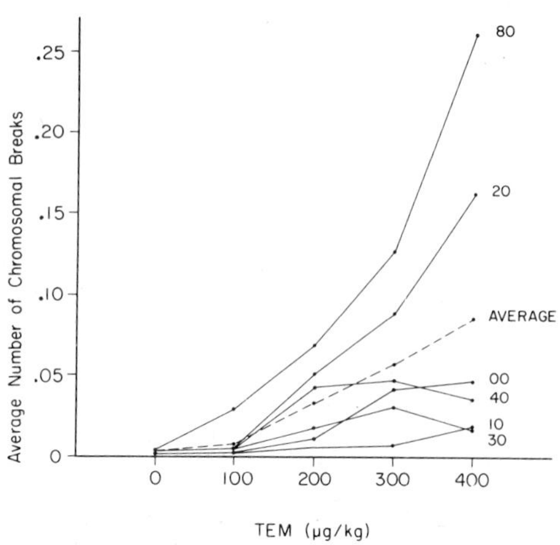

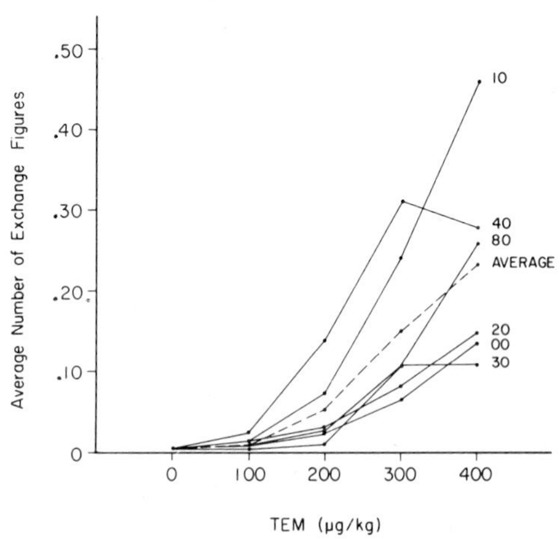

Fig. 8. Graph showing the average number of chromosomal breaks per non-severely damaged cell according to laboratory and dose level of TEM.

Fig. 9. Graph showing the average number of exchange figures per non-severely damaged cell according to laboratory and dose level of TEM.

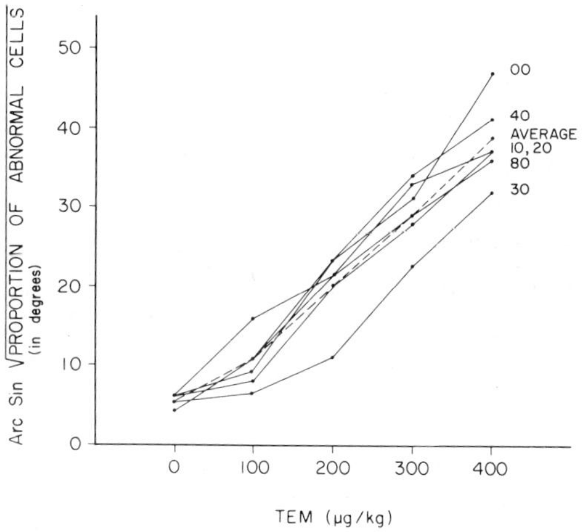

Fig. 10. Graph showing the average arc sin proportion of abnormal cells according to laboratory and dose level of TEM.

cell was considered abnormal when it contained at least one aberration. This variable, of all those considered, appeared to be the most consistent, both for the rate of increase across doses and for the overall agreement among laboratories.

The fifth aberration index, TR SV CL, deals with the proportion of *severely damaged cells.* TR SV CL was derived by using the transformation applied to the proportion of abnormal cells, above. A cell was scored as severely damaged when it contained at least ten aberrations. Results are graphed in Fig. 11. The dose-response patterns by laboratory are similar to those found with the vari-

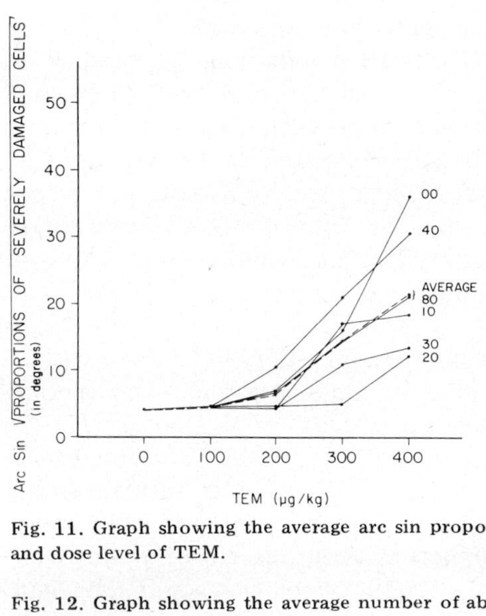

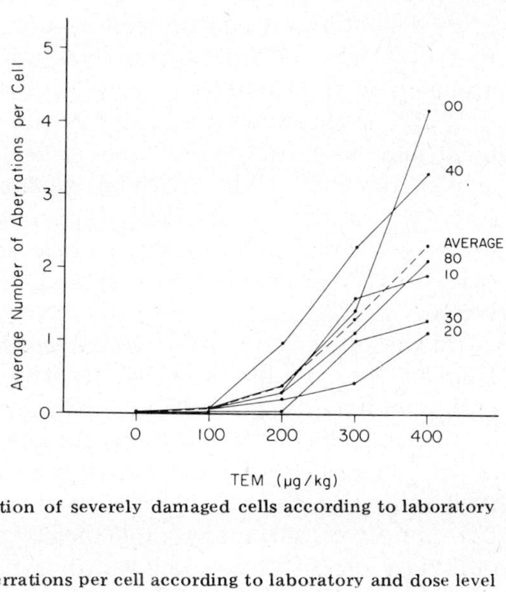

Fig. 11. Graph showing the average arc sin proportion of severely damaged cells according to laboratory and dose level of TEM.

Fig. 12. Graph showing the average number of aberrations per cell according to laboratory and dose level of TEM.

ables discussed above, and there is a tendency for the overall dose-response to be laboratory-dependent, despite the positive increase shown by most laboratories across doses.

The final variable, A ABS, was defined as the *average* number *of aberrations* per cell scored. The total number of aberrations was calculated as the sum of chromatid breaks, chromosomal breaks, dicentrics, rings, exchange figures, and aberrations in severely damaged cells — those with at least 10 aberrations — which were counted as exactly 10.

Fig. 12 shows a graphic representation of the results for A ABS. As could be expected, results are nearly identical with those for TR SV DM.

In summary, for each of the six indices of cell aberration presented here, it was found, using analysis of variance procedures, that significant ($P < 0.001$) dose effects existed overall, that some laboratories were significantly different ($P < 0.001$) from others, and that some laboratories showed significantly different ($P < 0.001$) dose-response curves. Generally the differences among laboratories were found to increase as the doses of TEM increased. No laboratories were significantly different at the control dose. It was also found that the overall consistency in laboratory dose-response patterns was greater for some variables than for others. The most consistent dose-response pattern among laboratories was shown by the variable dealing with the proportion of abnormal cells (TR AB CL).

Discussion

This collaborative research project was *not* designed to demonstrate that TEM is a known clastogen, but, rather, to utilize the well-recognized properties of this compound to test the extent of variation among laboratories using standardized, jointly devised methods. The data were analyzed to show the consistency, or lack of consistency, among the laboratories across doses.

In the interpretation of these results, however, it is important to recognize that there are conceptual problems to be examined regarding the validity or accuracy of the variables in measuring the extent of chromosomal aberration.

It can be shown that a variable that contains a constant error in its measurement may be entirely acceptable under certain circumstances. For example, in an experimental study in which all animal weights are *consistently* overmeasured by exactly 5 g, the bias of overmeasurement will not affect adversely the analyses and interpretations of average weight differences among groups. In this case, the *differences* remain unbiased, although each group's average weight is biased.

On the other hand, if the animal weights are overmeasured in an *inconsistent* manner, the variable may be of little value in making comparisons among groups and in estimations of each group's average weight.

This concept has implications for the results reported here. The category of severely damaged cells was established because of the difficulty in identifying each and every aberration when there had been marked injury, and when there were, in effect, too many anomalies to tabulate without benefit of a photographic enlargement. It was dedided, therefore, to define a cell as severely damaged when the scorer found at least 10 aberrations of any type and to assign an arbitrary value of *exactly* 10 non-specific aberrations for use in calculations of the variable for the average number of all aberrations (A ABS).

Use of this procedure causes concern about the validity of results calculated for each of the primary indices of cell aberration, except those dealing with the proportion of abnormal cells (TR AB CL) and the proportion of severely damaged cells (TR SV CL).

To illustrate, the variable dealing with chromatid breaks (AA CT) was measured as the total number of chromatid breaks observed in all cells scored. Some cells, however, were defined as severely damaged and these, it is assumed, contained chromatid breaks. Since the true number of chromatid breaks is unknown there must be a question about the accuracy of this count as a true measure of what exists. We find, in fact, the variable dealing with chromatid breaks reflects only those found in the non-severely damaged cells.

Calculation of the average number of chromatid breaks per cell scored could be done in terms of (1) the total number of cells scored or (2) the number of non-severely damaged cells scored. If we choose to use (1), the numerator does not relate appropriately to the denominator. With the use of (2), however, the numerator does relate to the denominator, but the resulting average is not a valid reflection of the actual average we are seeking.

Use of the number of non-severely damaged cells to calculate the average will also produce underestimation of the actual average sought, since the number of severely damaged cells (each of which is assumed to contain an increasing but unknown number of chromatid breaks) increases as the dose increases. It can be seen, therefore, that as the dose increases, the more underestimated the average becomes. Furthermore, it is difficult, if not impossible, to determine how much this underestimate is. Consequently, one could observe a small dose-response effect when the true slope is large, the truth being masked by those chromatid breaks that are never measured, yet which may exist in the severely damaged cells.

This problem also complicates the interpretation of results dealing with chromosome breaks, dicentrics, rings, exchange figures, and the average number of *all* aberrations per cell scored. Recognition of these problems does not, of course, negate use of these indices as indicators of chromosome damage, but in the event of severely damaged cells, caution should be used in the interpretation of results.

In summary, we find that final results show highly significant overall dose effects with the test compound (TEM) and that, while significant differences exist among laboratories, most had a similar pattern of dose-response. The most significant index dealt with the percentage of abnormal cells. The other variables and related transformations showed greater degrees of insensitivity.

It should be noted that four of the authors were participants in the initial collaborative study [1]. In the initial study, in which preparations from the same animals were scored by different investigators, the variable representing the number of aberrations per cell appeared to be the most sensitive index of damage. There was also fairly close agreement among the laboratories for the category of percentage of cells with aberrations. In the second study, however, the proportion of abnormal cells to total cells scored seemed the best indicator.

A comparison of the two studies was made, and a review of original data from the first study showed differences with respect to scoring and tabulation. In the initial study, cells were scored only once on the basis of the most severe aberration. Therefore, a cell with three chromatid breaks, two chromosome breaks, and one exchange figure would have been counted as an exchange figure only. In the tabulation of data, all breaks were included in a single category; chromatid breaks were counted as one and chromosome breaks were counted as two. There was difficulty in statistical analysis when only totals of gaps, breaks, rearrangements, and cells with greater than ten aberrations were submitted. In view of this, data from the first study were retabulated to give totals of abnormal cells. Reviewed on this basis, it appeared that, even in the initial study, the best agreement between laboratories was achieved in the category reflecting the proportion of abnormal cells.

In contrast to the initial study, this present investigation was conducted using several concentrations of TEM, rather than the single dose of a positive control, trimethylphosphate. The data would indicate that the most accurate range in which to achieve correlation between laboratories will be found at concentrations of test compounds that show moderate activity in producing chromosomal abnormalities. It may be that at higher concentrations animal variation and scorer bias are amplified and many of the severely damaged cells are eliminated, thus confounding the results.

In comparing these results with other collaborative studies in the field of toxicology, it appears that the evaluation of chemically induced chromosomal abnormalities is one of the more consistent toxicological procedures available. This study illustrates that rat cytogenetic analysis, when used with caution, is an effective test system for evaluation of a compound for mutagenic potential, particularly for the index reflecting the proportion of abnormal cells.

References

1 Legator, M.S., K.A. Palmer and I.D. Adler, A collaborative study of in vivo cytogenetic analysis. I. Interpretation of slide preparations, Toxicol. Appl. Pharmacol., 24 (1973) 337—350.

2 Mosteller, F. and C. Yourty, Tables of the Freeman-Tukey transformations for the binomial and Poisson distributions, Biometrika, 48 (1961) 433—440.

3 Nichols, W.W., P. Moorhead and G. Brewer, Chromosome methodologies in mutation testing; ad hoc committee report. Toxicol. Appl. Pharmacol., 22 (1972) 269—275.

4 Shaw, M.W., The uses of banding techniques for the identification of human diseases of cytogenetic origin, Environ. Health Perspect., 6 (1973) 151—156.

5 Weil, C.S. and R.A. Scala, Study of intra- and interlaboratory variability in the results of rabbit eye and skin irritation tests, Toxicol. Appl. Pharmacol., 19 (1971) 276—360.

6 Weil, C.S., C.P. Carpenter, J.S. West and H.F. Smythe, Jr., Reproducibility of single oral dose toxicity testing, J. Am. Ind. Hyg. Assoc., 27 (1966) 483—487.

Appendix

Animal Card

Definition of Fields and Field Codes

ANIMAL NUMBER: Six-digit identification number; right justified.

DATE TEST STARTED: Six-digit date in the form of day/month/year; e.g., 011274 for December 1, 1974.

SPECIES CODE: Two-digit number code denoting the species.
00 = Cat
01 = Chicken
02 = Cow
03 = Dog
04 = Frog
05 = Hamster
06 = Horse
07 = Mouse
08 = Pig
09 = Rabbit
10 = Rat
11 = Sheep
12 = Shrimp
13 = Turkey
14—99 = Breed, Variety or Strains

STATUS CODE: One-digit denoting the status of the animal.
0 = Intact Male
1 = Intact Female
2 = Castrated Male
3 = Castrated Female
4 = Pregnant
5 = Virgin
6 = Axenic
7 =
8 =
9 = Other

AGE OR WEIGHT CODE: One-digit denoting age or weight measurements to be recorded in card columns 17—19.
1 = Age in Days
2 = Age in Weeks
3 = Age in Months
4 = Weight in Grams
5 = Weight in Dekagrams
6 = Weight in Hectograms
7 = Weight in Kilograms
9 = When age *and* weight are desired; use card 2

AGE OR WEIGHT: Three-digit number denoting the age or weight according to above code.

DAYS ON TREATMENT: Three-digit number denoting the number of days the animal has been on the treatment.

PRIMARY CHEMICAL NUMBER: Six-digit chemical compound code to which the animal has been exposed.

SOLVENT CODE: One-digit code denoting the solvent used.
0 = Water
1 = Vegetable Oil
2 = Mineral Oil
3 = Ethanol

4 = DMSO
5 = Xylene
6 = Other

FORMULATION CODE: One-digit number code denoting the formulation.
0 = Solution
1 = Emulsion
2 = Suspension
3 = Polymer

9 = Other

DOSE QUANTITY: Three-digit number denoting the quantity of dose received.

DOSE UNITS CODE: One-digit number denoting the units the dose quantity is given.
0 = ppm (parts per million) in air
1 = ppt (parts per thousand) in air
2 = % (parts per hundred) in air
3 = ppm in feed
4 = micrograms/kilogram body weight/day
5 = milligrams/kilogram body weight/day
6 = grams/kilogram body weight/day

9 = Other

ROUTE OF ADMINISTRATION: One-digit number denoting the route of administration.
0 = Oral
1 = Inhalation
2 = IM = Intramuscular
3 = IP = Intraperitoneal
4 = IV = Intravenous
5 = SC = Subcutaneous
6 = Topical

9 = Other

PREFIX CODE: One-digit designed for use in coding special laboratory tests.

CULTURE NUMBER: Five-digit number of culture. (e.g., 00001—99999)

TISSUE TYPE: One-digit code denoting the type of tissue type in vivo.
1 = 72 Hour PHA Blood
2 = 48 Hour PHA Blood
3 = Fibroblast (Skin)
4 = Abortus
5 = Amniotic Fluid
6 = Bone Marrow Direct
7 = Gonadal
8 = Bone Marrow Culture
9 = Other/Combination

REPLICATION NUMBER: Two-digit number denoting the replication number of this particular subject (e.g., 01—99).

CULTURE DATE: Four-digit date (e.g., month/year, e.g., 0172 for January, 1972) that the culture was taken.

NUMBER OF CELLS ANALYZED: Three-digit number (e.g., 001–999).
DIPLOID CELLS: Two-digit number of diploid cells found in cells analyzed.
SEX CHROMOSOMES: One-digit code, assigned as follows:

0 = Not scored
1 = XY
2 = XX
3 = XXY
4 = XYY
5 = XO
6 = XXX
7 = Poly X
8 = Mosaic
9 = Other

HYPERDIPLOIDS: One-digit number of hyperdiploids found in cells analyzed.
CHROMATID BREAKS WITH FRAGMENTS: Two-digit number of chromatid breaks with fragments found in cells analyzed.
CHROMATID BREAKS WITHOUT FRAGMENTS: Two-digit number of chromatid breaks without fragments found in cells analyzed.
CHROMOSOME BREAKS WITH FRAGMENTS: Two-digit number of chromosome breaks with fragments found in cells analyzed.
CHROMOSOME BREAKS WITHOUT FRAGMENTS: Two-digit number of chromosome breaks without fragments found in cells analyzed.
DICENTRICS: One-digit number of dicentrics found in cells analyzed.
RINGS: One-digit number of rings found in cells analyzed.
EXCHANGES: Two-digit number of exchanges found in cells analyzed.
MISCELLANEOUS ABERRATIONS: One-digit. The assigned codes for this field are:

0 = None
1 = Endo Reduplication
2 = Polyploidy
3 = Aneuploidy (Hypodiploid)
4 = Dot Fragments
5 =
6 =
7 =
8 =
9 = Other

NUMBER OF ABNORMAL CELLS: Two-digit number of abnormal cells found in cells analyzed.
QUALITY: One-digit code. The assigned codes for this field are:

0 = Not Rated
1 = Excellent
2 = Good
3 = Fair
4 = Poor
5 = Not Reliable

6 = Not Analyzable
7 = No Growth
8 = Gross Contamination
9 = Other/Combination

LABORATORY: Two-digits for all participating laboratories. Codes assigned as needed.

SCORER: Two-digit number assigned to each scorer in each laboratory by each laboratory.

CARD CODE: One-digit number for computer use in denoting the type of score sheet. The assigned codes for this field are:

0 =
1 = Human Cytogenetics
2 = Animal Cytogenetics
3 = Semen

HUMAN PERIPHERAL BLOOD LYMPHOCYTES FOR THE ANALYSIS OF CHROMOSOME ABERRATIONS IN MUTAGEN TESTS

H. J. EVANS AND MAUREEN L. O'RIORDAN

Medical Research Council, Clinical and Population Cytogenetics Unit, Western General Hospital, Edinburgh (Great Britain)

I. SOME ADVANTAGES AND DISADVANTAGES

Studies on exposed individuals, and on cultured cells, have shown that the human peripheral blood lymphocyte is an extremely sensitive indicator of both *in vivo* and *in vitro* induced chromosome structural change. These changes in chromosome structure offer readily scored morphological evidence of damage to the genetic material. Although problems exist in the extrapolation from *in vitro* results to the *in vivo* situation, the lymphocyte offers several advantages as a test system:

(*1*) Easy availability of large numbers of *human* cells: a few ml of peripheral blood can be easily and repeatedly obtained from an individual and each 1 ml of blood can contain $1-3 \cdot 10^6$ small lymphocytes.

(*2*) The lymphocytes are distributed throughout the body, circulate in all tissues and a proportion are long-lived.

(*3*) Virtually all the peripheral blood lymphocytes are a synchronised cell population in the same G_0 or G_1 stage of mitotic interphase, and, in healthy individuals, these cells are only infrequently involved in mitotic proliferation *in vivo*.

(*4*) A proportion of the lymphocytes can be stimulated by mitogens to undergo mitosis in culture; they are easy to culture and thus provide a ready source of dividing cells for the scoring of chromosome aberrations.

(*5*) There are excellent techniques available for making chromosome preparations from lymphocytes and the cells have a low spontaneous chromosome aberration frequency.

Some disadvantages include:

(*1*) Although the lymphocytes are a synchronised population, there are different sub-populations of cells even within the same individual. These can have varying responses to different mitogenic agents and probably differing intermitotic times which may be further influenced by variations in culture conditions.

(*2*) As in other cell systems, there is the possibility of non-uniformity of exposure and of preferential growth in culture of less heavily damaged cells.

(*3*) It is now recognised that in man only T-lymphocytes, *i.e.* the thymic-derived cells, can be stimulated in culture by phytohaemagglutinin to undergo mitosis. The T-lymphocyte is closely concerned with the immune response, thus, where an analysis of an *in vivo* exposure is to be undertaken, any recent previous exposure of an individual to an immunological stimulus may positively or negatively alter the number of cells with chromosome aberrations, depending on when the cells are withdrawn from the individual.

These disadvantages relate largely to the use of lymphocytes for studying mutagen effects *in vivo* and are of less relevance to the use of lymphocytes for *in vitro* mutagen testing.

II. TYPES OF DAMAGE ASSAYED

The types of chromosome damage which can be cytologically distinguished at metaphase can be divided into two main groups: *chromosome-type* and *chromatid-type*. The circulating lymphocyte is in the G_0 or G_1 phase of mitosis and exposure to ionising radiations and certain other mutagenic agents during this stage produces chromosome-type damage where the unit of breakage and reunion is the whole chromosome (*i.e.* both chromatids at the same locus). However, cells exposed to these agents while in the S or G_2 stages of the cell cycle, after the chromosome has divided into two sister chromatids, yield chromatid-type aberrations and only the single chromatid is involved in breakage or exchange (Fig. 1). Other agents (*e.g.* some of the alkylating agents) will produce *only* chromatid-type aberrations although the cells are exposed to the mutagen whilst in G_0 or G_1.

(*A*) *Chromosome-type aberrations*

Studies on somatic metaphases show that seven classes of chromosome-type aberrations can be distinguished cytologically (Fig. 2). Aberration types 1–5 involve only a single chromosome and are known as *intrachanges* whereas types 6 and 7 involve the exchange of parts between chromosomes and therefore are classified as *interchanges*.

(*1*) *Terminal deletions* (Fig. 3). These are paired acentric fragments which have the appearance of resulting from a simple break across the chromosome and they are not associated with an obvious exchange aberration.

(*2*) *Minutes* (interstitial, isodiametric or dot deletions) (Fig. 4). These are pairs of acentric fragments, smaller in size than terminal deletions, characteristically appearing as paired spheres of chromatin. These are intercalary deletions.

(*3*) *Acentric rings* (Fig. 5). These are paired segments of chromatid without a centromere and which are joined to give a ring.

(*4*) *Centric rings* (Fig. 6). Ring structures containing a centromere. The centric ring can be easily distinguished morphologically from the acentric type and is generally accompanied by one acentric fragment.

(*5*) *Inversions* (Fig. 7). Chromosome inversions can be classified into two categories: (*a*) *paracentric inversions* where both points of breakage and reunion lie on the same arm of the chromosome; and (*b*) *pericentric inversions* where the points of breakage and inversion lie on opposite sides of the centromere.

In cells where the two points of pericentric exchange are of unequal distance from the centromere, the abnormal chromosome can be easily distinguished by the altered position of the centromere. However, where the exchange points are equally distant from the centromere, and in the case of paracentric inversions, there is no change in the relative position of the centromere. The detection of such inversions

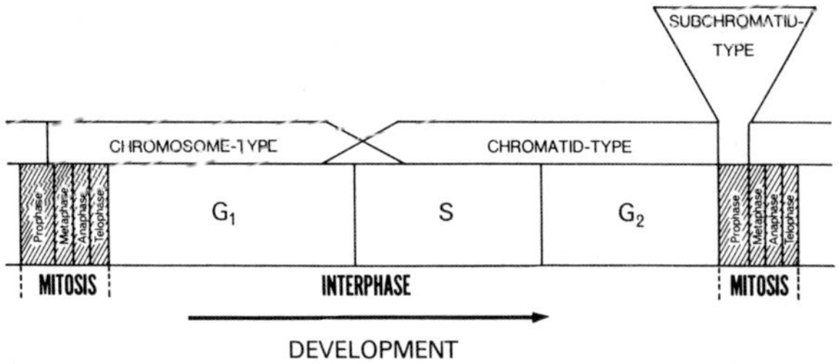

Fig. 1. Relation between type of aberration and stage in cell cycle at the time of exposure to irradiation or certain mutagens.

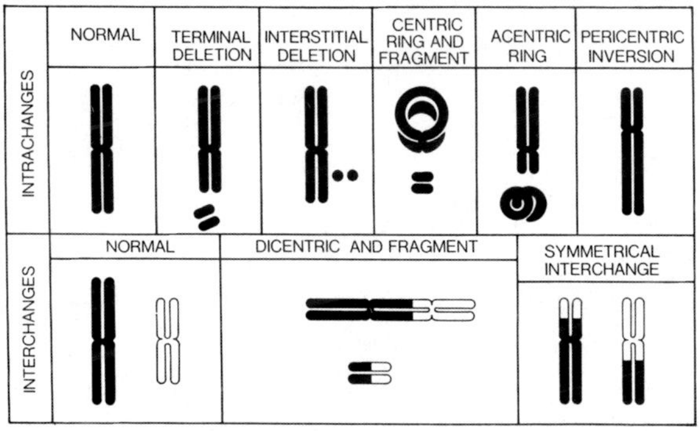

Fig. 2. Seven classes of chromosome-type aberrations cytologically distinguishable at mitotic metaphase.

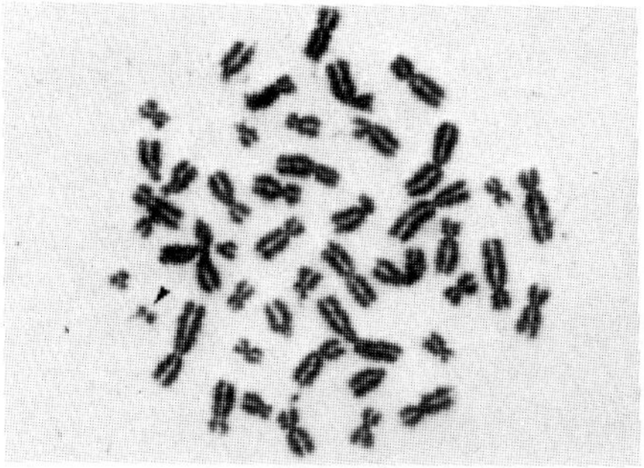

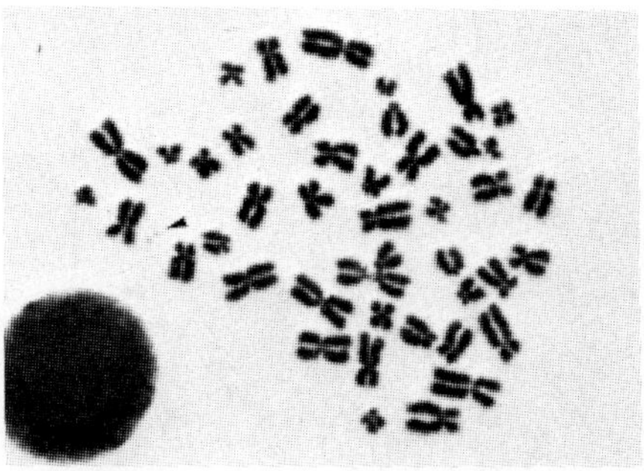

Fig. 3. Cell with a terminal deletion.

Fig. 4. Cell with a minute intercalary deletion.

may, therefore, be possible only through the use of chromosome banding techniques.

(6) *Reciprocal translocations* (symmetrical interchange) (Fig. 8). These are aberrations which involve breakage of two chromosomes and the reciprocal exchange of broken segments between these chromosomes. Conventional staining methods will not allow the detection of reciprocal translocations involving the exchange of two equal-sized segments, and banded chromosome preparations are required. Interchanges may, of course, also occur in the centromere regions of two chromosomes giving whole-arm exchanges which, if they involve acrocentric chromosomes, are sometimes referred to as centric fusions (Robertsonian translocations).

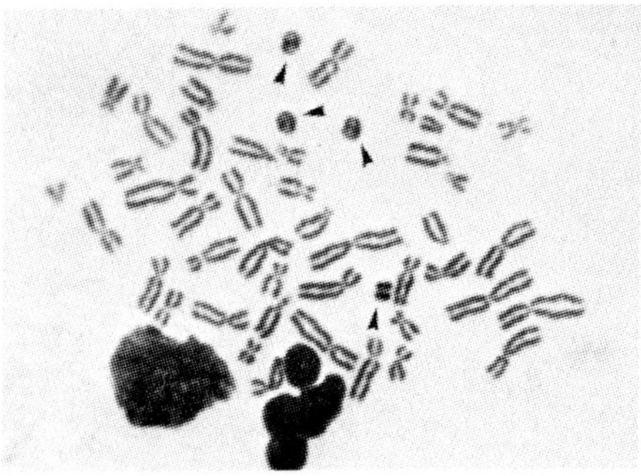 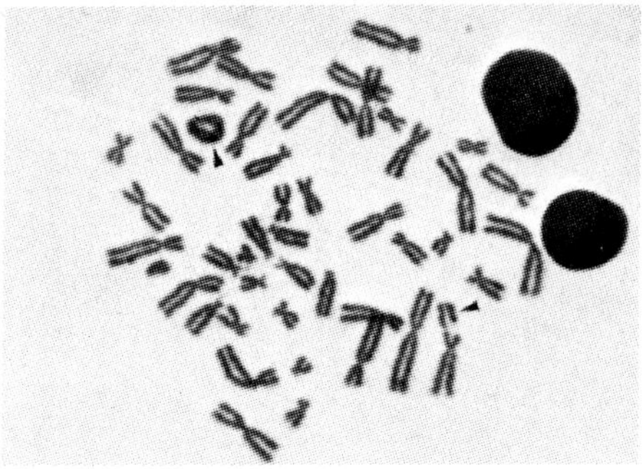

Fig. 5. Cell with acentric rings.
Fig. 6. Cell with centric ring and fragment.

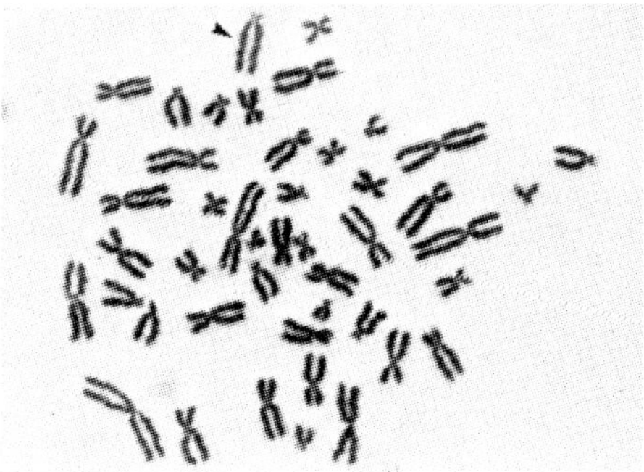 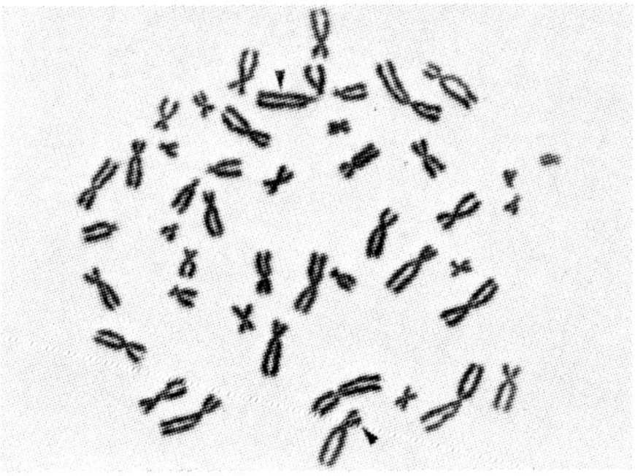

Fig. 7. Cell with a pericentric inversion in a C group chromosome.
Fig. 8. Cell with reciprocal translocation.

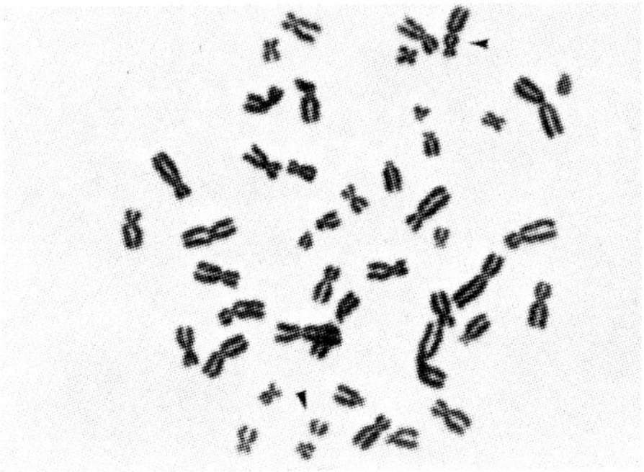

Fig. 9. Cell with dicentric chromosome and acentric fragment.

(7) *Dicentric or polycentric aberrations* (asymmetrical interchange) (Fig. 9). Aberrations which arise from an exchange between two or more chromosomes which results in the centric products reuniting in such a way to form a dicentric or polycentric structure and an associated acentric fragment.

Aberration types (*1*), (*2*) and (*3*) *i.e.* terminal deletions, minutes and acentric rings, are often loosely grouped together as acentric fragments which are not associated with any obvious exchange event resulting in a rearrangement. Fragments associated with an exchange event *e.g.* the fragment observed in association with a dicentric, are scored as part of the exchange and not as separate aberrations in their own right.

(B) Chromatid-type aberrations

Chromatid-type aberrations are induced by ionising radiations when cells are exposed in the S or G_2 stages of interphase, chromosome-type damage being produced in cells irradiated in G_1. However, many chemical agents and viruses cause only chromatid-type damage even though the cells are exposed to the agent whilst in the G_1 phase and examined in their first subsequent division. These aberrations are a consequence of errors in replication that occur at the DNA synthesis, or S, phase following exposure.

In contrast to the chromosome-type aberrations, chromatid breaks and particularly gaps are unreliable indicators of real damage to the genetic material. Not only can the scoring of gaps and breaks be extremely subjective, resulting in considerable observer differences, but also many gaps are caused by technical artefacts *e.g.* "poor" culture conditions, and the use of "drastic" processes during slide preparation. For this reason, if chromatid gaps and breaks are to be scored and included in the aberration yield, it is particularly important that "control" blood cultures are cultured in a similar way and at the same time as those exposed to the agent being tested.

(*1*) *Chromatid and isochromatid gap* (Fig. 10). The gap or achromatic lesion appears as a non-staining and constricted region in the chromatid arm and the apparently "broken" segments of the chromatid area are in alignment. Where the gap involves both chromatid arms at the same position, this is referred to as an isolocus or isochromatid gap.

(*2*) *Chromatid break* (Fig. 11). Where there is a discontinuity with displacement in the chromatid arm so that the broken chromatid ends are not aligned. An apparently "simple" break results in a terminal deletion.

(*3*) *Chromatid minutes*. Single, or unpaired, intercalary fragments.

(*4*) *Chromatid acentric rings*. Intercalary fragments joined to give ring structures.

(*5*) *Centric rings*. Intrachanges resulting in chromatid ring structures in which the centromere is included in the ring.

(*6*) *Inversions*. Paracentric inversions are not readily scorable, but pericentric inversions may be detected because of the close pairing of chromatids.

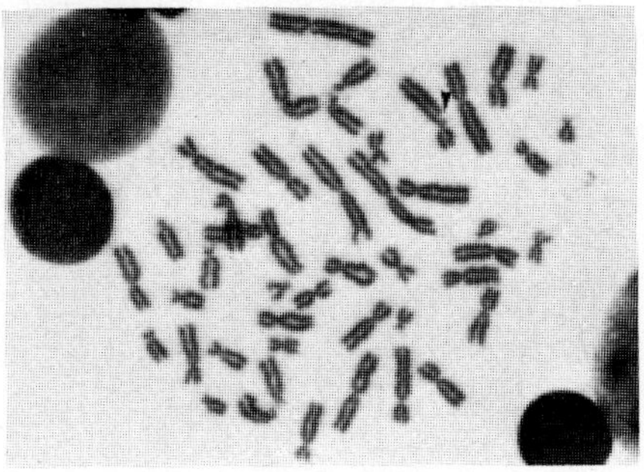

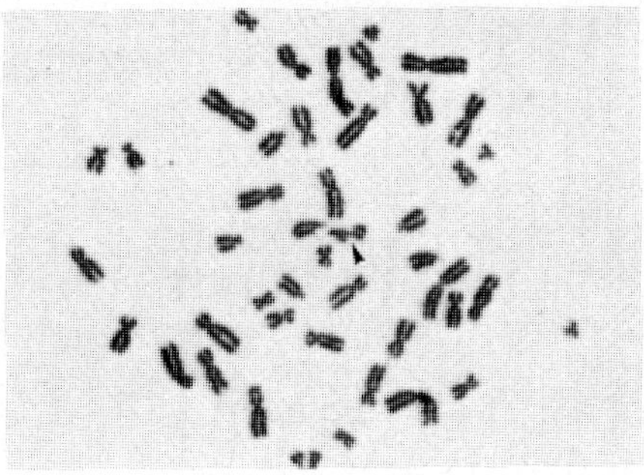

Fig. 10. Cell with chromatid gap.
Fig. 11. Cell with chromatid break.

(7) *Isochromatid aberrations* (Fig. 12). These involve exchanges between sister chromatids and may be confused with chromosome-type terminal deletions. They may be distinguished from chromosome-type aberrations, however, since the majority of isochromatid breaks involve a union between sister chromatids either proximal or distal to the point of breakage.

(8) *Symmetrical interchanges.* This type of aberration involves an exchange (exchanges) between two (or more) chromosomes. Simple exchanges between one chromatid in each of two chromosomes results in a configuration having four arms and referred to as a quadri-radial.

(9) *Asymmetrical interchanges* (Fig. 13). An exchange between two or more chromosomes resulting in the formation of one (or more) dicentric chromatids.

(C) *Aberration yield and type in relation to time in culture*

Probably the most important *in vitro* factor which influences the chromosome

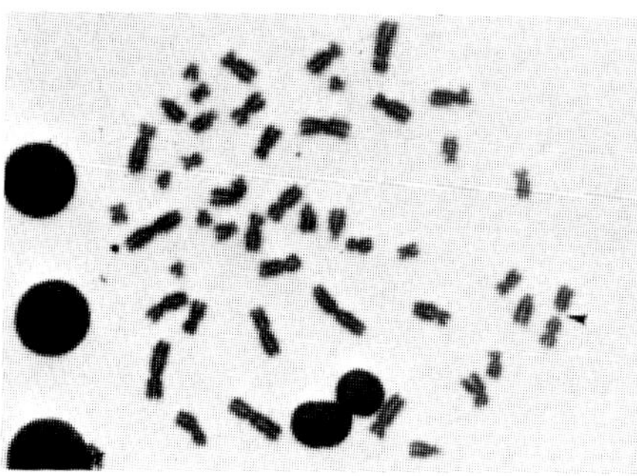

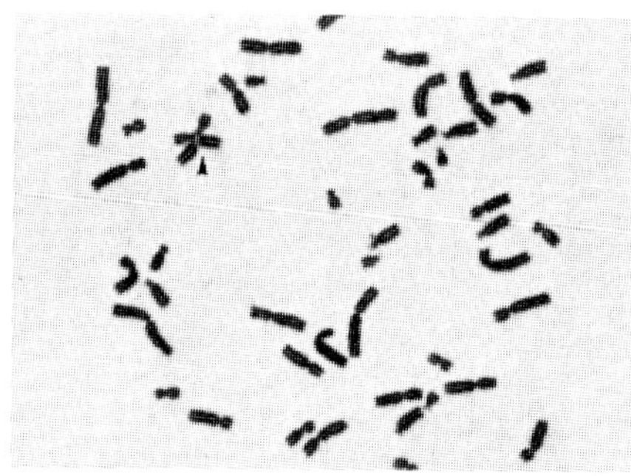

Fig. 12. Cell with sister chromatid (isochromatid) exchange.
Fig. 13. Cell with chromatid interchange between homologous chromosomes.

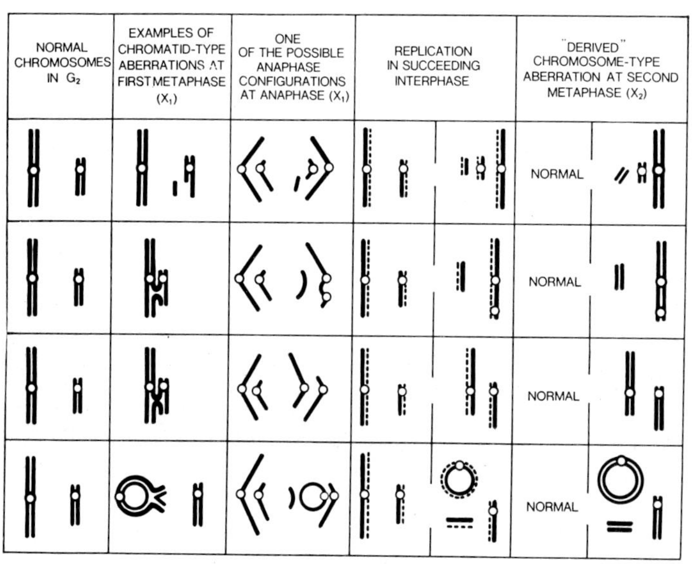

Fig. 14. Diagram of examples of "derived" chromosome type aberrations observed at the second mitosis (X_2) after exposure to irradiation or certain mutagenic agents.

aberration result is the time the cells are maintained in culture. After two days in culture at 37° the majority of transformed lymphocytes are undergoing their first mitosis. Preparations made from cells cultured for longer periods will contain increasing proportions of cells in their second or subsequent divisions. Consequently, if lymphocytes with chromosome-type damage fail to undergo repeated division *in vitro*, due to their genetic unbalance or mechanical difficulties at anaphase, examination of cells after 72 h of culture may not reflect the true *in vivo* aberration level. Furthermore, a proportion of the chromosome-type aberrations observed at this later time may be really of a "derived" type resulting from a duplication of aberrations that were initially chromatid-type (Fig. 14). Thus, for tests which are designed to detect the presence or absence of chromosome damage in lymphocytes following *in vivo* exposure of individuals to a suspected mutagenic agent, the blood samples should only be cultured for up to 50 h at 37°. The same restriction applies to *in vitro* studies, but consideration must also be given to the fact that the agent and the test may itself inhibit cell development and so prolong the time of culture necessary to obtain mitotic cells.

III. APPARATUS NEEDED

Small autoclave for sterilisation.
Temperature controlled water bath or incubator in which temperature can be controlled to ± 1°.
Bench centrifuge.
Microscope; good quality light microscope capable of magnifying 1500 times.
Refrigerator, preferably with deep-freeze.
Still or deionising equipment.
Filter sterilising system.
Sterile glass or plastic syringes in sizes between 1–20 ml.
Disposable needles to fit syringes with bore sizes between 19 gauge and 21 gauge.
Sterile glass or plastic containers with screw caps: (*a*) for collection of blood; (*b*) for the blood cultures; (*c*) for sterile liquids.
Glass or plastic containers for non-sterile liquids.
Centrifuge tubes with polythene or rubber stoppers.
Glass pipettes.
Glass microscope slides and coverslips.
Staining dishes and slide racks.

IV. CHEMICAL MATERIALS REQUIRED

Anticoagulant: Lithium heparin.
Mitogen: Phytohaemagglutinin.
Spindle inhibitor: Demecolcine (colcemid) 0.02% solution, or Colchicine or Vinblastine (velban).
Tissue culture media: Virtually any standard tissue culture medium, *e.g.* Hams F10; McCoy's; TC 199; Eagle's Minimum Essential Medium (MEM).
Serum: A wide variety of sera can be used. Human AB+ or bovine (calf, but not necessarily foetal calf) are amongst the most popular.

Antibiotics: Penicillin 100 IU/ml of medium; Streptomycin 100 IU/ml of medium.
pH control: Sodium bicarbonate; 5% CO_2 in air.
Hypotonic solutions: Potassium chloride 0.075 *M* solution; Sodium citrate 0.95% solution; Hanks' solution: diluted 1:4 with distilled water.
Fixatives: Methanol; Glacial acetic acid.
Stains: Lactic aceto-orcein, 1% or 2% solution in 1:1 lactic acid (70% aqueous)–glacial acetic acid; Giemsa 2% (aqueous solution).
Cleaning agents: Cellosolve; Xylene; Euparal essence.
Mountants: Euparol; De Pe X.
For "in vitro" tests: Known chemical mutagen as a positive control, *e.g.* nitrogen mustard or triethylenemelamine.

V. COSTING

A very approximate figure for the cost of a single test culture and its analysis can be estimated. Each test will involve a number of cultures at different dose levels, positive and negative controls and replicates. Other expenses which should be considered include the expenditure on basic equipment (III), accommodation and electricity costs. In addition it is not easy to assess the cost involved in obtaining blood samples.

An estimated cost for *one culture* is given based on using 10 ml of culture medium and the analysis of 100 cells.

Reagents

Culture medium, mitotic inhibitor, hypotonic solution, fixative, stain: Estimated cost $ *1.50*

Equipment

Disposable syringes and needles, disposable containers, microscope slides and coverslips, pipettes: Estimated cost $ *1.00*

Time

Culture handling, slide preparation:	*0.5 h*
Analysis of cells:	*4.0 h*
	4.5 man hours

VI. PREPARATION OF CULTURE MEDIUM

It is preferable to prepare the culture medium in advance and then store at 4° until required. Each new batch of medium should be tested for sterility and compared with previous batches for ability to support cell growth.

In practice it is often easier to prepare sufficient medium in bulk for a whole experiment rather than prepare small individual aliquots. This also helps to minimise technical artefacts which can arise from variation in different batches of medium.

A suitable starting quantity of culture medium, *i.e.* sufficient for 45 cultures and using 10 ml medium for each culture, consists of:

400 ml proprietary tissue culture medium.

5.0 ml phytohaemagglutinin.
70 ml serum.
optional – streptomycin, 100 IU/ml of medium.
– Penicillin, 100 IU/ml of medium.

A variety of tissue culture medium can be used with success (see IV) but, in our hands, Ham's F10 has been found to yield the highest and Eagle's MEM the lowest, number of dividing cells in a 48-h culture.

The pH of the final culture medium should be between 6.8 and 7.2; sodium bicarbonate or 5% CO_2 in air may be used to adjust the pH.

After preparation, 10-ml or 5-ml aliquots of culture medium should be dispensed into sterile glass or plastic screw-capped vials. Where the larger quantity is used, a 30-ml container is recommended, and a 15-ml vial is suitable for the 5-ml aliquots. The choice between the two largely depends on the number of cells which are finally required, but the 10-ml amount generally yields a superior final result.

VII. COLLECTION OF BLOOD SAMPLES

(a) *Choice of subject for study*

The relationship between the frequency of chromosome aberrations detected in an individual's cultured lymphocytes and the level of his *in vivo* exposure to the chemical being specifically tested, may be complicated by other environmental factors: for example, previous or current exposure to prescribed and unprescribed drugs, diagnostic or therapeutic radiation exposure, viral infections. It is therefore particularly important to collect adequate data for each individual being examined on: age, occupational history, radiation and drug histories, exposure to toxic substances, *e.g.* organic solvents, insecticides.

For *in vivo* studies, a similarly sized control population is required to provide background data on the frequency of chromosome aberrations. Ideally, control individuals should have a closely similar environmental background and be sex- and age-matched with the "exposed" persons. Any batch of blood samples should contain a mixture of samples from both control and exposed groups and these should be processed together. In this way any differences between groups that may be due to, or influenced by, technical artefacts, *e.g.* variations in culture medium, fluctuations in incubation temperature and cell fixation, will be minimised.

Where persons are being selected to provide blood samples for an *in vitro* test, medical data should also be collected, but it is preferable to exclude individuals suffering from virus or other infections or who have received excessive doses of radiation or drugs, or recently received immunisation. Wherever possible, it is also advisable to use blood samples from several individuals to minimise the possibility of a varying response between the lymphocytes of different individuals and the chemical being tested.

(b) *Method*

Blood samples (5–10 ml) for both *in vivo* and *in vitro* test systems are normally taken from adults by venipuncture using a 10–20 ml sterile syringe. The blood is then immediately transferred to a suitable sized sterile glass, or plastic, vessel containing lithium heparin (preservative-free) at concentrations of between 10–100 IU of heparin/1 ml of blood. The specimen is then gently mixed to prevent clotting.

If the blood is not immediately required for culture, it should be stored at between 4–25°. It is advisable to set up the cultures within 24 h of withdrawal, since delay may adversely affect cell viability.

To prevent observer bias in the *in vivo* tests, the *blood samples from the control and the exposed persons should be randomised and coded by someone who is not concerned in the study*. Coding of samples in the *in vitro* test system should be *done after slide preparation*. It is most important that the scorer has no prior knowledge of which slides are from exposed and which from control cultures, so that all slides are coded and then randomised before being presented to the scorer.

VIII. PERIPHERAL BLOOD LYMPHOCYTE CULTURE TECHNIQUE

Two methods of culturing the peripheral blood lymphocytes can be used.

(*a*) *Macroculture method*

This is not a recommended technique since it requires relatively large quantities of blood. After gently centrifuging the blood sample, the plasma is separated off from the red cells and cultured as below.

(*b*) *Microculture method*

(*1*) The heparinised blood sample, unless freshly withdrawn, should be gently shaken to allow proper mixing of the cells and plasma.

(*2*) 0.8 ml of whole blood is added to 10 ml of culture medium in a 30-ml sterile glass or plastic vial, which should be then tightly closed by a screw cap. Where 5 ml of medium is used, only 0.4 ml of blood should be used for each culture.

(*3*) The culture is then incubated at 37° in a water bath or incubator for 45–50 h. The cultures are *not* shaken during incubation.

(*4*) During the last 3 h prior to harvesting the cells, a spindle inhibitor should be added to the culture. Since spindle inhibitors can be toxic to cells in high concentration, it is advisable to ascertain in advance the most suitable concentration. Many laboratories use demecolcine to give a final concentration in the culture of 0.2 μg/ml, or colchicine to give 0.25 μg/ml.

IX. HARVESTING THE CULTURE

(*1*) Time of harvesting, see II(C).

(*2*) The cells and culture medium should be removed from the incubator, gently mixed and then the contents emptied into a centrifuge tube.

(*3*) The cells are spun down by gentle centrifugation for 5 min at approx. 2000–2500 rev./min (approx. $60 \times g$). The cells should be then in a loose button at the bottom of the tube and the supernatant clear.

(*4*) The supernatant is gently poured off and the button of cells shaken and resuspended in approx. 10 ml of the hypotonic solution (we routinely use the 0.075 *M* potassium chloride hypotonic solution). The suspension should then be allowed to stand at either room temperature or at 37° for 8 to 10 min.

(*5*) The culture should be centrifuged as before and the supernatant decanted off. The cells are resuspended and 5 ml of fixative (3:1 methanol–glacial acetic acid) should be slowly added using a small pipette, whilst agitating the cells.

(6) After centrifugation two further changes of fixative are recommended and the cells should be exposed to fixative at least 15 min before the slides are made. The required cells can be satisfactorily stored in the second or third change of fixative for several days before slides are made, although this is not recommended. If cells are stored in fixative a further change of fresh fixative will be necessary immediately prior to slide preparation.

The fixative should always be freshly made up and not stored for periods of time longer than one hour.

X. SLIDE PREPARATION

The cells are suspended in fresh fixative to give a suitable, slightly cloudy suspension. One or two drops of suspension should be dropped on to a clean grease free microscope slide and left to dry. The ambient temperature and humidity affects the degree of cell spreading. A relative humidity of 15–65% is found to give best results. It may be necessary to speed up the air-drying process by vigorously waving the slide through a spirit flame. The fixative on no account should be allowed to ignite, otherwise the cells tend to break.

The slides should then be stained with aceto-orcein, or Giemsa, cleared and finally mounted.

XI. *In vitro* TESTS

(*a*) *General aspects*

The major problem in the *in vitro* testing of suspected mutagenic agents is the assessment of the result and the extrapolation of either the positive or negative cytogenetic findings to the *in vivo* situation. It is well known that a number of chemicals which are apparently inactive mutagens *in vitro* have a cytogenetically positive effect *in vivo*, due to their metabolic activation. Conversely, other compounds produce a positive *in vitro* effect even at small dosage levels, but have a reduced *in vivo* activity. *In vitro* tests can therefore only provide a limited guide to the possible *in vivo* effects of a suspected mutagen.

(*b*) *Method*

(*1*) *Blood samples and cultures.* If possible, blood samples from several individuals should be used for *in vitro* testing. This ensures that any variability in results due to possible inherent differences between individuals is taken into account.

Two kinds of control are recommended and these should be cultured in parallel, using the same batches of medium and incubation times, to ensure the adequacy of the system and to exclude the possibilities of false negatives or the presence of unknown contaminants. For the "positive" control, a known chromosome breaking agent, *e.g.* triethylenemelamine should be added to the cultures at various concentrations. "Negative" controls are provided by cultures to which "carrier" (saline) but no mutagen, has been added.

Three dose levels of the agent under test are recommended: (*i*) the maximum tolerated concentration which allows the cells to undergo mitosis, (*ii*) a concentration comparable to the expected *in vivo* physiological level, (*iii*) an intermediate concentration.

(*2*) *Exposure and harvesting time.* Many chemical agents produce "delayed" effects so that aberrations are only evident in the second or subsequent mitosis after the cells have been exposed to the mutagen. In addition, the different stages of the cell cycle may show large variations in sensitivity to the induction of chromosome damage, and some agents cause mitotic delay. For example a good positive control involves exposure of cultures to nitrogen mustard (Mustine hydrochloride), at a final concentration of $3 \cdot 10^{-6}\,M$ in the culture, for 30 min at 37°, 24 to 36 h after culture initiation. Such an exposure, however, results in a delay in cell development, so that the maximum yields of first division cells are observed in cultures incubated for 65 to 75 h and in 72-h cultures some 40 to 60% of the cells contain chromatid structural changes.

(*3*) *Chromosome aberration analysis.* In good preparations, generally those with a high mitotic index, the cells selected for analysis tend to be of a superior quality than those from poorer preparations. To avoid bias in cell selection, certain criteria should be defined and adhered to irrespective of the overall quality of the preparations. The following suggestions are given as guide lines:

(*a*) The coded slides should be methodically scanned.

(*b*) With the low power objective, a metaphase spread which appears to be in an unbroken cell should be selected for analysis.

(*c*) Under high power, the chromosomes should be well defined and should not be in an early C-anaphase state with completely separated chromatids.

(*d*) To avoid the analysis of cells with random chromosome loss due to technical artefacts, only cells with a minimum number of 45 centromeres should be scored.

(*e*) If the cell is suitable for analysis the vernier reading should be recorded to locate the cell for possible future reference.

(*f*) The chromosomes should be then counted and any aberrations noted.

Where difficulty is found in the interpretation of an aberration, the cell should not be rejected but the difficulty recorded and the opinion of a second observer sought.

The cells can be fully counted and analysed in detail (karyotyped) but this is an extremely time-consuming procedure and may be not necessary to obtain a valid estimate of aberration yield. Alternatively, the chromosomes in each cell should be counted and the presence of any chromosome- or chromatid-type damage recorded.

All unstable types of chromosome aberrations, *e.g.* dicentrics, rings and fragments, are particularly good parameters of damage since they are all easily identifiable. It may not be profitable to spend time attempting to analyse reciprocal translocations, since they are more difficult to detect. It is more profitable and informative to spend that time scoring more cells for the readily visible fragments and asymmetrical exchanges.

A recommended minimum number of 100 cells (see XII) should be scored from each of the duplicate blood culture treatments. A record should be kept of all cell references, whether or not aberrations were found.

XII. STATISTICAL TREATMENT

The chromosome aberrations induced by many mutagens may have a random distribution *between* cells, although they may be non-randomly distributed *within* chromosomes. Consequently the number of cells with 0, 1, 2------n aberrations may conform with a Poisson distribution, where the mean and its variance are approximately equal. The mean frequency of aberrations may be small and the smaller the mean the larger will be the number of cells to be scored to obtain good estimates of aberration yield for meaningful comparisons with both negative and positive controls. However, since slides are scored blind, observers usually aim to score a fixed number of cells per slide *e.g.* 100 metaphases with say two slides per culture (or 50 metaphases on each of four slides). Replication is important to estimate "between slides", "between culture" and "between treatment" variances.

A simple guide for determining the approximate numbers of cells to be scored (or aimed for) at anticipated, or known, aberration frequency levels in order to detect a significant result, may be obtained by simply calculating (or looking up in tables) binomial confidence limits. For example: if P is the proportion of abnormal cells (cells with aberrations) and $Q = 1-P$ then the approximate 95% confidence limits of P are given by:

$$P \pm t\sqrt{\frac{PQ}{n}}$$

where t is the upper 2.5% point on the normal curve and n is the number of cells to be scored. For 95% limits $t = 1.96$; for 80% limits $t = 1.28$ *etc. e.g.* for a frequency of 0.20 (20% of cells abnormal) after scoring 100 cells, the 95% confidence limits of that mean estimate are:

$$0.2 \pm 1.96(0.04),\ \text{or}\ 0.2 \pm 0.08 = 0.28\ \text{to}\ 0.12$$

If 400 cells are scored, and $P = 0.20$, the approximate limits (95%) are 0.24 to 0.16.

Standard significance tests are employed to test the null hypothesis of a difference between results from negative control and treated cultures. In this way the conclusion that a given substance is, or is not, mutagenic in the test described can therefore be made in a quantitative fashion with defined levels of probability.

In some cases, for instance in studying the effects of different dose levels of a known mutagen, such as ionizing radiations, it may be required to express the relationship between dose (D) and aberration yield (y), where y will be some function of D, in the form of the parameters of an equation representing the dose–response curve. For example this could be a linear function where

$$y = a+b_1D$$

where a is the spontaneous aberration yield in control cultures and b is a constant. Alternatively, the response may be defined by a quadratic:

$$y = a+b_1D+b_2D^2$$

where b_1 and b_2 are constants.

Or by a power law:

$$y = a+b_1D^n$$

here, assuming a to be negligible, the logarithm of the aberration frequency is linearly related to the logarithm of the dose, *i.e.* $\log y = \log b_1+n \log D$.

The constants for these equations for any particular set of dose–response data are calculated using least squares regression methods and various computer programmes are available for a number of dose–response models.

XIII. LITERATURE

BUCKTON, K. E., AND H. J. EVANS (Eds.), *Methods for the Analysis of Human Chromosome Aberrations*, World Health Organisation, Geneva, 1973.

COHEN, M. M., AND K. HIRSCHHORN, Cytogenetic studies in animals, in *Chemical Mutagens. Principles and Methods for their Detection*, A. HOLLAENDER (Ed.), Vol. 2, Plenum, New York, 1971.

DE SERRES, F. J., AND W. SHERIDAN (Eds.), The evaluation of chemical mutagenicity data in relation to population risk, in *Environmental Health Perspectives*, Exp. Issue No. 6, US Department of Health, Education and Welfare, Washington D.C., 1973.

EVANS, H. J., Chromosome aberrations induced by ionizing radiation, *Int. Rev. Cytol.*, 13 (1962) 221–231.

EVANS, H. J., Population cytogenetics and environmental factors, in *Pfizer Medical Monographs 5*, Edinburgh University Press, 1970, pp. 192–216.

KIHLMAN, B. A., *Actions of Chemicals on Dividing Cells*, Prentice Hall, Englewood Cliffs, N.J., 1966.

IN VITRO AND IN VIVO ANALYSIS OF SISTER CHROMATID EXCHANGE FORMATION

SAMUEL A. LATT, JAMES W. ALLEN, WILLIAM E. ROGERS and LOIS A. JUERGENS

Clinical Genetics Division, Mental Retardation Program, Children's Hospital Medical Center and the Department of Pediatrics, Harcard Medical School, Boston, Mass. 02115 (U.S.A.)

Summary

5-Bromodeoxyuridine-dye techniques have stimulated renewed interest in sister chromatid exchanges as sensitive indices of certain types of chromosome damage and repair. Sister chromatid exchange frequencies can be increased by treatment of cells with a variety of agents, many of which are mutagen-carcinogens. In addition, abnormalities in sister chromatid exchange formation appear to be involved in two inherited diseases characterized by chromosome fragility and a predisposition for neoplasia. Sister chromatid exchange can now be detected both in vitro, in different types of cultured cells, and in vivo, in intact animals. Methods which can be used for sister chromatid exchange analysis in different in vitro and in vivo systems are described and illustrated with specific examples.

Introduction

Sister chromatid exchanges represent the interchange of DNA between replication products at apparently homologous loci. These exchanges, which are generally detected in cytological preparations of metaphase chromosomes, presumably involve DNA breakage and reunion, although the molecular basis of sister chromatid exchange (S.C.E.) formation remains unknown.

Sister chromatid exchanges were first described by J. Herbert Taylor [63], who utilized autoradiography to detect differentially labeled sister chromatids in cells which had undergone one cycle of [^{3}H]dT incorporation followed by a repiication cycle in non-radioactive medium. Reciprocal alterations in labeling (S.C.E.s) were detected along the chromatids of a number of metaphase chromosomes. Consistent with Taylor's early observation [62], most, if not all exchanges were induced by the radioactivity necessary for their detection [5, 18,48].

Analysis of the correlation between exchange formation in metaphase chromosomes which were derived from a common precursor, i.e. twin versus single exchanges [62], provided strong evidence that sister chromatid exchange formation was constrained by the polarity of the DNA double helix. The occasional observation in autoradiographic emulsions of grains at homologous loci of both sister chromatids ("isochromatid labeling") led to some transient confusion [67] about the number of DNA helices in a single chromatid. However, subsequent detection of exchanges by 5-bromodeoxyuridine (BrdU)-dye techniques (vide infra) corroborated an earlier suggestion [9] that isolabeling was really pseudoisolabeling originating from the inability to resolve closely spaced exchanges by autoradiography.

Kato employed autoradiography to demonstrate an increase in the S.C.E. frequency in cells treated with agents such as mitomycin C, 4-nitroquinoline-oxide [27], or ultraviolet light [26]. In some of these experiments the S.C.E. frequency appeared to level off at high treatment doses. This effect, like isolabeling, presumably reflects the resolution limits of autoradiography. Autoradiographic analysis was sufficient to establish that S.C.E. formation required cell traverse through the S phase [69] and that photoreactivation prevented S.C.E. induction by UV light [28]. However, the tedious nature of autoradiography probably discouraged more extensive investigations.

Analysis of S.C.E. formation has been facilitated by BrdU-fluorochrome and Giemsa techniques for detecting DNA synthesis in cytological chromosome preparations. The dye 33258-Hoechst was observed to fluoresce less efficiently when bound to DNA substituted with BrdU than when bound to unsubstituted DNA [37,44]. If cells were allowed to incorporate BrdU for one replication cycle followed by a cycle in the presence or absence of this analogue, and the chromosomes were then stained with 33258-Hoechst, sister chromatids exhibited differential fluorescence [38,39]. Sister chromatid exchanges were apparent as sharp reciprocal alterations in fluorescence along these chromosomes. Sister chromatid exchanges in BrdU-substituted chromosomes have subsequently been detected by changes in acridine orange fluorescence [15, 29] or Giemsa stain, the latter with [35,54,68] or without [36] photosensitization by bound dye. More recently, exchanges have been detected by an immunological technique which utilizes antibodies directed against BrdU [20].

BrdU-33258-Hoechst methodology was used to show that low doses of alkylating agents such as mitomycin C or nitrogen mustard induced large numbers of S.C.E.s at concentrations well below those causing significant numbers of chromosome breaks [40], and S.C.E. analysis was proposed as a highly sensitive means of detecting chromosome damage and repair. BrdU itself was observed to induce exchanges [29,40,54], although this effect can be minimized by working at low BrdU concentrations. Subsequent reports confirmed these observations [3,6,31] and extended them to include other agents known to damage chromosomes [6,22,30,33,34]. Multiple closely spaced exchanges were resolved [30,40] and presumably accounted for the isolabeling observed with autoradiography. Perry and Evans surveyed a number of exchange-inducing agents [53]. Since many of these are mutagens and/or carcinogens, they suggested that S.C.E. analysis could be used as an assay for mutagenesis and carcinogenesis.

Analysis of S.C.E. formation has also been used to differentiate between a number of inherited diseases characterized by chromosome fragility and a predisposition for the development of neoplasia. These diseases include Bloom's syndrome, Fanconi's anemia, and ataxia telangiectasia [17]. Chaganti et al. detected an elevated S.C.E. frequency in BrdU-cultivated cells from patients with Bloom's syndrome [6]. In contrast, normal S.C.E. frequencies were detected in Fanconi's anemia, ataxia telangiectasia [16,23] and xeroderma pigmentosum [70]. Lymphocytes from Fanconi's anemia patients, while exhibiting essentially normal S.C.E. frequencies in the present of BrdU, did not respond to mitomycin C treatment with a normal increase in sister chromatid exchange formation [45]. Cells from patients with this disease are highly susceptible to killing [13] or to chromosome breakage [58,59] by bifunctional alkylating agents. The reduced stimulation by mitomycin C of S.C.E. formation in Fanconi's anemia was associated with an increased number of chromatid breaks [45]. Approximately half of these breaks occurred at sites of incomplete sister chromatid exchange formation, an observation compatible with the hypothesis that the break increment and exchange deficit were causally related. The results were interpreted [45] to suggest that Fanconi's anemia cells were defective in a form of DNA repair perhaps akin to the recombinational repair [7,56] of bacteria.

The scope of S.C.E. studies has been extended from in vitro to in vivo systems. Bloom and Hsu described the formation of S.C.E.'s in ovo in chick embryos [4]. More recent reports [1,64] have described the induction by alkylating agents of S.C.E. in marrow or spermatogonia of mice which received repeated doses of BrdU. The host-mediated aspects of this system, its applica-

TABLE I

AGENTS CAPABLE OF INDUCING SISTER CHROMATID EXCHANGES

Radiation			Miscellaneous Chemicals	
X-ray	S > G1, not G2	[53]	BrdU	[29,30,40,54]
			Adriamycin	[53]
Visible light (cells with BrdU-substituted DNA)		[25]	33258-Hoechst	[53]
			Bleomycin	[53]
			7,12-Dimethylbenz(a)anthracene	[61]
UV light		[26]	Proflavine	[27]
Alkylating Agents			**Negative**	
Mitomycin C (in vitro)		[31,40]	Caffeine	[33,34]
	(in vivo)	[1]	8-Ethoxycaffeine	[33,34]
Trinemon		[3,22]	Lead acetate	[3]
Ethylmethane sulfonate		[30,46]	Maleic hydrazide	[53]
MNNG		[53]		
Nitrogen mustard		[40,53]		
4NQO		[53]		
Diepoxybutane		[53]		
β-Propiolactone		[53]		
Triaziquon (in vivo)		[64]		
Cyclophosphamide (requires in vivo activation)		[2,64]		
Psoralen plus light		[43]		
Thiotepa		[33,34]		
Chlorambucil		[60]		
Quinacrine mustard		[53,60]		

bility to multiple cell types in the same organism and the obvious relevance of spermatogonial damage to germ cell formation make this approach unique for studying environmental mutagenesis. In addition, the spermatogonial system can potentially be extended to an analysis of DNA replication kinetics and of chromatid interchange in meiosis. Agents thus far reported to induce S.C.E. in vivo or in vitro are listed in Table I.

Implicit in many of the above studies is the assumption that S.C.E. formation bears a direct relationship to DNA repair and mutagenesis. A number of lines of evidence support this idea: (1) alkylation of DNA by psoralen plus light, a powerful inducer of S.C.E. formation in human lymphocytes [43] and Chinese hamster ovary (CHO) cells [41], is known [8] to stimulate DNA strand interchange in recombination-proficient but not in recombination-deficient (REC A) bacteria; (2) recombinational repair in bacteria appears to be error-prone. Specifically, induction of mutations by UV light [65,66] or psoralen plus near-UV light [24] requires a functional REC A system; (3) as mentioned previously, there is a correlation between the ability of a chemical to act as a mutagen and/or carcinogen and its ability to induce S.C.E. formation [53], and abnormalities in S.C.E. formation appear to be involved in some of the chromosome fragility diseases [6,45]. However, direct demonstration of a cause and effect relationship between S.C.E. formation and mutagenesis has not been reported, and the precise relationship of S.C.E. formation to DNA repair and carcinogenesis remains to be elucidated.

BrdU-Dye methodology for S.C.E. detection

Sister chromatid exchange analysis in BrdU-sunstituted chromosomes has now been performed with a variety of cell types. Exchange detection can be effected by fluorescent or modified Giemsa techniques, the latter being preferable for large scale routine analyses. The following section describes some of the cell culture, BrdU-substitution, slide preparation, and photomicroscopy protocols which can be used for S.C.E. analysis in vitro or in vivo.

In vitro S.C.E. analysis

Peripheral lymphocyte cultures

Blood is collected in preservative-free heparin (e.g. Abbott's "Panheparin"), 10 units/ml, and allowed to settle 90 min before removal of the buffy coat. Aliquots of plasma containing 5—8 $\times 10^6$ leukocytes are added to 16 $\times$ 125 mm culture tubes (e.g. Falcon No. 3033) together with the following:

(1) Medium: Eagle's MEM with Earle's BSS and 2 mM L-glutamine (e.g. Microbiological Associates No. 12-611), 100 U/ml penicillin, 100 μg/ml streptomycin, and 20% fetal bovine serum. Total volume used is 5 ml.

(2) Phytohemagglutinin: either 0.2 ml of a crude saline extract of red kidney beans or 0.1 ml of Burroughs-Wellcome phytohemagglutinin (HA-15, reagent grade).

(3) Nucleosides: we routinely employ BrdU (Sigma), fluorodeoxyuridine (FdU), and uridine (U) (Calbiochem) at final concentrations of 10^{-5} M, 4 $\times 10^{-7}$ and 6 $\times 10^{-6}$ M. Other workers have omitted the FdU and U. Addition of

deoxycytidine (dC) (10^{-4} M) improves cell growth [50] but also results in higher baseline S.C.E. frequencies [45,46]. It is convenient to add the BrdU together with the cells, although an 18—24 h delay in BrdU addition may improve the yield of mitotic cells [10].

Cells are cultured at 37°C in the absence of light. Tubes are sealed tightly to prevent CO_2 escape and inclined at an angle of approximately 45°. Good yields of second division metaphases are generally obtained after 72—74 h total culture time. Retardation of cell growth by various clastogens may necessitate the use of a longer culture time before cell harvest [60].

Agents to be tested for potential S.C.E. induction are typically added at the start of culture or 24 h before harvest. The latter protocol reduces bias due to selection of those cells which are able to reach the second metaphase after addition of BrdU. Exposure of cells for only the last few hours of culture should further reduce this selection, although the chromosomal distribution of exchanges might then be altered. All manipulations of cells after addition of BrdU to the cultures should be done in subdued light and preferably in the absence of light with wavelength less than 500 nm.

Lymphocyte harvest consists of the following steps: (1) addition of 0.1 μg/ml Colcemid (Calbiochem) for 2 h; (2) removal of medium and resuspension in 0.075 M KCl for 12 min; and (3) fixation in at least two changes of a 3 : 1 mixture of methanol and acetic acid (for 16 min and 10 min). Minor variations in mitotic arrest or hypotonic treatment might be appropriate in individual circumstances. For example, metaphase preparations of lymphoblasts appear to be improved by reducing the Colcemid concentration to 0.023 μg/ml and the time in KCl to 8 min. At each step, the cells are centrifuged, all but about 0.2 ml liquid is removed, and the cell pellet is gently resuspended before adding the next solution. Careful cell resuspension appears to increase the probability of obtaining good metaphase spreads.

Cultured cell lines

Analogous procedures are employed when examining cells other then lymphocytes. The type of medium and container (e.g. petri dish vs. T-flask) varies with the cell type. BrdU is generally added a few hours following cell transfer and cell harvest is timed to occur at the second metaphase following BrdU addition. As with lymphocyte cultures, it is generally convenient to allow cells to incorporate BrdU for two replication cycles, although sister chromatid differentiation can also be effected if BrdU is present for the first but not the second cycle. Examples of specific conditions that have been employed for different cell types are given in Table II. Cell harvest begins with proteolytic release of cells (0.05% trypsin plus 0.02% EDTA, or 0.1% Pronase) followed by the steps described for lymphocytes.

Slide preparation

Cells are centrifuged and resuspended in fresh fixative. Clay Adams Goldseal slides (3 in. × 1 in. × 1 mm, No. 3010) are washed in chromic acid, rinsed with water, dipped in methanol and air dried. The cell suspension in fixative is applied dropwise to the slides. A covered plastic box (e.g. 12 in. × 8 in. × 4 in.) containing warm water can serve as a humidity chamber to control drying of

TABLE II

CONDITIONS FOR S.C.E. ANALYSIS WITH DIFFERENT CELL TYPES

Cell	Medium	Time	Colcemid	Ref.
Human fibroblast	Ham's F-10 or Dulbecco's MEM + 20—25% fetal calf serum (FCS); 5% CO_2	48—72 h	0.45 μg/ml for 4 h	[45]
WI-38	Ham's F-10 + 10% FCS; 5% CO_2	48 h	0.45 μg/ml for 2 h	[43]
HeLa	Eagle's MEM + 10% FCS; 5% CO_2	48 h	0.05 μg/ml for 2 h	[43]
3T3	Eagle's MEM + 10% FCS; 5% CO_2	45 h	0.45 μg/ml for 2 h	[46]
CHO	Modified Ham's F-10 plus 10% calf serum, 5% fetal calf serum, HEPES buffer	30—32 h	0.6 μg/ml for 2 h	[21]

material added to slides. Slides containing cells are placed on a rack inside this chamber and the top removed after a time (e.g. 15 sec) determined to give satisfactory metaphase spreads. For example, if chromosome spreading (examined under a phase microscope) at low ambient relative humidity is found to be insufficient, the time before removal of the chamber top is increased. The chamber top can be used to fan the slides to effect even drying in a few seconds.

Slide staining

1. 33258-Hoechst

Slides are successively dipped in 0.14 M NaCl, 0.004 M KCl 0.01 M phosphate, pH 7.0 (PBS) (5 min), 0.5 μg/ml 33258 Hoechst in PBS (10 min), PBS (1 min) and PBS (10 min). Stock solutions of dye (50 μg/ml) in H_2O can be stored at 4°C in the dark for at least 2 weeks. Samples of 33258-Hoechst [47] were originally obtained from Dr. H. Loewe, Hoechst AG, Frankfurt, G.F.R., although the dye can now be purchased from American Hoechst Co., Chemical Department, Route 202-206N, Somerville, N.J. 08876.

2. Giemsa staining

A. Following fluorescence photomicrography of 33258-Hoechst-stained slides (mounted in pH 7 or pH 7.5 buffer), the slides can be washed, incubated in 2 × SSC (0.3 M NaCl, 0.03 M Na citrate, pH 7) at 60—65°C for 15—30 min, and restained in Giemsa (e.g. 4% Gurr's R-66 in 5 mM pH 6.8 phosphate buffer). Sister chromatid differentiation previously apparent only with fluorescence can then be observed with transmitted light [35].

In this procedure, the illumination used to excite 33258-Hoechst fluorescence also presumably promotes selective destruction of the chromatid which has been highly substituted with BrdU.

B. Direct photosensitization of chromosomes can be achieved [54] by mounting slides previously stained with 33258-Hoechst in a 10^{-4} M solution of the dye [19] in one-third strength phosphate-buffered saline (PBS) (or a comparable buffer with a pH near 7). A coverslip is applied and the slides are then placed in a petri dish and exposed to light for a period of time which depends on the illuminating conditions. For example, approximately 6 h exposure is sufficient after positioning the slides 6 cm from a Sylvania 20 W cool white bulb. The slides are then rinsed in H_2O, incubated at 60—65° C in 2 × SSC (0.30 M NaCl, 0.03 M sodium Citrate (pH 7)) for 15 min, rinsed with H_2O and stained with Giemsa as before. Contrast can be increased by increasing the time during which slides are exposed to light. The 2 × SSC step can in principle promote G-banding of chromosomes. However, this can apparently be minimized by "aging" the slides for a few days before use [3,51].

Photography

1. Fluorescence

Slides are mounted at pH 7.0 or pH 7.5 (e.g. in McIlvaine's buffer, a mixture of 0.1 M citric acid and 0.2 M Na_2HPO_4), and fluorescence is recorded on Tri-X film. Since success in photographing fluorescence patterns depends on the efficiency of detection of the fluorescence, the use of an automatic camera system which removes a large fraction of light before it reaches the film is not recommended. Use of a buffer with appreciably lower ionic strength or pH reduces contrast due to BrdU-dependent dye quenching [44,46].

Fluorescence excitation of the dye-DNA complex peaks at approximately 360 nm, while emission peaks at approximately 475 nm [37,44]. Excitation of 33258-Hoechst fluorescence is thus carried out with illumination from a high pressure mercury lamp which traverses filters transmitting maximally in the range of 360—400 nm. For example, this can be accomplished with a Leitz Orthoplan microscope with incident illumination, a UG-1 filter and a TK 400 dichroic mirror. Emission is best observed through filter combinations excluding light below 460 nm.

If incident illumination is used, the lenses should not have a large number of correcting elements, since many of these attenuate light intensity below 400 nm and thus reduce the excitation of illumination in the range absorbed by the Hoechst dye. Achromat objectives have proved to be satisfactory for fluorescence microscopy of 33258-Hoechst-stained chromosomes.

The fluorescence of 33258-Hoechst fades rapidly under the conditions in which it is sensitive to the incorporation of BrdU. Thus, for routine detection of S.C.E., BrdU-Giemsa techniques are more convenient than fluorescent methods. Fluorescence analysis can often help, however, to locate the source of problems which might arise in setting up Giemsa methods.

2. Giemsa

Slides can be mounted in a standard embedding medium, or immersion oil can be applied directly to the slide without a coverslip. A 544 nm interference filter can be used to enhance contrast, and the optical image is recorded on high contrast copy film. Highly corrected microscope objectives can be employed.

In vivo S.C.E. analysis

Male CBA mice (Jackson Lab., Bar Harbor, Me.), 3—5 months old, are administered hourly intraperitoneal injections of BrdU (10^{-2} M) over the approximate duration of one DNA synthesis period. Deoxycytidine (5×10^{-3} M) can be mixed with BrdU to reduce toxicity of the latter [50]. The series of injections serves to replenish nucleoside which is though to undergo rapid depletion in vivo [57].

In germ cell studies, 14 (0.2 ml) hourly injections of the BrdU-dC solution are given, and spermatogonial cells are collected 44 h following the last injection. In bone marrow studies, 9 hourly (0.4—1.0 ml) injections of the BrdU-dC solution are given. Marrow cells are examined 11—19 h following the last BrdU injection. Drugs (e.g. mitomycin C or cyclophosphamide), when administered, are given as a single intraperitoneal (i.p.) injection following the last BrdU injection — 12 h later in the case of spermatogonia, and 15 min later in the case of bone marrow. Animals are killed by ether narcosis 2—5 h following intraperitoneal injection of 0.6 mg/kg Colcemid.

Spermatogonial chromosomes are prepared in accordance with the method of Meredith [49]. Following orchidectomy, both testes are placed in P-60 culture dishes, half filled with hypotonic 1% sodium citrate at room temperature. The tunica is removed from each testis and seminiferous tubules are gently teased apart. The tubules are then transferred to two successive dishes of 1% sodium citrate for further teasing and washing. The total time for tissue exposure to hypotonic solution in the 3 dishes is 12 min. Tubules are then transferred to 12 ml of fixative (methanol : acetic acid, 3 : 1) for 15 min, at which time they are either used immediately for slide preparation or stored at —20° C for slide preparation at a subsequent time (ranging up to a month later). For preparation of slides, approximately 2—3 drops of tubule suspension (collected with a forceps) are placed in a small test tube containing approximately 0.5 ml of 60% acetic acid. Release of the cells from the tubule is aided by gentle tapping of the tube and mixing with a Pasteur pipette. Cells are withdrawn with a Pasteur pipette and applied dropwise to warm (60° C) slides. To avoid regional clumping of cells, the slide is tilted after every droplet, allowing the solution to roll across the surface, and the fluid is then immediately withdrawn with the pipette. Two slides per tube are prepared within about 5 min of the time tubules were initially exposed to 60% acetic acid.

For preparation of bone marrow cells [11,14] both femurs are excised and the epicondyle tips removed. Marrow is expelled by aspiration with 0.075 M KCl through a 26-gauge needle into a centrifuge tube. Then 0.075 M KCl is added to give a total of 6 ml. The cell solution is mixed and incubated for 12 min at 37°C, and extraneous matter removed. The cells are then centrifuged and fixed twice in 3 : 1 methanol : acetic acid.

In vivo S.C.E. analyses are more difficult to carry out then are in vitro studies because of the need for multiple BrdU injections. Alternative, more convenient approaches for BrdU administration are being explored. We have recently utilized subcutaneous implantation of BrdU tablets for this purpose (Allen, J., Shuler, C., Mendes, R. and Latt, S., Cytogenet. Cell Genet., in press). In addition, spermatogonial preparations yield cells with widely varying quality, as well as polyploid cells. In scoring the latter cells, SCE counts are normalized for the number of genomes present as determined by chromosome count.

Sister chromatid exchange scoring

At present, S.C.E.s are counted manually on photomicrographs of metaphases containing chromosomes with sister chromatid differentiation. Using this approach, a range of baseline S.C.E. frequencies has been reported from different laboratories (Table III). A number of factors might contribute to this variation, for example (1) cell type, (2) culture conditions, in particular BrdU concentration, (3) slide preparation and staining and (4) S.C.E. scoring procedures. These will each be considered in turn.

Cell type

Differences between the S.C.E. frequencies of lymphocytes and human fibroblasts are no greater than the variations in the frequencies reported by different laboratories for each cell type, although fibroblasts appear to be more responsive to S.C.E. induction by alkylating agents [22,45]. Modest day-to-day variations also occur in the S.C.E. frequency of lymphocytes from different individuals [43].

S.C.E. frequencies determined in vivo appear to be significantly less than those reported in vitro. Bloom and Hsu [4] observed fewer than one S.C.E./

TABLE III

BASELINE SISTER CHROMATID EXCHANGE FREQUENCIES

Cell type	BrdU(M)	S.C.E. frequency	Detection[a]	Comments	Ref.
Human lymphocyte	10^{-5}	7	H + G		[31]
	10^{-5}	10	H + G	Lymphoblasts	[11]
	10^{-5}	4–5	H + G		[3]
	10^{-5}	10–13	H	4×10^{-7} M FdU 6×10^{-6} M U	[27,38,45]
	10^{-5}	15	H	FdU, U, 10^{-4} M dC	[45,46]
	6.5×10^{-5}	14	H, H + G		[35]
	10^{-4}	7	H + G		[26]
10	10^{-4}	80	H + G	Bloom's syndrome	[26]
	10^{-4}	15–17	H	FdU, U	[18,69]
	10^{-4}	3–6	H, H + G		[60]
	1.6×10^{-4}	14[b]	H + G		[16]
	6.5×10^{-4}	27	A.O.		[12]
Human fibroblast	10^{-5}	9	H + G		[22]
	10^{-5}	9[b]	H + G	Normal or xeroderma pigmentosum cells	[23]
	10^{-5}	12[b]	H + G	SV40-transformed cells	[70]
	10^{-5}	10	H		[45]
	$\leqslant 3 \times 10^{-6}$	4	A.O.		[32]
Human embryonic lung					
	10^{-5}	6	H + G	1 cycle BrdU	[51]
	10^{-5}	8	H + G	2 cycles BrdU	[51]
Chinese hamster					
CHO	10^{-5}	12–13	H + G		[53]
	10^{-5}	12–14[b]	H + G		[54]
D-6	2×10^{-6}	2.4	A.O.		[31]
Mouse	10^{-4}	25	H		[46]

[a] H: 33258-Hoechst fluorescence; H + G: Giesma following 33258-Hoechst staining and illumination, A.O.: acridine orange fluorescence.

b Computed from values of S.C.E. per chromosome.

cell in chick embryos after two cycles of BrdU incorporation. The S.C.E. frequency in mouse bone marrow was determined to be 4/cell [2,64], while a value of 2/cell was observed in mouse spermatogonia [1] after a cycle of BrdU incorporation followed by a cycle without BrdU. In contrast, mouse 3T3 cells in vitro exhibit an S.C.E. frequency of approximately 25 [46].

Cell culture conditions

The most obvious variation in cell culture conditions influencing the S.C.E. frequency relates to BrdU incorporation. Growth of cells in the absence of light is assumed. The S.C.E. frequency increases gradually as cells are grown in medium containing increasing BrdU concentrations in the range 10^{-5} to 10^{-4} M. At even higher BrdU concentrations the S.C.E. frequency in many cell types increases markedly [43]. Use of a single cycle of BrdU incorporation followed by one division in the presence of deoxythymidine (dT) but not BrdU results in a somewhat lower exchange frequency [51] than if the cells incorporate BrdU for two cycles. Inclusion of FdU in the culture medium probably increases the extent and uniformity of BrdU incorporation by suppressing endogenous dT synthesis, although cell growth might be reduced.

Deoxycytidine, which has been added to lymphocyte cultures to reduce BrdU toxicity [23], can increase the S.C.E. frequency without significantly altering BrdU incorporation [46]. One explanation for this effect, which is associated with improved cell growth [45], is that cells which have experienced greater damage are still able to reach the second metaphase at which S.C.E. are scored. In this regard, other factors altering cell viability might alter the S.C.E. frequency by affecting the sample of cells which are scored for exchanges.

Other components of the culture medium might in principle influence the exchange frequency, although details here are scant. We have found no consistent difference in the peripheral lymphocyte S.C.E. frequency when using purified phytohemagglutinin or a crude red kidney been extract containing phytogemagglutinin [43]. However, variations in culture conditions should be minimized when surveying S.C.E. frequencies in different individuals.

Slide preparation and staining

This category has not been extensively explored. Inefficient cell harvest could lead to a loss of metaphase cells. Cells with chromosomes in various states of contraction might differ in exchange frequency and be selectively lost. Increased effectiveness in sister chromatid differentiation would reduce the necessary BrdU concentrations and, indirectly, the S.C.E. frequency.

S.C.E. scoring procedures

The majority of S.C.E. can be scored without great difficulty (Fig. 1). However, at least four sources of variation in S.C.E. scoring can be identified. (1) Choice of cell to be scored might alter the frequency by biasing the count to cells with better contrast and greater BrdU incorporation. (2) S.C.E. located near the centromere might be difficult to differentiate from 180° chromosome arm twists (Fig. 2). (3) S.C.E. located near the tips of chromosomes may be difficult to detect (Fig. 3). (4) Multiple, closely spaced S.C.E. might be diffi-

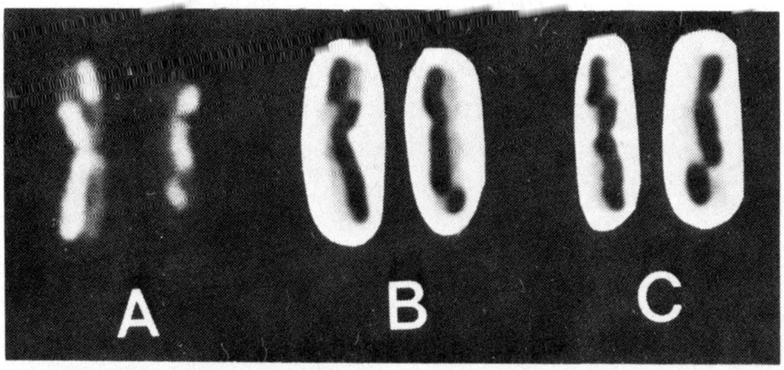

Fig. 1. Sister chromatid exchanges. The chromosomes in this figure are from human lymphocytes which replicated twice in medium containing 10^{-5} M BrdU, 6 × 10^{-6} M U, and 4 × 10^{-7} M FdU. Those in (A) were stained with 33258-Hoechst and photographed under conditions described for fluorescence microscope. Chromosomes in (B) were previously photographed to record fluorescence, as in (A), and then washed with H_2O, incubated 15 min at 60—65°C in 2 × SSC, and stained with Giemsa. Chromosomes in (C) were exposed to fluorescent light while mounted in buffer containing 10^{-4} M 33258-Hoechst as described in the methods section, incubated in 2 × SSC, and stained with Giemsa. Chromosomes shown were chosen to demonstrate relatively unambiguous sister chromatid exchanges.

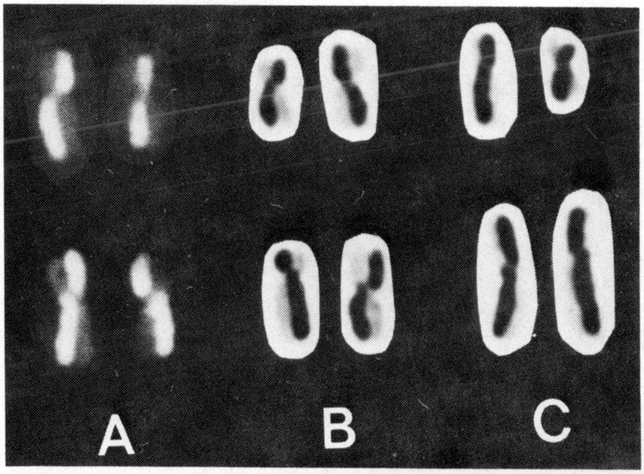

Fig. 2. Sister chromatid exchanges vs. 180° twists. Cell culture, chromosome staining and photography are as described in the captain to Fig. 1. The top row of chromosomes demonstrates 180° centromeric twists, while the bottom row consists of chromosomes containing an S.C.E. which resembles a twist. Differentiation between twists and S.C.E.s is based on observation of dark or light staining chromatid continuity in the former but not the latter, which exhibit a discernible discontinuity in staining. Differentiation between 180° twists and S.C.E.s is not always this apparent.

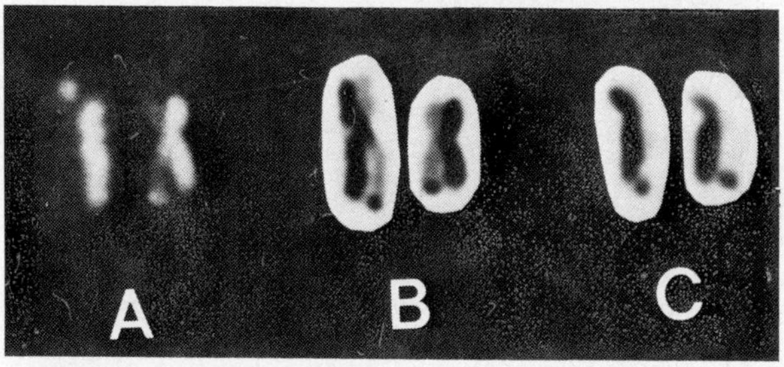

Fig. 3. Sister chromatid exchanges near chromosome tips. Cell culture, chromosome staining, and photography are as described in the caption to Fig. 1. These chromosomes were chosen to demonstrate S.C.E.s near chromosome tips. Some show terminal dark areas on one chromatid without a discernible faint area on the other. S.C.E.s more terminal than those shown would be difficult to detect.

cult to resolve completely (Fig. 4). Additional information of interest is a description of the frequency of chromosome breaks and of the relative location of breaks and incomplete exchanges. The relative importance of breaks and exchanges, standardized for the time at which chromosome treatments are applied, should further distinguish patterns of chromosome damage and repair.

Additional examples

Analysis of S.C.E. in response to chromosome damage is illustrated with some data on the response of human lymphocytes to varying concentrations of mitomycin C (Fig. 5). In this particular experiment, mitomycin C was present for only the second cycle rather than both cell cycles in order to reduce cell selection effects. The S.C.E. values shown utilize 33258-Hoechst fluorescence data, although S.C.E. in these cells could also be detected by Giemsa staining after fluorescence microscopy or directly after exposure of slides stained with 33258-Hoechst to light (see Fig. 1—4). We have made relatively little use of acridine orange methodology, although this has been effectively employed by others [15,29].

The 30 control cells for the experiment demonstrating the induction of sister chromatid exchanges by mitomycin C were scored by six different individuals. The inter-individual variation in the mean exchange frequency observed was

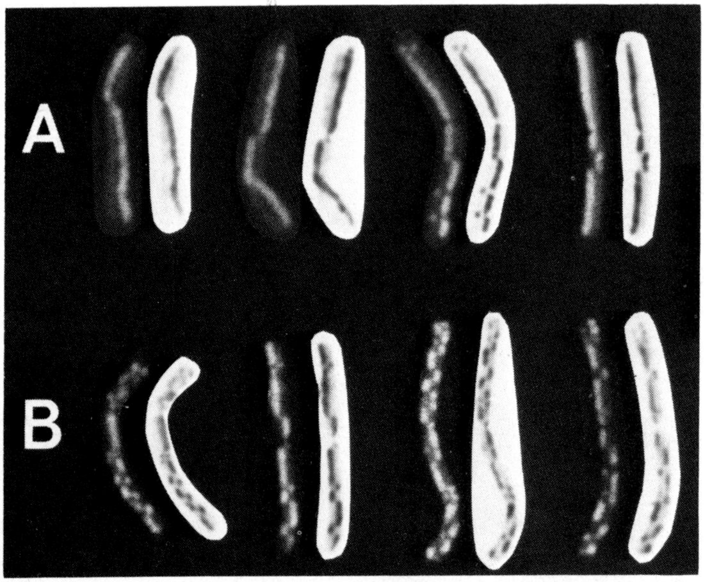

Fig. 4. Extensive S.C.E. induction by an alkylating agent. These chromosomes are from CHO cells grown 42 h in medium containing 5×10^{-5} M BrdU. The left hand picture of each chromosome reflects 33258-Hoechst fluorescence. The right hand picture was obtained with transmitted light after the same slides were washed, incubated 15 min in 2 × SSC at 60—65°C and stained with Giemsa. The top row of chromosomes (A) represents the control, while those in the bottom row (B) are from cells exposed to 10^{-5} M 8-methoxypsoralen and near-UV light (1 or 2 min of approximately 100 ergs/mm^2/sec, principally at 365 nm) before addition of BrdU [42]. The chromosomes in this figure are less than 25 μm long, and several S.C.E.s in (B) are less than 1 μm apart.

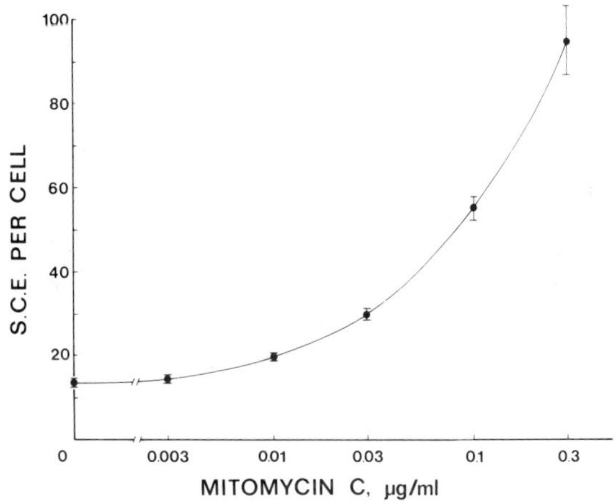

Fig. 5. Induction of S.C.E.s in human lymphocyte chromosomes. Human lymphocytes were cultured 75 h in medium containing 10^{-5} M BrdU, 6 × 10^{-6} M U, and 4 × 10^{-7} M FdU. Twenty-five hours before harvest, mitomycin C was added to the final concentration shown. Slides were stained with 33258-Hoechst and S.C.E.s were scored by one person from fluorescence photomicrographs. The horizontal lines bracketing the data points represent the standard error of the mean S.C.E. frequency of 20—54 cells.

less than 1 S.C.E./cell. This number in effect sets a lower limit on the number of cells which should be scored before the statistical error in the S.C.E. frequency is less than the counting error. At a mean of N S.C.E./cell, the standard error for one S.C.E. count should approximate $\sqrt{N}$ (assuming a Poisson distribution of exchanges). For N cells the error of the mean would then be 1 S.C.E./cell, e.g. a mean of 25 S.C.E./cell is adequately determined by counting 25 cells. When chromosome damage and S.C.E. induction is extreme (e.g. data

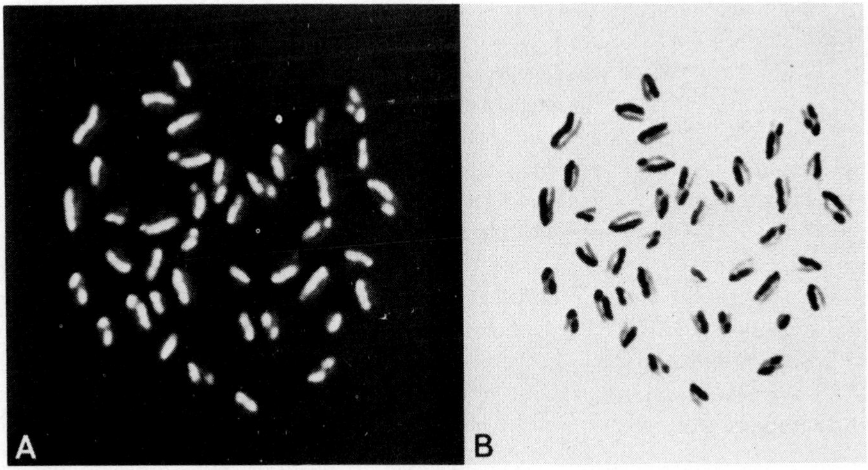

Fig. 6. Induction by mitomycin C of S.C.E. formation in vivo in mouse bone marrow. A 25 g mouse received nine 0.4 ml injections of 10^{-2} M BrdU plus 5 × 10^{-3} M dC. Fifteen minutes later, 7.5 μg of mitomycin C was injected intraperitoneally. The animals were sacrificed 19 h later, and chromosome preparations made as described in the Methods section. The cell was stained with 33258-Hoechst, the fluorescence under incident illumination was photographed (A), the slide was washed with xylene, methanol and water, stained with Giemsa and rephotographed (B). An average of 13 S.C.E. was observed in cells following mitomycin C treatment, as compared with fewer than 4/cell without mitomycin C [2].

at 0.1 μg/ml and 0.3 μg/ml mitomycin C, Fig. 5), the standard error of the exchange frequency estimate is greater. This may reflect fluctuations in the survival of cells with varying degrees of damage.

Analysis of S.C.E. induction is now being extended to in vivo systems. For example, BrdU can be injected into mice [1] and S.C.E. frequencies in different tissues, such as spermatogonia [1] or bone marrow [2,64] measured after injection of animals with clastogens such as mitomycin C (Fig. 6).

Future directions

1. Use of a limited number of standardized cell lines would improve the prospects of comparing data from different laboratories. Employment of standard cell culture medium preparations would be similarly desirable. Use of a battery of human mutant cell strains, e.g. those with different DNA repair defects, might be used to diffentiate between agents causing various types of DNA damage.

2. BrdU concentrations should be minimized. At 10^{-5} M BrdU appears satisfactory for exchange detection by all cell culture techniques. This level is sufficiently low that neither BrdU-induced exchanges nor the rate at which these exchanges vary with BrdU is a problem. Immunochemical detection of BrdU substitution [20] might permit the use of lower BrdU concentrations. A cell growth protocol consisting of one cycle of BrdU incorporation followed by a cycle in the absence of BrdU will also reduce the number of BrdU-induced exchanges [51]. Problems anticipated with this latter variation include accurate determination of the timing at which medium is changed and the need for more vigorous chromosome pretreatment for sister chromatid differentiation if a Giemsa staining method is used.

3. Automatic exchange detection has been accomplished with images on photographic negatives [71] and is being developed for Giemsa-stained slides [72]. This will standardize exchange scoring, although a somewhat greater uncertainty in identifying exchanges might be anticipated. Successful automation would simplify exchange analyses on large numbers of cells.

4. Biochemical detection of S.C.E.s appears feasible [52,55]. While more complicated than current cytological methods, biochemical assays would be less susceptible to the bias introduced by the need for cells to reach second metaphase to be scored.

5. Perhaps the major development needed for effective use of S.C.E. measurements is a classification of the relationship between S.C.E.s and mutagenesis or carcinogenesis. If S.C.E. formation accurately reflects the other two processes, S.C.E. analysis might become an essential component of environmental mutagenesis and carcinogenesis testing.

Acknowledgements

This work was supported by research grants from the National Institutes of Health (GM 21121), National Foundation March of Dimes (1-353) and American Cancer Society (VC-144). S.A.L. is the recipient of a Research Career Development Award (GM 00122), and J.W.A. is supported by funds from a postdoctoral training grant (GM 00156). The technical assistance of Ms. Susan Brefach and Ms. Gayle Forbes is greatly appreciated. 33258-Hoechst and 8-methoxypsoralen were generous gifts from Dr. H. Loewe, Hoechst, G.F.R., and the Paul Elder Co., Bryan, Ohio, respectively.

References

1 Allen, J.W. and S.A. Latt, Analysis of sister chromatid exchange formation in vivo in mouse spermatogonia as a new test system for environmental mutagens, Nature, 260 (1976) 449—451.

2 Allen, J.W. and S.A. Latt, In vivo BrdU-33258 Hoechst analysis of DNA replication kinetics and sister chromatid exchange formation in mouse somatic and meiotic chromosomes, Chromosoma, 58 (1976) 325—340.

3 Beek, B. and G. Obe, The human leukocyte test system VI. The use of sister chromatid exchanges as possible indicators for mutagenic activities, humangenetik, 29 (1975) 127—134.

4 Bloom, S.E. and T.C. Hsu, Differential fluorescence of sister chromatids in chicken embryos exposed to 5-bromodeoxyuridine, Chromosoma, 51 (1976) 261—267.

5 Brewen, J.G. and W.J. Peacock, The effect of triatiated thymidine on sister-chromatid exchange in a ring chromosome, Mutation Res., 7 (1969) 433—440.

6 Chaganti, R.S.K., S. Schonberg and J. German, A manyfold increase in sister chromatid exchanges in Bloom's syndrome lymphocytes, Proc. Natl. Acad. Sci. (U.S.A.), 71 (1974) 4508—4512.

7 Clark, A.J., Recombination deficient mutants of *E. coli* and other bacteria, Ann. Rev. Genet., 7 (1973) 67—86.

8 Cole, R.S., Repair of DNA containing intrastrand crosslinks in *Escherichia coli*: sequential excision and recombination, Proc. Natl. Acad. Sci. (U.S.), 70 (1973) 1064—68.

9 Comings, D.E., Isolabelling chromosome strandedness, Nature New Biol., 229 (1971) 24—25.

10 Craig-Holmes, A.P. and M.W. Shaw, Cell cycle analysis in asynchronous cultures using the BUDR-Hoechst technique, Exptl. Cell Res., 99 (1976) 79—87.

11 Declève, A., G.B. Gerber, A. Leonard, M. Lambiet-Collier, A. Sassen and J.R. Maisin, Regeneration of thymus, spleen, and bone marrow in X-irradiated AKR mice, Radiation Res., 51 (1972) 318—332.

12 Dutrillaux, B., A.M. Fosse, M. Prieur et J. LeJeune, Analyse des Echanges de chromatides dans les cellules somatiques humaines, Chromosoma, 48 (1974) 327—340.

13 Finkelberg, R., M.W. Thompson and L. Siminovich, Survival after treatment with EMS, γ rays, and mitomycin C of skin fibroblasts from patients with Fanconi's anemia, Am. J. Hum. Genet., 26 (1974) 30A (abstract).

14 Ford, C.E. and J.L. Hamerton, A colchine, hypotonic citrate, squash sequence for mammalian chromosomes, Stain Technol., 31 (1956) 247—251.

15 Franceschini, P., Semiconservative DNA duplication in human chromosomes treated with BUDR and stained with acridine orange, Exptl. Cell Res., 89 (1974) 420—421.

16 Galloway, S.M. and H.J. Evans, Sister chromatid exchanges in human chromosomes from normal individuals and patients with ataxia telangiectasia, Cytogenet. Cell Genet., 15 (1975) 17—29.

17 German, J., Genes which increase chromosomal instability in somatic cells and predispose to cancer, Progr. Med. Genet., 8 (1972) 61—101.

18 Gibson, D.A. and D.M. Prescott, Induction of sister chromatid exchanges in chromosomes of rat kangaroo cells by tritium incorporated into DNA, Exptl. Cell Res., 74 (1972) 397—402.

19 Goto, K., T. Akematsu, H. Shimazu and T. Sueiyma, Simple differential Giemsa staining of sister chromatids after treatment with photosensitive dyes and exposure to light and the mechanism of staining, Chromosoma, 53 (1975) 223—230.

20 Gratzner, H.G., A. Pollack, D.J. Ingram and R.C. Leif, Deoxyribonucleic acid replication in single cells and chromosomes by immunologic techniques, J. Histochem. Cytochem., 24 (1976) 34—39.

21 Hamlin, J.L. and A.B. Pardee, S-phase synchrony in monolayer CHO cultures, Exp. Cell Res., 100 (1976) 265—275.

22 Hayashi, K. and W. Schmid, The rate of sister chromatid exchanges parallel to spontaneous chromosome breakage in Fanconi's anemia and to trenimon-induced aberrations in human lymphocytes and fibroblasts, Humangenetik, 29 (1975) 201—206.

23 Hook, E.B., B. Pollard and N.H. Hatcher, Chromatid exchange rates in ataxia telangiectasia, Pediat. Res., 9 (1975) 313 (abstract).

24 Igali, S., B.A. Bridges, M.J. Ashwook-Smith and B.R. Scott, Mutagenesis in *Escherichia coli*. IV. Photosensitization to near ultraviolet light by 8-methoxypsoralen, Mutation Res., 9 (1970) 21—30.

25 Ikushima, T. and S. Wolff, Sister chromatid exchanges induced by light flashes to 5-bromodeoxyuridine- and 5-iododeoxyuridine-substituted Chinese hamster chromosomes, Exptl. Cell Res., 87 (1974) 15—19.

26 Kato, H., Induction of sister chromatid exchanges by ultraviolet light and its inhibition by Caffeine, Exptl. Cell Res., 82 (1973) 383—390.

27 Kato, H., Induction of sister chromatid exchanges by chemical mutagens and its possible relevance to DNA repair, Exptl. Cell Res., 85 (1974) 239—247.

28 Kato, H., Photoreactivation of sister chromatid exchanges induced by ultraviolet irradiation, Nature, 249 (1974) 552—553.

29 Kato, H., Spontaneous sister chromatid exchanges detected by a BUDR-labelling method, Nature, 251 (1974) 70—72.

30 Kato, H., Is isolabelling a flase image?, Exptl. Cell Res., 89 (1974) 416—420.

31 Kato, H. and H. Shimada, Sister chromatid exchanges induced by mitomycin C: a new method of detecting DNA damage at chromosome level, Mutation Res., 28 (1975) 459—464.

32 Kato, H. and H.F. Stich, Sister chromatid exchanges in aging and repair-deficient fibroblasts, Nature 260 (1976) 447—448.

33 Kihlman, B.A., Sister chromatid exchanges in *Vivia faba*. II. Effects of thiotepa, caffeine and 8-ethoxycaffeine on the frequency of SCE's, Chromosome, 51 (1975) 11—18.

34 Kihlman, B.A. and D. Kronborg, Sister chromatid exchanges in *Vicia faba*. I. Demonstration by a modified fluorescence plus Giemsa (FPG) technique, Chromosoma, 51 (1975) 1—10.

35 Kim, M.A., Chromatid exchange and heterochromatin alteration of human chromosomes with BUDR-labelling; demonstration by benzomidazol fluorochrome and Giemsa stain, Humangenetik, 25 (1974) 179—188.

36 Korenberg, J. and E. Freedlender, Giemsa technique for detection of sister chromatid exchanges, Chromosoma, 48 (1974) 355—60.

37 Latt, S.A., Microfluorometric detection of DNA replication in human metaphase chromosomes, Proc. Natl. Acad. Sci. (U.S.), 70 (1973) 3395—3399.

38 Latt, S.A., Localization of sister chromatid exchanges in human chromosomes, Science, 185 (1974) 74—76.

39 Latt, S.A., Microfluorometric analysis of DNA replication kinetics and sister chromatid exchanges in human chromosomes, J. Histochem. Cytochem., 22 (1974) 478—491.

40 Latt, S.A., Sister chromatid exchanges, indices of human chromosome damage and repair: detection by fluorescence and induction by mitomycin C. Proc. Natl. Acad. Sci. (U.S.), 71 (1974) 3162—3166.

41 Latt, S.A., J.W. Allen, C. Shuler, K.S. Loveday and S.H. Munroe, The detection and induction of sister chromatid exchanges, in 6th ICN-UCLA Symp., Molecular Human Cytogenetics, Academic Press, New York, 1977.

42 Latt, S.A., in preparation.

43 Latt, S.A. and L. Juergens, Determinants of sister chromatid exchange frequencies in human chromosomes, in E. Hook and I. Porter (eds.), Population Cytogenetics, Academic Press, New York, 1976, pp. 217—236.

44 Latt, S.A. and J.C. Wohlleb, Optical studies of the interaction of 33258 Hoechst with DNA, chromatin, and metaphase chromosomes, Chromosoma, 52 (1975) 297—316.

45 Latt, S.A., G. Stetten, L.A. Juergens, G.R. Buchanan and P.S. Gerald, The induction by alkylating agents of sister chromatid exchanges and chromatid breaks in Fanconi's anemia, Proc. Natl. Acad. Sci. (U.S.), 72 (1975) 4066—4070.

46 Latt, S.A., G. Stetten, L.A. Juergens, H.F. Willars and C.D. Scher, Recent developments in the detection of DNA synthesis by 33258 Hoechst fluorescence, J. Histochem. Cytochem., 23 (1975) 493—505.

47 Loewe, H. and J. Urbanietz, Basisch Substitutient 2,6-Bis-Bensimidazole-derivate, Eine Neue Chemotherapeutische Active Korperklasse, Arzneimittel-Forsch., 24 (1974) 1927—1933.

48 Marin, G. and D.M. Prescott, The frequency of sister chromatid exchanges following exposure to varying doses of ^{3}H-thymidine or X-rays, J. Cell Biol., 21 (1964) 159—167.

49 Meredith, R., A simple method for preparing meiotic chromosomes from mammalian testis, Chromosoma, 26 (1969) 254—258.

50 Meuth, M. and H. Green, Induction of a deoxycytidineless state in cultured mammalian cells by bromodeoxyuridine, Cell, 2 (1974) 109—112.

51 Miller, R.C., M.M. Aronson and W.W. Nichols, Effects of treatment on differential staining of BrdU labeled metaphase chromosomes: three-way differentiation of M_3 chromosomes, Chromosoma, 55 (1976) 1—11.

52 Moore, P.D. and R. Holliday, Evidence for the formation of hybrid DNA during mitotic recombination in Chinese hamster cells, Cell, 8 (1976) 573—580.

53 Perry, P. and H.J. Evans, Cytological detection of mutagen-carcinogen exposure by sister chromatid exchange, Nature, 258 (1975) 121—124.

54 Perry, P. and S. Wolff, New Giemsa method for the differential staining of sister chromatids, Nature, 251 (1974) 256—258.

55 Rommelaere, J. and A. Miller-Faures, Detection by density equilibrium Centrifugation of recombinant-like DNA molecules in somatic mammalian cells, J. Mol. Biol., 98 (1975) 195—218.

56 Rupp, W.D. and P. Howard-Flanders, Discontinuities in the DNA synthesized in an excision-defective strain of *E. coli* following ultraviolet irradiation, J. Mol. Biol., 31 (1968) 291—304.

57 Russev, G. and Tsanev, R., Application of precursors adsorbed on activated charcoal for labeling of mammalian DNA in vivo, Meth. Cell Biol., IX (1975) 115—122.

58 Sasaki, M., Is Fanconi's anemia defective in a process essential to the repair of DNA cross links?, Nature, 257 (1975) 501—503.

59 Sasaki, M.S. and A. Tonomura, A high susceptibility of Fanconi's anemia to chromosome breakage by DNA cross-linking agents, Cancer Res., 33 (1973) 1829—1836.

60 Solomon, E. and M. Bobrow, Sister chromatid exchanges — a sensitive assay of agents damaging human chromosomes, Mutation Res., 30 (1975) 273—278.

61 Sugiyama, T., Specific vulnerability of the largest telocentric chromosome of rat bone marrow cells to 7,12-dimethylbanz(α)anthracene, J. Natl. Cancer Inst., 47 (1971) 1267—1275.

62 Taylor, J.H., Sister chromatid exchanges in tritium-labelled chromosomes, Genetics, 43 (1958) 515—529.

63 Taylor, J.H., P.S. Woods and W.L. Hughes, organization and duplication of chromosomes as revealed by autoradiographic studies using tritium-labelled thymidine, Proc. Natl. Acad. Sci. (U.S.), 43 (1975) 122—128.

64 Vogel, W. and T. Bauknecht, Differential chromatid staining by in vivo treatment as a mutagenicity test system, Nature, 260 (1976) 448—449.

65 Witkin, E.M., The mutability toward ultraviolet light of recombination-deficient strains of *Escherichia coli,* Mutation Res., 8 (1966) 9—14.

66 Witkin, E.M., Mutation-proof and mutation-prone modes of survival in derivatives of *Escherichia coli* B differing in sensitivity to ultraviolet light, Ann. Rev. Genet., 3 (1969) 17—55.

67 Wolff, S., Strandedness of chromosomes, Int. Rev. Cytol., 25 (1965) 279—296.

68 Wolff, S. and P. Perry, Insights on chromatid structure from sister chromatid exchange ratios and the lack of both isolabelling and heterolabelling as determined by the FPG technique, Exptl. Cell Res., 93 (1975) 23—30.

69 Wolff, S., J. Bodycote and R.B. Painter, Sister chromatid exchanges induced in chinese hamster cells by UV irradiation at different stages of the cell cycle: the necessity of cells to pass through S, Mutation Res., 25 (1974) 73—81.

70 Wolff, S., J. Bodycote, G.H. Thomas and J.E. Cleaver, Sister chromatid exchange in xeroderma pigmentosum cells that are defective in DNA excision repair or post-replication repair, Genetics, 81 (1975) 349—355.

71 Zack, G.W., J.A. Spriet, S.A. Latt, G.H. Grandlund and I.T. Young, Automatic detection and localization of sister chromatid exchanged, J. Histochem. Cytochem., 24 (1976) 168—177.

72 Zack, G.W., W. Rogers and S.A. Latt, J. Histochem. Cytochem., in press.

TESTS FOR HERITABLE TRANSLOCATIONS IN MALE MAMMALS

A. LÉONARD

Laboratory of Genetics, Department of Radiobiology, C.E.N.–S.C.K., B-2400 Mol (Belgium)

(Received December 9th, 1974)

INTRODUCTION

The observation of dividing spermatocytes for the presence of reciprocal translocations is, together with the specific locus method and the dominant lethality test, routinely used to estimate genetic hazards in mammalian germ cells. The term of reciprocal translocation designs exchanges of terminal segments between non-homologous chromosomes (Fig. 1).

At the meiotic synapsis, heterozygosity for a reciprocal translocation results in production of: a quadrivalent configuration in form of a ring (RIV) when the non-centromeric ends of the chromosomes maintain association through terminalyzing chiasmata; a chain of four figure (CIV) by failure of association of chromosomes in one arm; a chain of three plus an univalent figure (C III+I) by failure of association of two adjacent arms; two bivalents, or one bivalent and two univalents or, even, four univalents. More complex configurations such as hexavalents, octovalents or decavalents can be observed if two, three or four translocations occur.

Using an appropriate interval between treatment and observation (50–100 days) one can estimate the rate at which reciprocal translocations are induced in spermatogonia (the main germ cell type at risk) of the treated animals by scoring multivalent configurations in the derived spermatocytes (spermatocyte test on treated males). The production of reciprocal translocations in the postmeiotic male germ cells is generally ascertained by analysing the dividing spermatocytes of the F1 males (F1 translocation test). This test can also be used to compare the differential sensitivity of various germ cell stages and to test female germ cells.

Refs. 7, 8, 15 discuss the problems related to the induction of heritable reciprocal translocations in mammals by environmental mutagens and their importance for the evaluation of genetic hazards to man.

THE TECHNICAL PROCEDURE

Instruments and chemicals required to perform tests for heritable translocations:

(1) A centrifuge giving exact speeds from 14 to 88 *g* and which is equipped with 4–8 carriers for 3-ml centrifuge tubes.

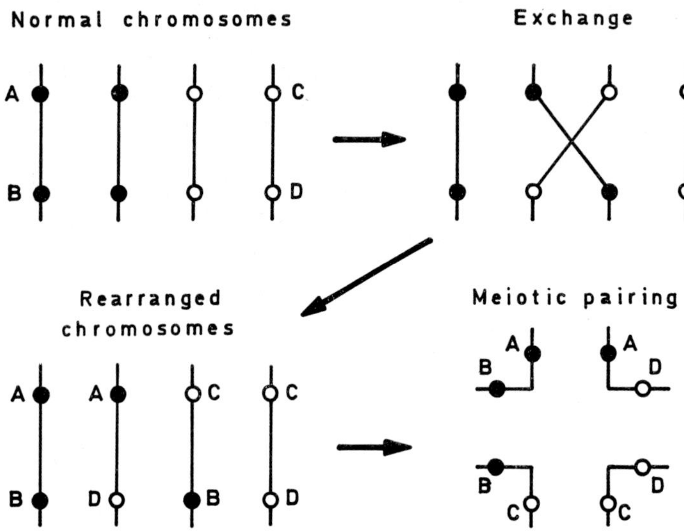

Fig. 1. Diagram of a reciprocal translocation between two acrocentric chromosomes.

(2) A microscope.

(3) A warming plate.

(4) Dissecting instruments (small scissors and small straight and curved forceps).

(5) Pyrex petri dishes (6 cm diameter), conical centrifuge tubes of 3 ml capacity, conventional glass pipettes, Pasteur pipettes drawn out to a fine tip, propipettes of 2 ml, coverglasses, grease-free microslides.

(6) Sodium citrate ($C_6H_5Na_3O_7 \cdot 2H_2O$), acetic acid ($CH_3COOH$), ethyl alcohol ($CH_3CH_2OH$), chloroform ($CHCl_3$), toluidine blue.

TEST ANIMALS

The forty chromosomes of the mouse are acrocentric. Therefore the meiotic association configurations are rather simple for these species (Fig. 2) and it is relatively easy to identify the translocation figures in the form of a ring or of a chain in the dividing spermatocytes (Figs 3–6). Due to the presence of metacentric or submetacentric chromosomes the meiotic pairing results in rats and Chinese hamsters in the formation of bivalents of a much more complex nature than the 20 bivalents of the mouse. Consequently the mouse is the most convenient animal on which to perform the tests for heritable translocations in male mammals. The frequency of spontaneous translocations among young adult control mice is extremely low. Since a slight increase has been reported in old animals[12], young 3–4-month old adult mice are to be preferred.

EXPERIMENTAL DESIGN

The general outline of the tests for heritable translocations in male mouse is shown in Table I. In general, only very few translocation configurations are recovered in dividing spermatocytes when mouse spermatogonia are treated with mutagenic

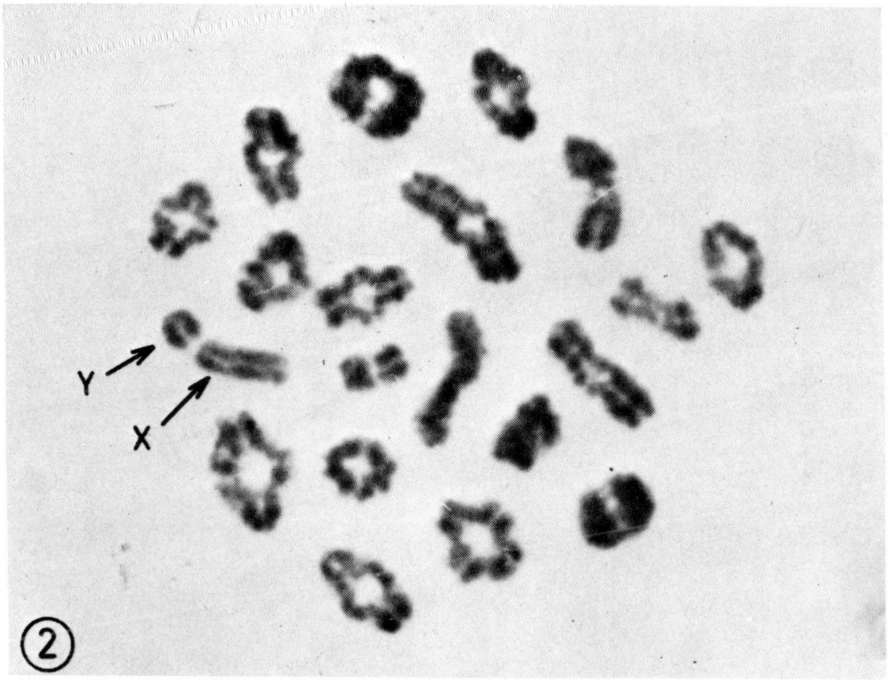

Fig. 2. Mouse dividing spermatocyte I with 20 bivalents.

compounds. Moreover, the data on ionizing radiations have often revealed great variations among animals given the same treatment. A minimum of 10 animals with an analysis of 100–200 cells per animal is therefore required for each treatment. To allow the treated spermatogonia to differentiate into dividing spermatocytes, an interval of 50–100 days between treatment and examination is suitable. To ascertain that a compound does not produce heritable translocations in postmeiotic germ cells a great number of male offspring must be investigated. In general, spermatids are the most sensitive stage with respect to the induction of chromosome damage by chemical mutagens but, as indicated by the example of mitomycin C which yields higher dominant lethality in spermatocytes[2], the possibility also exists that certain compounds act on spermatocytes but not on other postmeiotic germ cell stages in producing heritable translocations. Depending on the fertility of the strain 10 to 25 males are usually used. To avoid fertilisation before the compound could reach the germ cells, the animals are usually treated in the morning and caged with virgin females in the afternoon. The females are exchanged for fresh ones every seven days in order to compare the sensitivity of the different male germ cell stages[13,14]: 1–7 days spermatozoa, 7–14 days spermatids, 15–21 days early spermatids, 22–35 days spermatocytes. The F1 females are discarded at weaning and the F1 males are examined when mature. Experiments with ionizing radiation[9] and cyclophosphamide[16] have shown that in a translocation heterozygote the incidence of multivalent configurations can vary from 12 to 100%. Based on these results it is generally considered that 25 cells are enough to detect heterozygosity for a reciprocal translocation in the F1 males. If the 25 analysed cells have been found with 20 bivalents the animal is considered as normal.

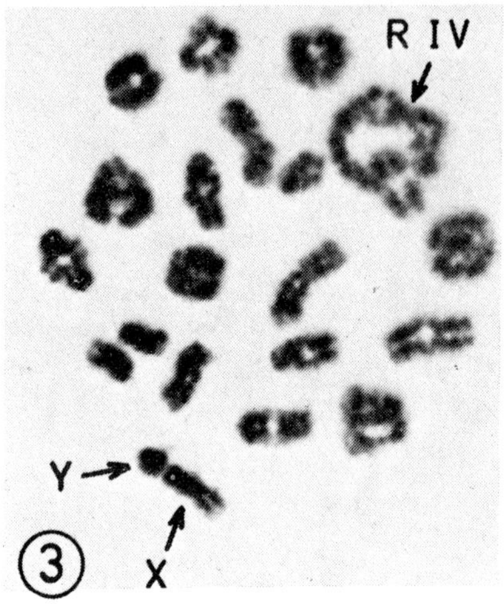

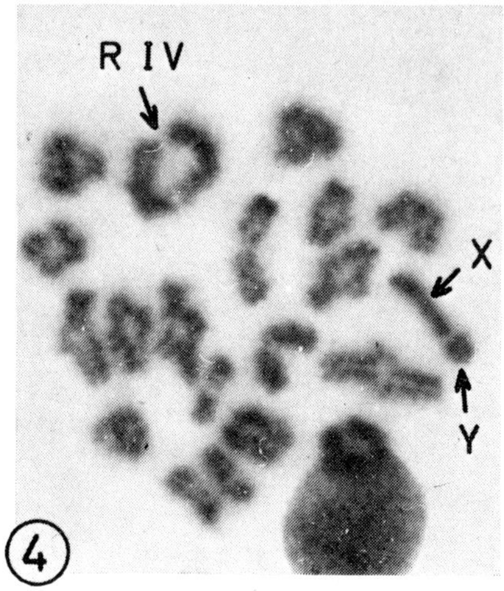

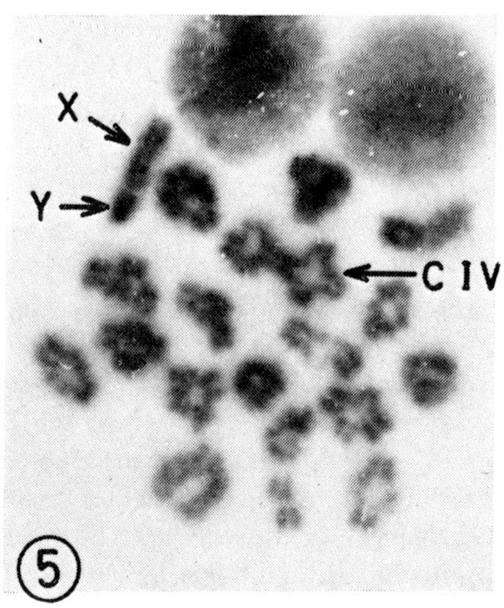

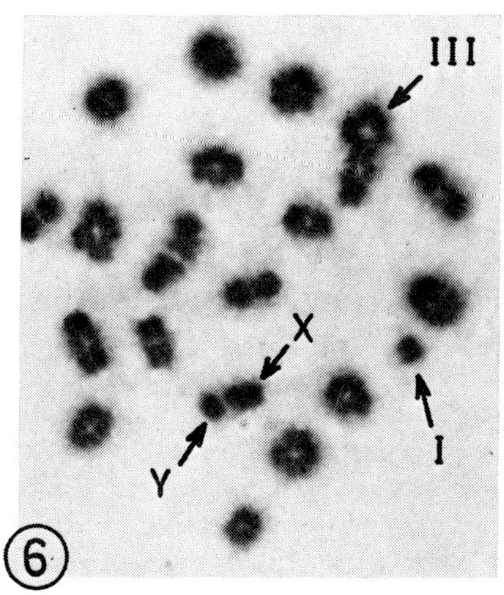

Fig. 3. Mouse dividing spermatocyte I with 18 bivalents and a ring qudrivalent.
Fig. 4. Mouse dividing spermatocyte I with 18 bivalents and a ring quadrivalent.
Fig. 5. Mouse dividing spermatocyte I with 18 bivalents and a chain quadrivalent.
Fig. 6. Mouse dividing spermatocyte I with 18 bivalents and a III+I configuration.

If some multivalents have been detected 100 cells are examined for the presumed translocated male.

DESCRIPTION OF METHODS

Two main techniques have been described for the preparation of male mammalian meiotic chromosomes (ref. 28 for review). The air-drying method developed by Evans *et al.*[3] is more appropriate for the mouse spermatocytes. This method takes very little time and provides a large number of scorable cells per animal. Since it involves a thorough mixing of the testis population the cells scored can be con-

TABLE I

GENERAL OUTLINE OF THE TESTS FOR THE HERITABLE TRANSLOCATIONS IN MALE MICE

Test	*Number of males to be*	*Number of cells to be analyzed per*	*Interval of time (days) between treatment and*				
Spermatocyte test on the treated males	*Examined* 10	*Treated male* 100	*Observation* 50–100				
F1 translocation test	*Mated* 10–25	*F1 male* 25	*Mating*				
			Spermatozoa	*Mature spermatids*	*Young spermatids*	*Spermatocytes*	*Spermatogonia*
		F1 translocated male 100	1–7	8–14	15–21	22–35	35–

sidered as a random sample, representative of the whole spermatocyte population. The method of MEREDITH[11] is rather time-consuming and requires the examination of many preparations to avoid sample heterogeneity but is preferable for those F1 males which are sterile and have very small testes yielding very few dividing spermatocytes. The two techniques will be described separately.

The air-drying method[3]

(1) The mice are sacrificed by cervical dislocation and the testes removed immediately. In order to wash away fat they are quickly placed in 2.2% (w/v) trisodium citrate ($C_6H_5Na_3O_7 \cdot 2H_2O$) solution in a 6-cm diameter Petri dish maintained at room temperature. It is important to use fresh citrate solutions prepared once a week with water double-distilled in a quartz apparatus and stored in a refrigerator.

(2) The washed testes are then transferred into 2.5 ml of a 2.2% citrate solution in fresh petri dishes and the tunica is removed. The contents of the tubules are gently teased out with curved forceps. In the F1 translocation test, the contents of the two testes are generally mixed together whereas in the spermatocyte test the two testes are prepared separately in order to detect any differences between testes in the frequency of induced reciprocal translocations.

(3) The cell suspension produced is transferred into a 3-ml conical tube with a Pasteur pipette and centrifuged at 22 *g* for 5 min.

(4) The supernatant is discarded, and the pellet is resuspended in 2 ml of a 1% (w/v) citrate solution added drop by drop with a Pasteur pipette while flicking the tube with a finger to mix well.

(5) After 10 min the suspension is centrifuged for 5 min at 22 *g*, and as much if the supernatant as possible is removed. 2 ml of fixative (3 parts ethyl alcohol to 1 part glacial acetic acid) is added dropwise flicking the tube vigorously after each drop.

(6) The fixative is changed after 5 min and again after 10 min.

(7) Three to four droplets of the final suspension are allowed to fall from a fine tipped Pasteur pipette on to a grease-free slide and blown upon to dry. According to the cell concentration the process is repeated 5 to 10 times to cover the whole slide with a homogeneous layer of cells.

Meredith's method

(1) The first step is the same as in the air-drying method.

(2) After removing the testis tunica the tubules are gently washed in 1% citrate solution, the total duration of the hypotonic treatment being 12 min.

(3) The tubules would then be transferred to fixative (3 parts ethyl alcohol to 1 part acetic acid) for a least 15 min. The tubules can be stored in a stoppered glass tube for several days in refrigerator if necessary.

(4) Three or four pieces of the tubules (about 1 cm long) are transferred to a glass tube containing 1 ml of 60% acetic acid, and the tubes are gently flicked until the tubules become transparent.

(5) With a fine tipped Pasteur pipette, one drop is put on a grease-free slide kept at 60° on a warming plate. The drop is drawn into and expelled from the pipette several times until the liquid has disappeared. This procedure is repeated with three or four drops.

MICROSCOPIC EXAMINATION

For immediate examination, the preparations are normally stained with toluidine blue but a lactic–acetic–orcein stain may be used for finer examination and for slides. The preparations are scanned at low power magnification (10×15), and the dividing spermatocytes are examined for quality of spreading. The selected cells are analysed at higher magnification (10×40 or 10×100).

DRAWBACKS

It must be pointed out that selective elimination may occur before the spermatogonia bearing reciprocal translocations reach the spermatocyte stage thus decreasing the frequency of detectable multivalents. Failure of certain translocations to form multivalent configurations also reduced the frequency observed.

A review of the literature shows that, for compounds giving positive results in the spermatocyte test on treated males, the frequency of reciprocal translocations is generally very low ($< 1\%$) when compared with the rate of reciprocal translocations induced in premeiotic male germ cells by ionizing radiation[7]. It is noteworthy that most of the chemicals that fail to induce reciprocal translocations in spermatogonia have been shown to be potent mutagens in other organisms or other test systems and are even capable of causing chromosome breakage in the post-meitoic male germ cells of the mouse as demonstrated by the dominant lethality test. Furthermore, some compounds (*e.g.* mitomycin C) which fail to produce chromosome rearrangements in premeiotic male germ cells[1,6] are able to produce chromosome breaks resulting in the killing of spermatogonia as shown by the induction of a temporary sterile period[10].

ACKNOWLEDGEMENT

This work has been performed with a grant from the Fonds National de la Recherche Fondamentale Collective.

REFERENCES

1 Adler, I. D., Comparative cytogenetic study after treatment of mouse spermatogonia with mitomycin C, *Mutation Res.*, 23 (1974) 369–379.

2 Ehling, U. H., Comparison of radiation- and chemically-induced dominant lethal mutations in male mice, *Mutation Res.*, 11 (1971) 35–44.

3 EVANS, E. P., G. BRECKON AND C. E. FORD, An air-drying method for meiotic preparations from mammalian testes, *Cytogenetics*, 3 (1964) 289–294.

4 FORD, C. E., A. G. SEARLE, E. P. EVANS AND B. J. WEST, Differential transmission of translocations induced in spermatogonia of mice by irradiation, *Cytogenetics*, 8 (1969) 447–470.

5 GERBER, G. B., AND A. LÉONARD, Influence of selection, non-uniform cell population and repair on dose-effect curves of genetic effects, *Mutation Res.*, 12 (1971) 175–182.

6 GILLIAVOD, N., AND A. LÉONARD, Tests for mutagenic effects of chemicals in mice, I. Effects of mitomycin C on spermatogonia, *Mutation Res.*, 13 (1971) 274–275.

7 LÉONARD, A., Radiation induced translocations in spermatogonia of mice, *Mutation Res.*, 11 (1971) 71–88.

8 LÉONARD, A., Observations on meiotic chromosomes of the male mouse as a test of the potential mutagenicity of chemicals in mammals, in A. HOLLAENDER (Ed.), *Chemical Mutagens: Principles and Methods for their Detection*, Vol. 3, Plenum, New York, 1973, pp. 21–56.

9 LÉONARD, A., AND GH. DEKNUDT, The sensitivity of various germ-cell stages of the male mouse to radiation induced translocations, *Can. J. Genet. Cytol.*, 10 (1968) 496–507.

10 LÉONARD, A., AND N. GILLIAVOD, Testicular changes in mice after treatment with mitomycin C, *Toxicology*, 1 (1973) 217–223.

11 MEREDITH, R., A simple method for preparing meiotic chromosomes from mammalian testis, *Chromosoma*, 26 (1969) 254–258.

12 MURAMATSU, S., Frequency of spontaneous translocations in mouse spermatogonia, *Mutation Res.*, 24 (1974) 81–82.

13 OAKBERG, E. F., Duration of spermatogenesis in the mouse and timing of stages of the cycle of the seminiferous epithelium, *Am. J. Anat.*, 99, (1956) 507–516.

14 OAKBERG, E. F., AND R. L. DIMINNO, X-ray sensitivity of primary spermatocytes of the mouse, *Int. J. Radiat. Biol.*, 2 (1960) 196–209.

15 SEARLE, A. G., Chromosome damage and risk assessment, in J. DE GROUCHY, F. J. G. EBLING AND I. W. HENDENSON (Eds.), *Proc. 4th Int. Congr. Hum. Genet., 1971*, pp. 58–66.

16 SOTOMAYOR, R. E., AND R. B. CUMMING, Induction of translocations by cyclophosphamide in different germ cell stages of male mice: cytological characterization and transmission, *Mutation Res.*, 27 (1975) 375–388.

CYTOGENETIC ANALYSIS OF PRE- AND POSTOVULATORY OOCYTES AND PRE-IMPLANTATION EMBRYOS IN MUTAGENESIS OF MAMMALS

G. RÖHRBORN [1], I. HANSMANN [2] and U. BUCKEL [3]

[1] *Institut für Anthropologie und Humangenetik, Universität Düsseldorf;* [2] *Institut für Humangenetik, Universität Göttingen und* [3] *Institut für Anthropologie und Humangenetik, Universität Heidelberg (G.F.R.)*

Introduction

Chromosome analysis in pre- and postovulatory oocytes and also in early cleavage stages, gives information about the amount and type of induced and spontaneous chromosomal anomalies long before modifying factors such as the prenatal elimination can interfere with the primary mutagenic effects.

Only few data are available from animal experiments dealing with the elimination of chromosomally abnormal embryos [4,7,9]. From cytogenetic studies of newborns and of induced abortions in man we do known that the incidence figure of chromosomally abnormal fetuses is about 10 times higher than the figure among newborns (e.g. refs. 15 and 23). The others are aborted in between. What is the incidence, however, earlier in the course of prenatal development, before implantation?

From these above-mentioned studies in man we known that aneuploidies prevail among chromosomally abnormal newborns and spontaneous abortions. Less than 30% are newborns carrying a balanced or unbalanced structural chromosome anomaly (e.g. ref. 15). This fact is often not considered in mutagenesis in mammals. The result of spontaneous and induced non-disjunction during oogenesis as well as during spermatogenesis can be analyzed already in metaphase II oocytes (for events in oogenesis) and in pronucleus stages or in the following cleavage stages (both for spermatogenesis and oogenesis) immediately after non-disjunction occurred. Also, polyploidies and all types of structural anomalies can be detected by these methods.

Paternal and maternal chromosomes are usually separated at the pronucleus stage [5,8] and the origin of an anomaly can thus be traced to oogenesis or spermatogenesis. Moreover, if a strain of mice with a meta- or submetacentric marker chromosome is used, this possibility of analysis can still be extended.

Analyzing chromosomes in postovulatory oocytes at the stage of metaphase II [18] is a feasible method to pick up all aneuploidies just a few hours after the first meiotic division when non-disjunction occurred. This becomes even more important, as it was shown recently that in man only about 30% of trisomy 21 (Down syndrome) originates from non-disjunction during spermatogenesis and that non-disjunction during the first meiotic division of oogenesis should be about 10 times more frequent than during the second meiotic division [12].

Restricting for the application of all these methods in pre- and postovulatory oocytes and in early cleavage stages, however, is the relatively time-consuming preparation of these cells. These techniques, therefore, cannot be recommended for screening or testing a large number of chemicals. However, to study special problems, above all the induction of non-disjunction in mammals, these methods are of high relevance.

An example may be the investigation of the chromosome-damaging effect of hormones used as oral contraceptives: it might hardly be advisable to test such compounds, for example, in bone marrow cells, lymphocytes or spermatocytes since female germ cells are exclusively exposed. In these cases, therefore, the investigation of ovulated metaphase II oocytes and/or of cleavage stages after acute, subacute or chronic treatment is strongly suggested.

Types of damage which can be looked for

1. Aneuploidy

The dictyotene, a characteristic stage of oogenesis, rests in oocytes from about birth up to just a few hours before ovulation. The first meiotic division takes place shortly before ovulation and as the oocytes stay at metaphase II for some hours until fertilization, the result of a chromosomal mis-segregation can already be analyzed immediately after its manifestation. Non-disjunction during the second meiotic division in oogenesis as well as in spermatogenesis can be observed one day after its formation in the first mitotic division at the pronucleus stage (Fig. 3). Fig. 1 schematically shows periovulatory stages.

2. Polyploidy

As the chromosomes of the first polar body differ very quickly from the corkscrew-like twisted metaphase II chromosomes because of their fuzzy-like appearance, the suppression of the first meiotic division can be analyzed direcly after ovulation, or even more simply by determining the number of pronuclei during the first mitotic interphase after fertilization.

One should, however, bear in mind that triploidy or tetraploidy represent such a serious detriment to the fetus that those carrying this chromosomal anomaly are aborted relatively early.

3. Structural anomalies

The analysis in preovulatory oocytes at diakinesis/metaphase I gives the same important information about induced structural anomalies, for example translocations and deletions as they occur in spermatogenesis (e.g. refs. 6 and 20). And the chromosomes of the oocyte I are even more clearcut (Fig. 2a) at this stage than those of the spermatocyte. It is advisable to use a species like that

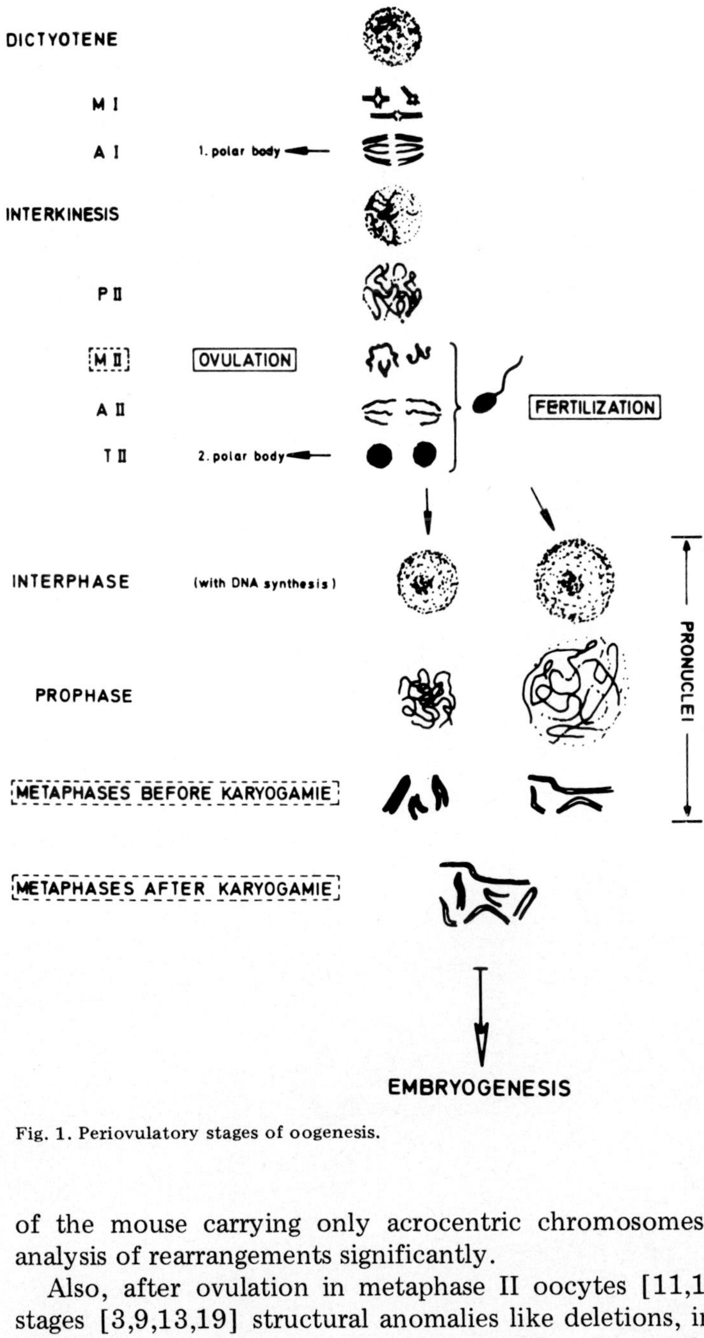

Fig. 1. Periovulatory stages of oogenesis.

of the mouse carrying only acrocentric chromosomes, which facilitates the analysis of rearrangements significantly.

Also, after ovulation in metaphase II oocytes [11,17,18] and at cleavage stages [3,9,13,19] structural anomalies like deletions, inter- and intrachanges, fragments and breaks can be observed (Figs. 2, 4 and 5). As far as cleavage stages are concerned, however, we would like to stress very strongly that in the mouse it is impossible to detect all structural anomalies of the chromosome type, which should usually be observed in cleavage stages when germ cells are treated. Only those which show (a) a changed centromere position, (b) a very large or (c) a very small chromosome can be observed in the mouse. If, how-

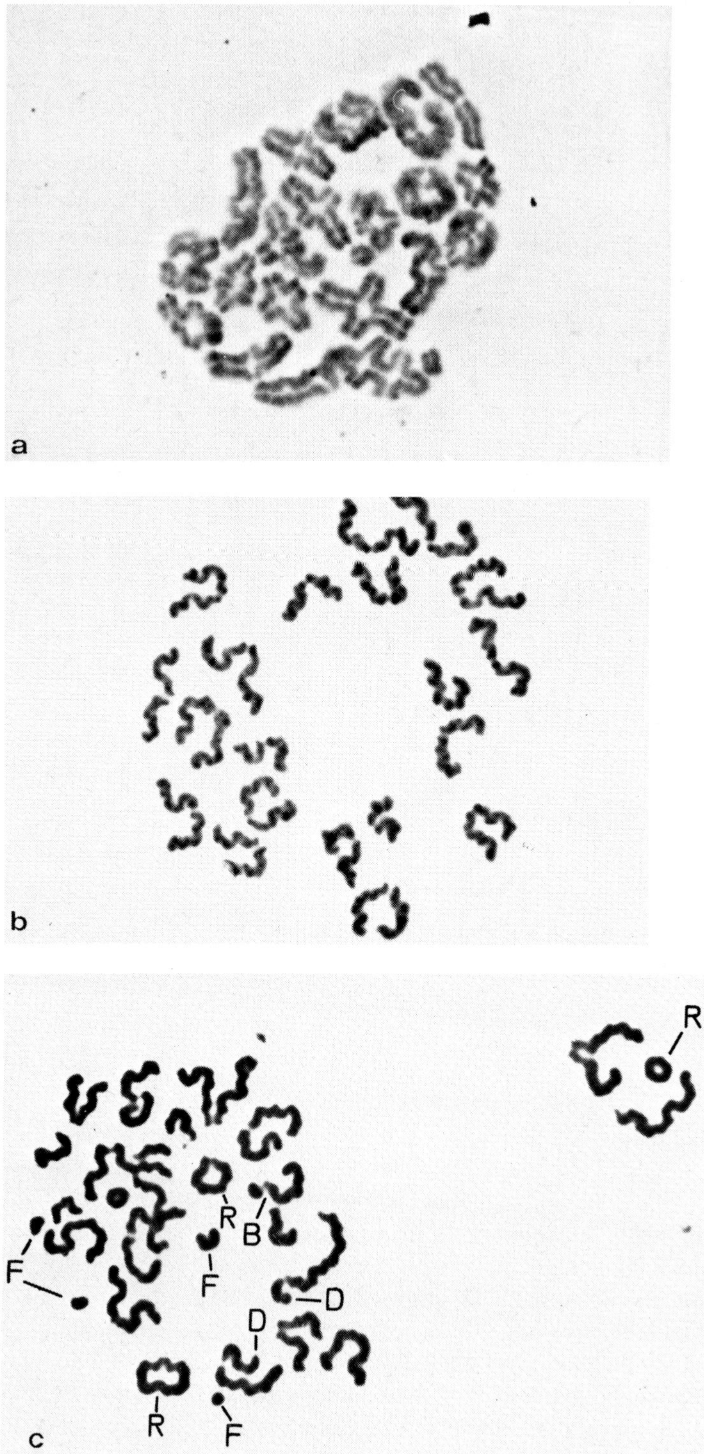

Fig. 2. a: metaphase I of female mouse after irradiation of oogenesis. b: normal metaphase II of mice. c: metaphase II chromosomes after exposure to 600 R with more than 4 anomalies. Rings (R), fragments (F), a deletion (D) close to the centromere and an abnormal chromosome resulting most probably from an interchange (♀).

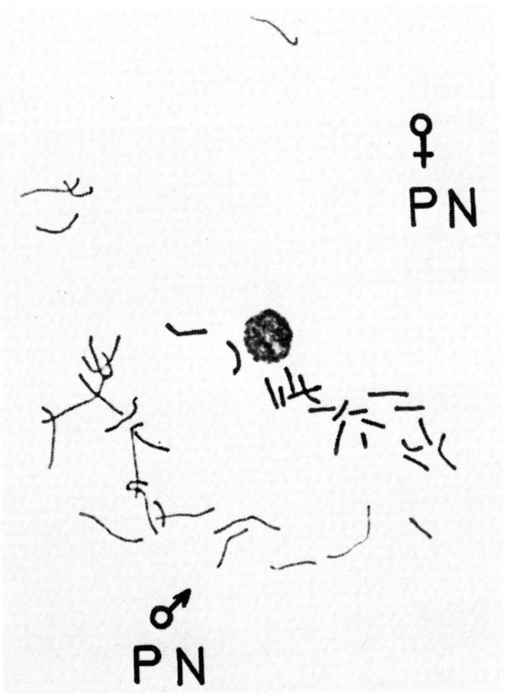

Fig. 3. Male and female pronucleus chromosomes of mice.

ever, specific banding techniques well known from human cytogenetics are used, then the analysis may even be improved. Without the application of these staining methods the results of chromosome analysis represent only minimal yields and it would be not correct to correlate these minimal data with, for example, dominant lethal effects obtained with the same compound and experimental procedure.

Abnormalities of the chromatid-type in cleavage stages which should not originate from the primary effect of treatment at a definite germ-cell stage but from a delayed effect can, of course, be observed without the use of banding techniques.

4. *Subnuclei*

Subnuclei were observed in preparations from cleavage stages after the previous treatment of oogenesis [9] and of spermatogenesis [13] with chemicals indicating the induced loss of chromosomes or parts of them one or several mitotic cycles after fertilization and treatment. These observations might be useful to develop a "quick-test' for chromosome loss similar to the micronucleus test in bone marrow preparations [2,14]; however, with the advantage of different target cells, for example oocytes, spermatocytes or cleavage stages.

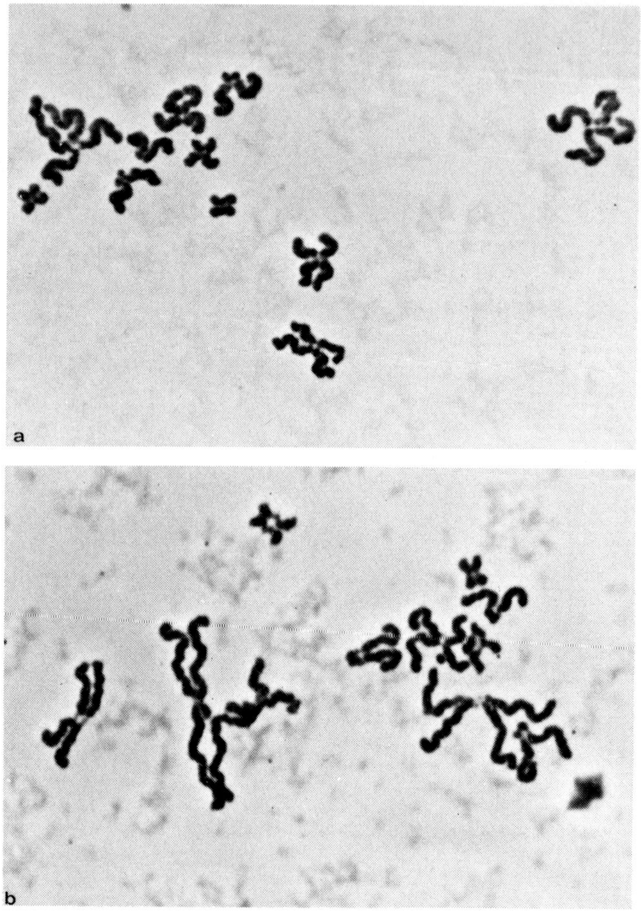

Fig. 4. Metaphase II chromosomes of Chinese hamsters (*Cricetulus griseus*). a: normal metaphase II oocytes. b: chromatid-type interchange after preovulatory treatment with triaziquone.

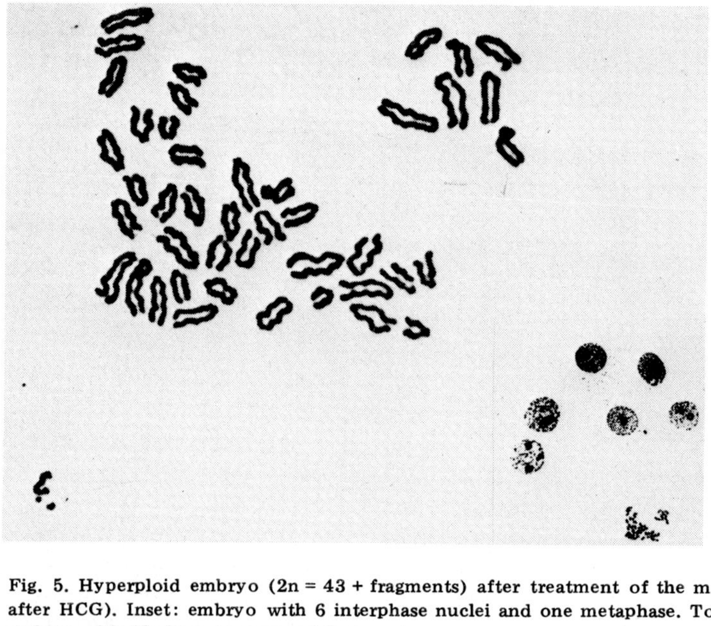

Fig. 5. Hyperploid embryo (2n = 43 + fragments) after treatment of the maternal prometaphase I (3 h after HCG). Inset: embryo with 6 interphase nuclei and one metaphase. Top left: metaphase from this embryo with 43 chromosomes and fragments.

Technical requirements

A. Apparatus

Microscope, if possible with phase-contrast.
Binocular, with 10—20-fold magnification.
Small scissors and forceps, preparation needles for dissection.
Water bath.
Embryological watch glasses.
Glass pipettes which can be drawn out to a very fine tip.
A glass marker, to mark the position on the slide where the oocyte settled down after fixation.
Container for hypotonic, fixation, and balanced salt solutions.
Staining jars.

B. Chemicals

Balanced salt solution (e.g. Hanks).
Sodium citrate, methanol, acetic acid.
Aceto-orcein (2%) or Giemsa (2%).
Fetal calf serum.
Xylene, Euparal, Eukitt.
Colchicine.
Hyaluronidase.
Ringer solution.
PMS (pregnant mares serum) and HCG (human chorionic gonadotrophin), both for stimulated ovulation.

Animals

Because of the favorable karyotype and the available data from numerous mutagenicity studies, the mouse is the most convenient species for this purpose and therefore it has been used nearly exclusively for these investigations up to now. But the Chinese hamster is also very suitable because of its well distinguishable chromosomes and the low chromosome number [1,10]. Because of the smallness of many chromosomes, the rat is less convenient for investigations of the oocytes (unpublished).

Methods

All methods for chromosome preparation of the preovulatory (metaphase I) and postovulatory — as well as cleavage — stages described here are based on the air drying method reported first by Tarkowski [21]. As laid down below, modifications have proved to be important especially for the preparation of metaphase II chromosomes.

1. Diakinesis/metaphase I in preovulatory oocytes

Sacrifice females by cervical dislocation.
Place ovaries in embryological watch glasses containing about 1—2 ml culture medium.

Culture medium: (a) Hanks balanced salt solution supplemented with 30% fetal calf serum; (b) total calf serum (Evans, unpublished).

Perform subsequent manipulations under binocular with 10—20-fold magnification.

Liberate oocytes by puncturing follicles.

Remove follicular cells mechanically or by treatment with hyaluronidase (150 IU/ml Ringer solution).

Place oocytes in medium with or without equilibrated paraffin oil in an embryological watch glass or a petri dish and incubate for 4—6 h in a humidified 5% CO_2 incubator at 37° C.

Transfer oocytes with a fine glass pipette to another embryological watch glass containing sodium citrate (1%) for 15—30 min.

Pick up single oocytes and transfer them to a grease-free slide in a small drop of hypotonic solution.

Fix with 2—5 small drops methanol : acetic acid (3 : 1) and then air dry.

Mark immediately during or after fixation the location of the oocyte on the slide by use of a glass marker.

Stain with aceto-orcein (2%) or Giemsa (2%) and mount.

2. Metaphase II in postovulatory oocytes

To obtain ovulated oocytes females can be pretreated with PMS (e.g. 1.5 IU) and 48 h later with HCG (e.g. 2 IU) to stimulate ovulation or a second possibility is to mate them with vasectomized males. Fifteen hours after HCG or at 09.00 h on the morning of vaginal plug (when the dark period lasts from 18.00—06.00 h) the females are sacrificed. Spontaneous ovulation is preferred, when the timing of the experiment allows it, to avoid possible secondary effects of hormonal pretreatment and/or synergistic effects with the compound tested.

Dissect females and place ovary, tube and part of the uterus in an embryological watch-glass containing prewarmed Hanks solution (37° C).

The following manipulations are all carried out under a binocular microscope.

Look out for a clear-cut ampulla in the upper third of the tube and destroy the tubal wall with the tip of a needle.

Oocytes surrounded by the corona radiata are released in the Hanks solution.

Pick up oocytes with small forceps or by using a fine pipette and transfer them to an embryological watch glass containing some drops of Hyaluronidase (150 IU/ml Ringer Solution).

Treat oocytes at least for 30 min at room temperature until the oocytes are free from all corona cells.

Transfer oocytes in a fine glass pipette to an embryological watch glass containing sodium citrate as hypotonic solution. Concentration and time of treatment should be adapted to the conditions in the experimenter's own laboratory. Convenient are 1.5% for at least 45 min or 1.0% for 30 min.

Hypotonic treatment should be performed in a refrigerator at 4° C.

Fixation and staining are the same as with metaphase I oocytes.

8 Cleavage stages

Timing of cleavage division proves to be variable between strains and species. Stages should be checked in each strain individually, to increase the frequency of cleavage stages at metaphase. Two hours before dissection females receive 0.008 mg/kg body weight Colcemid to arrest metaphases.

After dissection of the females ovaries, together with the tube and part of the uterus, are transferred to an embryological watch glass containing prewarmed (37° C) Hanks solution.

Separate tube from ovary and most of the uterus and flush out cleavage stages by the use of a fine glass pipette filled with warmed Hanks solution. The tube should be flushed from the infundibulum.

Transfer cleavage stages to another embryological watch glass containing sodium citrate.

The concentration should be adapted to the cell stage investigated: pronucleus 1.5%, blastocyst 1.0%.

Hypotonic treatment for about 20—30 min at 4° C.

Fixation, staining and mounting are the same as in methods to prepare oocytes.

In all these preparations, however, grease-free slides, chilled overnight in distilled water, are used.

Conclusions

As already discussed earlier, these methods are well suited to test the mutagenic effect of compounds acting especially only in oogenesis. By using a similar experimental set-up like that of the dominant lethal method, however, the mutagenic effect in spermatogenesis can also be investigated by analyzing cleavage stages. The significant advantage of these methods is based on the fact that the effects are observed directly in the germ cells, i.e., in the oocytes, or in cleavage stages a long time before the prenatal elimination known in man can interfere with the primary effects.

Because of the relatively time-consuming preparation which is in contrast to other methods, application of these techniques for screening seems not to be advisable. These are special techniques, which, as already pointed out earlier, should only be used if special problems arise, but then they are highly significant.

We therefore encourage all efforts to modify these cytogenetic methods in a way to avoid the time-consuming handling of single cells [16].

The relatively few investigations which have already been performed demonstrate that it is possible to investigate problems of dose-effect relations, of stage-sensitivity in oogenesis, of the different effects of various chemical compounds (hormones, antimetabolites, alkylating agents), as well as comparing chemically and X-ray-induced chromosomal anomalies [1,6,9—11,17,18,20, 22].

Acknowledgement

Sponsored by the Deutsche Forschungsgemeinschaft.

References

1 Binkert, F. and W. Schmid, Types of chromosome aberration in preimplantation embryos of chinese hamster after treatment of the sires with Trenimon, Mutation Res., 29 (1975) 215.

2 Boller, K. und W. Schmid, Chemische Mutagenese. Das Knochenmark des chinesischen Hamster als in-vivo Testsystem. Hämatologische Befunde nach Behandlung mit Trenimon, Humangenetik, 11 (1970) 35—54.

3 Brewen, J.G., H.S. Payne, K.P. Jones and R.J. Preston, Studies on chemically induced dominant lethality. I. The cytogenetic basis of MMS-induced dominant lethality in post meiotic male germ cells, Mutation Res., 33 (1975) 239—250.

4 Buselmaier, B. and I. Hansmann, Elimination of induced chromosome aberrations in embryogenesis of mice, Mutation Res., 29 (1975) 215.

5 Donahue, R.P., Fertilization of the mouse oocyte: chromosomal behaviour up to the first cleavage division, in 4th Int. Congr. of Human Genetics, Paris, 1971, Excerpta Medica, Amsterdam, 1971, p. 233.

6 Gilliavod, N. and A. Léonard, Dose relationship for translocations induced by X-irradiation in mouse oocytes, Mutation Res., 25 (1974) 425—426.

7 Gropp, A., D. Giers and U. Kolbus, Trisomy in the fetal backcross progeny of male and female metacentric heterozygotes of the mouse. I, Cytogenet. Cell Genet., 13 (1974) 511—535.

8 Hansmann, I., Induced chromosomal aberrations in pronuclei, 2-cell stages and morulae of mice, Mutation Res., 20 (1973) 353—367.

9 Hansmann, I. and G. Röhrborn, Chromosome aberrations in preimplantation stages of mice after treatment with triazoquinone, Humangenetik, 18 (1973) 101—109.

10 Hansmann, I., J. Neher and G. Röhrborn, Chromosome aberrations in metaphase II-oocytes of Chinese hamster (*Cricetulus griseus*). I. The sensitivity of the preovulatory phase to triaziquone, Mutation Res., 25 (1974) 347—359.

11 Jagiello, G.M., Action of phleomycin on the meiosis of the mouse ovum, Mutation Res., 6 (1968) 289—295.

12 Langenbeck, U., I. Hansmann, B. Hinney and U. Hönig, On the origin of the supernumerary chromosome in autosomal trisomies with special reference to Down syndrome. A bias in tracing non-disjunction by chromosomal and biochemical polymorphisms. Humangenetics, 33 (1976) 89—102.

13 Matter, B.E. and J. Jaeger, Premature chromosome consendation, structural chromosome aberrations and micronuclei in early mouse embryos after treatment of paternal post-meiotic germ cells with triethylenemelamine. A possible mechanism for chemically induced dominant lethal mutations, Mutation Res., 33 (1975) 251—260.

14 Matter, B.E. and W. Schmid, Trenimon-induced chromosomal damage in bone-marrow cells of six mammalian species evaluated by the micronucleus test, Mutation Res., 12 (1971) 417—425.

15 Nielsen, J., Chromosome examination of newborn children. Purpose and ethical aspects, Humangenetik, 26 (1975) 215—222.

16 Payne, H.S. and K.P. Jones, Techniques for mass isolation and culture of mouse ova for cytogenetic analysis of the first cleavage mitosis, Mutation Res., 33 (1975) 247—249.

17 Reichert, W., I. Hansmann and G. Röhrborn, Chromosome anomalies in mouse oocytes after irradiation, Humangenetik, 28 (1975) 25—38.

18 Röhrborn, G. and I. Hansmann, Induced chromosome aberrations in unfertilized oocytes of mice, Humangenetik, 13 (1971) 184—198.

19 Röhrborn, G., O. Kühn, I. Hansmann and K. Thon, Induced chromosome aberrations in early embryogenesis of mice, Humangenetik, 11 (1971) 316—322.

20 Searle, A.G. and C.V. Beechey, Cytogenetic effects of X-rays and fissions neutrons in female mice, Mutation Res., 24 (1974) 171—186.

21 Tarkowski, A.K., An air-drying method for chromosome preparations from mouse eggs, Cytogenetics, 5 (1966) 394—400.

22 Uchida, I.A. and C.P.V. Lee, Radiation induced non-disjunction in mouse oocytes, Nature, 250 (1974) 601—602.

23 Yamamoto, M., R. Fujimori, T. Ito, K. Kamimura and G. Watanabe, Chromosome studies in 500 induced abortions, Humangenetik, 29 (1975) 9—14.

THE SPECIFIC LOCUS TEST IN THE MOUSE*

A. G. SEARLE

Medical Research Council, Radiobiology Unit, Harwell, Oxon. OX11 ORD (U.K.)

INTRODUCTION

The specific locus test of mutagenicity in the mouse was independently developed over 25 years ago by T. C. CARTER in Edinburgh and W. L. RUSSELL in Oak Ridge[1]. It has been used extensively since then, especially by RUSSELL and shows no signs of obsolescence[23]. Although it requires the production of a large number of mice yet the technique itself is simple and no special skills are needed apart from those of animal husbandry and good powers of observation. Briefly, the test involves the crossing of wild type exposed or control mice with homozygotes for a particular set of recessive genes and the external examination of immediate offspring for divergences from the expected wild type.

We shall not consider in detail certain other tests for mutation at specific loci in the mouse, in which the methodology differs from that of the usual specific locus experiment. VALCOVIC AND MALLING[25] have described such a test for mutation induction at a set of nine biochemical loci, in which two inbred strains of mice (differing with respect to alleles at these loci) are crossed together and electrophoretic methods are used to determine whether the expected enzymic patterns are found in F_1 blood. Crosses between inbred strains have also been used in experiments on mutation induction at histocompatibility loci[9] with skin-graft tests between progeny which allow presumptive mutations to be scored as gains, losses or gains+losses.

TYPES OF DAMAGE ASSAYED

By use of this test it is possible to detect, in the generation after exposure, recessive mutations at specific loci as well as visible dominant mutations at any locus. The test will also reveal certain deletions and translocations as well as some examples of the complementation of aneuploid gametes.

Recessive mutations

These will include small inter- or intragenic deletions affecting a specific locus

*Dedicated to Dr. W. L. Russell, pioneer and chief exponent of this method, in recognition of his outstanding achievements in mutational research.

as well as gene mutations at the base-pair level which lead to a change from the dominant wild-type to a recessive allele at the locus concerned.

Some translocations with a break-point at one of the specific loci also behave like recessive mutations.

In the first specific locus (T) stock[16] the seven loci at which recessive mutations can be detected at weaning age are non-agouti (*a*), brown (*b*), albino (*c*), pink-eyed dilution (*p*), Maltese dilution (*d*), short-ear (*se*), and piebald (*s*). These are located on five different chromosomes, since the *c* and *p* loci are linked as well as the *d* and *se* loci. In the second specific locus (HT) stock[12] the six loci concerned are non-agouti (*a*), brachypodism (*bp*), pallid (*pa*), fuzzy (*fz*), leaden (*ln*) and pearl (*pe*). The *a*, *bp* and *pa* loci are linked, as are the *fz* and *ln* loci. Fig. 1 shows the positions of all these loci on the mouse karyotype.

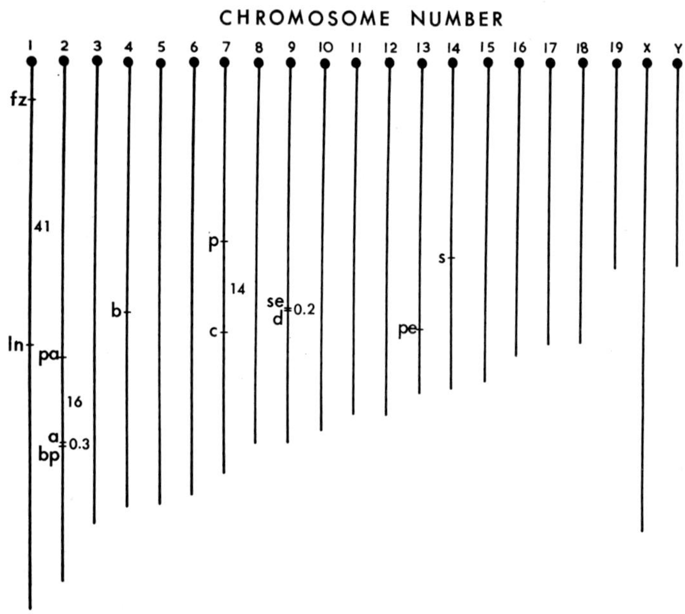

Fig. 1. Map of mouse karyotype, showing approximate positions of the twelve specific loci used so far in mutagenicity tests, with distances in map units between linked loci.

Six of the first set of seven loci control various aspects of coat and eye colour in the mouse, while the seventh (*se*) controls ear length. In the second set, one locus is concerned with hair texture (*fz*) while another (*bp*)helps to control limb development. The others are again mainly concerned with pigmentation, although *pa* also affects the inner ear.

Dominant mutations

In principle, any mutation with a visible heterozygous effect can be detected among the F_1 offspring from homozygous wild type X recessive matings. The number actually detected in a specific locus experiment will, of course, depend on the powers of observation of those examining the mice. Mutations affecting the coat or causing gross changes in limbs, tail or ears will normally be noticed, but behaviour differences or small changes in individual digits etc. may be missed. The detection of dominants is a more subjective process than is the detection of specific locus recessives since, with the latter, the observer has a much better idea of what to expect.

Deletions

It has already been mentioned that the recessive class of mutations may include deletions. If these cover the closely linked *d* and *se* loci (0.1 units apart) in the T stock or the *a* and *bp* loci (0.3 units apart) in the HT stock they will be revealed by the occurrence of an apparent simultaneous mutation affecting both loci. However, not all of these double mutations are deletions, as explained in the next section.

Nullosomic and disomic gametes

An apparent specific locus mutation will result if a wild type gamete which lacks one of the chromosomes containing a specific locus fuses with a specific locus tester stock gamete which has an extra copy of the same chromosome, so that a balanced zygote is formed[18]. This type of complementation will give what appears to be a mutational reoccurrence of the recessive allele present on the tester stock chromosome concerned. Its true nature can only be recognised when a chomosome with more than one marker locus (*i.e.* chromosomes 7 or 9 in the first stock, and 1 or 2 in the second) is involved, since a double or triple "mutation" will appear. This will not be lethal in the homozygote and thus can be distinguished from a deletion covering the loci concerned.

Translocations

As already mentioned reciprocal translocations with a break-point at a specific locus may have a phenotypic effect in the F_1 generation like that of true recessive mutations, but can be distinguished from them by semi-sterility in the heterozygote and cytological effects. Moreover, certain X-autosome translocations affecting chromosomes carrying specific locus coat colour mutations have been detected among F_1 progeny because they produced variegation in females heterozygous for the translocation and the coat colour gene concerned[12].

For references to specific locus tests, see refs. 3, 5, 12, 16, 23, 27.

APPARATUS NEEDED

The only apparatus needed is the breeding mouse cage with food-hopper and water bottle. Many designs are available, none of which are of obvious superiority for the purpose on hand.

Mouse strains needed

In nearly all specific locus experiments the treated mice have been F_1 hybrids between the inbred strains C3H/He and 101, the males being of the latter strain. This hybrid stock will be called 3H1. Each of the parent strains should be maintained by sib-mating pairs. A grading scheme should be adopted, so that the less productive lines are quickly recognised and culled. Table I gives examples of such schemes, based on production of weaned young per cage-month, but the right scheme for particular animal house conditions can only worked out *in situ*. It should be noted that strain

TABLE I

A. GRADING SYSTEM USED ON STRAIN 101/H MATINGS, TO DECIDE WHETHER MATINGS SHOULD BE KEPT AND WHETHER OFFSPRING SHOULD BE USED FOR STRAIN MAINTENANCE.

	Months in cage							*Diagnosis*
	3	4	5	6	7	8	9	
Number weaned	0–4	0–9	0 –14	0–19	0–24	0–29	Any	Poor-cull
	5–9	10–14	15–19	20–24	25–29	30–34	–	Fair
	10–14	15–19	20–24	25–29	30–34	35–39	–	Good
	15+	20+	25+	30+	35+	40+	–	Very good

B. GRADING SYSTEM USED ON STRAIN C3H/HeH MATINGS

	Months in cage						*Diagnosis*
	3	4	5	6	7	8	
Number weaned	0–13	0–21	0–29	0–32	0–45	Any	Poor-cull
	14–19	22–27	30–35	38–43	46–51	–	Fair
	20–25	28–33	36–41	44–49	52–57	–	Good
	26+	34+	42+	50+	58+	–	Veiy good

C3H is particularly susceptible to virus-borne infantile diarrhoea, which can be controlled by placing filter-tops on the cages[10,11].

The 3H1 matings can be in the form of trios. Again, a grading scheme is desirable.

The multiple recessive T stock is homozygous for non-agouti (*a*), brown (*b*), chinchilla (c^{ch}), pink-eyed dilution (*p*), Maltese dilution (*d*), short-ear (*se*) and piebald (*s*). This stock seems the most appropriate for general mutagenicity testing, since average radiation induced mutation rates per locus are about three times as high as for the second set of specific loci of the HT stock[12,13].

In T stock matings, inbreeding should be avoided as far as possible. As in 3H1 stock matings, trios can be used and grading should be practised.

Costing

The only published estimate of the cost of a specific locus experiment seems to be that by SANDERS[20], also quoted by EHLING[5]. This stated that a test to determine the mutagenicity of triethylenemelamine (TEM), carried out at Oak Ridge in 1965–66, cost more than $10,000. The test took about 18 months, in which period about 24,000 offspring of 500 pairs of mice were examined.

It is probably better to cost in terms of manpower and cage-space needed per 10,000 weaned offspring produced for classification. Let us assume male mice are being tested for spermatogonial mutations, as is usually the case. Productivity per cage is decidedly higher if trios rather than pairs are used in the specific locus matings, in those of tester stock mice and in the C3H♀×101♂ matings. A rough estimate of production in the specific locus matings is given by $N = 6(t-1)$ where N is the total number weaned per female and t is the number of months the cage is occupied, up to 9. This presupposes some replacement of non-breeding or poorly breeding females. Thus 100 cages, containing 200 females at any one time should produce nearly 10,000 mice in 9 months, while 200 cages should achieve this in 5 months. To keep matings for a short period would be uneconomic, since normally the first weanings won't take place until nearly 6 weeks after the matings are put up. If treated males are thereby temporarily sterilised the breeding programme will be correspondingly delayed.

The cost of preparatory matings must also be included. If the hundred trios are to be mated up over 10 weeks, then 10 3H1 males will be required per week. Since the productivity of these matings (when in trio form) is roughly given by $N = 8(t-1)$, 7 cages (each occupied for 4 months) should suffice, making due allowance for fluctuations in the sex ratios. This in turn will require 2 cages of strain 101 paired sib-matings (productivity given by $N = 5(t-1)$) and 3 cages of strain C3H paired sib-matings ($N = 8(t-1)$), each over 3 months.

For each 100 cages of the test matings, 200 T stock females will be needed initially, with perhaps an extra 12 per month for replacement purposes. If the T stock is mated in trios its productivity is given $N = 6(t-1)$. If again we assume a 10 week mating period, around 12 cages of T-stock trios will be needed, each occupied over 4 months, followed by a cut to 3 cages to supply the replacements. Thus it seems likely that the preparatory and main matings will need 1000–1100 cage months per 10,000 weaned F_1 mice produced in a specific locus experiment if the production cages are occupied for 9 months, with more cage months for shorter periods of production. If the specific locus test matings were spread over 10 weeks the main part of the experiment would occupy a year, with preparatory matings over the previous 6 months, and some genetic tests to follow.

It is impossible to estimate the number of extra cage months needed for genetic tests on presumptive mutants, since this will depend on the number of these recovered and the loci involved. However the maximum number of cages needed per mutant is unlikely to be more than 15 or so and the period of occupation of some of these will be short. Separate estimates will have to be made if tests are planned on germ-cells other than spermatogonia, *e.g.* oocytes in the female or post-meiotic stages in the male. If the breeding span of the treated individual is limited, either because of sterility or because the germ-cells under test are of limited duration, the number of preparatory matings producing mice for testing will have to be correspondingly increased.

It is difficult to estimate the manpower required, since this will depend on a number of factors, such as the extent of automation in the animal house and the thoroughness of scrutiny of each F_1 offspring. Thus some laboratories might consider it unnecessary to count, sex and classify offspring at birth as well as at weaning. In general, one efficient animal technician can look after and change about 400 breeding cages with reasonable thoroughness, providing other staff clean and sterilize the cages and prepare them for occupation. This cage cleaning and preparation would occupy about half the time of one ancillary worker. In addition, about one-quarter of the time of a higher grade technician or scientific assistant would be required to help with the examination of offspring at weaning (done routinely by 2 persons working together), maintenance of stocks and mutants, grading procedures, book-keeping etc. Scientific responsibilities will include planning, general supervision and decisions on what to do with particular known or suspected mutants. In general this would only take up a small proportion of a scientist's time, say 5 h per week.

In summary, a specific locus experiment with spermatogonial treatment and with a target of 50,000 weaned mice would probably require around 5500 cage-months, if 400 cages are used and if males are kept in full production. It might well employ 1 animal technician full-time, 1 ancillary worker half-time, 1 scientific assistant quarter-time, with about 5 h per week of supervision by a scientist.

Controls

Estimates of spontaneous mutation frequencies at specific loci obtained by different workers and laboratories are in good agreement[4,14,19,24] and there have been no reports so far of heterogeneity associated with seasonal or other factors in the normal environment. However, any test for mutagenicity should include concurrent controls, so as to check on the genetic and mutational behaviour of the stocks in the absence of treatment. It is also very desirable for any laboratory embarking on specific locus experiments to build up a large volume of data on spontaneous mutation frequencies which, in the absence of heterogeneity, can be used *en masse* in statistical calculations. As is shown later, the narrow confidence limits associated with such large control series are of considerable help to the investigator in his attempt to achieve a clear-cut positive or negative result.

Minimum permissible size

Calculations of the minimum permissible size of an experiment on purely theoretical considerations (*e.g.* to give a high probability of detecting a true doubling of the overall mutation frequency) are difficult and have not been attempted yet, as far as we know. On practical grounds and from past experience, it would seem desirable to aim for at least 30,000 classifiable progeny at each dose level in the experimental series and to have at least two dose levels.

DESCRIPTION OF METHOD

The method is best outlined in a flow diagram (Fig. 2). Before the experiment starts it is necessary to decide how many offspring are needed over what period of time, so that the required number of matings can be calculated, in the way already described. More details of the different steps in the test are given below.

Preliminary matings

Mice of the inbred strains C3H and 101 and of the multiple recessive T stock should be obtained several months in advance and should be mated at the age of 7–8 weeks. The inbred strains should be maintained by pairs of brother-sister matings while the T stock can be maintained by pairs or trios (the latter being more productive) in which the mating of close relatives is avoided, in order to keep the inbreeding coefficient low.

Strain C3H females are then mated to strain 101 males (again at 7–8 weeks and in pairs or trios) to give the experimental and control 3H1 mice. If the germ-cells of only one sex are under study the other sex is usually discarded at weaning age, although this could sometimes be done at birth if sexing is accurate and sufficient young are left so that maternal lactation is unaffected.

The 3H1 male and/or female mice should normally be treated with the known or suspected mutagen as soon as they reach maturity, *i.e.* at about 7 weeks or so. By this time all germ-cell stages are present in the gonads. Treatment can be started immediately after weaning if it is going to be protracted and is unlikely to have any severe effects. Sometimes it may be desirable to expose embryonic germ-cells also, by treating the mother from the time of conception of her litter until the litter is weaned.

Controls should be sibs of treated animals.

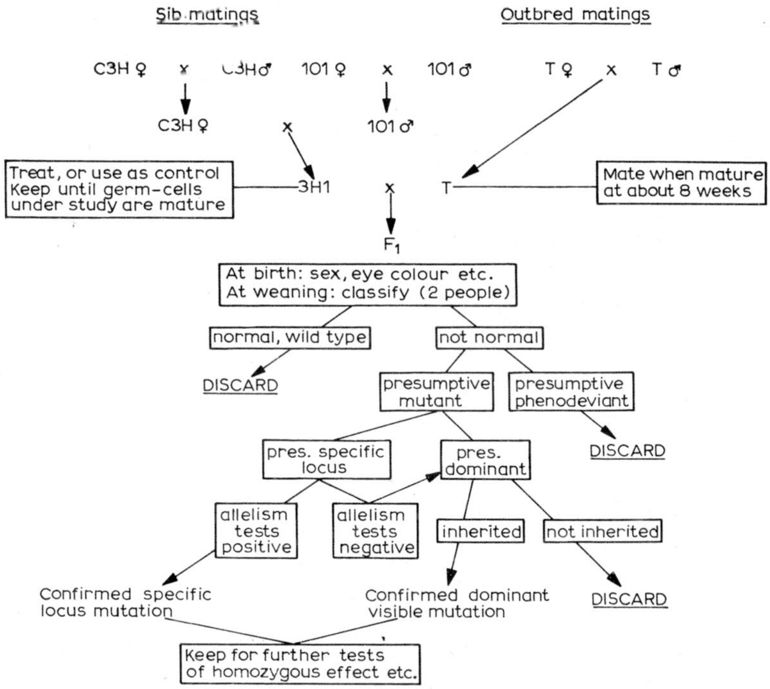

Fig. 2. Flow diagram of specific locus test procedures. T is the 7-locus test stock. Note that a few specific locus mutants (at *a* and *b* loci) are also dominant.

Matings with tester stock

If the 3H1 mice were mature when treated they can be mated at once to mature T stock mice, in pairs or trios. Trios have the disadvantage that (1) there tends to be greater post-natal mortality, which may affect offspring with deleterious specific locus mutants *e.g.* at the *s* locus; (2) with treatment of females, wrong assignment of a litter might make it look as if two independent mutations had occurred rather than a cluster of two mutants. However, they allow higher productivity per cage.

Matings can be delayed if the investigator is interested only in the response of an immature germ-cell stage. It should be remembered that in the weeks after treatment of males, germ-cell stages being sampled are as follows: 1st week, mature spermatozoa in vas deferens and epididymis; 2nd and 3rd, immature spermatozoa and mature and immature spermatids; 4th and 5th, spermatocytes; 6th, later generations of spermatogonia; 7th and later, stem-cell spermatogonia. This is the normal timing and may not apply where the treatment kills a high proportion of germ-cells at particular stages.

For maximum productivity, it is desirable to keep records of the young born and weaned to each female and replace those with poor performance. No females should kept with a male for more than 9 months, but many males will remain fertile for much longer than this, unless adversely affected by the treatment.

Classification of young

Pregnancies should be recorded and young examined soon after birth. They should be counted, sexed and carefully examined with respect to (1) eye colour (normal dark, light or pink) (2) normal pink body-colour or signs of anaemia (3) tail length and straightness (4) normality of limbs. Any differences from normal should be recorded.

Litters should be examined again when cages are changed, to see if abnormal newborn have survived and if other differences have become visible, *e.g.* changes in coat colour. Final examination at weaning age (usually 18–19 days) should be by two persons together, each checking on the normality of the following.

(1) Coat colour, including colour of underfur (blow up fur).
(2) Coat texture (whether wavy, shiny etc.).
(3) Eye colour.
(4) Ear colour and size.
(5) Vibrissae (whether normally straight, kinked or wavy etc.).
(6) General size and shape of head, trunk, limbs and tail (look and feel for kinks).
(7) Number and relative size of digits on fore- and hind-feet.
(8) External genitalia (when sexing): whether normal in appearance and pigmentation.
(9) Postural reflexes; hold mouse up by tail and see if back is hollowed and head held back (normal) or tucked under, also if fore-limbs are pushed forward when the tail-held mouse is quickly lowered ("landing reaction").
(10) Locomotion (no circling etc.).

Keep any deviants from normal for genetic tests, except those which are clearly polygenic or non-genetic.

Genetic tests

These are dealt with in the Appendix. Before carrying them out it is important to decide whether any presumptively mutant mouse is likely to have a specific locus mutation, so that it can be crossed to a member of the appropriate stock. Diagnostic features of mutants at the T stock specific loci are as follows (see also refs. 8, 21).

a Will usually be non-agouti (solid black) or possibly dark agouti or umbrous, with a coat intermediate between agouti and black. A yellow or mottled yellow and agouti coat colour are other possibilities (though very unlikely).

b May lighten the coat colour to cinnamon, or something between cinanamon and wild type agouti, and may also lighten eye colour (visible at birth). However, some *b* locus mutants may do little more than lighten the underfur.

c Most mutants will be very light (cc^{ch} phenotype) but some may be chinchilla ($c^{ch}c^{ch}$) with yellow pigment removed from coat and pinna-hairs so that the coat looks silvery and the ears look white inside. Intermediates between wild type and chinchilla or between chinchilla and cc^{ch} phenotypes are possible, as well as mottled effects.

p Pink-eyed animals are almost certainly mutants at this locus. They may well show yellowish fur, due to removal of black pigment from the agouti coat. However, some *p* alleles don't produce pink eyes, even in combination with *p*. Therefore, mutants with slightly diluted coats, in which black pigment is affected rather than yellow, may be at this locus. Again, mottled coats with patches of wild type and yellowish fur, may result from mutation at this locus.

If an almost white-coated and pink-eyed mutant occurred one would suspect the induction of a deficiency covering both the *c* and *p* loci or, more likely, combination of a gamete from the tester stock which was disomic for chromosome 7 with a complementary nullosomic gamete from the wild type stock.

d Mutants at this locus usually lead to a rather streaky grey phenotype on the agouti background, but something intermediate between this and normal intense agouti might be found.

se Gives short and crinkled external ears, but again an intermediate phenotype is possible.

s Gives irregular white-spotting, which may be high or low grade. However, a small belly-spot is more likely to be either a heterozygous effect of *s* or the result of a dominant visible mutation, *e.g.* at the splotch (*Sp*) locus.

Calculations and statistics with examples

The mutation frequencies per locus in the control and treated series are given by total No. specific locus mutations/(No. classified $\times$No. loci). Members of a clone of mutants, derived from a single original mutation, are normally all counted, since there must be similar clones among the non-mutant mice. In order to obtain the induced mutation rate, where there is a mutagenic effect, the control or known spontaneous frequency is subtracted from that found after treatment and the result is divided by the number of dose units administered to give a mutation rate per unit of dose, if linearity is assumed.

For instance, RUSSELL[19] found 15 mutations in 49, 569 offspring after 300R γ-ray exposures. The mutation frequency per locus per gamete was 15/(49,569 $\times$ 7), since there were 7 specific loci in the stock. This frequency is $4.32 \cdot 10^{-5}$. Since the control frequency was $0.75 \cdot 10^{-5}$ the induced mutation rate is $(4.32-0.75) \cdot 10^{-5}/300 = 11.9 \cdot 10^{-8}/R$/locus.

The statistical treatment of specific locus result depends on the fact that these mutations are very rare events, for which a Poisson distribution can be assumed. Therefore the confidence limits can be calculated from the appropriate tables such as those in FISHER AND YATES[7]. For example LYON *et al.*[15] found only 1 specific locus mutant in 32,463 offspring of male mice which had been given 0.1% caffeine in their own (or their parents') drinking water from conception until the age of 10 weeks. Where 1 is a Poisson variable, its lower and upper 95% confidence limits are 0.0253 and 5.57. Dividing by 7$\times$32,463 (since 7 loci are under test) we can estimate the mutation frequency per locus to be $0.44 \cdot 10^{-5}$, with 95% confidence limits of $0.01 \cdot 10^{-5}$ and $2.45 \cdot 10^{-5}$. The spontaneous mutation frequency for these 7 loci, based on examination of 157,421 offspring at Harwell (site of the caffeine experiment) is $1.0 \cdot 10^{-5}$ (see ref. 23), which is well inside the limits of error for the caffeine experiment and higher than the frequency actually found. We can conclude, therefore, that there is no evidence for mutagenicity of caffeine in mice from this result.

The data of CATTANACH[2] on specific locus mutation induction by triethylenemelamine (TEM) illustrate a positive result. In all, with various dose regimes and mainly with treatment of spermatogonia he recovered 7 specific locus mutations (2 untested) in 23,998 F_1 mice. Thus the frequency per locus was $4.17 \cdot 10^{-5}$, *i.e.* 7/(7$\times$ 23,998). Because 95% confidence limits of 7 are 2.81 and 14.42, the confidence limits are $1.67 \cdot 10^{-5}$ and $8.58 \cdot 10^{-5}$. CATTANACH compared this with the spontaneous frequency of 28 mutations in 531,500 offspring obtained by RUSSELL at the same laboratory. This frequency per locus was $0.75 \cdot 10^{-5}$, with 95% confidence limits of $0.50 \cdot 10^{-5}$

and $1.09 \cdot 10^{-5}$. There was no overlap between these two sets of confidence limits, from which one can conclude that the induced frequency after TEM was significantly higher than the spontaneous frequency.

The significance of the difference can best be calculated by FISHER's "exact treatment" from the 2×2 table:

	With mutations	*Without*	*Total*
Treated	7	23, 991	23, 998
Control	28	531, 472	531, 500
Total	35	555, 463	555, 498

Here, the exact probability is 35! 555,463! 23,998! 531,500!/555,498! (1/7! 28! 23,991! 531,472!, 1/8! 27! 23,990! 531,473! *etc.*), that is, the combined probability of this particular distribution and all more extreme ones. This probability is 0.00065, *i.e.* the difference between the two distributions is highly significant. In this experiment the smallest *expectation* is less than 5, so Yates' correction must be applied[6] if a χ^2 test is used, with 0.5 *subtracted* from the smallest entry and the others corrected accordingly, to make the heterogeneity less extreme. χ^2 is 17.5, $P < 0.001$. It should be noted that if numbers in control and experimental series are fairly similar, the table of VOGEL[26] can often be used to give a direct estimate of the probability that the two frequencies are really the same.

Recording

This is best done in large lined record books, with about forty lines to the page. With treatment of males, each one can then be given two facing pages. The name of the male (stock, parental mating No., litter No., and position in litter) and date of birth should be given at the top of the page, as well as his own mating No. Details of the tester stock females (A, B, C, D etc.) mating to him should also be given, with date of mating. One line can be devoted to each of his litters and should give such details as the number of each sex born and weaned (with classification) date of birth, the running total of weaned and particulars of any abnormal variant. It may be useful to keep a separate running total of all those weaned and classified in each series. It is also useful to keep a separate record of all variant progeny kept for genetic tests, with details of presumptive locus involved, matings and their results etc.

Where females are treated and mated up as a trio to a tester stock male, details of the progeny of both females should be given on the same page, headed by the name of the tester stock male to which both are mated.

Further potentialities of specific locus experiments

We have seen that the specific locus experiment can give information not only on recessive mutations at specific loci but also on visible dominants at any loci, and certain other types of mutation such as deletions involving specific closely linked loci. From littersize changes it can also give a rough estimate of dominant lethal mutation induction, and on the induction of mutations killing between birth and weaning. Moreover, the surviving F_1 progeny are a valuable source of further information. For instance, they can be (and have been) used to screen for biochemical and skeletal mutation. They can also be tested for sterility or semi-sterility and in this way can show whether or not transmissible reciprocal translocations are being induced by the treatment. Thus the specific locus method can easily be expanded into a whole battery of very informative tests.

Simplifying the specific locus procedure

If the experimenter is only interested in specific locus mutations and if he or she is an experienced mouse geneticist, who can recognise with confidence the action of mutant alleles at the specific loci involved, then certain aspects of the experiment can be simplified or omitted. Thus experienced mouse workers have found that a very high proportion of mutants presumptively assigned to a particular specific locus have in fact proved to belong to that locus when confirmatory genetic tests are made. Under these circumstances, the error introduced by not confirming the presumptive classifications is small, so that a case can be made for omitting genetic tests. However, this means that the nature of the homozygote remains in doubt. Another step which is sometimes omitted is the recording, sexing and classification of young at birth, this being done only at weaning. This increases the number of cages which an animal technician can look after, but means that certain information pertaining to dominant lethals and sub-lethals, and the pre-weaning mortality of certain presumptive specific locus recessives, is not obtained. Moreover, certain *p*-locus alleles can be recognised by pink eyes at birth but at weaning have normal-looking dark eyes.

If specific locus mutants alone are wanted the classification procedure at weaning can also be simplified, examination being restricted to coat colour and ear length.

CONCLUSION

The great advantages of the specific locus test are that (1) the information it provides is particularly relevant to human genetic hazards, since it concerns transmitted mutational damage in a mammal, (2) it permits the detection of both dominant and recessive mutations in the first generation after treatment, (3) for mutation detection, the F_1 mice merely have to be observed with care, rather than being processed (4) the new mutations recovered can then be studied in detail by genetic, biochemical and other means, (5) the test can easily be combined with others so as to provide further information of mutational interest. Its disadvantages are that (i) only six or seven out of many thousands of loci are scanned for recessives, so a large number of progeny are needed, (ii) it only detects those forward mutations at the specific loci which have visible phenotypic effects in combination with the tester stock recessive allele, (iii) the nature of the mutations is not easily discovered, unless they simultaneously involve the *d* and *se* loci, (iv) certain non-mutational events may be scored as mutations, (v) specific locus mutations with adverse heterozygous effects may not survive long enough to be recognised or tested.

It seems likely that there will be many occasions in the future, as in the past, when the advantages will outweigh the disadvantages and the test will be performed. It is hoped that this detailed account of what it involves will make it easier for those concerned with mutation assays to make a realistic cost-benefit assessment for this important test.

SPECIAL TESTS AND TESTER STOCKS

It is desirable to test the presumptive specific locus and dominant mutations for inheritance and for allelism at the locus thought to be involved. They can also be tested to see if they are lethal in the homozygote.

Allelism tests at specific loci

Allelism can be confirmed by crossing the suspected mutant to a recessive homozygote for the locus concerned, to test if all offspring have a recessive phenotype with respect to that locus. The specific locus stock itself can be used for suspected *p* alleles, but since the effect of *p* tends to mask effects of the other loci it is best to test the others for allelism by crossing to a multiple recessive stock not carrying *p* (*e.g.* LIVA of Harwell which is *aa bb* c^{ch} c^{ch} *dd sese ss wa-1 wa-1*). A dd^1 stock will enable simultaneous tests for allelism and lethality at the *d* locus since lethal *d* alleles will give the dilute lethal phenotype when combined with d^1. Note that some *p* alleles do not affect eye colour.

When testing for allelism, raise at least 8 young which can be classified for mutant effects at the locus concerned. If all 8 are mutant, the alteration can be considered heritable and belonging to the locus under test.

Dominant tests

Cross to a wild type stock to see if the deviant phenotype reappears in the F_1. If fully penetrant there should be a 1:1 ratio of affected and normal.

Tests for lethality

If the mutant is a male, or if the presumptive mutation is thought to be at the *d* locus, these can be carried out at the same time as the allelism test. If the mutant is a non-dilute female they should be done after the allelism tests.

Basically, tests for lethality involve outcrossing the mutant and then intercrossing progeny, with classification of F_2 offspring. Suppose the original mutant is aa^x, where a^x is the newly arisen specific locus mutation. F_1 progeny will be either $+a$ or $+a^x$, and will have the wild phenotype. Intercrosses of these will be $+a\times+a$, $+a\times+a^x$ or $+a^x\times+a^x$ in the expected ratio of 1:2:1. The last type will produce a^xa^x animals. If these are lethal then about 3 in every 12 intercrosses at random will fail to segregate at the *a* locus (because the homozygote dies before classification) or will produce a distinctive lethal phenotype. Thus at least 12 intercrosses should be put up.

Those segregating in the specific locus mutant can be discontinued when two viable mutant young have been obtained. Those not segregating should be continued until 16 non-mutant young have been obtained, when lethality is indicated.

Special stocks can shorten the procedure with mutants at the following loci.

(1) *c* and *p*. A $c^{ch}c^{ch}$ non-pink-eyed dilute stock is useful. If the mutant looks like cc^h (very light fur) it should be crossed to this to test for allelism. Also, phenotypically cc^{ch} offspring can then be intercrossed (4 matings) to see if they produce albino (*cc*) or not. If only chinchilla or cc^{ch}-type offspring are produced it can be assumed that the

newly induced mutation is lethal. If the mutant is a confirmed *p*-allele it should be crossed to $c^{ch}c^{ch}$ and non-chinchilla offspring should be kept. These will mainly carry the newly induced *p*-allele (*e.g.* $p^{x}+/+c^{ch}$). Six intercrosses of these should be made to determine the nature of p^{x}. Note that some recombination will be expected between the *c* and *p* loci.

(2) *d* and *se*. A $d^{1}+/d\ se$ stock enables one more simply to test *d* and simultaneous $d+se$ mutants for lethality, since if they are lethal they will produce dilute lethal offspring with typical opisthotonus and convulsions when crossed to members of this stock. In this way one can distinguish the deficiency-type *d se* mutants from others. If short-ear mutants ($+se^{x}/d\ se$) are crossed to $d^{1}+/d\ se$, then non-dilute non-short-eared offspring will nearly all be $d^{1}+/+\ se^{x}$ and can be intercrossed to find the nature of $se^{x}se^{x}$.

Note that deficiency carrying offspring may be rarer than expected because of reduced transmission[18].

REFERENCES

1 CARTER, T. C., M. F. LYON AND R. J. S. PHILLIPS, Induction of mutations in mice by chronic gamma irradiation; interim report, *Brit. J. Radiol.*, 29 (1956) 106–108.

2 CATTANACH, B. M., Chemically induced mutations in mice, *Mutation Res.*, 3 (1966) 346–353.

3 CATTANACH, B. M., Specific locus mutations in mice, in (A. HOLLAENDER, ed.) *Chemical Mutagens*, 2 (1971) 535–9. New York–London, Plenum.

4 EHLING, U. H., Differential spermatogenic response of mice to the induction of mutations by antineoplastic drugs, *Mutation Res.*, 26 (1974) 285–295.

5 EHLING, U. H., The multiple loci method, in (F. VOGEL AND G. ROHRBORN eds.) *Chemical Mutagenesis in Mammals and Man* (1970) Berlin, Springer–Verlag, 156–161.

6 FISHER, R. A., *Statistical methods for research workers*, 10th edition (1946) Edinburgh, Oliver and Boyd.

7 FISHER, R. A. AND F. YATES, *Statistical tables for biological, agricultural and medical research*, 3rd ed. (1948), Edinburgh, Oliver and Boyd.

8 GREEN, M. C., Mutant genes and linkages, in (E. L. GREEN, ed.) *Biology of the Laboratory Mouse*, New York, McGraw–Hill (1966) 87–150.

9 KOHN, H. I., H-gene (histocompatibility) mutations induced by triethylenemelamine in the mouse, *Mutation Res.*, 20 (1973) 235–242.

10 KRAFT, L. M., Observations on the control and natural history of epidemic diarrhoea of infant mice (EDIM), *Yale J. Biol. Med.*, 31 (1958) 121–137.

11 KRAFT, L. M., R. F. PARDY, D. A. PARDY AND H. ZWICKEL, Practical control of diarrheal disease in a commercial mouse colony, *Lab. Animal Care*, 14 (1964) 16–19.

12 LYON, M. F. AND T. MORRIS, Mutation ıates at a new set of specific loci in the mouse, *Genet. Res. Camb.*, 7 (1966) 12–17.

13 LYON, M. F. AND T. MORRIS, Gene and chromosome mutation after large fractionated or unfractionated radiation doses to mouse spermatogonia, *Mutation Res.*, 8 (1969) 191–198.

12 LYON, M. F. AND T. MORRIS, Mutation rates at a new set of specific loci in the mouse, *Genet. Res. Camb.*, 7 (1966) 12–17.

13 LYON, M. F. AND T. MORRIS, Gene and chromosome mutation after large fractionated or unfractionated radiation doses to mouse spermatogonia, *Mutation Res.*, 8 (1969) 191–198.

14 LYON, M. F., R. J. S. PHILLIPS AND H. J. BAILEY, Mutagenic effects of repeated small radiation doses to mouse spermatogonia. I. Specific-locus mutation rates, *Mutation Res.*, 15 (1972) 185–190.

15 LYON, M. F., R. J. S. PHILLIPS AND A. G. SEARLE, A test for mutagenicity of caffeine in mice, *Z. Vererbungslehre*, 93 (1962) 7–13.

16 RUSSELL, W. L., X-ray induced mutations in mice, *Cold Spring Harb. Symp. Quant. Biol.* 16 (1951) 327–336.

17 RUSSELL, L. B. AND J. W. BANGHAM, Variegated-type position effects in the mouse, *Genetics*, 46 (1961) 509–525.

18 RUSSELL, L. B. AND W. L. RUSSELL, Genetic analysis of induced deletions and of spontaneous non-disjunction involving chromosome 2 of the mouse, *J. Cell. Comp. Physiol.*, 56 suppl. 1 (1960) 169–188.

19 Russell, W. L., The effect of radiation dose rate and fractionation on mutation in mice, in (F. H. Sobels, ed.) *Repair from Genetic Radiation Damage*, Oxford, Pergamon (1963) 205–218.
20 Sanders, H. J., Chemical mutagens, *Chem. Eng. News*, 71, May 19 (1969) 50–65.
21 Searle, A. G., *Comparative Genetics of Coat Colour in Mammals*, London, Logos/Academic (1968).
22 Searle, A. G., Spontaneous frequencies of point mutations in mice, *Humangenetik*, 16 (1972) 33–38.
23 Searle, A. G., Mutation induction in mice, *Adv. Radiation Biol.* 4 (1974) 131–207.
24 Selby, P. B., X-ray-induced specific-locus mutation rate in newborn male mice, *Mutation Res.*, 18 (1973) 63–75.
25 Valcovic, L. R. and H. V. Malling, An approach to measuring germinal mutations in the mouse, *Environmental Health Perspectives*, 6 (1973) 201–205.
26 Vogel, F., Eine Tafel für den Vergleich zweier kleiner Häufigkeitsziffern bei seltenen Ereignissen, *Acta Genet.*, 9 (1959) 314–319.
27 Vogel, F., G. Röhrborn, E. Schleiermacher and T. Schroeder, *Strahlengenetik der Sänger*, Vol. 1 of *Fortschritten der allgemeine und klinischen Humangenetik*, Stuttgart, Georg Thieme (1969).

THE DOMINANT LETHAL ASSAY IN THE MALE MOUSE

A.J. BATEMAN

Paterson Laboratories, Christie Hospital and Holt Radium Institute, Manchester M20 9BX (Great Britain)

Following my experience with the induction of dominant lethal mutations by known mutagens, I recommended the use of this technique as a screening method for mutagenicity [2]. Since then, various industrial laboratories have acquired a great deal more practical experience than myself, in the use of the assay as a means of assessing unknown substances. It is hoped, nevertheless, that the following account may prove useful, not only to laboratories who have not yet used the assay, but also to those with much experience, who may still have problems over the interpretation of their data.

The nature of the dominant lethal mutation

A dominant lethal mutation is a genetic change in a gamete, egg or sperm, which kills the zygote produced by that gamete. To have such an effect the change must not incapacitate the gamete in which it arises. In a plant the dominant lethal will reduce seed set, in Drosophila it will lead to unhatched eggs, and in a mammal it will reduce litter size. In the mouse the reduction in litter size can be due to failure of the fertilized egg either to implant, or to develop after implantation.

Death before implantation is manifest as a discrepancy between the number of eggs shed, counted as corpora lutea, and the number of implants. If the corpora lutea are uncounted pre-implantation losses can only be inferred from a reduction in the average number of implants per uterus in comparison with control matings. Unless there has been microscopic examination of the pre-implantation eggs to determine the rate of fertilization, pre-implantation losses through dominant lethal mutation cannot be distinguished from failure of fertilization, which may result from many factors, including inadequate quality or quantity of sperm. Pre-implantation lethals are therefore unsuitable for our assay. Fortunately for the assay, failure of some fertilized eggs to implant always seems to accompany failure of some implants to develop and this can be recognized with little difficulty or ambiguity.

It is therefore advisable to ignore pre-implantation deaths and concentrate on the fate of implantations, examined in mid-term pregnancy. They can then be classified as live embryos, late deaths (where there is some evidence of growth of a foetus which has then died) and early deaths, which are a very standard product, the deciduoma, commonly and aptly called a "mole". For simplicity, we shall refer in the rest of this paper to these early deaths as "moles". The deciduum is a growth of maternal uterine tissue under the stimulus of the implanting egg. Its growth is then autonomous, whatever the fate of the egg, until the 11th day of pregnancy, when it stops. Where there is a live embryo, foetal tissues overwhelm the deciduum, which is nevertheless still there. If the embryo has died at or soon after implantation the deciduum is the total growth and from the 11th day it is recognizable as a mole. This mole persists unaltered, apart from necrosis of the upper part, the deciduum capsularis, throughout pregnancy, being shed at parturition with the rest of the uterine contents.

As a consequence of this behaviour the proportion of implants which are moles can be recorded at any stage of pregnancy from the 11th day onwards. The mole is sometimes pseudoscientifically referred to as a "resorption site". There is no evidence however that there was ever anything there beyond what is seen at dissection: nothing has been resorbed.

Background

Dominant lethal mutations in a mammal were first recognized in 1937 by Brenneke [3]. They were first used experimentally as an indicator of mutation by Kaplan and Lyon [7] in 1953 and by Russell et al. [8] in 1954. In 1966 I recommended their use in screening for mutagens [2]. Since then they have been widely used in academic and industrial laboratories.

From the start it was assumed that the prime cause of dominant lethality was chromosome breakage, but cytological confirmation of this has only recently been obtained by Brewen et al. [4] and by Bateman (unpublished). It has been shown that the incidence of broken chromosomes at the first cleavage metaphase corresponds to the incidence of dominant lethal eggs. The broken chromosomes will be lost, leading to monosomy, which is probably the genetic cause of the death of the embryo. It was argued on theoretical grounds [1] that preimplantation losses arose from multiple hits. Bateman (unpublished) has obtained cytological confirmation of this.

Materials

Nothing is required beyond the normal equipment of an animal laboratory. Any mouse strain that breeds freely and is free of t-lethals and translocations or anything else causing high and erratic levels of lethal embryos, can be used. The greater the normal litter size the more information is obtained per mating. It is therefore recommended that F1 hybrids or outbred strains can be used. The only other requirements are mouse boxes, food and water, some standard dissecting instruments and skilled hands to dissect the females. Syringes and fine needles (18-gauge) will be needed for commonest (i.p.) route of administration of the test compound. Other routes which have been used are ingestion, gavage, and inhalation.

Costing

The unit of costing is the tested male, to be mated to 3 females per week for 8 weeks, each mated female to be dissected. The rationale of the 8 weeks of mating is that following a short exposure to the test substance 8 weeks are necessary to sample all the stages of male germ cells which have been exposed to the substance. The first week will sample epididymal sperm, the second and third weeks late and early spermatids respectively, the fourth and fifth weeks late and early spermatocytes, and the sixth, seventh and eighth weeks successively earlier generations of spermatogonia back to the stem cell.

For each test male there should be one control male mated over the same period. It is also necessary to include a positive control, a male treated with a known mutagen to ensure that the assay is working properly. Since the positive control serves only to yield positive results it is pointless to mate this male beyond the period known to produce a positive response. Thus, if TEM * is used as the positive control, mating need not extend beyond 3 weeks.

For reasons that will be discussed below, the recommended number of males to include in the test is 10 test males, 10 controls and 5 positive controls. At the above mating protocol this will produce 525 females to be dissected. At 10 min per dissection (a generous allowance) this makes 87 h. The cost of the test can be calculated on the basis of the following units of expenditure:

25 males + 525 females prior to test	550 mice
3 weeks housing of each female during the test	1575 mouse-weeks
Technicians' dissection time	87 man-hours

The cost of an assay is of course dependent upon the sensitivity required. The above protocol is devised to detect an increase of dominant lethal incidence from the control level of 5% to 10% in any one week of mating.

It is common practice to include more than one dose level of the test substance, but we are not interested at this stage in dose dependence. Indeed we hope the results of the assay will be negative. One dose level should suffice, provided it is chosen to maximize the chance of a response.

A method has been devised [6] which aims to survey the whole spermatogenic system without 8 weeks of mating. Instead, the males are exposed to the test substance for 8 weeks and then mated for 2 weeks. (Actually those authors used rats, which necessitates 10 weeks treatment owing to their longer maturation period.) This would have the same sensitivity as the protocol recommended above for detecting a response confined to a single maturation stage provided the *dose per week* was the same in the two methods, which means an 8-fold total dose (10-fold in rats). The method would fail to identify which germ cell stage was responsible for any positive response.

The prime determinant of the sensitivity of a test is the total number of pregnancies dissected. The 8-week treatment schedule will, therefore, only be as sensitive as the 8-week dissection schedule if the same total of pregnancies is analyzed. In that case there would seem to be no saving in time, animals or labour. Experience may show the 8-week treatment schedule to be superior to

* Triethylene melamine.

the 8-week mating schedule. Meanwhile, the latter is recommended.

Until recently one might have argued against the 8-week mating period on the grounds that experience had shown that the response always occurred in postmeiotic cells (detectable in the first 3 weeks). This is a dangerous philosophy since it assumes that all unknown mutagens will operate in the same way as known mutagens. The unsoundness of such an assumption has recently been exposed. Mercaptopurine [5] has been shown to produce dominant lethals only in the fifth week after treatment (early spermatocytes).

The Protocol

Assuming that preliminary toxicity tests have been made, the dose of test substance which will maximize the chances of detection will be the highest dose that permits the male to produce fertile matings over an 8-week period. For the purposes of this test, this dose will be the maximum tolerated dose (MTD). It is convenient to sample the matings on a weekly basis. It is appropriate, therefore, to administer the dose over one week, say in equal daily portions over 5 consecutive days.

Mutation is of course equally important and equally undesirable in both sexes (i.e. in eggs and sperm), but it is more convenient to treat one male and mate him to many control females, than to treat many females each of which, dissected after mating, provides a single litter for analysis. There is also the possibility that systemic responses of the treated female may interfere with the pregnancy in such a way as to be confused with dominant lethality.

All males to be mated should be sexually mature (8—10 weeks old) at the beginning of the experiment, and should be of proven fertility. Assuming that the selected route is intraperitoneal, the negative controls would receive 1 ml of isotonic saline on 5 consecutive days, the test animals would receive the MTD of the test substance in 1 ml saline over the same period, and the positive controls would receive TEM at 0.05 mg/kg in 1 ml saline over the same period. A lower dose might be used, if one wished to use the positive control as a critical test of the sensitivity of the system. Any chemical known to be mutagenic by the selected route is suitable.

Each male is then caged with 3 virgin females aged 8—10 weeks for one week with a change of females every week for 8 weeks (3 weeks following TEM). The females are dissected 17 days after the introduction of the male. The pregnancies will therefore vary in age from 10 to 17 days, with a mode at 14 days. Daily recording of vaginal plugs in mice (or sperm-positive vaginal smears in rats) enables one to distinguish sterile matings from non-matings, to determine the day of the week of each mating (in case there should be variation in response to treatment during the week), and to dissect all pregnancies at the same age. These refinements could be very valuable in scientific research. The extra labour involved would not seem to justify their use in screening tests.

The females are killed by dislocation of the neck. The uterus is removed and stretched across a dissecting board covered with white paper for good visibility. Two sizes of implantation will be seen through the uterine wall: normal and moles, with rarely some late deaths of intermediate size. The late deaths occur sporadically and their incidence seems to have no relation to mutation rates.

The normal foetuses will vary in size according to the age of the pregnancy. The moles are constant in size from 10 days to parturition (19th day). In late pregnancies care must be taken not to overlook the moles because of their small size, especially when close to the relatively enormous live foetuses. Record the numbers of total implants and moles in each pregnancy. That completes the practical side of the test.

Statistics

Experience has shown this to be the most difficult part of the test, because there are so many possible parameters. It is desirable to decide which is the best, both in sensitivity to detection of positive results, and in discrimination against false-positive results, and to get this generally adopted for different laboratories using the assay.

The simplest parameter to measure is litter size. It is, however, subject to the influence of many variables unconnected with mutation: ovulation, fertilization and the cannibalism of newborns by their mother. Dissection of the pregnant female to simply count live embryos involves much more work and only eliminates cannibalism. This does not justify the extra effort, yet the number of live embryos has been widely used in the dominant lethal assay.

The other extreme in terms of completeness and effort would be to subtract the number of live implants from the number of fertilized eggs. Unfortunately these parameters cannot be obtained from the same pregnancy. Counting fertilized eggs is also a very laborious procedure and subject to a large bias through incomplete harvesting of the eggs.

The nearest one could approach to this in a single pregnancy would be to compare the number of live implants with the total number of eggs shed, as represented by the corpora lutea. This involves careful dissection of the ovaries and their examination under a dissecting microscope for corpus luteum counts. A little experience will enable this to be done accurately. There is still, however, the possibility of confusing pre-implantation dominant lethals with unfertilized eggs. Since substances under test for mutagenicity may well affect both the quality and quantity of sperm of the treated male, pre-implantation losses are a very unreliable indication of mutation, though they will frequently accompany it [1].

The solution is to ignore pre-implantation loss, thus avoiding corpus luteum counts. All that is required, as given in the protocol, is the total of implants per pregnancy and the number of moles. We do not risk obtaining false-positives by this procedure. At low mutation rates pre-implantation losses are rare, and at high mutation rates, when pre-implantation deaths become more common, moles also become an increasing proportion of the total implants [1].

With only two parameters, total implants and moles, statisticians have yet found a remarkable variety of statistical treatments:

(1) live foetuses/pregnancy;
(2) moles/pregnancy;
(3) pregnancies with moles/total pregnancies;
(4) pregnancies with more than one mole/total pregnancies;
(5) frequency distribution of moles/pregnancy; and
(6) moles/implant.

The first five of the six statistics use the pregnant female as the statistical unit instead of the foetus (≡ gamete). A little consideration of the biological principles establishes that the mutation rate is a property of the gametes, not of the mated females. The first five statistics also make no allowance for the variation in the total number of implants per female by which they will be in part determined. Indeed, for a given mutation rate, statistics 1—4 will be *directly* proportional to the number of implants per female, which is itself dependent upon the ovulation rate and the fertilization rate, both of which we wish to exclude from our assay.

So we will settle for statistic 6 as the only one appropriate to the assay. The *operational* unit of sampling is of course the female. If, however, there is homogeneity between the females mated in one week to the same class of male, all their implants can be pooled to give moles/implant/treatment week. This greatly simplifies the statistics. For each week of sampling we can construct at 2×2 contingency table between the controls and the treated matings and obtain a χ^2 with 1 degree of freedom for the difference between them. The 8 weeks will provide 8 such χ^2. If we are to use the 5% probability level for judging the significance of our comparisons, it follows that out of 8 such comparisons there will be a 40% chance that one of these will achieve the 5% level for significance in *either* direction (with treatment above *or* below the control). If, therefore, this significance level is *just* reached in one and not more than one of the 8 comparisons we can ignore it, or, to be on the safe side, if it shows an excess of treated over controls, we can repeat the whole assay.

If, however, as frequently happens, there is heterogeneity between pregnancies, the χ^2 test is not appropriate. Each pregnancy will contribute a value (transformed or not) of moles/implant to an estimated mean, the standard error of which is calculated from the data. The Student's *t*-test can then be applied to a comparison of means.

The χ^2 approach provides us with a simple way of estimating the sample size needed to achieve a desired sensitivity of the test. Let us suppose that the control incidence of moles is 5% (a common level) and that we wish to detect a doubling of that level, to 10%. We can express the situation in the following table with the Control and Treated sample sizes each being equal to n.

	Moles	Other implants	Total implants	χ^2
Control	0.05n	0.95n	n	$\dfrac{(0.95n \times 0.10n - 0.90n \times 0.05n)^2 \times 2n}{0.15n \times 1.85n \times n \times n}$
Treated	0.10n	0.90n	n	
Total	0.15n	1.85n	2n	

This table allows us to calculate χ^2 for heterogeneity, for 1 degree of freedom, in terms of n: n/55.5. We can now equate this to the χ^2 corresponding to a significance at the 5% level, which is 3.84. This gives a value of n = 213, which is the sample size to distinguish between 5% moles and 10% moles at the 5% level of significance. With 10 implants per pregnancy we would need 21.3 pregnancies per week. This should be more than achieved with 3 females per week for each of 10 males, which is why these numbers were recommended in the above section on costing. If the control data are homogeneous over the 8 week period the sensitivity can be greatly increased by pooling the control data.

If the treated data are homogeneous in the same way (as would happen if the results were negative) they could be similarly pooled achieving even greater sensitivity. If n = 213 this will give a 50% chance of a P value <0.05% if there is a true increase in the proportion of dead implants from 0.05 in the control to 0.10 in the treated series. (The chance will be greater if the treated value exceeds 0.10.) Larger sample sizes are advocated by Ehrenberger in a later chapter of this volume.

No statistical treatment should be necessary for the positive control, whose sole purpose is to validate the assay. If the positive control fails to give a clear positive result the assay is void and must be repeated, modified as seems appropriate, until the positive control yields positive results.

Experience with the assay will soon establish what is the minimum size of positive control to guarantee positive results and this would be all that was required in future assays. If, however, the route of administration was changed, for example from i.p. to drinking water, it would be necessary to reinstate a full scale positive control in order to establish that results with the new route were reliable.

One might, however, wish to use the positive control as a standard for comparison, expressing the treatment response as "not more than 1/n of the positive control". In that case the positive control value must be estimated with the same accuracy as the treatment and negative control values.

Sensitivity

It is clear from the above discussion that the protocol proposed would be adequate to detect a mutation rate of 5% dominant lethals in any one of the 8 weeks of the test, induced by the MTD level of the test substance against a background of 5% in the controls. To put it another way, it will detect mutagens whose doubling dose is less than the MTD, in any maturation stage of the spermatogenic cycle, in mice with a spontaneous incidence of dominant lethals not less than 5%. The control level of dominant lethals is crucial to the sensitivity of the test. The higher the control level the smaller the sample needed to detect a *doubling* of the mutation rate, but the larger the sample needed to detect a given mutation rate (say 5%).

The sensitivity would be increased if more than one maturation stage was sensitive or, of course, if more animals were used.

It should be borne in mind that valid comparisons of the sensitivity of different assays can only be made in terms of equal expenditure of time or money or effort. No test is intrinsically more sensitive than another, since any test can be made more sensitive by increasing the size of sample.

Limitations

The assay can only detect mutations which do not affect the fertilizing ability of the sperm carrying them. This probably explains why it detects mutations in post-meiotic stages rather than earlier stages. The mercaptopurine results [5] reveal however that dominant lethal mutations *can* be inherited by sperm without loss of function. The dominant lethal assay detects chromosome breakage rather than point mutations, though it would probably detect high levels of non-disjunction.

Appendix

I am very grateful for being allowed to use a practical example from an industrial source, which may assist in the appreciation of the problems involved in the dominant lethal assay. I present a summary of the results (with my comments in parentheses).

The chemical being tested was 5-butyl-2-ethylamino-6-methylpyrimidin-4-yl dimethylsulphamate, which I will abbreviate to BEMD. It was insoluble in water and was administered orally in suspension in 2% "Cellofas" (carboxylated methyl cellulose). The test animals were the Charles River (CD-1) strain of mice. In preliminary tests it was found that concentrations corresponding to a dose greater than 3000 mg/kg became too viscous to administer, so the maximum practical dose was 3000 mg/kg.

Prior to testing, the males were placed with 2 females each for one week and only those males producing 2 litters from 2 females were entered in the test. At the time of treatment the males were 10—12 weeks old. The females were 8—10 weeks old at the time of mating.

The test substances was given at three dose levels: 3000, 300 and 30 mg/kg/day for 5 consecutive days prior to test mating. (According to my schedule only one dose, the highest, is necessary.) The positive control received 100 mg/kg/day of EMS for 5 days.

Fifteen males were included in each test and mated to 2 females per week for 8 weeks. (This is equivalent to the 10 males mated to 3 females per week in my schedule. The positive controls were mated for 8 weeks compared to the 3 weeks regarded as sufficient in my recommendation.)

Females were dissected 15—16 days after introduction to the males, i.e. 8—9

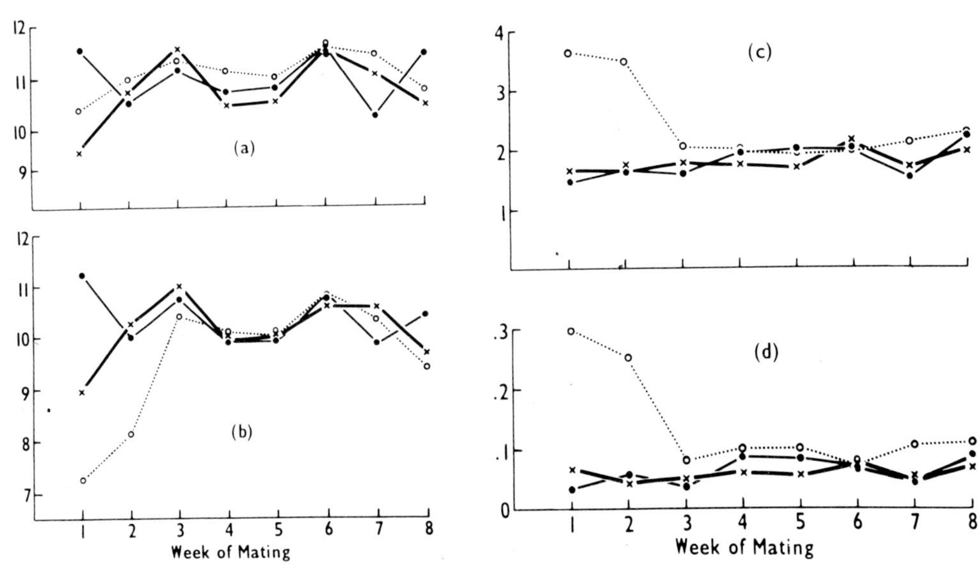

Fig. 1. a: implants/pregnancy. b: live implants plus late deaths/pregnancy. c: moles/pregnancy (Y), transformed to $\sqrt{Y} + \sqrt{Y+1}$. d: moles/implant. ●———●, control males (5 × 10 ml/kg/day of 2% Cellofas, orally); X———X, males receiving 5 × 3000 mg BEMD/kg/day, orally; ○- - - - - -○, males receiving 5 × 100 mg EMS/kg/day, orally.

days after the removal of the males. Matings during the last two days of the week would be infrequent but the pregnancies would be only 8—9 days old at dissection, making it difficult to score with confidence. There seems no good reason not to dissect at 17 days, when all pregnancies will be at least 10 days old. The uterine contents were then scored as live implants, early deaths and late deaths. (The early deaths are the moles of my schedule. The "late-deaths" were pooled with the live implants for the analysis.)

The data were analyzed on the basis of the following statistics on a week-by-week comparison. (Some are more relevant than others to mutagenicity screening. Effects on fertility are useful for a general assessment of a compound. They are not critical evidence of mutation.)

1. *Frequency of non-pregnant females.* No effect. (This confounds failure to mate with pseudopregnancy, i.e. mating with no fertilization or with early death of the entire litter. It is of little relevance to a mutagenicity test.)

2. *Implantations per pregnancy* (Fig. 1a). The highest dose of BEMD and the EMS had reduced numbers in weeks 1 and 8, but only the BEMD was significant. (Not strictly relevant to mutagenicity test. The effects are probably real, however, i.e. due to the test compound.)

3. *Pregnancies with moles.* EMS showed the expected increase in weeks 1 and 2. BEMD showed increases in weeks 1 and 3, but these were surprising because they were confined to the *lowest* dose. (This is not a very informative statistic. It will be correlated with mutation, but fails to take account of variation in number of moles per pregnancy.)

4. *Live implants plus late deaths per pregnancy* (Fig. 1b). The expected highly significant reductions were found in weeks 1 and 2 with EMS. Highest dose of BEMD gave reductions in week 1. (This is a direct result of the reduced number of total implants per pregnancy: item 2 above. The dominant lethal test might appear to be more sensitive using this statistic since dominant lethal agents can cause both pre-implantation losses and early deaths, but since non-mutagenic factors also cause pre-implantation losses, this statistic, though more sensitive in one way, is also liable to produce false-positives, as it seems to have done here with BEMD, week 1.)

5. *Proportion of implants recorded as early deaths* (Fig. 1d as moles/implant). Only EMS gave positive results, in weeks 1 and 2 as expected. (The accompanying graph shows that for the remaining 6 weeks EMS gave consistently higher values than the controls, none individually significant. I believe this represents a true mutagenic response, revealed by the sensitivity of the test. Some laboratories have argued against this statistic on the grounds that a high mutation rate, which causes pre-implantation losses, will automatically reduce the implantation rate, reducing the denominator of the fraction. A reduced implantation rate will equally reduce the absolute number of dead implants per pregnancy, thus biassing the next statistic. The purpose of this assay, however, is not to measure the mutation rate but to detect mutation. The best statistic will combine high sensitivity to mutation with a low possibility of false-positives due to low ovulation rate or low fertilization rate.)

6. *The final statistic* (Fig. 1c) is the most sophisticated. It is the number of early deaths per pregnancy (Y) transformed, to stabilize the variance, into $\sqrt{Y} + \sqrt{Y+1}$. Apart from the highly significant effect expected with EMS in weeks

1 and 2, BEMD at the low and medium doses, but *not* the highest, showed a significant reduction in week 1. (The low level of total implants with the highest BEMD dose will automatically reduce this statistic also, and thus explain the inconsistency between the 3 dose levels with BEMD. The trouble seems to arise from the exceptionally low incidence of moles in the controls in week 1. These unexpected and inconsistent results must raise doubts of the value of this statistical test. As already stated the number of early deaths/pregnancy is inevitably correlated with implants/pregnancy and no transformation can alter that fact.)

This material shows fluctuation in total implants per female over the 8 weeks (Fig. 1a). It is more obvious if all levels of BEMD are included. All 5 treatments then show the fluctuation. It is obviously irrelevant to the matter of the test. It may be due to fluctuation in ovulation rate (an explanation which I would favour) or in fertilization rate (which in normal matings is very close to 100% and invariable), but it will be the direct cause of fluctuation in both live implants per pregnancy and moles per pregnancy, transformed or untransformed. It will *not* affect the incidence of moles per implant.

The above example not only serves to illustrate the many ways in which the dominant lethal data can be analyzed. I suggest that it vindicates the moles/implant statistic as the one most relevant to the mutagenicity test.

Acknowledgements

This work was supported by grants from the Medical Research Council and the Cancer Research Campaign.

References

1 Bateman, A.J., The partition of dominant lethals in the mouse between unimplanted eggs and deciduomata, Heredity, 12 (1958) 467—475.
2 Bateman, A.J., Testing chemicals for mutagenicity in a mammal, Nature, 210 (1966) 205—206.
3 Brenneke, H., Strahlungschädigung von Mäuse- und Rattensperma, beobachtet an der Frühentwicklung der Eier, Strahlentherapie, 60 (1937) 214—238.
4 Brewen, J.G., H.S. Payne, K.P. Jones and R.J. Preston, Studies on chemically induced dominant lethality. I. The cytogenetic basis of MMS-induced dominant lethality in post-meiotic germ cells, Mutation Res., 33 (1975) 239—250.
5 Generoso, G.M., R.J. Preston and J.G. Brewen, 6-Mercaptopurine, an inducer of cytogenetic and dominant lethal effects in pre-meiotic and early meiotic germ cells of male mice, Mutation Res., 28 (1975) 437—447.
6 Green, S., F.M. Moreland and G.W. Flamm, A more refined approach to dominant lethal testing, Mutation Res., 31 (1975) 340.
7 Kaplan, W.D. and M.F. Lyon, Failure of mercaptoethylamine to protect against the mutagenic effects of radiation. II. Experiments with mice, Science, 118 (1953) 777—778.
8 Russell, W.L., L.B. Russell and A.W. Kimball, The relative effectiveness of neutrons from a nuclear detonation and from a cyclotron in inducing dominant lethals in the mouse, Amer. Naturalist, 88 (1954) 269—286.
9 Sirlin, J.L. and R.G. Edwards, Sensitivity of immature mouse sperm to the mutagenic effect of X-rays, Nature, 179 (1957) 725—727.

DROSOPHILA AS ASSAY SYSTEM FOR DETECTING GENETIC CHANGES

F.E. WÜRGLER [1], F.H. SOBELS [2] and E. VOGEL [2]

[1] *Department of Zoology, Swiss Federal Institute of Technology, Universitätstrasse 2, CH-8006 Zürich (Switzerland) and* [2] *Department of Radiation Genetics and Chemical Mutagenesis, University of Leiden, Leiden (The Netherlands)*

Contents

Abbreviations: DEN, *N*-nitrosodiethylamine; DMSO, dimethylsulfoxide; EMS, ethylmethanesulfonate; MMS, methylmethanesulfonate; NaF, sodium fluoride; PDMT, 1-phenyl-3,3-dimethyltriazene; TEB, 2,3,5,6-tetra-ethyleneimino-1,4-benzoquinone; and Trenimon, 2,3,5-tris(1-aziridinyl)-*p*-benzoquinone. The genetic symbols are described in ref. 41.

Introduction

During the last few years, the evaluation of genetic damage resulting from exposure to chemical mutagens in the human environment has received considerable attention. Recent studies have revealed a remarkable overlap between the carcinogenic and mutagenic potential of chemical agents. Consequently, mutagenic activity, apart from constituting a risk to future generations, poses an additional threat of inducing malignancy in the presently living one.

An identification of the potential hazards involved first of all requires adequate detection of mutagenic activity relevant to man. A great number of different assay systems have now been developed, microbial test systems constituting the most sensitive and fastest ones.

Regulatory measures aimed at restricting exposure to genetically active chemicals require careful weighing of risks against benefits. In view of the far-reaching consequences associated with regulation and restriction of any particular chemical or group of compounds, microbial data should, in our opinion, require the confirmation from a battery of different test systems.

In this context, we believe that tests with Drosophila can fulfil an extremely useful function. Drosophila has a short generation time of 10—12 days, is cheap and easy to breed in large numbers with simple facilities; in brief, it provides the quickest in vivo assay system for the recovery of mutations available in animals today. Genetic testing methods make it possible to provide unequivocal answers about the whole spectrum of relevant genetic damage ranging from recessive lethal (or visible) mutations, deletions, translocations, chromosome loss, dominant lethals to non-disjunction and recombination. In this respect, Drosophila compares favourably with most other assay systems, where only one class of genetic changes can be detected. When Drosophila is used in screening programmes employing a battery of different test systems, it will be too cumbersome to assay for all genetic changes that can be detected. Instead, a proper choice between the various techniques will have to be made. A comparison of the detection capacity of the various assays indicative for different kinds of genetic damage, such as chromosome breaks, chromosome rearrangements and mutations, at different concentration levels by Vogel and co-workers [38,85,87] revealed that various substances, like DEN, are highly effective in inducing mutations, but do not produce chromosome breakage effects at all; or if so, as with MMS and TEB, only at considerably higher concentrations than those required for mutation induction. Of the different test systems available, the classical sex-linked lethal test therefore deserves priority in view of its superior power to discriminate between mutagenic and non-mutagenic substances. This test and the possibility of combining the recessive lethal test with assays for chromosome loss and translocations will be described in detail. Testing for recessive lethals is also important in the light of considering what particular kind of genetic damage might be most relevant to man. Auerbach [10] recently concluded that small deletions, which are known to occur at high frequency after exposure to chemical mutagens, presumably constitute the greatest risk, possibly more so than missense and nonsense mutations. Since most deletions will act as recessive lethals, assays designed to detect recessive lethals have the additional advantage that they will provide information on the induction of this important class of genetic damage. The simplest and most commonly employed assay systems test for the induction of recessive lethals in the X-chromosome, which represents one-fifth, and thus a sizeable section, of the total Drosophila genome.

In addition, tests for the detection of dominant lethals, chromosome loss and translocation will be described. Dominant lethals, although constituting a less precise estimate (see ref. 88 for details) have the advantage of quick ascertainment of whether any chromosome breakage effects have been induced. Recessive lethal tests may be supplemented by tests recovering induced losses of the X- or Y-chromosomes or translocations between the second and third chromosomes. Translocations provide information on both breakage of chromosomes and misrepair of chromosome breaks. The possibility of storing sperm is another advantage Drosophila has to offer. Storing of treated sperm in inseminated females greatly increases the frequency of translocations in cases of delayed effects which are characteristic for many chemical mutagens [9; for reviews see refs. 70,72]. The recovery of translocations entirely depends on genetic methods, that is on the all-or-non-recovery of clearly marked classes of Mendelian segregation. When Drosophila is employed in this way, it may provide guidance on the kinds of questions to which answers should be sought in more laborious mammalian systems. For example, observations in Drosophila that a particular effect can be modified or exhibits pronounced stage specificity, can then be followed up in mammalian test systems.

Indirect mutagens and carcinogens require activation by the microsomal enzyme systems present in the mammalian liver. In microbial test systems, mutagens of this kind register as negative, unless host-mediated assays or plating on microsomal extracts from mammalian tissues are employed. Extensive studies on the metabolism of insecticides, performed during the last 15 years, have revealed that insect microsomes are capable of facilitating similar enzymic reactions to those from the mammalian liver (see the review of Wilkinson and Brattsten [91]). The endoplasmic reticulum is considered the center for drug and pesticide metabolism and the enzymes involved are mixed-function oxidases. Unlike the mammalian liver, insects do not have any specific organ in which those enzymes are predominantly located [91]. Biochemical characterization of microsomal enzymes in insects has demonstrated that a large number of foreign compounds are oxidatively metabolized by isolated microsomes or other subcellular fractions. The nature and diversity of the reactions catalyzed indicate that insect microsomes exhibit a similar degree of metabolic versatility and substrate-non-specificity as those from the mammalian liver [15]. Most of the more direct biochemical information is derived from studies with the housefly, *Musca domestica.* The evidence that Drosophila is capable of facilitating similar reactions is based on mutation studies, carried out mainly by Vogel during the last few years (for details see ref. 88). Some 50 compounds, falling into 9 different groups, that all require metabolic activation for the manifestation of their mutagenic and carcinogenic properties have now been tested in Drosophila and yielded positive response for the induction of recessive lethals. The genetic activity of these widely different substances convincingly demonstrates the versatility and lack of substrate-specificity of the Drosophila enzyme systems.

An additional advantage Drosophila has to offer is the facility with which mutagenic activity can be tested in a wide variety of different germ cell stages. This is of importance because one of the characteristic features of chemical mutagens is their specificity of action. In cases of very pronounced stage-specificity, testing of only one germ cell stage, e.g. only mature sperm, can lead to false negative results. In Drosophila sensitivity differences between germ cell stages can be ascertained by mating treated males to fresh females in a succession of different broods, or by setting up successive subcultures with treated females.

The great potential of this brooding technique for the understanding of fundamental questions is illustrated by the studies on the mode of action of indirect mutagens by Vogel [84]. They show that, despite considerable differences between these various chemicals, the substances tested exhibit a similar stage-specific sensitivity pattern; that is, spermatids and spermatocytes appear to show greatest response. From studies with the electron microscope by Tates [79], it is known that in these stages the endoplasmic reticulum is most highly developed. Since this system is regarded as the site of metabolic activation, these observations provide an indication for intracellular activation within these particular stages of sperm development. Thus, even in situations where short-lived activation products are involved, with mutagenic effects restricted to the cells in which activation has occurred, Drosophila will still permit their detection. Such effects may well be missed in host-mediated or in vitro microsome-faciliated assay systems.

As a final point we have to mention the efforts presently undertaken in a number of laboratories to study the genetic control of enzymatic repair systems in Drosophila. In addition to the in vivo studies of repair using whole animals [43,54,57,74,94] some biochemical approaches became possible [12, 80,82]. It is promising that already several mutagen-sensitive stocks are known

[24,26,52,67,68,83]. As with microbial systems, mutants present in such stocks may be used to enhance the sensitivity and by this the detection capacity of the Drosophila test systems (Vogel and co-workers in prep.). As soon as the genetic control of the different repair pathways is understood, repair mutants may be used for the characterization of mutagens and their mode of action in Drosophila.

Strategy of testing

Before starting an experiment a number of decisions have to be made. If there is already some information available concerning the mode of the possible mutagenic activity of a compound (activity expected in resting cells or only in connection with DNA synthesis) this information has to be taken into account already in the planning stage. Based on experience it is reasonable to start with testing the possible effects of the compound on spermatogenesis.

In all those cases where it is to be expected that the compound will not be degraded to a large extent in the digestive tract, one will start by feeding the compound to adult males. If there are strong indications that the digestive tract has to be excluded it is possible, but quite laborious, to inject small amounts of the substance into adult males. The choice of the route of application might be a crucial point in the case of indirect mutagens and directly acting chemicals of short half-life, e.g. β-propiolactone.

The next problem to be considered is the concentration of the compound to be used. One might intend to start using the highest possible concentration which permits the testing for transmissible genetic effects. Toxicity and sterility effects may then be the limiting factors. Solubility problems may be overcome by using DMSO, alcohol, acetic acid, etc.

After the route of application has been chosen, pilot experiments have to give information on the toxicity and possible sterilizing effects of the compound in adult males. The procedures for the toxicity test and the sterility test are described below. If partial sterility is observed an appropriate dominant lethal test might indicate whether genetic or non-genetic damage is responsible for the induced sterility.

The main experiment will be a test for sex-linked recessive lethals. The most economic procedure is a brood pattern analysis which includes mature sperm, spermatids, spermatocytes and possibly spermatogonia. In some special cases the conventional test may be supplemented with a test for lethals manifested in F_3. This can reveal the induction of mosaics resulting from delayed effects.

If the recessive lethal test turns out to be positive it will be interesting to test specifically for phenomena connected with chromosome breakage. Loss of sex chromosomes can be recorded in a short one-generation test. Heritable, reciprocal translocations need two test generations. But this test has several advantages: (1) in order to get a recoverable translocation, breakage and reunion of chromosomes must have taken place; (2) the rate of spontaneous translocations is extremely low; (3) the rate of induced translocations can be increased by storing the treated sperm in females.

The test organism

Life cycle

The fruit fly (*Drosophila melanogaster*) undergoes complete metamorphosis. The succession and duration of the different stages in the life cycle are: embryonic development, 1 day; first larval instar, 1 day; second larval instar, 1 day; third larval instar, 2 days; prepupa, 4 hours; pupa, 4.5 days. Therefore the duration of the whole cycle under standard culture conditions (in a room or incubator at 25°C and about 60% relative humidity) is 9—10 days.

Culture medium

There exist numerous recipes for culture media on which *Drosophila melanogaster* flourishes. The preparation of the medium used at Northwestern University is described by King [36] as follows: 13 liter water, 710 g yellow corn meal, 710 g brewers yeast, 110 g agar, 38 g methyl-*p*-hydroxybenzoate, 80 ml propionic acid. The last two components serve as fungicides. Mold growth will rapidly destroy cultures lacking fungicides. Methyl-*p*-hydroxybenzoate is marketed by Teneco Chemicals, Inc., 290 River Drive, Garfield, New Jersey 07026 under the trade name Methyl Parasept. The powdered components are added to the water, all lumps are broken up, and then the mixture is brought to a boil and cooked for 15 min. The propionic acid is stirred into the medium just prior to dispensing it into culture bottles.

At the Swiss Federal Institute of Technology in Zürich we use the following recipe (the amount of the different components is given in ml because this is the way it is used for daily work in the kitchen): 3 liter water, 54 ml agar, 255 ml sugar, 660 ml corn meal, 150—200 g live yeast, a point of a knife (about 1—2 g) Nipasol-M-sodium (a trade name for propyl-*p*-hydroxybenzoate; Nipa Laboratories Ltd., Treforest Industrial Estate, South Wales, Great Britain) as a fungicide and 10 ml of a 0.1% aqueous solution of streptomycin.

A list of sources of equipment and supplies used in setting up a kitchen and a laboratory for Drosophila studies was provided by Merriam [44].

Glassware

Bottles of 200 ml volume are used for stock cultures. For smaller cultures, e.g. pair matings, vials of about 40 ml are used. These are glass tubes, about 8 cm high and with a diameter of about 3 cm. Glass is preferable because scoring of normal cultures (see below) is then possible without taking out the flies.

The cultures are stoppered with plastic foam plugs or with plugs made of tightly clumped cotton-wool.

With the bottles, one uses about 40 ml medium and with the vials about 10 ml medium per culture.

Laboratory techniques

Stock cultures

Newly hatched flies are transferred into culture bottles supplemented with live yeast on the surface of the medium. For starting new cultures one needs actively egg-lying, well fed, fully mature flies. Therefore the newly hatched flies are kept up to 4 days in these "feeding" bottles. Then new stock cultures are started. To avoid overcrowding in the new bottles only 20—50 flies (about half females and half males) are brought into every new bottle. After one or two days the parental flies are shaken into fresh bottles or discarded by throwing them into a glass containing 70% alcohol. Depending on the culture conditions in the laboratory it might be useful to add some live yeast and a piece of filter paper (regulation of humidity within the bottle) to the bottles from which the flies were removed and which contain eggs and larvae of the next generation.

For simply maintaining the stocks, the cultures can be raised at 18°C. At this low temperature, which is tolerated by most stocks, the generation time is increased to nearly 4 weeks. In general, two to three bottles are sufficient for maintaining a stock.

For enlarging a stock in the preparation of an experiment the cultures are raised at the standard temperature of 25°C.

Stock cultures can be sent from one laboratory to another by ordinary mail or for longer distances by air mail. Cultures in small vials or small plastic tubes have to be used. Before sending away the cultures have to be kept in the laboratory for some 3—4 days. This is to ensure that in addition to the adult flies some eggs and larvae are present. These young developmental stages may arrive in better conditions than adults which accidently might stick to the medium. In general Drosophila can be mailed all over the year, but problems may arise during extremely cold winters.

Handling of adult flies

Adult flies can be transferred from one bottle to another just by putting the empty bottle upside-down on the bottle containing the flies (with stoppers removed), then turning around and shaking the flies down into the new bottle. As long as the lower bottle is slightly shaken, the flies will not move upwards and the bottle can easily be stoppered with the foam rubber stopper.

For a close inspection of flies under a binocular microscope (see Fig. 1) the flies can easily be immobilized by narcosis using either ether or carbon dioxide.

The principle of etherization is to keep the flies for a short time in ether fumes prepared within a bottle. An etherizer may be constructed by using a culture bottle, e.g. a half-pint milk bottle, a small plastic funnel (obtainable in many hardware stores, a cork (not rubber, because rubber is destroyed by ether fumes) and a plastic tube (in earlier times gelatin veterinary capsules were used, but these are difficult to obtain). With a heated needle about a dozen holes are pierced into the closed end of the plastic tube. The holes should be large enough to admit ether fumes but not large enough for the flies

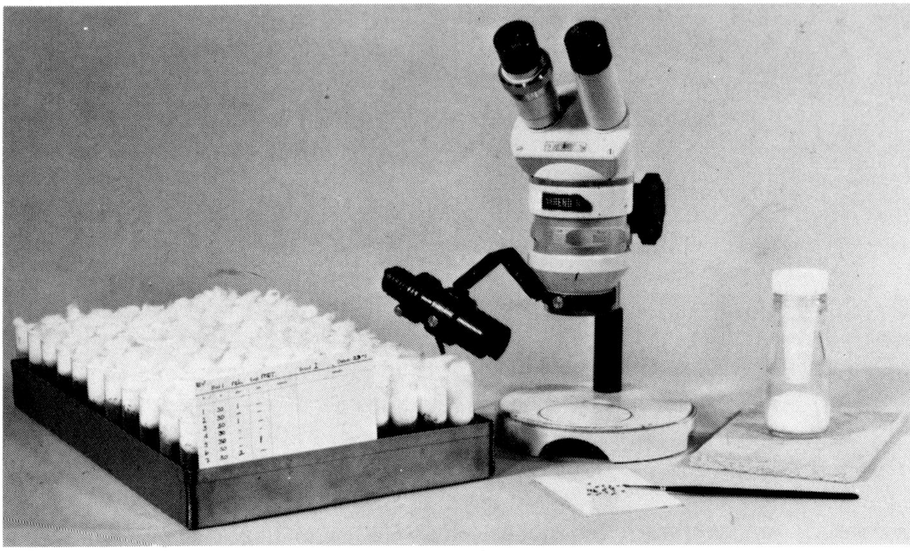

Fig. 1. Basic equipment for experiments with Drosophila.

to fall through. The closer end of the funnel is cut down until the funnel fits on the plastic tube. The plastic tube with the holes at the lower end is fitted into a hole bored in the cork, the funnel is fitted on top and all three parts are glued together. Cotton is put into the bottom part of the bottle and a little ether is added to the cotton. Putting the cork on the bottle the plastic tube will now be located in the space which is immediately filled with ether fumes. Additional descriptions of etherizers are found in Demerec and Kaufmann [21] and Seecof et al. [58]. Flies are shaken into the funnel and will fall down to the tube. It is easy to determine when the flies are anesthetized just by looking into the tube and then uncork the etherizer and shake the flies out. Now the flies can be inspected under a binocular microscope. If the flies should begin to wake up before examination is completed, they can be re-etherized without being returned to the etherizing bottle. The re-etherizer usually consists of a flat dish (section of a small Petri dish) with a strip of absorbent paper or foam rubber fastened to the inside. A few drops of ether are placed on the paper, and the flies are covered with the re-etherizer until they are immobilized again. Care should be taken not to overetherize flies. In flies killed by overetherizing, the wings are extended at a 45° angle or are turned upwards.

A device for easily anesthetizing large numbers of flies with carbon dioxide was described by Würgler et al. [97].

Fig. 2 shows a scheme of the device. A bunsen burner equipped for a pilot flame (1) is connected by plastic tubing to a CO_2-cylinder (2). The air inlet (3) of the burner is sealed with Araldit. If the valve of the burner (4) is in position "flame", the carbon dioxide flows through the large tube of the burner (5) and enters a plastic funnel (6) fixed to the top of the burner. A foam rubber stopper (7), through which the gas can penetrate, is placed in the neck of the funnel. Bottles or vials (8) containing flies are put upside down

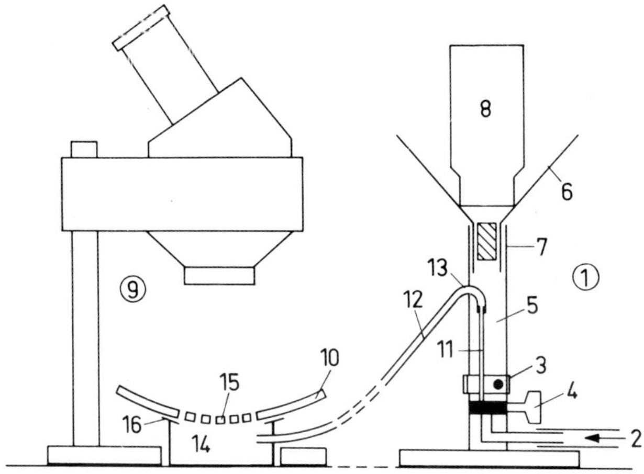

Fig. 2. A device for easily anesthetizing large numbers of flies.

into the funnel. Upon gentle shaking the flies fall into the funnel and are immediately immobilized by the CO_2 and accumulate on the foam rubber stopper. For inspection of the flies under a microscope (9) they are — by removing the funnel — transferred to a concave plastic dish (10). In order to keep the flies continuously in a CO_2-atmosphere, the valve (4) is switched to position "pilot". Now the gas stream passes through the thin tube (11) of the burner. On top of the shortened thin tube a plastic tubing (12) is fixed. This plastic tube passes through a hole (13) in the wall of the large tube (5) and is connected to a cylindrical box (14). From this box the gas stream reaches the flies through a large number of very small holes (15) in the plastic dish. This dish is made by cutting out a circular piece from a conventional plastic bowl. The edge of the box (14) is covered with a strip of rubber (16) to avoid electrostatic loading of the plastic dish resulting from its movements. The use of a heat-shielded microscope lamp is recommended. With this device large numbers of flies can be inspected over a very long period without the interruption of work for reanesthetizing and without danger of killing the animals.

In general the carbon dioxide method might be preferred, because there is practically no possibility to kill the flies by a prolonged treatment. Ether treatment can have adverse effects on the flies such as clumping of mature sperm in males as observed by B. Leigh (personal communication).

Collection of virgins

At the beginning of most experiments virgin females have to be used. In order to collect virgin females, the culture bottles containing dark brown pupae ready to hatch (in the morning of the 9th or 10th day after starting the culture at 25°C) are freed from all adult flies. Depending on the speed of the sexual maturation of the males in the stock, all flies that have hatched within the next 6—12 h are shaken out of the bottles. If collection of virgins has to continue, the bottles have to be carefully cleaned from every living fly. The flies obtained from the culture bottles are anesthetized and the females separated from the males under a low magnification microscope using a needle or a small dry brush. The following criteria can be used to identify the females: absence of the sex comb on the first pair of legs, the pattern of coloration of the abdomen, the larger wings, the shape of the abdomen which terminates in a point or the structure of the external genitalia. For the identification of newly hatched females the absence of the sex combs and the structure of the external genitalia are the best criteria.

Groups of 25 females are then brought into individual culture vials containing standard medium with live yeast. Sticking of the flies to the medium can be avoided by placing the females on the glass wall and keeping the vials, after stoppering, in horizontal or upside-down position. Restricting the number of flies to no more than 25 avoids overcrowding and thus provides good feeding conditions, if the females are shaken over at least every 3 to 5 days to fresh culture medium. For the development of the oocytes, the substances contained in live yeast are a first requirement. Consequently, to guarantee sufficient egg-laying, it is recommended to add some extra yeast to cultures to be employed for further crosses. Since this may occasionally result in stickiness of the medium, females should be transferred to freshly-prepared vials every three days. Optimal fertility and most reproducible results are obtained if females 3—5 days after hatching are used for experimental crosses. Limiting the number of females has an additional advantage. If inadvertently non-virgin females have been included, such vials, which can be recognized by inspection for hatched eggs and young larvae after 2—3 days, can be discarded.

Collection of males

For the collection of males no such rigorous time scale has to be observed. On the other hand, the age of the males used for treatment with a chemical should not vary too much. A simple method is to shake out the flies from the culture bottles once a day, transfering them to new bottles and to separate the males from the females the day before the treatment is planned. Especially for the sex-linked recessive lethal test this procedure is recommended because by mating with the females in the same bottles the first sperm batches — which show an increased rate of spontaneous lethals (for review see ref. 25) — will not be included in an experiment. For reasons of standardization it is recommended to use 3—4-day-old males.

Treatment of adult flies

For general screening purposes, it is recommended to use only males for chemical treatment, since females are more easily sterilized by chemicals and have, so far, proved more refractory to the induction of genetic changes (Clark and Sobels, unpublished). Moreover, so far no chemicals have been found with effects specifically restricted to female germ cell stages.

A number of methods to treat adult flies with chemicals have been developed: feeding [23,40,42,86], injection [46], inhalation (F. Verburgt, unpublished), aerosol treatment [59,89], vaginal douches [27] and topical application.

Fig. 3. Set up for feeding adult flies.

From a technical point of view the simplest method is the feeding of solutions. In order to ensure immediate uptake in any feeding procedure the flies to be fed should be kept in empty bottles for several hours before feeding begins. This is particularly important in case of handling unstable chemicals. The method published by Vogel and Lüers [86] is a modification of the technique described by Lüers [42]. This simple but effective procedure (Fig. 3) consists of a 30 ml glass filter (1D3 of Schott, Mainz, G.F.R.) being large enough to treat about 30 flies. The filter is placed into a 100 ml beaker which contains 15—20 ml test solution. (For smaller amounts, the 15 ml filter 10D3 or the 8 ml filter 63D3 might be used.) The test solution should nearly contact the filter plate's lower part. A wick of cellulose wadding is used to ensure contact between the test material and the plate. A cotton ring around the filter's upper part prevents water evaporation. Filters contaminated with a mutagen are decontaminated and cleaned successively by hot water containing a detergent, distilled water, 70% ethanol, ethanol and acetone. This procedure is particularly essential to clean the filters from insoluble ingredients.

Depending on the toxicity of the chemical and its ability to induce dominant lethals exposure times between a few hours up to several days can be used. With unstable chemicals (such as MMS, see ref. 87) the test solution may have to be replaced, e.g. after 24 h of feeding.

The method described by Lewis and Bacher [40] is excellent and very convenient for mutagenizing flies with EMS. For routine screening of unknown chemicals feeding on paper may result in separation of solvent and substances. Flies might well feed selectively in different regions on the paper and in this way avoid even slightly repellent substances. We therefore do not recommend this feeding procedure for routine testing. The same holds for the method proposed by Félix [23].

A simple but efficient inhalation procedure was used by Verburgt (unpublished) for exposing flies to gaseous vinyl chloride for up to 17 days. Flies were treated in 1 liter blood plasma bottles which had been sealed air-tight with rubber caps. The bottles contained a few grams of sterilized food, killed yeast, and a known quantity of vinyl chloride gas. None of the methods listed here, however, is sufficient to quantify the amount of active material at the genetic target cells (gonads), unless exact dosimetry by labeled compounds is done, for example, according to the procedure developed by Sega et al. [60].

Handling of test cultures

For test-crosses, cultures have to be started with individual females. In this case one uses vials. As with bottles it is convenient to use foam rubber stoppers. Provided that the surface of the medium and the live yeast are fairly dry, the anesthetized flies are picked up with a sharp forceps (at a wing or a leg), with one finger the foam rubber stopper is pushed to one side and the fly thrown into the vial. With a little bit of experience this simple procedure is also possible in the case that already one or a few non-anesthetized flies are in the vial. Therefore females and males have not to be transferred to the vial at the same time. If thousands of tests have to be set up such a time-saving procedure is important.

In order to avoid admixture of the parents used to start the culture with their progeny, it is recommended to eliminate the paternal flies about 5—6 days after starting the culture. Depending on the experience in a given laboratory it might be desirable to add at the same time a small piece of filter paper to regulate humidity in the vial and/or to add some additional live yeast.

Brooding

Since chemicals often exhibit stage-specificity, i.e. exert more or less pronounced mutagenic effects in different stages of germ-cell development, it is recommended to examine at least a number of different germ-cell stages for induced genetic effects. To this end, the brooding technique is applied which, in fact, consists of translating the spatial pattern of spermatogenesis into a temporal pattern of successive broods. Treated males are therefore mated at regular intervals (2—3 days) to a succession of, for example, 3 fresh virgin females. With 1500—2000 R X-rays, 2-day brood intervals with 5 females per male, and 1—2-day-old males at the start, the succeeding broods represent cells treated on different stages of spermatogenesis: mature sperm in the first brood (A), late spermatids in the second brood (B), early spermatids with maximum radiosensitivity in the third brood (C), meiotic stages, characterized by high sterility, in the fourth brood (D), and spermatogonia in brood E. This last brood is sampled 8—10 days after exposure of the males to irradiation. A rough verification of these stages can be obtained by simultaneous tests for induced non-disjunction and the appearance of single cross-overs (indicative for spermatocytes) and bunches of identical or complementary cross-overs and germinal selection, indicating that proliferating spermatogonia have been sampled (for details see ref. 8), or by means of applying radioactive labeling [16,53]. The sampling of the different stages in the successive broods is different for different stocks, and depends on the ability of the males to mate, the speed of sperm release and on the dose of X-rays or the concentration of the chemical applied.

Irrespective of the fact whether stages with maximum sensitivity can be identified, the application of the brooding procedure at least safeguards against the possibility that chemicals with more pronounced effects in earlier stages of spermatogenesis are not erroneously dismissed as false negatives. A few examples may serve to illustrate stage-specificity. All indirect carcinogens, requiring metabolic activation, that were tested with the brooding procedure exhibited peak activities in broods corresponding to early spermatids and late spermatocytes [84,88], EMS at low concentration exhibits greatest activity in early spermatids (Sobels, unpublished), hycanthone peaks in the second brood, corresponding to late spermatids [37], chloroethylmethanesulfonate only affects spermatogonia [7,27] and formaldehyde and dehydroxydimethylperoxide, when injected into adult males, produces mutations in mature sperm and spermatogonia [69]. When larvae are treated, the first brood already represents earlier stages of spermatogenesis, depending on the age of the larvae at the time of treatment. With larval feeding the pronounced mutagenic activity of formaldehyde, which is restricted to a growth stage of the primary spermatocytes, has been discovered by Auerbach.

In the case of females being used for treatment, successive stages of egg cell development can be obtained by simply transfering the treated females after fertilization to successive subcultures. However, for reasons outlined above, we do not particularly propose using females for chemical treatment in routine assay procedures.

Scoring of progeny

If Drosophila is used for routine testing, the time needed for scoring the progeny in the test vials can present a bottle neck. Some of the tests to be described below only need the observation of the presence or absence of a certain easily detectable phenotype. This is true for the following tests: in the Basc test for sex-linked recessive lethals the presence of males with round red eyes indicates a non-mutated chromosome and with the yB-test it is the presence of Bar males; with the translocation test a culture which contains light-orange eyed flies (males or females) indicates normal non-translocated chromosomes. In these three cases the progeny in a vial with more than about 20 flies can easily be inspected with the naked eye. If at least two animals of the critical phenotype are found, the vial can safely be classified as normal. This is best done when the vial is illuminated from the side with a table lamp.

Vials with few progeny or vials which seem to contain none or only one fly of the critical phenotype should be inspected very carefully later on, by anesthetizing the flies and inspecting under a low-power dissecting microscope. Often the reinspection of the critical vials can be combined with setting up a new set of vials needed for verification of a mutation or the retest of non-decisive cultures.

Egg collection and registration of dominant lethals

The egg collection method described by Sankaranarayanan [56] (Fig. 4) may be used to study the rate of dominant lethals expressed during embryonic development. Each egg-collection unit is hexagonal in shape and is made from a set of 19 cylindrical plastic subunits (8 cm × 2 cm) glued together and open at both ends. At one end of the unit is a nylon netting held taut by cementing the netting to the wall of the individual cylinders. One inseminated female can then be introduced into each tube which is sealed with a cotton plug. The unit is then placed gauze side downwards in a large petri dish (15 cm in diameter) containing about 0.5 cm thickness of corn-meal sugar-agar medium darkened with powdered charcoal and smeared with a fairly thick yeast suspension (Fig. 4). The females can easily oviposit through the gauze and can also feed readily. Each subunit is numbered (1—19) as is its position on the petri dish (the latter ba marking an imprint of the rims of the whole unit on the medium using a spare egg-collecting unit, and labeling them). It is thus possible to keep an accurate record of the eggs laid by the individual females. Approximately every 12 h the units are removed to nutrient-containing petri dishes and the eggs are counted. The unhatched eggs are counted again after another 24—28 h of incubation. It is possible to score eggs over up to 10 successive 12-h egg-laying periods. Eggs laid by unfertilized females can be excluded from the calculations. The egg-collecting units and the petri dishes with the eggs are kept in fly-proof rectangular plywood boxes provided with a metal

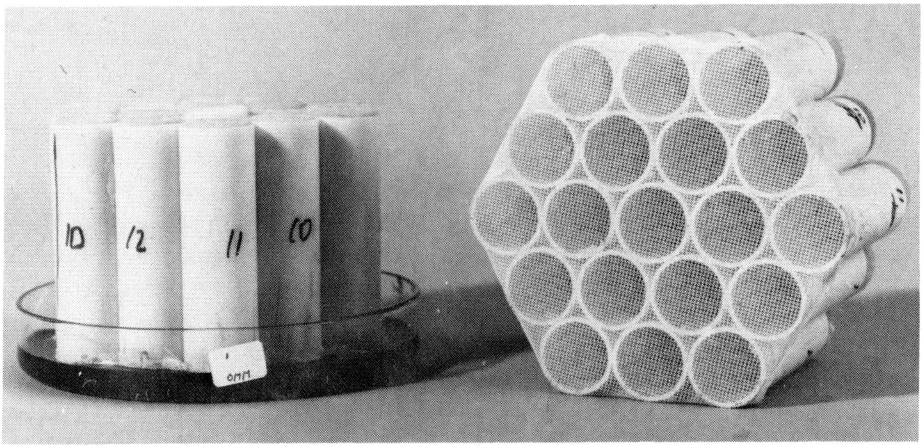

Fig. 4. Egg-collection unit developed by Sankaranarayanan [56] for dominant lethal assays.

Fig. 5. Egg-collection unit developed by Bürki [14].

gauze at the top, to allow for proper ventilation, in an incubator maintained at 25°C.

If one intends to determine dominant lethals expressed in embryonic as well as in postembryonic stages one has to collect eggs and incubate them in such a way that the hatching larvae are picked up and can be raised to the adult stage. In this case the egg collection method described by Büchi and Bürki [14] (Fig. 5) may be used. Groups of about 150 flies (males:females about 2:1) are transferred into egg collection units. Each collection unit consists of a plastic beaker (5 cm in diameter, 8 cm high) with a nylon gauze covering the opening. A fine gauze through which the eggs will not pass is used. A small opening at the bottom of the beaker allows easy removal of the flies and circulation of air (see Fig. 5). The beaker is placed with the nylon gauze downwards on a small petri dish filled with fermenting egg laying medium (see below). The liquid of the medium is immediately soaked up by the gauze. Therefore the females deposit eggs only on the gauze and not on the wall of the dry beaker. At the end of the egg laying period the gauze can easily be removed and the eggs counted.

The egg laying medium is prepared by boiling a mixture of 500 ml water, 5 ml acetic acid, 100 ml malt extract, 85 g sugar and 135 g corn meal. Upon cooling of this medium 20 ml of a concentrated suspension of live yeast in water are added. After standing at room temperature for one day the fermenting medium is ready to be used [99].

The eggs deposited on the gauze are transferred on to strips of agar (2% agar; 5 mm thick) lying on small glass plates (7 cm long, 1 cm wide and 1 mm thick, e.g. halves of microslides). A fine brush is used for handling the eggs. The number of eggs on each agar strip is counted under a dissecting microscope by placing them in groups of 10 beginning about 1.5 cm from the tip of the agar strip. The glass plate with the agar strip and the eggs is then brought into a vial or a special tube such as described by Würgler et al. [99]. The vials with the eggs are kept in a room or an incubator with a temperature of 25°C and 60% relative humidity. After about 21 h the surviving larvae hatch. Attracted by the yeast and following a humidity gradient they crawl into the medium. At 48 h after egg collection practically all surviving larvae have reached the medium. Now the glass plate with the agar strip is removed from the culture vial, and surviving larvae found on the agar are brought back into the vial. The dead embryos and the empty egg shells lying on the agar strip are gently dried in a stream of heated air and covered with paraffin oil. Since the egg shells are now transparent the dead embryos can be classified under a dissecting microscope and counted according to early embryonic death (white mottled eggs) or late embryonic lethality (embryos with a brownish colour and/or larval structures present). For further details on different lethal syndromes see Würgler and Ulrich [95]. The vials with the surviving larvae are brought into a culture room at 25°C and a relatively high humidity. When the flies hatch the surviving adults are counted (if needed separately for different phenotypes).

Literature on the biology and genetics of Drosophila

The book "Biology of Drosophila" edited by M. Demerec [20] is still a very useful source, especially concerning the morphology and histology of Drosophila.

The "Genetics and Biology of Drosophila" has a much wider scope. The first volume, edited by M. Ashburner and E. Novitsky [4] was published in 1976 and two further volumes are in preparation. Introductory texts with further details on culturing methods are: Demerec and Kaufmann [21] *, Strickberger [78], Wheeler [90] and Shorrocks [61].

All the information on genetic variations (gene mutations and chromosome rearrangements) available until 1967 is compiled in the book of Lindsley and Grell [41]. More recent information is found in recent issues of Drosophila Information Service (DIS). DIS, published annually, is edited by E. Novitsky (Department of Biology, University of Oregon, Eugene, Oregon). It includes research notes, technical notes, stock lists, recent literature and from time to time a geographical directory. The literature published up to 1972 has been compiled in six bibliographies (see refs. 29—33 and 48).

Predominantly technical aspects of mutagenicity studies in Drosophila are presented in the following publications: Muller and Oster [51], Auerbach [8], Abrahamson and Lewis [2], Crow and Abrahamson [18], Zimmering [101] and Mollet and Würgler [45].

Actual results and general problems of mutagenicity testing using Drosophila are discussed in Sobels [70,72], Vogel [84], Sobels and Vogel [75,76] and Vogel and Sobels [88]. For the technical nomenclature and terminology special to the field of genetics the reader may consult "The Glossary of Genetics and Cytogenetics" [55].

The toxicity test

The first pilot experiment should give an approximation of the possible toxicity of the test compound. For this purpose groups of males are fed with different concentrations of the test compound. After an overnight feeding period the males are removed from the feeding arrangement and placed into vials with standard medium and live yeast. Twenty-four hours later we count the number of flies still alive and compare this with controls fed with identical solutions

---

* The introduction by Demerec and Kaufmann contains a lot of illustrations on how to handle flies.

without the test compound. Based on our experience with fungicides and related compounds, such a short-time experiment provides sufficient information on toxicity for further experiments to be adequately planned. In general, relatively high concentrations of the chemical should be included in this quick pilot experiment. If solubility problems turn up DMSO, acetic acid or alcohol might be used for first dissolving the substance. An additional possibility is to use suspensions. In certain cases, for example if insecticides have to be tested, it is necessary to determine the highest concentration tolerated by the flies.

The sterility test

By the time a rough impression about the toxicity of a compound has been obtained, a somewhat more time-consuming test should be carried out to provide information on the potential sterilizing action of a compound. This test should already include a brood-pattern analysis, since this will help to organize later genetic tests in the most economical way.

The easiest way to perform such a test is the following: groups of males are fed with reasonably selected concentrations of the compound during an overnight feeding period. The males of a group are brought together with about twice the number of virgin females (of any genotype) into one culture bottle.

After one or two days, depending on the procedure adopted, the flies from the bottle are anesthetized and the males separated from the females. The females are brought back to the bottle for another few days (4—5 days) and then discarded. The males are given a new set of virgin females and placed in a new bottle. This procedure is continued until 10—14 days after the end of the feeding period of the males. The possibility of reduced fertility or complete sterility induced by certain concentrations of the compound or by lower concentrations in certain broods can now be simply analyzed by looking at the number of progeny in the different bottles. Depending on the situation encountered it might be sufficient to note the result of the experiment in a qualitative way only. For a quantitative estimate the number of progeny per bottle are counted. These figures are then compared with those obtained in a control run with males that did not receive the test compound. If some reduction of fertility is observed it might be interesting to continue with a dominant lethal test in order to check whether this results from genetic or toxic effects of the test compound.

The dominant lethal test

To start with, we have to stress that the term "dominant lethal" is generally used as an operational term. The problem is that true dominant lethals (that are genetic alterations which, in heterozygous condition, do not allow a zygote to develop into an adult fly) can be mimicked by other phenomena such as unfertilized eggs, etc.

In the dominant lethal test the eggs deposited by the females are collected and tested for developmental potential. As an endpoint for "surviving" progeny one can choose either the hatching of the larvae from the egg or the hatching of the adult fly from the puparium.

For the first type of experiments, in which only embryonic lethality is measured, the technique of Sankaranarayanan [56] can be used successfully. It is recommended that the scoring of dead "eggs" is done in such a way that "white" lethal eggs and "brown" lethal embryos are counted separately. This is important because situations will be encountered in which only the amount of white lethal eggs is increased, but the brown lethal eggs remain at the control level. The class of white lethal eggs, identified under a dissecting microscope without cytological analysis represents embryos killed during the first few hours of embryonic development (cleavage divisions, gastrulation) as well as unfertilized eggs and eggs inseminated by non-functional sperm. An increase of only white lethal eggs strongly suggests that the substance under test does not induce true lethals in the chromosomes but interferes either with fertilization, gametogenesis or physiological functions of the gametes. On the other hand, an increase of the amount of brown lethal embryos represents the induction of damage leading to late death (comparable to the post-implantation lethality in mammalian dominant lethal tests). In most cases this damage results from chromosomal alterations which do not interfere with the cleavage mitoses. An increase of white lethal eggs *and* brown lethal embryos strongly suggests that the test compound does damage the chromosomes. In insects non-genetic damage in the cortical cytoplasm of the egg or early embryo may also lead to late lethality, to brown lethal embryos and in some cases to larval lethality [3,95]. Up to now these phenomena have been observed after X- and UV-irradiation; with chemicals this problem has not yet been studied in detail.

If dominant lethality is observed after treatment with unknown chemicals, the question always exists whether genetic or non-genetic damage contributes mostly to this lethality. In order to answer this question directly a special type of experiment can be carried out. The basic idea is that one expects the amount of genetic damage induced in developing and mature gametes to be related to (e.g. proportional to) the amount of genetic material treated within the cell.

If flies are treated which produce at least two types of gametes differing in the amount of chromosomal material included in the nuclei, the amount of true dominant lethals should be higher in the type with more chromosome material treated. In Drosophila a number of chromosome rearrangements are available which can be used for such purposes. Especially suitable are some of the rearrangements between the sex chromosomes. For a test with male germ cells we treat XY/O males and mate these to wild-type females. The XY/X females represent the progeny resulting from XY-bearing sperm and the X/O males the progeny from null-X null-Y bearing sperm. If true dominant lethals are induced, less females than males will be found in the progeny. For a test with female germ cells XX/O females are treated and mated to wild-type males. In the XX/Y females (from XX-eggs) more lethality is expected than in the X/O males (from null-X eggs). A number of successful tests were run with these systems using X-rays [39,92,93,98,100], Trenimon, PDMT and sodium monofluorophosphate [13].

Since it might be argued that sex chromosomes (especially the Y chromosome) are a non-representative part of the genome, a test system based on autosomal rearrangements was developed. Experiments with Trenimon indicate that the same conclusions are reached as with sex-chromosome studies [13].

Of particular importance in these experiments is that the registration of the pre-imaginal lethality is in fact noted. Since genetic markers are used which have to be identified (or are only easily identifiable) in the adult stage, the hatching of the surviving flies from their puparium has to be chosen as an end-point. Therefore, the egg collection method of Sankaranarayanan is less suitable than the techniques developed by Würgler and co-workers. The technically simplest test to study the correlation between lethality and amount of chromosome material treated is the following.

(1) Feeding of XY/O males with the test compound.

(2) Crossing these males to X/X females and collection of eggs according to Büchi and Bürki [14].

(3) Count groups of eggs and place them on the surface of the standard medium in a vial.

(4) Raise these cultures under standard conditions and count the surviving flies after 10—12 days. Males and females are recorded separately.

(5) Calculate the lethality separately for males and females.

An example may illustrate these calculations: in the control group a number of eggs, E, were put into vials and F females and M males survived. In the treated group with e eggs tested f females and m males survived. From e eggs one would expect e · F/E females and e · M/E males. The lethality within the female progeny in the treated group can therefore be estimated as f/(e · F/E) and the lethality of the male progeny as m/(e · M/E). If the value for the females is significantly higher than the one for the males it can be assumed that the chemical treatment of the fathers induced more damage in the XY-bearing sperm (leading to female progeny) than in the null-X null-Y sperm (leading to male progeny). And this is strong evidence indicating that the chemical under test can damage the chromosomes.

The sex-linked recessive lethal test

This is a test for genetic changes which in hemizygous and homozygous, but not in heterozygous, conditions kill the developing individual somewhere between the zygote and the adult stage. Such genetic factors, called recessive lethals, can be induced on all chromosomes, but the detection of factors induced on the X-chromosome needs only two test generations whereas the autosomal factors need three generations to be detected. For this reason sex-linked recessive lethals are preferred in routine testing procedures. An additional advantage is that, if males are used for treatment, the treated X-chromosome does not contain pre-existing lethals, with the exception of those arising during the development of that particular male fly.

There is also a very important practical reason for choosing recessive lethals as the genetic change to be scored for: the criterion used to decide whether a mutation is present or not is very objective. The decision is based on the observation whether a certain Mendelian class is present or not. This is — with the genetic markers available — a criterion which is used by different workers throughout the world in the same way.

The sex-linked lethal test is very sensitive because the X-chromosome tested represents about 20% of the whole genome. The X-chromosome corresponds to about 1000 bands in salivary gland giant chromosomes. It contains, therefore, a large number of loci which can mutate to recessive lethals.

Types of damage assayed

Gene mutations, small deletions as well as certain types of chromosome aberrations, can lead to the phenotype of a recessive lethal. This statement is based on the analysis of single gene inheritance in Drosophila (examples are found in ref. 41) and cytological analysis [19,28,62—66,81]. The studies by Slizinska provide evidence that among recessive lethals and chromosome aberrations induced by chemicals in Drosophila, small deletions occur with much higher frequency than among the same genetic changes produced by X-irradiation. Since most deletions will act as recessive lethals, substances that produce no or few recessive lethals are not likely to produce many small deletions.

Furthermore, it has to be stressed that the recessive lethal test is the most sensitive test available in Drosophila at present. This is based on the fact that point mutations as well as genetic changes connected with chromosome breakage are picked up by the test. With certain mutagens positive results are obtained at concentrations which do not induce dominant lethals or chromosome aberrations [85,87].

The genetic basis of the sex-linked recessive lethal test

The test for sex-linked recessive lethals requires an understanding of the basic mechanism underlying X-linked inheritance. That is: (1) the X-chromosome of the father is passed on to the daughters, whereas sons receive the X-chromosome from the mother; (2) recessive genes, located on the X-chromosome come to expression in males in a hemizygous condition, because the Y-chromosome does not contain the dominant, wild-type alleles to suppress their manifestation.

All tests for recessive sex-linked lethals rely on the fact that among the progeny of females, carrying a recessive lethal on one of her X-chromosomes, i.e. being heterozygous for a recessive lethal, half of the sons will die. Consequently, progeny of such females will contain half as many males as females. The first test for recessive lethals by Muller and Altenburg [47], actually consisted of determining the sex ratio among progeny of females suspected of carrying a lethal on one of their X-chromosomes. By providing suitable genetic markers, the class of males carrying the X-chromosome of the treated grandfather can easily be determined. If a lethal was induced, this class will be missing and this is easily scored for. Another prerequisite for this assay system is the prevention of crossing-over in the females, heterozygous for the lethal-bearing chromosome, because transfer of the lethal from the paternal to the maternal X-chromosome by genetic recombination would restore viability of the chromosome under test and thus lead to the erroneous result that males receiving this X-chromosome survive. Since combinations of suitable inversions effectively inhibit the occurrence of crossing-over, the females used for the test carry both: an inversion covering the whole X-chromosome, sc^{S1} sc^{8}, and a smaller inversion, i.e. *S* in the *Basc* and *dl-49* in the *Inscy* chromosome.

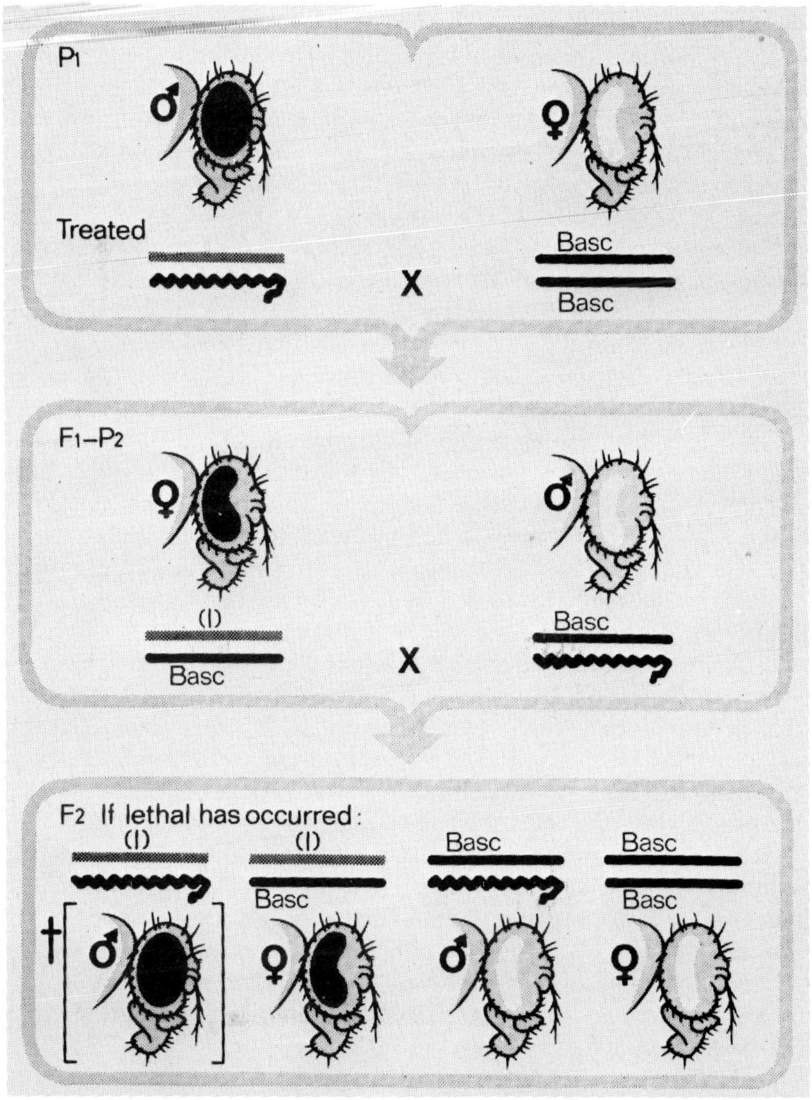

Fig. 6. The Basc test.

Two tests will be described: (1) the *Basc* or *Muller-5* test, and (2) the *yellow-Bar* test with *Oster* females. The most commonly employed *Basc* or *Muller-5* (Fig. 6) test is a variant of the classical *ClB* test, developed by Muller in the late forties for the experimental work by Stern and co-workers [77]. Wild-type males, characterized by normal, round, red eyes are used for treatment, and mated to females homozygous for the *Basc* chromosome. Care should be taken that only virgin *Basc* females are used for this cross. The *Basc* chromosome carries the genetic markers *Bar* and *white-apricot*. *Bar* produces a narrow eye shape in homozygous or hemizygous conditions and a kidney-shaped eye when heterozygous in females. *White-apricot* flies markedly differ from wild-type, in that the eye pigment is changed to light orange. The *Basc*-flies homozygous for *Bar* and *white-apricot* are thus characterized by narrow-shaped, orange eyes. In addition to these visible markers, the *Basc* chromosome contains a combination of two *scute* inversions, i.e. the left-hand part of sc^{S1} and the right-hand part of sc^{8}, covering the entire X-chromosome, and a smaller inner inversion, *In-S;* the precise genetic description is $In(1)sc^{S1L}sc^{8R} + S$, $sc^{S1}sc^{8}w^{a}B$ (for refs. see ref. 77).

The progeny of the cross treated wild-type male × *Basc* females now consists of females heterozygous for the treated X-chromosome, characterized by kidney-shaped red eyes, and males of the *Basc* phenotype having received the X-chromosome from their *Basc* mother (see Fig. 6). *Each F_1 female thus represents one paternal X-chromosome, treated in the male gametes; the understanding of this principle is absolutely essential for a proper test procedure.* It means that the vials to breed the F_2 generation each correspond to one treated male gamete, and for that reason never should contain more than one F_1-P_2 female. If this precaution is not taken erroneous results will be obtained, because of mixing of lethal-carrying with lethal-free females, and thus a lethal may be missed. The vials of the F_2 generation are then individually inspected for the presence of males with round red eyes. If this class is missing from a vial containing at least 20 flies, the conclusion can be drawn with $P < 0.05$ that the treated male gamete contained a recessive lethal. The test thus relies on the disappearance of a whole Mendelian class, i.e. half of the F_2 males, characterized by round red eyes. Since this is a criterion on which different observers cannot disagree, personal bias is reduced to a minimum in the scoring procedure and this is indeed a great advantage of all tests for recessive lethals.

A note concerning *Basc* stock cultures: if one is rearing a homozygous *Basc* stock over a long period of time, one will observe that occasionally the marker *Bar* is lost in some flies. The eye color of such flies is still *white-apricot*, but they have round eyes or heterozygous females have kidney-shaped eyes. Since the "loss" of *Bar* results from unequal crossing-over during female meiosis, the "accident" concerns not the stock as such but only the few aberrant chromosomes. Therefore the stock may be "cleaned" just by elimination all flies which do not have a narrow eye shape.

The practical procedure of the Basc (Muller-5) test

1. P_1 mating-schedule and rearing of the F_1 generation

Following treatment of 3—4-day-old males of a standard wild-type strain like *Oregon-K*, the males are mated individually in single vials to 3 virgin *Basc* females per male. The culture vials are numbered. It is convenient to keep the vials in an upside-down position until the flies have recovered from anesthesia. It is recommended to mate the males individually in single vials and to record the data in such a way that at the end of the experiment the origin of every tested chromosome can be traced back to the individual male used initially. Such a procedure has the advantage over mass-mating that bunches of mutations (clusters) derived from individual males can be identified in the F_2. Such clusters of mutations can originate from a spontaneous mutation in pre-meiotic germ cells of that particular male. They should be eliminated when the final calculations are made.

After 2—3 days, depending on the brooding procedure adopted, the paternal flies are anesthetized and males and females are separated. The females of the first mating are placed in a new vial, so as to increase the number of progeny. This vial is attached to the first vial with a rubber band and these two vials then constitute the first brood (roughly corresponding to treated mature sperm). Every parental male is given 3 fresh virgin females in order to raise the second brood. The same procedure is repeated 2—3 days later, for changing over from the second to the third brood, and this can then be repeated according to design.

As a rule two vials will produce sufficient progeny from the females. In the case of highly sterilizing treatments, the females can be shaken over to a third subculture. Normally they are removed from the second subculture after six days so as to avoid mixing of parental females with progeny.

About 10—12 days after the first cultures have been started the progeny are shaken over to fresh vials supplemented with some extra yeast, and aged for 2—3 days, so as to promote rapid maturation of the ovaries of the F_1 females.

2. The test matings for the F_2 generation

When sufficient progeny of the P_1 parental females have been collected and aged, the flies are anesthetized and females separated from males. Careful inspection of F_1 females will reveal whether, inadvertently, non-virgin P_1 females have been used or a P_1 female has been left in the vial. In that case, *Basc* females (with narrow, orange eyes) are present. These exceptional *Basc* females should never be used for the subsequent cross. The F_1 females with kidney-shaped red eyes are now individually mated to one or more males. For this mating their brothers can be used. If these are not available, as in the *yellow Bar* test when the P_1 males are used for a translocation test, one takes males from the stock generating the P_1 females. If trays are used with a width to accomodate 10 vials, the simplest procedure is to mate, for example, 20 F_1 females per treated male. The numbers of the treated P_1 males can be indicated on paper partitions, and it is recommended to mark both this number, and the number of F_1-P_2 females (i.e. the number of tested chromosomes) on a card which is stuck into that particular tray of vials, so that right from the beginning a record is kept of the number of test crosses and an identification of the individual P_1 male from which the progeny is derived. These cards can later be used for scoring the F_2 progeny and marking off of sterile cultures.

Optimal rearing conditions are achieved by placing the vials with anesthetized flies upside-down. When a tray containing, say, 200 vials is filled up, it is put away until the flies have recovered from anesthesia; this will prevent sticking of the flies to the culture medium. By placing an empty tray on top the whole tray can then be turned over (with the food in a downward position) and placed in an incubator at 25°C, so as to obtain standardized results. The F_1-P_2 parents do not have to be removed but the trays should be left at 25°C for 12—14 days; this will guarantee also that males with semi-lethals characterized by a longer development, have hatched. Mixing of F_2 with F_3 progeny should, in any case, be avoided. It is recommended to set up 600 cultures per brood, i.e. three trays, each containing 200 vials.

3. Scoring of the F_2 generation

The F_2 progeny is most rapidly inspected for the occurrence of sex-linked recessive lethals by taking out 3 vials at a time and simply looking for the presence or absence of wild-type males with red, round eyes at low power (4—6 ×) under a dissecting microscope. If these appear to be missing the culture vial is marked with a felt pen, indicating the number of the treated grandfather, and set aside. The presence of at least 2 wild-type males can be used as a criterion for a non-lethal culture and all these are discarded right away on a separate tray. If scoring of a tray of vials is completed, bundles of cultures suspected of lethals are kept together with a rubber band, also containing the card which had accompanied that particular tray of vials. On these cards the number of sterile cultures and lethal-containing cultures are indicated.

If, in post-meiotic broods, large clusters of lethals are observed among progeny derived from the same P_1 male, it is recommended to eliminate these from the final score, because they may reflect spontaneous mutations that arose in dividing spermatogonia during development of that particular P_1 male.

When scoring of one brood has been completed, the cultures suspected of containing lethals are anesthetized and individually inspected for further verification under the dissecting microscope. For further identification of recessive lethals the following criteria can be applied. (1) If more than 20 progeny (males and females together) are present, and no wild-type males, the culture can be considered as indeed containing a lethal; the chance of missing one being $(1/2)^5$, i.e., <0.05. Further retesting then becomes unnecessary. (2) If less than 20 progeny are present, the particular culture may be kept for a couple of days in order to wait for late hatching flies. (3) If after this time less than 20 progeny, or if one wild-type male, is present, retesting is required. To this end 3 (or more) females heterozygous for the treated and the *Basc* chromosome (kidney-shaped, red eyes) are mated individually with a few *Basc* males, and to ensure adequate ageing, shaken over to fresh subcultures after a few days. These females are, with regard to their X-chromosomes, identical to the F_1-P_2 female under test, and this extension to an F_3 should establish without doubt whether a lethal is indeed present in the treated X-chromosome. In a few cases all the F_3-vials might give less than 20 progeny or the females were sterile. Such cases have to be eliminated from the final calculations.

In this context, it is appropriate to point out that cases may be observed where only 1 or 2 cultures used for retesting may indicate the occurrence of a lethal and the others do not. Such results are obtained if the gonads of the P_2 female were mosaic for both mutated and non-mutated cells, and these cases can be counted as lethals. Characteristic for chemical mutagens is their delayed effect, often resulting in mosaicism. An impression of the extent of such delayed effects can be obtained by setting up F_3 cultures with F_2 females from non-lethal F_2 cultures. But for screening purposes this procedure increases the amount of work too much to be used routinely.

When the retesting has been completed, the final score can be made up. The number of sterile cultures are subtracted from the total number of cultures set up. When, for example, 600 F_1-P_2 cultures had been set up, and 27 were sterile or retesting proved impossible as a consequence of small numbers of progeny, the total number of tested X-chromosomes was 573. If 12 cultures containing verified recessive lethals were obtained, the frequency of lethals is 12/573 × 100 = 2.1%. In this context it should be pointed out that among the 12 lethals recovered some may have arisen by spontaneous mutation. When the control frequency of spontaneous mutation in the wild-type stock used is well known, as is, for example, the case for *Oregon-K*, an adequate correction can be made. The final evaluation of whether a particular substance under test significantly raises the frequency over that of the control and thus can be considered as a mutagen, will be treated separately.

4. The problem of clusters in progeny of individual males

Clusters of mutations among F_2 progeny from the same treated P_1 male may

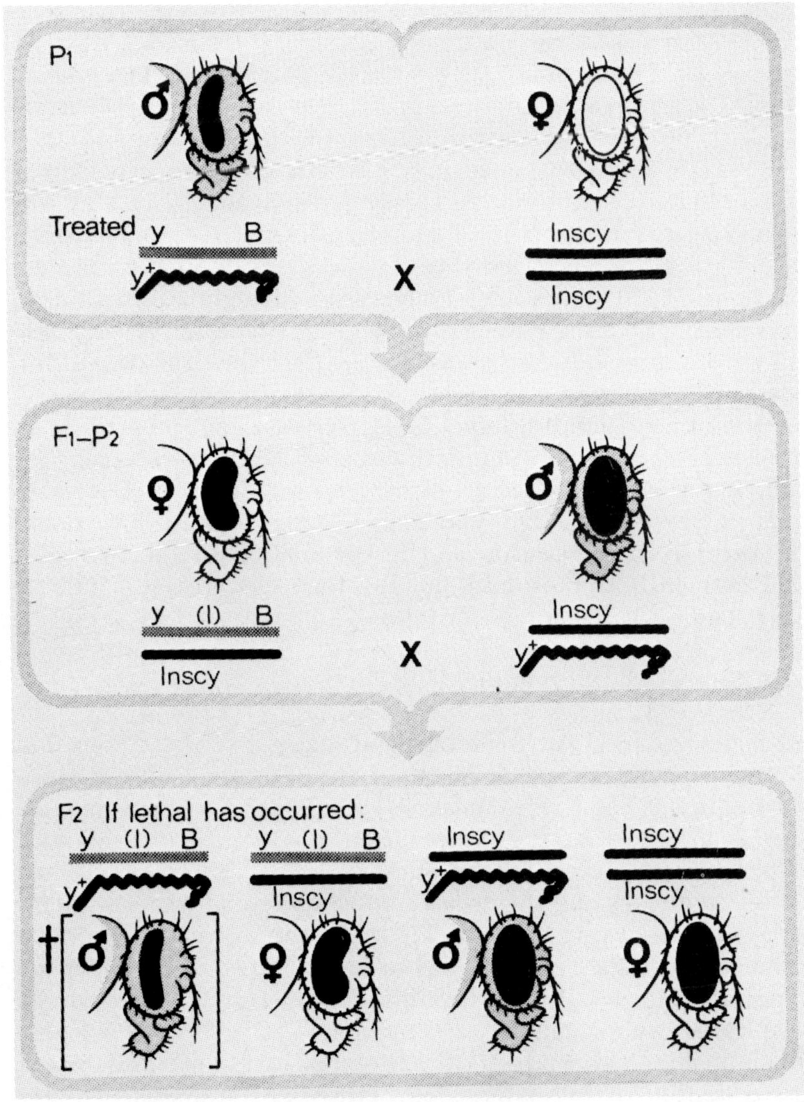

Fig. 7. The yellow-Bar test.

have different origins. If occurring in post-meiotic broods, they may originate from spontaneous mutation during development of that individual, and are better eliminated from the score. If they are observed in progeny derived from treated spermatogonia they should be counted as individual mutational events, because non-mutated cells also have undergone division and a correction according to Muller [50] can be applied. If they occur regularly, one may be dealing with a mutagen that for some uncontrolled reason affects some individuals more than others, as has been reported by Clark [17] and Auerbach [6] for pyronin B.

The yellow-Bar test

In the *yellow-Bar* test males of the genotype $yB/y^{+}Y$ are used for treatment and mated to "Oster" females of the genotype *Inscy; bw; st* p^{P}.

In the males the gene determining yellow body color does not come to expression because it is covered by the wild-type allele for *yellow*, present on the Y-chromosome. The genetic constitution of the *Inscy* chromosome is $In(1)sc^{S1L}\ sc^{8R} + dl\text{-}49,\ y\ sc^{S1}\ sc^{8}$. Since inversion delta-49 is longer than inversion S, it provides even better inhibition of crossing-over in the F_1 females. The *yellow-Bar* test has the advantage that, if required, loss of X- or Y-chromosomes can be scored in F_1 progeny and that F_1 males can be used to test for induced translocations. Fig. 7 shows the scheme for the detection of sex-linked recessive lethals. The genetic principles involved and the practical procedure do not differ basically from those outlined above for the *Basc* test. The only difference is that in this test *Bar*-eyed males are used for treatment. Consequently the criterion for the occurrence of a lethal in scoring of F_2 progeny relies on the presence or absence of males having narrow-shaped *Bar* eyes. The gene for yellow body color is suppressed by the wild-type allele of y^{+} on the Y-chromosome, and is thus not expressed. Eye colors may vary from red, brown and orange to white, because of segregation and recombination of the eye color genes on the 2nd and 3rd chromosomes, but this does not greatly encumber the scoring for *Bar*-eyed males.

The test for loss of X- or Y-chromosomes

When $yB/y^{+}Y$ males are used for treatment and mated to "Oster" females, the presence of the wild-type allele (y^{+}) on the Y-chromosome makes it possible to test for loss of the X- or Y-chromosomes (see Fig. 8). Chromosome loss can be considered as an indication for the occurrence of breaks, not followed by restitution or rejoining. According to our experience, chromosome loss occurs at only a low frequency and the decision of whether one wants to incorporate this test is a personal one. In view of the fact, however, that for the sex-linked recessive lethal test F_1 progeny has to be examined anyway for the separation of males and females, it takes very little extra effort to make counts and note the exceptional yellow males.

From the scheme in Fig. 8 it can be seen that non-disjunction in the treated fathers, i.e., non-separation of the paternal chromosomes in meiosis, can also give rise to *yellow* males. This class of exceptions cannot be distinguished from

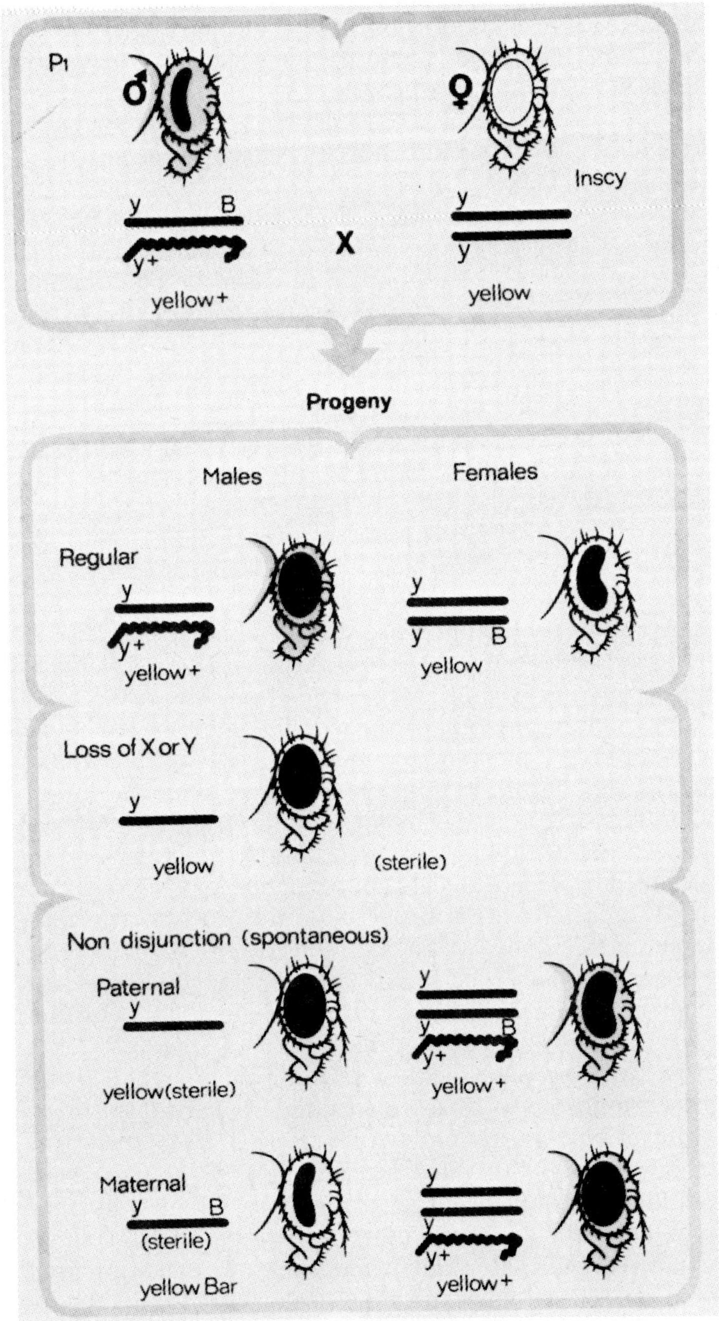

Fig. 8. The test for loss of X- or Y-chromosome.

yellow males originating from loss of X- or Y-chromosomes. If desired, a correction becomes possible by counting the complementary class of exceptions, i.e. non-yellow (y^+) females with kidney-shaped eyes ($B/+$). This class of females is expected to arise at equal frequency to the *yellow* males originating from paternal non-disjunction. Consequently, by simple subtraction of the number of non-yellow females, a more precise estimate of the frequency of yellow males derived from chromosome loss can be obtained.

The test for translocations between the second and third chromosomes

Translocations, i.e. exchanges of fragments derived from broken chromosomes, usually do not arise at high frequency after chemical treatment, though the yield, at least after treatment with certain chemicals, can be considerably increased by applying storage of sperm in the females (see below). An advantage of the testing for translocations is that, in consequence of their very low rate of spontaneous occurrence, controls become practically superfluous. Moreover, data on the induction of translocations do provide a clear indication of the capacity of a chemical to break chromosomes. Such information may be useful as a pointer to be followed up in more laborious cytogenetic tests employing mammalian assay systems.

The genetic principle involved

The presence of markers on the second and third chromosomes in the "Oster" stock makes it possible to test genetically for the induction of heritable reciprocal translocations. These marker genes are *brown* (*bw*) on the second chromosome, producing a phenotype characterized by light-brown eyes, because the development of red eye pigment is suppressed. The third chromosome carries the gene *scarlet (st)*, resulting in bright red eyes, since the formation of the brown component of the eye pigment is suppressed. In addition, the *pink* allele of the gene *peach* (p^p) has been incorporated. In combination with p^p the *scarlet* eye color is changed to light-orange, and this facilitates the scoring of F_2 progeny for the presence or absence of a translocation. The combination of the genes *brown* and *scarlet* within the same fly results in a white-eyed phenotype, because, in homozygous condition, *brown* suppresses the formation of red eye pigment and *scarlet* that of brown eye pigment, and thus no color is developed.

An experiment starts by treating males with the wild-type eye color genes on the second and third chromosomes. These males are crossed to *bw; st* p^p females which show white eye color. F_1 individuals derived from this cross will be heterozygous for the genes *bw* on their second and *st* p^p on their third chromosomes. Since crossing-over only occurs in Drosophila females, but not in males, and we want to keep the treated wild-type paternal chromosomes separate from the untreated marked maternal chromosomes, F_1 males are used to generate the F_2 generation. These males are then back-crossed individually to 2 virgin females of the maternal genotype (characterized by white eyes).

This test designed for the detection of translocations by a genetic procedure, simply relies on Mendels second law (see Fig. 9). By independent reassortment of the chromosomes at meiosis an individual heterozygous for two genes located on different chromosomes will produce four different types of gametes. These can be visualized by back-crossing such a "double" heterozygote to flies homozygous for the marker genes under consideration. The scheme for detect-

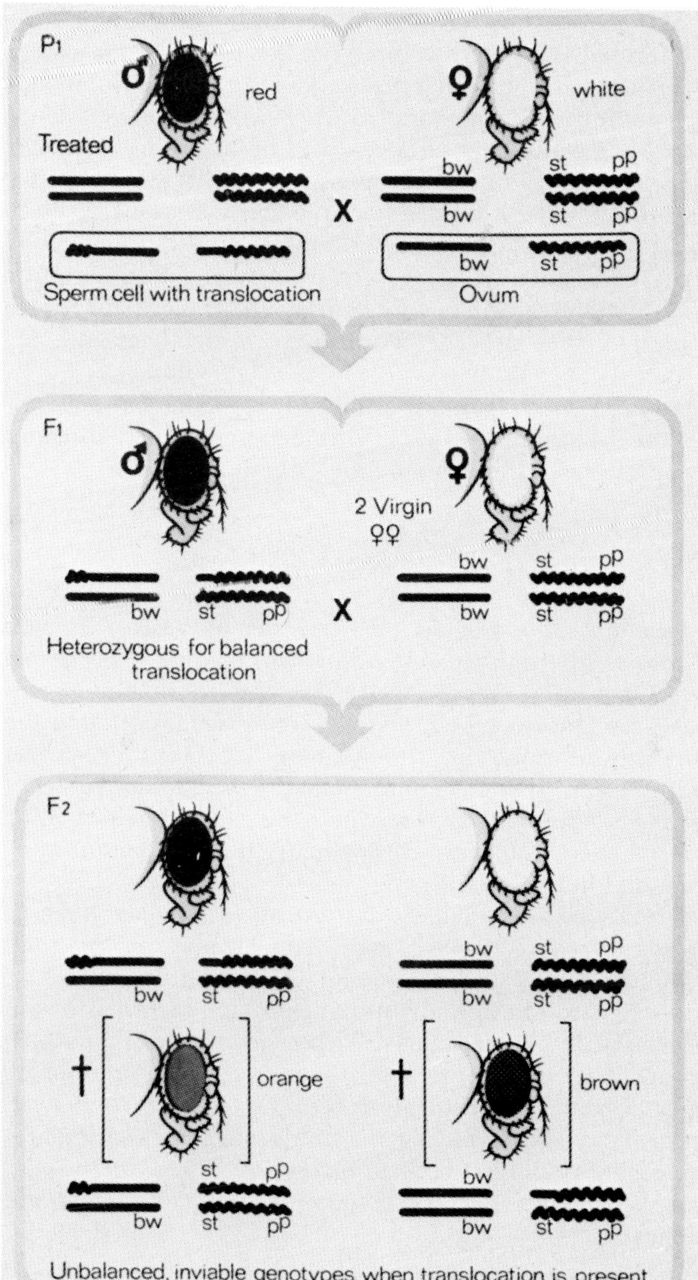

Fig. 9. The translocation test.

ing translocations using the brown-scarlet-pink-peach system is shown in Fig. 9.

In the absence of a II-III translocation the F_2 progeny will contain four different phenotypes with red, white, orange and brown eyes, respectively. However, in the case that the P_2 male is heterozygous for a translocation, two unbalanced genotypes will be formed. The zygotes receiving the second chromosome with a translocated piece of the third attached to it and a *st* p^p chromosome are unbalanced. Instead of a complete chromosome set they will carry a deficiency for a part of their second and a duplication for a part of their third chromosomes. Similarly, genotypes receiving the third chromosome with a translocated piece and the second chromosome with *bw* are unviable, because of a deficiency for part of the third and a duplication of part of the second chromosomes. As a result, individuals with these unbalanced genotypes die during development, and instead of four Mendelian classes with red, white, orange, and brown eyes, only two types of flies with red and white eyes, respectively, are observed.

Translocation thus results in suppression of independent Mendelian reassortment, but instead genes on the second and third chromosomes of a II-III translocation heterozygote segregate as if they were linked together.

Practical procedures

1. P_1 mating schedule

The P_1 cross involves the mating of treated yB/y^+Y males to "Oster" females of the genotype *Inscy; bw: st* p^p (yellow body color, white eyes). With a test for II-III translocations we are not concerned with the markers on the X- and Y-chromosomes. They can be ignored when the test is executed. Scoring in the F_2 generation will simply rely on observing the eye color of the progeny. The use of the above-mentioned genotypes is recommended, because flies of these stocks are also used for the sex-linked lethal test and the test for loss of X- or Y-chromosome. The use of such multipurpose stocks reduces the number of different stocks to be cultured in the laboratory.

The procedure to be followed for a genetic translocation test has much in common with that for sex-linked recessive lethals. The F_1 may be obtained by mass-mating treated males with "Oster" females or by starting with pair matings which, as described with the lethal test, will provide information on genetic effects in germ cells of individual males. With respect to brood pattern analysis the practical procedures are identical to those described for the sex-linked recessive lethal test. Ordinary translocation tests may be complemented by experiments concerning delayed effects. The reasoning and the experimental protocol for storage experiments will be described below.

Within the progeny of the first generation cross, each F_1 male represents one gamete of the treated father.

2. P_2 mating schedule

The second test generation is obtained by crossing the double heterozygous F_1 males to homozygous virgin "Oster" females. Since each F_1-P_2 male represents one treated gamete of the father, the F_1-P_2 males *have to be mated individually* in single vials to "Oster" females. In order to ensure sufficient progeny every male is mated to at least 2 well-fed virgin females. A special technique facilitating the collection of large numbers of virgins will be described below.

With this second test generation the parental flies have not to be removed from the vials because identical genotypes to the P_2 flies will be present in the progeny.

3. Scoring and retesting of the F_2 generation

Scoring of the F_2 generation is exceedingly simple in that one looks for the presence or absence of flies (males or females) with brown or orange eyes. The absence of these classes can be considered as an indication that the father had been heterozygous for a II-III translocation. In cases of doubt with insufficient F_2 progeny the test cross is repeated, three-fold, by taking red-eyed males, which are genotypically identical to their father, and back-crossing these again to virgins of the "Oster" stock.

4. Sperm storage

Extensive investigations of Auerbach (for reviews see refs. 5, 9 and 70) have shown that chemical mutagens, alkylating agents in particular, are characterized by delayed effects. Chemical damage to the genetic material does often remain latent for several cell generations before it comes to expression as chromosome breaks or mutations. If breaks open up in different cell cycles, the chances that a translocation can be formed are greatly reduced because, obviously, the formation of a translocation requires the simultaneous presence of two open breaks.

In Drosophila, the mature sperm represent a higly specialized cell type. Experiments of Muller [49] and Sobels [71,73] indicated that X-ray-induced chromosome breaks do not rejoin before the sperm has inseminated an egg. With sperm exposed to chemical mutagens, Auerbach observed a shortage of translocations in comparison to X-ray doses producing comparable frequencies of sex-linked recessive lethals. That the observed shortage of translocations does not result from a shortage of chromosome breaks, but rather from delayed effects, could be shown by storing the treated sperm for 6 days in the inseminated females. By storing of the sperm, more and more breaks become available simultaneously and thus the probability of the formation of a translocation is increased.

It is recommended, therefore, to incorporate this storing procedure in the experimental protocol if tests for translocations are desired. Storage simply consists of preventing the females from depositing their eggs and thus retaining the sperm in their storage organs: the spermathecae and ventral receptacle. This is done by placing the inseminated females on sugar-agar medium, composed of 2% agar and 10% sugar (or honey). The omission of fresh yeast will then prevent development of the ovaries and egg deposition. To avoid stickiness, the inseminated females should be transferred to vials containing fresh sugar-agar medium every 2—3 days. If this precaution is taken, inseminated females can be kept for several weeks at a temperature of 25°C.

5. "Automatic virgins"

As discussed above, the test matings for the translocation test require large scale collection of virgin females from the "Oster" stock. That is, if one plans to test 800 gametes of the treated P_1 males, one has to collect at least 1700 virgins, so as to allow for some loss through the ageing period. Moreover, in translocation tests one always has to reckon with considerable sterility in the test matings, so that more cultures have to be set up than the planned number of gametes to be tested.

The labor involved in the collection of virgins can be greatly reduced by making use of stocks that permit the collection of so-called "automatic virgins". To this end virgin females of the "Oster" stock are mated to males having both arms of the Y-chromosome attached to the X-chromosome of the following genetic composition Y^S *X.*Y^L*,InEN,dl 49,y; bw; st* p^p (a stock recently made up by Professor E. Novitski). Males derived from this cross have no Y-chromosome, and, in consequence, are sterile, so that all female progeny are not fertilized and thus remain automatically virgin. When 20 bottles are set up of this cross with 20 pairs of flies each, this only requires the collection of some 400 virgins of the "Oster" stock and then thousands of automatic virgins of the desired genetic composition are generated, which can be collected every few days for further ageing and feeding in vials.

Some statistical considerations

The basic problem of the statistical analysis of mutagenicity data is to decide whether the mutation frequency in the treated group is significantly larger than the frequency observed with an appropriate control group. With such a significance test the size of the groups analyzed plays an important role: the larger the groups the smaller the difference between the two mutation frequencies that can be shown to be significant.

With the sex-linked recessive lethal test, which is a two-generation test, performing of an "ample" number of tests poses some problems. These will be discussed in some detail below. With the dominant lethal test and the test for chromosome losses, which are both one-generation tests, it is much easier to get fairly large samples. With the translocation test it is also laborious to perform large numbers of tests, but here the spontaneous frequency is negligibly low and therefore a straightforward interpretation of the data is generally possible.

The significance test

The aim of an experiment is to answer the following question: is the mutation rate found in the treated group significantly larger than the spontaneous rate? The spontaneous mutation rate may be determined in the same experiment in an appropriate non-treated control group, or the rate normally found with the stocks used in a particular laboratory is already known based on experience with preceding experiments.

Let us first consider the simultaneous registration of the control rate and the experimental rate. With respect to statistical significance the most economic way to conduct the experiment is to test about equal numbers of chromosomes in the control and the treated group. More detailed evidence for the validity of this proposal is found in Würgler et al. [96]. After the sterile and non-decisive cultures have been eliminated the experimental result consists of the following numerical values:

Nc = number of tested chromosomes in the control group;
Mc = number of lethals found in the control group;
Nt = number of tested chromosomes in the treated group; and
Mt = number of lethals found in the treated group.

The question to be answered can be stated as follows: is the rate pt = Mt/Nt found in the treated group significantly larger than the rate pc = Mc/Nc found in the control?

The best way to find the answer is to apply the significance test developed by Kastenbaum and Bowman [34]. Berchtold [11] has shown that this test is nearly equivalent to Fisher's exact test. In certain regions the Kastenbaum-Bowman test is slightly more conservative than Fisher's test. It is highly recommended to use the simple Kastenbaum-Bowman test instead of the much more laborious exact test of Fisher. Appropriate tables are found in ref. 35 and their use described by Kastenbaum and Bowman [35] and Würgler et al. [96].

In the second case, if an experimentally determined rate is to be compared with a standard spontaneous rate observed over a long period of time in a certain laboratory, we recommend — in order to be on the safe side — the following treatment of the problem. The experiment led to the values of Nt and Mt, and from experience pc is known. We recommend to use Nc = Nt to calculate Mc as Mc = Nt.pc. With the four values Nc, Mc, Nt, Mt we then perform the Kastenbaum-Bowman test.

The frequency of induced mutations

Especially for comparisons of different sets of data it might be convenient to calculate the induced mutation frequency (= pi). In order to calculate pi from the two experimentally determined values pc and pt we apply Abbott's correction [1]. An example is:

control: Mc = 21	Nc = 2441	pc% = (Mc/Nc) · 100 = 0.86%
treated: Mt = 39	Nt = 1917	pt% = (Mt/Nt) · 100 = 2.03%

Assuming that the frequency of spontaneous mutations in the treated group is the same as in the control group we calculate that S = Nt · pc = 1917 · 0.0086 = 16.5 spontaneous mutations are present in the treated group. A test in which a spontaneous mutation was already present when the group was treated could not be tested for a mutation induced by the treatment (with standard procedures an X-chromosome with two recessive lethals cannot be distinguished from an X-chromosome containing only one recessive lethal). Therefore, we have to eliminate the tests assumed to contain spontaneous mutations from the treated group in order to get an estimate of the induced mutation frequency. For the frequency pi of induced mutations we find:

$$pi = (Mt-S)/(Nt-S) = (39-16.5)/(1917-16.5) = (22.5)/(1900.5) = 0.0118$$
$$pi\% = 1.18\%$$

It should be stressed that the new pi has a larger variance than the original pt. In connection with the Kastenbaum-Bowman test we interpret this result as follows: based on the Kastenbaum-Bowman test the original data show that the mutation frequency in the treated group pt% = 2.03% is significantly higher

TABLE I

SAMPLE SIZES BASED ON THE KASTENBAUM-BOWMAN TEST

Alpha (one-sided) = 0.05. The figures in the table give the smallest sample size (N/2 = Nt = Nc) which is needed to prove a significant difference between pt% (to the left) and pc% (heading of table).

pt%	pc%								
	0.00	0.05	0.10	0.15	0.20	0.25	0.30	0.35	0.40
0.05	16,000	0	0	0	0	0	0	0	0
0.10	8,000	28,000	0	0	0	0	0	0	0
0.15	5,334	11,334	40,667	0	0	0	0	0	0
0.20	4,000	6,500	14,000	50,500	0	0	0	0	0
0.25	3,200	5,200	8,000	18,400	74,800	0	0	0	0
0.30	2,667	3,667	5,667	9,334	20,334	88,667	0	0	0
0.35	2,286	3,143	4,572	6,572	10,858	23,429	104,858	0	0
0.40	2,000	2,500	3,250	4,750	7,000	12,250	26,250	120,750	0
0.45	1,778	2,223	2,889	3,778	5,334	8,223	13,556	34,445	134,889
0.50	1,600	2,000	2,600	3,200	4,000	5,600	9,200	14,800	38,400
0.55	1,455	1,819	2,000	2,546	3,455	4,364	6,546	9,637	16,182
0.60	1,334	1,667	1,834	2,167	2,834	3,834	4,667	6,834	10,167
0.65	1,231	1,539	1,693	2,000	2,462	3,077	3,847	5,077	7,385
0.70	1,143	1,429	1,572	1,858	2,286	2,715	3,286	4,000	5,429
0.75	1,067	1,334	1,334	1,734	1,867	2,267	2,667	3,600	4,400
0.80	1,000	1,250	1,250	1,375	1,625	2,000	2,375	2,875	3,876
0.85	942	1,177	1,177	1,295	1,530	1,883	2,236	2,715	3,177
0.90	889	1,112	1,112	1,223	1,445	1,556	1,889	2,223	2,667
0.95	843	1,053	1,053	1,158	1,369	1,474	1,685	2,000	2,422
1.00	800	800	1,000	1,100	1,300	1,300	1,600	1,900	2,000

than the frequency pc% = 0.86% in the control group. Based on Abbott's correction we estimate that the significant increase resulting from the treatment represents an induced mutation frequency pi% = 1.18%. It should be stressed here, that for statistical reasons, the significance test described above has to be based on pc and pt and not on pi.

The sample size

Especially with the sex-linked recessive lethal test the determination of the "ample" sample size is a serious problem that should be considered carefully before starting an experiment. The problem results from the fact that within a given time only a limited number of tests can be performed. With bacteria, 10^5 treated chromosomes are easily tested on one single petri dish, and the test is completed within a few days. With Drosophila testing of 10^5 treated X-chromosomes would keep a laboratory (as assumed in the following section) busy for about 30 weeks. This clearly demonstrates that the precision with which Drosophila experiments can answer a given question will always be limited by the number of tests that can be performed. It is highly recommended that, before starting an experiment, this statistical problem is considered carefully in order to avoid attacking questions which cannot be properly answered based on the capacity of the laboratory.

The sample size table prepared by Würgler et al. [96] will help to study the problem of the "ample" sample size. This table (Table I) may be used in the following way: choose an estimate of the spontaneous mutation rate pc% to be expected with your experiment. In most cases an estimate of about 0.25% might be reasonable. Now we choose the column headed 0.25, and the values in that column tell us how many tests have to be performed in order to be able to find a significant difference between the control and the treated group if the mutation rate in the treated group (pt%) is equal to or larger than the value given in the left most column of the same line of the table. For example, with pc% = 0.25% and 6000 tests performed one will be able to find a significant increase in pt% provided that value is 0.50% or larger. Or, in other words, in order to be able to prove statistically that a doubling of the mutation rate from pc% = 0.25% to pt% = 0.50% has been found, at least 5600 tests have to be performed in the treated group (and, if simultaneously run, also in the appropriate control group).

Such considerations will lead to the decision how large an experiment has to be planned in order to be able to answer the actual question. It is absolutely mandatory to realize that performing a few hundred tests will only detect strong mutagens which increase the mutation rate four, five or more times.

Cost estimates

The following calculation is based on the assumption that recessive lethal assays form the main part of mutagenicity testing using Drosophila. The estimate is based on a personnel cost of two technicians and one laboratory helper (for cleaning purposes and preparing the food).

From our experimental experience with rapid screening testing on Drosophila, two experiments consisting of three successive broods each, can easily be handled per week. If 600 vials are set up per brood, then a testing capacity of 1800 tested X-chromosomes results per experiment. One repetition experiment under "normal" circumstances should be sufficient to detect with a reasonable certainty the mutagenic activity of a given test substance. Complicated borderline cases will need more experiments, of course. The total annual capacity will then reach 40, to a maximum of 48, substances.

With such a testing schedule time would still be available each week to run one dominant lethal test or one experiment on chromosome loss, using up to four concentrations per test. Without difficulty an average sample size of 5000 should be feasible in both assays. Particularly for dominant lethal assays this is rather an underestimate. At the Leiden laboratory we frequently scored about 20 000 eggs from 7 different groups (concentrations) side by side with recessive lethal assays.

References

1 Abbott, W.S., A method for computing the effectiveness of an insecticide, J. Econ. Entomol., 18 (1925) 265—267.

2 Abrahamson, S. and E.B. Lewis, The detection of mutations in *Drosophila melanogaster*, in A. Hollaender (ed.), Chemical Mutagens, Vol. 2, Plenum Press, New York, 1971, pp. 461—487.

3 Amy, R.L. and R.C. von Borstel, The effects of different wave lengths of ultraviolet light on the Habrobracon egg, in Proc. 2nd Int. Photobiol. Congr., Turin 1957, Ediz. Minerva Med., Minerva Fisioterapica (Collana Monografica), Vol. 2, 1957, pp. 419—422.

4 Ashburner, M. and E. Novitsky (eds.), Genetics and Biology of Drosophila, Vol. 1, Academic Press, New York, 1976.

5 Auerbach, C., After-effects of radiation and chemicals on chromosomes and genes, Brit. J. Radiol., 27 (1953) 122—124.

6 Auerbach, C., The mutagenic action of pyronin-B, Amer. Naturalist, 89 (1955) 241—245.

7 Auerbach, C., The production of visible mutations in Drosophila by chloroethyl methylsulphonate (CB 1506), Genet. Res. (Camb.), 3 (1962) 461—466.

8 Auerbach, C., Mutation: An Introduction to Research on Mutagenesis. Part I: Methods, Oliver and Boyd, Edinburgh, 1962.

9 Auerbach, C., The chemical production of mutations, Science, 158 (1967) 1141—1147.

10 Auerbach, C., The effects of six years of mutagen testing on our attitude to the problems posed by it, Mutation Res., 33 (1975) 3—10.

11 Berchtold, W., Comparison of the Kastenbaum-Bowman test and Fisher's exact test, Archiv Genetik, 48 (1975) 151—157.

12 Boyd, J.B. and J.M. Presley, Repair replication and photorepair of DNA in larvae of *Drosophila melanogaster*, Genetics, 77 (1974) 687—700.

13 Büchi, R., Induced dominant lethal mutations and cytotoxic effects in germ cells of Drosophila treated with Trenimon, PDMT and sodium monofluorophosphate, in preparation.

14 Büchi, R. and K. Bürki, The origin of chromosome aberrations in mature sperm of Drosophila: influence of sodium fluoride on the treatments with Trenimon and 1-phenyl-3,3-dimethyltrazene, Archiv Genetik, 48 (1975) 59—67.

15 Casida, J.E., Insect microsomes and insecticide chemical oxidations, in: J.R. Gillette, A.H. Conney, G.J. Cosmides, R.W. Estabrook, J.R. Fouts and G.J. Mannering (eds.), Microsomes and Drug Oxidations, Academic Press, New York, 1969, pp. 517—531.

16 Chandley, A.C. and A.J. Bateman, Timing of spermatogenesis in *Drosophila melanogaster* using tritiated thymidine, Nature, 193 (1962) 299—300.

17 Clark, A.M., The mutagenic activity of dyes in *Drosophila melanogaster*, Amer. Naturalist, 87 (1953) 295—305.

18 Crow, J.F. and S. Abrahamson, Drosophila methods for mutagenicity testing, in S.S. Epstein (ed.), Drugs of Abuse, MIT Press, Cambridge, Mass., 1971, pp. 136—139.

19 Demerec, M., Relationship between various chromosomal changes in Drosophila, Cytologie, Fujii Jub. Vol. (1937) 1125—1132.

20 Demerec, M. (ed.), Biology of Drosophila, John Wiley New York, 1950.

21 Demerec, M. and B.P. Kaufmann, Drosophila Guide, 7th edition, Carnegie Institution of Washington Publication, Washington, D.C., 1961.

22 Fahmy, O.G. and M.J. Fahmy, Mutagenicity of 2-chloroethyl methanesulfonate in *Drosophila melanogaster*, Nature, 177 (1956) 996—997.

23 Félix, R., A system for feeding liquids to adult flies, Drosophila Inform. Serv., 47 (1971) 129.

24 Ghelelovitch, S., La sensibilité des oeufs de la Drosophile (*Drosophila melanogaster*, Meig.) à l'action létale des rayons ultraviolets. II. Déterminisme génétique de la sensibilité des stades embryonaires postérieures à la gastrulation, Mutation Res., 5 (1968) 117—131.

25 Graf, U., Spontaneous mutations in *Drosophila melanogaster*, Humangenetik, 16 (1972) 27—32.

26 Graf, U. and F.E. Würgler, MMS sensitive strains in *Drosophila melanogaster*. Mutation Res., 34 (1976) 251—258.

27 Herskowitz, I.H., A new method for treating Drosophila gametes with chemicals, Evolution, 1 (1947) 111—112.

28 Herskowitz, I.H., The genetic basis of X-ray induced recessive lethal mutations, Genetics, 36 (1951) 356—363.

29 Herskowitz, I.H., Bibliography on the Genetics of Drosophila, Part 2, Commonwealth Agricultural Bureau, Farnham Royal, Slough, (1952).

30 Herskowitz, I.H., Bibliography on the Genetics of Drosophila, Part 3, Indiana University Press, Bloomington, Ind., 1958.

31 Herskowitz, I.H., Bibliography on the Genetics of Drosophila. Part 4, McGraw-Hill, New York, 1963.

32 Herskowitz, I.H., Bibliography on the Genetics of Drosophila, Part 5, Macmillan, New York, 1969.

33 Herskowitz, I.H., Bibliography on the Genetics of Drosophila, Part 6, Macmillan, New York, 1974.

34 Kastenbaum, M.A. and K.O. Bowman, The Minimum Significant Number of Successes in Binomial Sample, ORNL-3909, UC-32-Mathematics and Computers, 1966.

35 Kastenbaum, M.A. and K.O. Bowman, Tables for determining the statistical significance of mutation frequencies, Mutation Res., 9 (1970) 527—549.

36 King, R.C., *Drosophila melanogaster:* an introduction, in R.C. King (ed.), Handbook of Genetics, Vol. 3, Plenum, New York, 1975, pp. 625—652.

37 Knaap, A.G.A.C. and P.G.N. Kramers, Mutagenicity of hycanthone in *Drosophila melanogaster*, Mutation Res., 22 (1974) 55—62.

38 Kolar, G.F., R. Fahrig and E.Vogel, Structure-activity dependence in some novel ring-substituted 3,3-dimethyl-1-phenyltriazenes. Genetic effects in *Drosophila melanogaster* and in *Saccharomyces cerevisiae* by a direct and a host-mediated assay, Chem.-Biol. Interact., 9 (1974) 365—378.

39 Leuthold, U., Einfluss der Geschlechtschromosomen auf die Strahlenempfindlichkeit verschiedener Zelltypen von *Drosophila melanogaster*, Mutation Res., 14 (1972) 65—81.

40 Lewis, E.B. and F. Bacher, Method of feeding ethyl methanesulfonate (EMS) to Drosophila males, Drosophila Inform. Serv., 43 (1968) 193.

41 Lindsley, D.L. and E.H. Grell, Genetic Variations of *Drosophila melanogaster*, Carnegie Institution of Washington Publ. 627, Washington, D.C., 1967, 472 pp.

42 Lüers, H., Untersuchungen über die Mutagenität des Triäthylenmelanin (TEM) an *Drosophila melanogaster*, Arch. Geschwulst-Forschung, 6 (1953) 77—83.

43 Mendelson, D., The effect of caffeine on repair systems in oocytes of *Drosophila melanogaster*. II. The induction of chromosome aberrations in irradiated males, Mutation Res., 35 (1976) 91—100.

44 Merriam, J.R., On setting up a lab and kitchen, Drosophila Inform. Serv., 50 (1973) 196—197.

45 Mollet, P. and F.E. Würgler, An apparatus to inject large numbers of Drosophila with constant amounts of fluid within a short time, Drosophila Inform. Serv., 50 (1973) 202.

46 Mollet, P. and F.E. Würgler, Detection of somatic recombination and mutation in Drosophila. A method for testing genetic activity of chemical compounds, Mutation Res., 25 (1974) 421—424.

47 Muller, H.J. and E. Altenburg, The rate of change of hereditary factors in Drosophila, Proc. Soc. Exptl. Biol. Med., 17 (1919) 10—14.

48 Muller, H.J., Bibliography on the Genetics of Drosophila, Oliver and Boyd, Edinburgh, 1939.

49 Muller, H.J., An analysis of the process of structural change in chromosomes of Drosophila, J. Genet., 40 (1940) 1—66.

50 Muller, H.J., The standard error of the frequency of mutants some of which are of common origin, Genetics, 37 (1952) 608.

51 Muller, H.J. and I.I. Oster, Some mutational techniques in Drosophila, in W.J. Burdette (ed.), Methodology in Basic Genetics, Holden-Day, San Francisco, Calif., 1963, pp. 240—278.

52 Ogaki, M. and E. Nakashima-Tanaka, Inheritance of radioresistance in Drosophila I, Mutation Res., 3 (1966) 438—443.

53 Olivieri, G. and O. Olivieri, Autoradiographic study of nucleic acid synthesis during spermatogenesis in *Drosophila melanogaster*, Mutation Res., 2 (1965) 366—380.

54 Parker, D.R., On the nature of sensitivity changes in oocytes of *Drosophila melanogaster*, in F.H. Sobels (ed.), Repair from Genetic Radiation Damage, Pergamon Press, Oxford, 1963, pp. 11—30.

55 Rieger, R., A. Michaelis and M.M. Green, A Glossary of Genetics and Cytogenetics, Springer-Verlag, New York, 1968, 506 pp.

56 Sankaranarayanan, K., The effects of nitrogen and oxygen treatment on the frequency of X-ray-induced dominant lethals and on the physiology of the sperm in *Drosophila melanogaster*, Mutation Res., 4 (1967) 641—661.

57 Sankaranarayanan, K. and F.H. Sobels, Radiation genetics, in M. Ashburner and E. Novitsky (eds.), The Genetics and Biology of Drosophila, Academic Press, New York, 1976, pp. 1090—1250.

58 Seecof, R.L., W. Kaplan, P. Wong, W. Trout III and J. Donady, A versatile etherizer, Drosophila Inform. Serv., 46 (1971) 154.

59 Sega, G.A. and W.R. Lee, A vacuum injection technique for obtaining uniform dosages in *Drosophila melanogaster*, Drosophila Inform. Serv., 45 (1970) 179.

60 Sega, G.A., P.A. Gee and W.R. Lee, Dosimetry of the chemical mutagen ethyl methanesulfonate in spermatozoan DNA from *Drosophila melanogaster*, Mutation Res., 16 (1972) 203—213.

61 Shorrocks, B., Drosophila, Ginn, London, 1972.

62 Slizinska, H., Cytological analysis of formaldehyde induced chromosomal changes in *Drosophila melanogaster*. Proc. Roy. Soc. (Edinburgh) B, 66 (1957) 288—304.

63 Slizinska, H., Origin of repeats in Drosophila chromosomes, Genet. Res., 4 (1963) 154—157.

64 Slizinska, H., Mutagenic effect of X-rays and formaldehyde food in spermatogenesis of *Drosophila melanogaster*, Genet. Res., 4 (1963) 248—257.

65 Slizinska, H., The progressive approximation, with storage, of the spectrum of TEM-induced chromosomal changes in Drosophila sperm to that found after irradiation, Mutation Res., 8 (1969) 165—175.

66 Slizinsky, B.M., Salivary chromosome studies of lethals in *Drosophila melanogaster*, Genetics, 23 (1938) 283—290.

67 Smith, P.D., Mutagen-sensitivity of *Drosophila melanogaster*. I. Isolation and preliminary characterization of a methyl methanesulphonate-sensitive strain, Mutation Res., 20 (1973) 215—220.

68 Smith, P.D. and C.G. Shear, Mutagen sensitivity of *Drosophila melanogaster*. II. X-ray and ultraviolet light sensitivities of a methyl methanesulphonate sensitive strain, in R.F. Grell (ed.), Mechanisms in Recombination, Plenum, New York, 1974, pp. 399—403.

69 Sobels, F.H., Organic peroxides and mutagenic effects in Drosophila, Nature, 177 (1956) 979—982.

70 Sobels, F.H., The role of Drosophila in the field of mutation research, Arch. Gene., 45 (1972) 101—125.

71 Sobels, F.H., A dose-fractionation study to determine how long breaks induced in various stages of spermatogenesis of Drosophila stay open, Rev. Suisse Zool., 79 (1973) 143—152.

72 Sobels, F.H., The advantages of Drosophila for mutation studies, Mutation Res., 26 (1974) 277.

73 Sobels, F.H., The persistence of chromosome breaks in different stages of spermatogenesis of Drosophila, Mutation Res., 23 (1974) 361—368.

74 Sobels, F.H. and D. Mendelson, Caffeine treatment of the repair system and repair of chromosome breaks induced in Drosophila spermatids, Mutation Res., 28 (1975) 133—136.

75 Sobels, F.H. and E. Vogel, Assaying potential carcinogens with Drosophila, Environ. Health Perspect., 1976, submitted for publication.

76 Sobels, F.H. and E. Vogel, The capacity of Drosophila for detecting relevant genetic damage, Mutation Res., (1976) in press.

77 Spencer, W.P. and C. Stern, Experiments to test the validity of the linear r-dose/mutation frequency relation in Drosophila at low dosage, Genetics, 33 (1948) 43—74.

78 Strickberger, M.W., Experiments in Genetics with Drosophila, John Wiley, New York, 1962.

79 Tates, A.D., Cytodifferentiation during spermatogenesis in *Drosophila melanogaster*: an electron microscopic study, J.H. Pasmans, The Hague, 1971.

80 Trosko, J.E. and K. Wilder, Repair of UV-induced pyrimidine dimers in *Drosophila melanogaster* cells in vitro, Genetics, 73 (1973) 297—302.

81 Valencia, J.I. and McQuate, J.T., A cytogenetic analysis of 70 ultraviolet-induced lethals in Drosophila, Genetics, 36 (1951) 580—581 (abstract).

82 Valencia, J.I. and W. Plaut, X-ray-induced DNA synthesis in polytene chromosomes, J. Cell Biol., 43 (1969) 151a (abstract).

83 Vogel, E., Strain variations in response to certain indirect mutagens in *Drosophila melanogaster*, Drosophila Inf. Serv., 50 (1973) 138.

84 Vogel, E., Some aspects of the detection of potential mutagenic agents in Drosophila, Mutation Res., 29 (1975) 241—250.

85 Vogel, E., Mutagenic activity of cyclophosphamide, trofosfamide, and ifosfamide in *Drosophila melanogaster*. Specific induction of recessive lethals in the absence of detectable chromosome breakage, Mutation Res., 33 (1975) 221—228.

86 Vogel, E. and H. Lüers, A comparison of adult feeding to injection in *Drosophila melanogaster*, Drosophila Inform. Serv., 51 (1974) 113—114.

87 Vogel, E. and B. Leigh, Concentration-effect studies with MMS, TEB, 2,4,6,-triCl-PDMT and DEN on the induction of dominant and recessive lethals, chromosome loss and translocations in Drosophila sperm, Mutation Res., 29 (1975) 383—396.

88 Vogel, E. and F.H. Sobels, The function of Drosophila in genetic toxicology testing, in A. Hollaender, (ed.), Chemical Mutagens, Principles and Methods for their Detection, Vol. 4, Plenum Press, New York, 1976, in press.

89 Von Borstel, R.C., Differential response of meiotic stages in Habrobracon eggs to nitrogen mustard, Genetics, 40 (1950) 107–116.

90 Wheeler, M.R., The fruitfly, in The Universities Federation for Animal Welfare Handbook: The Care and Management of Laboratory Animals, 4th edition, Livingstone, Edinburgh, 1972, pp. 543–548.

91 Wilkinson, C.F. and L.B. Brattsten, Microsomal drug metabolizing enzymes in insects, Drug Metab. Rev., 1 (1972) 153–228.

92 Würgler, F.E. and U. Leuthold, Radiosensitivity of oocytes of Drosophila. II. Sensitivity of class-B oocytes of X/X and XY/XY females, Int. J. Radiat. Biol., 21 (1972) 465–473.

93 Würgler, F.E. and H.-U. Lütolf, Radiosensitivity of oocytes of Drosophila. I. Sensitivity of class-A oocytes of triploid and diploid females, Int. J. Radiat. Biol., 21 (1972) 455–463.

94 Würgler, F.E. and P. Maier, Genetic control of mutation induction in *Drosophila melanogaster*. I. Sex-chromosome loss in X-rayed mature sperm, Mutation Res., 15 (1972) 41–53.

95 Würgler, F.E. and H. Ulrich, Radiosensitivity of embryonic stages, in M. Ashburner and E. Novitsky (eds.), The Genetics and Biology of Drosophila, Vol. 1, Academic Press, New York, 1976, pp. 1269–1298.

96 Würgler, F.E., U. Graf and W. Berchtold, Statistical problems connected with the sex-linked recessive lethal test in *Drosophila melanogaster*. I. The use of the Kastenbaum-Bowman test, Arch. Genet., 48 (1975) 158–178.

97 Würgler, F.E., M. Lezzi and U. Graf, A device for easily anaesthetising large numbers of flies, Drosophila Inform. Serv., 46 (1971) 149.

98 Würgler, F.E., U. Petermann and H. Ulrich, A refined test for X-ray-induced dominant lethals in Drosophila, Experientia, 24 (1968) 1293–1294.

99 Würgler, F.E., H. Ulrich and H.W. Spring, An improved method for the collection of large numbers of inseminated eggs of *Drosophila melanogaster*, Experientia, 24 (1968) 1082–1083.

100 Würgler, F.E., U. Leuthold, U. Graf and H. Ulrich, Radiosensitivity of oocytes of Drosophila. III. Sensitivity of class-B oocytes of XX/O and X/X females. Int. J. Radiat. Biol., 22 (1972) 367–377.

101 Zimmering, S., A. Note on Drosophila as a mutagenicity test system, Environ. Health Perspect., 2 (1973) 111–113.

MEASURING DOMINANT LETHALITY IN HABROBRACON

R.C. VON BORSTEL and ROGER H. SMITH

Department of Genetics, University of Alberta, Edmonton, Alberta T6G 2E9 (Canada) and 902 Euclid Ave., Berkeley, Calif. (U.S.A.)

Introduction

Attention is beginning to be drawn to dominant lethality as a reflection of the most debilitating of birth defects. This is not only because sterility and spontaneous abortions are visible aspects of dominant lethality but because the lesions that result in dominant lethality have their viable corollary in the most enfeebling of the gross chromosomal aberrations recognizable in human populations. Therefore, it is essential that methods be developed and employed for detecting compounds in the environment which can induce dominant lethality. Since dominant lethality of one sort or another is the only genetic defect which can cause species extinction [19,30] then its manifestation from environmental mutagens should be studied and understood both explicitly and implicitly.

The parasitic wasps *Habrobracon juglandis* (Ashmead) = *Bracon hebetor* Say and *Habrobracon serinopae* (Ramkr.) = *Bracon serinopae* (Ramkr.) are the most useful living organisms for monitoring dominant lethality induced by environmental mutagens. Because the males are haploid and the females diploid, environmental agents preferentially affecting either the cytoplasm of the egg or motility or egg penetrability of sperm can be reliably and unequivocally identified as phenomena separate from dominant lethal actions of an agent on the egg or sperm nucleus. Moreover, Habrobracon is inexpensive to culture, has a large genetic literature for an operational base, has a short life span (at 35°C *H. serinopae* has an egg-to-adult development time of six days), and, because of male haploidy, Habrobracon has the entire genome available for identification of recessive lethal and visible mutations in one generation and viable chromosome aberrations in two generations.

We shall emphasize rearing methods and analysis of dominant lethality in this article because other reviews [2,18,31] have covered methods of introducing mutagens and aspects of analysis of a variety of genetic lesions in Habrobracon as well as analysis of genetic damage of all stages of oogenesis from the oogonium to the embryo. Also, it is likely that if Habrobracon is to be employed for identification of genotoxic agents in the environment, the unique feature for which it will be used most often is as a pre-eminent system for detecting dominant lethality.

Culture maintenance

In order to rear Habrobracon it is necessary to rear its host. The most commonly used host is the cornmeal moth *Ephestia kühniella* (Zeller) = *Anagasta kühniella* (Zeller). Fortunately both the wasp and its host are simple to rear and, in the properly equipped laboratory [25], one or two hours of attention weekly will suffice for culturing the organisms.

Rearing of host Ephestia

At room temperature the life cycle of Ephestia is approximately six weeks long. When Ephestia are sent from laboratory to laboratory, they are ordinarily shipped in small (15×17 cm) wax paper bags partly filled with corn meal. Young larvae are quite resistant to both high and low temperature shocks, and they can survive easily for the entire life cycle in these bags. The bags are normally used for rearing the larvae during the early part of the life cycle. Forty to fifty adult moths are collected in a shell vial by gently touching the moths at the anterior end (Fig. 1). Normally quiescent, the moth will fly to the bottom of the vial. The moths are then shaken into a wax paper bag containing 100 g of corn meal (Fig. 2) and the top twisted closed quickly before the moths fly or crawl out. The top of the bag is then bound with a rubber band.

If stored at a temperature of 28°C in approximately two weeks small Ephestia larvae appear in the bags of corn meal and can be seen next to the wax paper below the surface of the meal (Fig. 3). As can be seen in the figure, the bags are kept on a rack, the feet of which are in small metal wells. Mineral oil is kept in each well so that each rack is effectively isolated to prevent mite infestations.

When the small larvae can be seen in the bags, the contents of a bag are mixed with approximately 500 g of corn meal in a culture box such as those shown in Fig. 1. The culture boxes need be nothing more than commercial cardboard boxes in which shirts are packed. A moistened paper towel is placed over the corn meal as shown in Fig. 1. The culture boxes are stacked (Fig. 2) on mineral-oil-isolated racks at room temperature and checked thrice weekly. Corn meal and water are added as needed. When the Ephestia are nearly full grown, and before pupation, the paper towels on top of the corn meal are moistened and the culture boxes are placed in a refrigerator at 7—10°C. One or two boxes are retained at room temperature to build up stocks of Ephestia.

The Ephestia larvae can be collected from the same refrigerated culture box a number of times over a six-month period. There are usually from 2000 to 4000 larvae in a culture box. As can be seen in Fig. 4, the paper towel is turned back, and the full-grown Ephestia larvae are picked up with brass forceps of the sort used by insect taxonomists. The larvae, up to a maximum number of 500, are dropped in a collection dish (Fig. 4). The larvae are left to crawl about at room temperature for two or three hours; they spin a fine web and upon crawling through it they become free of corn meal dust. The collection dish is then placed in the refrigerator for use.

When the boxes are exhausted, they are sprayed with alcohol, placed in a

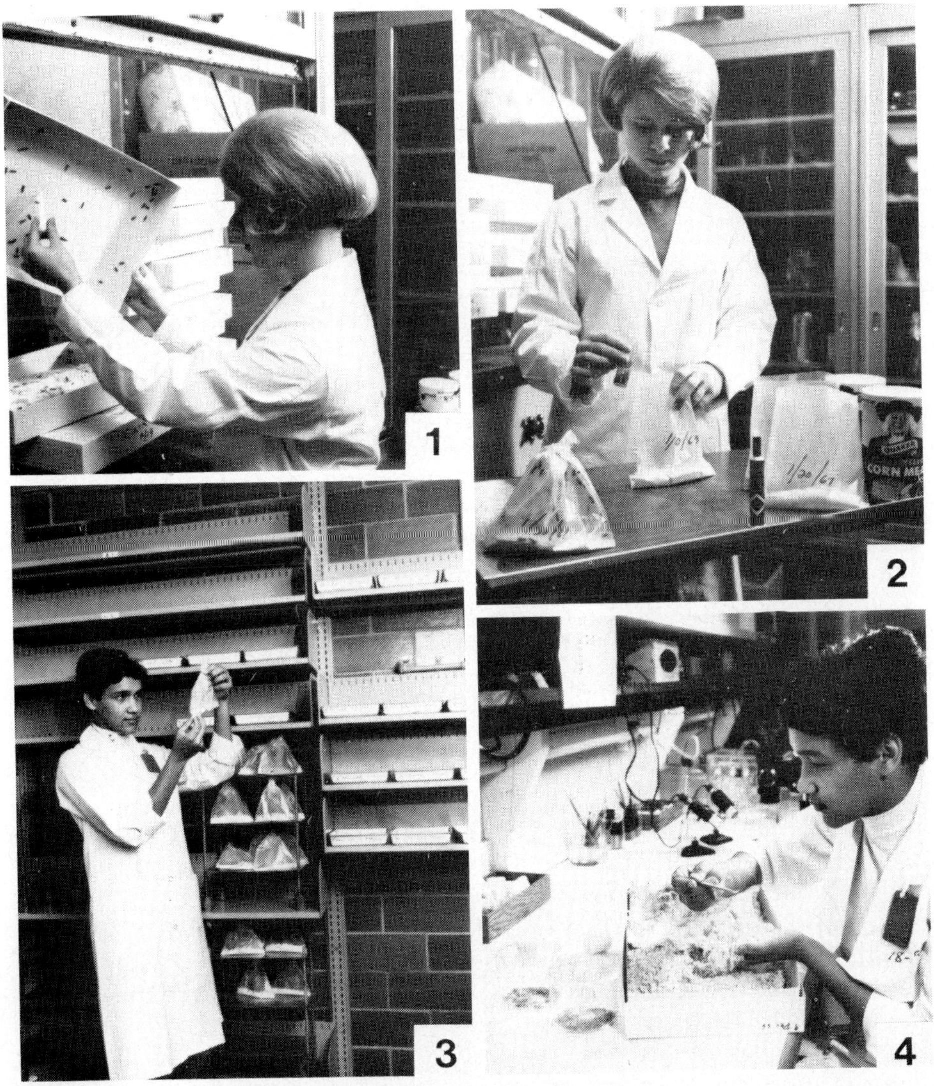

Fig. 1. Collection of adult Ephestia with a shell vial. (Martha S. Malling).

Fig. 2. Adult Ephestia are put into a wax paper bag containing corn meal. (Martha S. Malling).

Fig. 3. Small Ephestia larvae can be seen by examination of the bags of corn meal. (Margaret L. Yette).

Fig. 4. Full-grown Ephestia larvae are picked up gently with brass forceps and dropped into an Ephestia larvae collection dish. (Margaret L. Yette).

plastic bag and discarded. If the boxes ever become mite-infested, they are sprayed with toluene and sealed in a plastic bag before disposal.

Rearing of Habrobracon

Habrobracon are hardy and survive well from 3°C to 30°C. Indeed, Habrobracon can be stored easily for one or two months under refrigeration, and experiments are normally carried out at 28° C. Stock cultures are maintained at room temperature. *H. serinopae* can survive at higher temperatures than *H. juglandis*, is relatively disease-free and has a slightly shorter life cycle (7.5 days instead of 10 days at 28° C). Moreover, the females of *H. serinopae* lay twice as many eggs per day and live about three times as long at 28° C (two months instead of three weeks).

Habrobracon of both species are handled alike. The wasps are kept in 70 × 21 mm glass shell vials, usually three to five females to a vial, and two Ephestia larvae per female. The plugs are cotton wrapped in cheesecloth, tied with thread. The cotton plugs are used over and over again, and sterilized once or twice a year. Foam plastic plugs are useful for emergencies only, as they can be toxic to the wasps. The glassware is washed on the cold water cycle of a washer *without detergent* to remove webbing; a residue of detergent can remain in a vial and is powerfully insecticidal to Habrobracon.

The stock adults are moved to new vials every two or three days for three changes, and the process repeated about every two weeks.

For experimentation, individual females are kept in Stender dishes (37 × 19 × 24 mm) and the dishes are kept in aluminum dissecting pans (19 × 28 × 4 cm), 24 dishes per pan (Fig. 5). The wasps are changed twice daily, one Ephestia larva per *H. juglandis* female per change, and two Ephestia larva per *H. serinopae* per change. The pans of dishes are not stacked (Fig. 3) because stacking interferes with air exchange and the wasps die.

Anesthetization of Habrobracon

Habrobracon can be anesthetized with ether, but it has been found that carbon dioxide is best. The wasps quickly become immotile, recover quickly, and can be left under CO_2 anesthetization for up to an hour without being killed. A manifold system attached to a CO_2 line has been developed which permits anesthetization in a variety of ways (Fig. 5). The manifold is a lucite box containing a bubble flask half-filled with water. Two glass outlets extend from near the top of the flask and lead by Tygon tubing to a series of valves extending along both sides of the box. By turning these valves, one or all of the anesthetizing units can be activated. The bubble flask serves to humidify the CO_2 and acts as a visual indicator that the CO_2 is flowing.

There are three useful attachments for the manifold system so if enough valves are present (as the system shown in Fig. 5), then one manifold will suffice for two investigators. The three attachments are an anesthetizing dish (Fig. 5 and 6), an anesthetizing Stender dish cover (Fig. 5), and a needle (Fig. 6).

The anesthetizing dish is constructed from a funnel with a molded, fritted glass filter. The bottom of the funnel is cut off, sealed shut, and the bottom flattened. A tube is inserted from the side below the fritted glass filter into the middle of the dish. The outlet of the tube is pointed downward to ensure even distribution of the CO_2. A 1.5—2 cm slot is placed on the side of the funnel to the edge of fritted glass for ease in removal of the anesthetized Habrobracon. The height of the dish above the fritted glass is about 2.5 cm for the 10 cm dishes and about 1.5 cm for the 5 cm dishes.

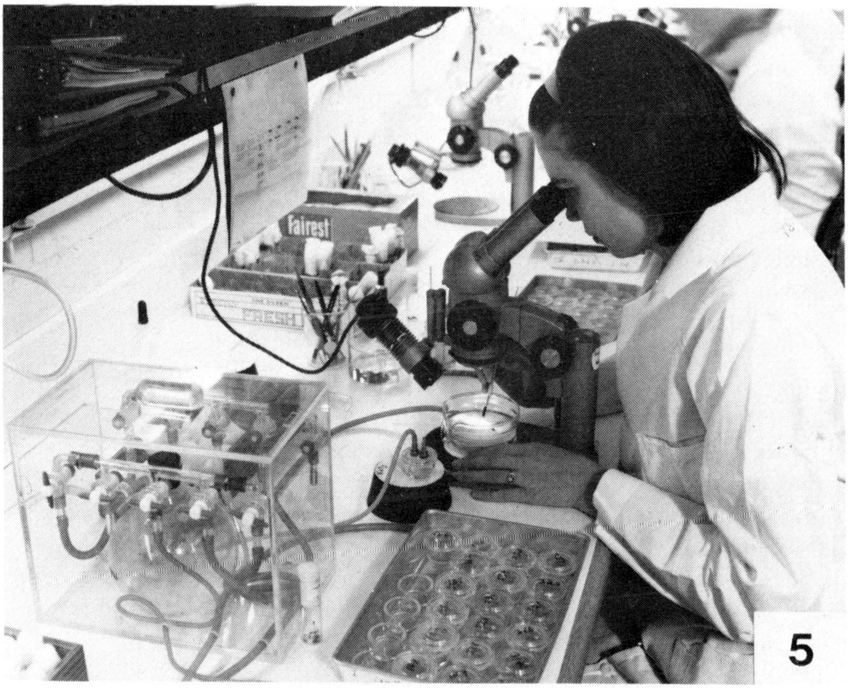

Fig. 5. The anesthetizing system in use under the microscope. (Katherine Taylor Cain).

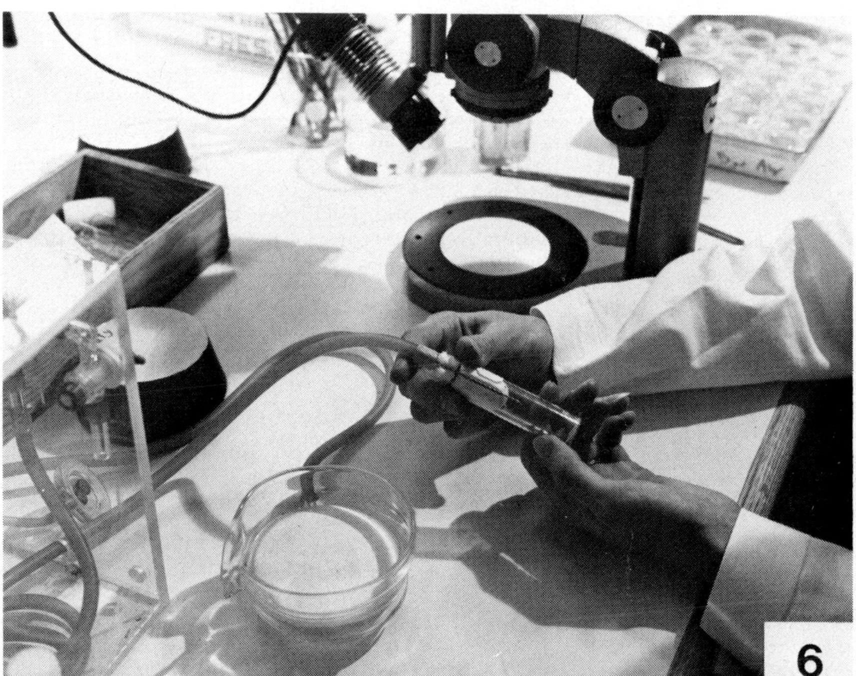

Fig. 6. A hypodermic needle is used for anesthetizing Habrobracon in the shell vials by sticking the needle into the vial beside the plug.

The anesthetizing Stender dish cover is made by drilling two holes in a Stender dish cover and attaching two metal tubes, each with a fine mesh screen, to the holes. One of the holes is the CO_2 inlet, the other an outlet; the screen prevents the wasps from escaping. The needle (No. 18 hypodermic) is used for anesthetizing the wasps in the vials by sticking the needle into the vial adjacent to the plug (Fig. 6).

Dominant lethality and cell death

Types of cell-lethal events

Before discussing the detection and analysis of dominant lethality in Habrobracon we shall review briefly the types of cell-lethal events to be found, and the genetic or molecular mechanisms shown or believed to underly them. The known types of dominant cell-lethal events were tabulated in 1963 (Table I) [23]. There has been little to add since then to explain the phenomenon of instantaneous cell death (Type IVa).

Cell death has been broadly separated into six different categories with some subdivisions. The Type I dominant lethality can be attributed to mitotic inhibi-

TABLE I

TYPES OF DOMINANT LETHAL, CELL KILLING AND SPERM INACTIVATION EVENTS INDUCED BY RADIATION OR CHEMICAL MUTAGENS IN MATURE GAMETES AND DURING GAMETOGENESIS (from ref. 22)

Type	Effect	Probable basis	References
Ia	Death very early in development	Mitotic inhibition through chromosome bridge formation	21
Ib	Death very early in development	Mitotic inhibition induced by ultraviolet and reparable by photoreactivating light	12, 29
Ic	Death very early in development	Mitotic inhibition reparable by presence of another nucleus (genetically nontransmissible damage, not dominant lethality sensu stricto)	11, 12, 22, 26, 28
Id	Death very early in development	Mitotic inhibition, non-reparable	11
II	Late embryonic death before hatching in unfertilized eggs of Habrobracon and after hatching in fertilized eggs	Genetic death, through loss of chromosome parts needed for normal development	1, 21, 27
III	Late embryonic death before hatching in both unfertilized and fertilized eggs of Habrobracon	Genetic death, through loss of chromosomes or drastic loss of chromosome parts	1, 21, 27
IVa	Causes sterile period. Cells die before going through division. Death at G_2 to early prophase.	Unknown	14, 16, 28
IVb	Causes sterile period. Cells die just after going through division. Death at anaphase or early telophase.	Unknown	16
V	Dominant lethality induced in gonial cells	Possibly destruction of gene function or enhancement of normal aging process	6, 8, 22, 28
VI	Sperm inactivation. Causes male sterility.	Unknown	3—5, 9, 10, 13 15, 32, 33

tion in large part from chromosomal bridges [21] when gametes are exposed to environmental mutagens. Type II dominant lethality can be attributed to the loss of chromosome parts one might expect from large deletions which provide too large a genetic imbalance to be completely compensated for by the normal haploid genome in a diploid nucleus [21,27]. Type III dominant lethality is from loss of chromosomes [21,27]. Genetic proof of the different phenotypic manifestations of these chromosomal events in Habrobracon and Drosophila has been provided by von Borstel [21] and von Borstel and Rekemeyer [27], using crossovers in a tandem metacentric of Drosophila to provide chromosomal bridges, translocation heterozygotes of both Habrobracon and Drosophila to provide deletion-duplications, and triploid Habrobracon virgins to provide simulations of loss of chromosomes.

Type IV cell death cannot be mimicked by genetic means because death takes place within two hours in these cells. This type of death is likely to be the same kind as the "lymphocyte death" from X-irradiation which has occupied the working hours of numerous radiation biologists for the past 50 years. Transitional gonial cells in mammals as well as insects are notably sensitive and respond to environmental mutagens by exhibiting this phenotype [14,16]. Excellent evidence has been provided that endomitosis in nurse cells can be inhibited [7], and some of the oocyte deaths can be attributed to this indirect effect [2] (days 3—5 in Fig. 7). Days 5—10 in Fig. 7 represent the true Type IV cell death.

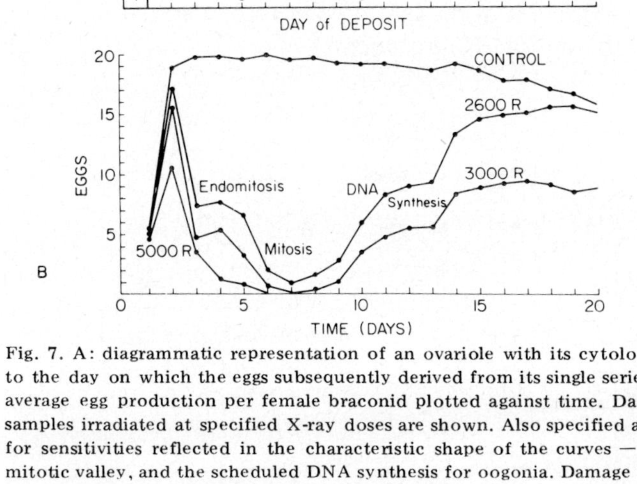

Fig. 7. A: diagrammatic representation of an ovariole with its cytological contents indicated and related to the day on which the eggs subsequently derived from its single series of cell types are deposited. B: the average egg production per female braconid plotted against time. Data from a control sample and from samples irradiated at specified X-ray doses are shown. Also specified are the cytological states responsible for sensitivities reflected in the characteristic shape of the curves — the endomitosis of nurse cells, the mitotic valley, and the scheduled DNA synthesis for oogonia. Damage to follicle cells as well as to oogonia may contribute to the mitotic valley (from ref. 2).

Type V dominant lethality presents a phenotype like that experienced from irradiating the cytoplasm of a mature egg with ultraviolet radiation [22,28]. Type V dominant lethality is seen after irradiation of oogonia, and the primary event is not manifest until late in embryonic development. Dominant lethality after irradiation of oogonia was discovered by LaChance [6]. As yet no genetic or molecular experiment has been undertaken which can test alternative explanations (e.g. sensitivity of cytoplasmic elements, enhanced aging of the oogenetic system, or loss of function of cells supporting egg formation) for Type V death.

Type VI cell death can be identified easily in any hymenopteran species where it has been sought. In diploid species special techniques must be used to distinguish sperm inactivation from dominant lethality; in hymenopteran species, dominant lethality induced in a sperm kills the egg, and sperm inactivation under extreme conditions permits the egg to still undergo haploid development as though mating of a male with a virgin had not taken place. Investigation has not been carried out to see whether the mechanism for sperm inactivation is from sperm immotility, from inability for a sperm to enter an egg, from inability for a sperm nucleus to combine with an egg nucleus, or from all of them.

Phenotypes of dominant lethal events in Habrobracon

Type I dominant lethality in Habrobracon is manifested as death early in embryonic development. Specifically, the syndrome is production of nuclei through five, six or seven replications; the nuclei enlarge dramatically and all but one or two of them are Feulgen-negative [20]. Fortunately the egg chorion is thin (2 μm) and transparent and consequently the embryo can be scored for stage of death without resorting to fixing and staining if the humidity is kept above 60% during incubation to avoid the egg becoming flaccid and wrinkled. Consequently, the Type I dominant lethal phenotype is seen as Stage 1 for *H. juglandis* and similar to Stage 2 for *H. serinopae* in Fig. 8 (see next section for precise description).

Type II dominant lethality was first described by Atwood et al. [1]. Death occurs late in embryonic development in the haploid embryo (Fig. 8, Stages 3–5) and in the larval stages of the diploid Habrobracon.

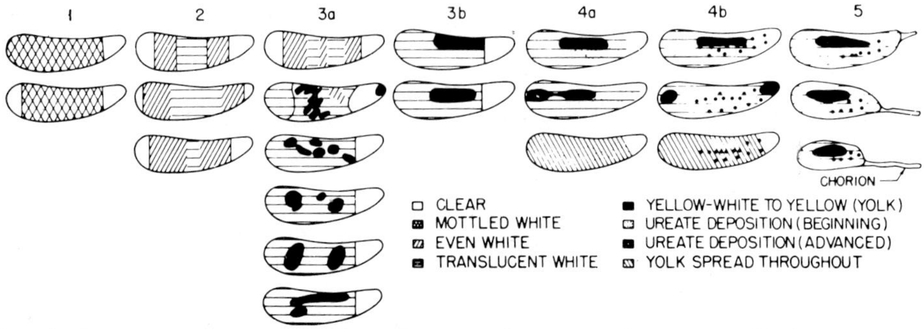

Fig. 8. Stages at which dominant lethality is expressed during the development of Habrobracon embryos (from ref. 27).

Type III dominant lethality is characterized by death at Stages 3 and 4 in the haploid embryo and at Stages 3—5 in the diploid embryo [27].

Type IV cell death is visualized and quantitated by the depression of egg production that takes place from days 5—10 of laying (see Fig. 7) [2,23].

Type V dominant lethality exhibits a curious fragility and milkiness of the developing embryo; if the development proceeds to later stages the developing gut usually bursts and yolk is spread throughout the embryo interior [22,28].

Type VI inactivation can be identified by treating males, mating them to untreated virgins, and examining the sex ratios in comparison with hatchability [13,32].

Stages of death

Embryos

The stages of death can be determined by following the diagram in Fig. 8. The embryos are placed in five categories, with Stages 3 and 4 being subdivided.

Stage 1 represents embryos in which death took place during karyokinesis before blastoderm formation; the embryos are mottled and white in appearance, and one or both ends is filled with a clear fluid.

Stage 2 is characterized by two white bands; it is seldom seen after treatment of *H. juglandis* and is commonly seen after treatment of *H. serinopae.* In *H. serinopae*, common practice is to classify them as Stage 1, so Stage 2 is seldom used in classification in mutagenesis experiments.

Stage 3 represents embryos where the yolk sac is changing from white to yellow and the yolk is becoming more centrally located.

Stage 4 is a nearly fully-developed embryo and is characterized by the presence of ureate cells. The gut occasionally bursts in Stage 4 embryos and the yolk becomes distributed throughout the embryo. Even so, Stage 4 embryos with ruptured guts appear different from the Type V dominant lethals described earlier.

Stage 5 represents embryos fully-developed and trying to emerge from the egg chorion.

Larval stages

Dead larvae are divided into five categories; although there are five larval instars, these categories do not represent the instars.

Stage 6 is a non-feeding larva which has emerged completely from the egg chorion but has often not moved far from it. These can be scored at the time hatchability is taken.

Stage 7 represents dead larvae which have just begun to feed but have died early in development. These are difficult to find after the adults emerge. Consequently, any deficiencies in scoring at time of emergence are recorded as Stage 7 deaths.

Stage 8 represents small to medium-sized dead larvae.

Stage 9 represents large dead larvae.

Stage 10 are large larvae which have begun to spin cocoons, but where a meconium has not been deposited.

Pupal stages

Four stages are used to categorize the pupae. The pupal stage begins when a meconium has been deposited.

Stage 11 represents larvae in cocoons, where the meconium has been deposited. The sexes cannot yet be distinguished.

Stage 12 pupae have distinguishable sexes; the pupae are lightly pigmented or non-pigmented.

Stage 13 pupae are well pigmented, but obviously are not ready to emerge; for example, the wings have not begun to expand.

Stage 14 pupae are nearly ready to emerge from the cocoons; the wings are partly or fully expanded.

Testing for dominant lethal action of mutagens

Procedures for testing

Virgins can be collected individually from cocoons or with the use of temperature-sensitive strains wherein the males all die and the females are heterozygous for the temperature-sensitive lethal [17]. The females should be tested for spontaneous mutations either directly or by their being collected through pedigrees before testing begins.

The methods for exposure of Habrobracon to environmental mutagens have been discussed elsewhere in detail [18]. In brief, mutagens can be applied by feeding, by injection, by aerosols and by topical application. Excellent dose-action data have been compiled after each of the four methods of application [18], so all are reliable.

After the virgins are exposed and the residual mutagen inactivated, usually by waiting for a day, one-half of them are mated to untreated males. After mating, usually observed individually, each female is placed in a Stender dish for six hours. Eggs laid in this time period were in the first meiotic metaphase during the treatment. If more than 20 eggs per Ephestia larva have been oviposited at this or any succeeding transfer, then the excess number are removed when the eggs are counted after the transfer. The females are transferred for a three-hour period during which a mixture of metaphase and late prophase eggs are laid. The next two transfers of females of six to twelve hours comprise eggs which were oocytes in the diplotene state when treated. Earlier oocyte stages are represented during the next two days, and at this time transfers are done twice daily. From days 5—10 the eggs represent treated transitional oogonia, and transfers are done once daily. After day 10 the eggs represent treated oogonia, and transfers may be done once or twice daily. A schematic and graphic representation of these stages is shown in Fig. 7.

In order to test effects of mutagens on sperm, males are treated with the presumptive mutagen and mated to females. Pair matings are observed. The females are transferred twice daily for two or three or even more days. Excess eggs over 20 per larva are removed when egg counts are made immediately after each transfer of females. More than 20 larvae per host will starve.

Since at 28°C the normal embryo hatches in 29 hours, hatchability and dead embryo classifications are done between 30 and 36 hours after each transfer. After the adults have emerged about 10 days later, dead larva and dead pupa classifications are performed, and sex ratios are taken. If further testing for recessive lethals and chromosomal aberrations is to be done, then the adult females are removed from the cocoons manually to avoid their mating with their brothers.

Analytical procedures

The estimation of genetic parameters is not straightforward because not all of the eggs are fertilized after an insemination. Equations have been developed which are precise [1,27,31]. The final development of these equations is presented here from Whiting et al. [31] because it is in a source that is often not accessible.

The frequency of embryo dominant lethal mutations, D, that depress hatchability when irradiated females are mated to unirradiated males, can be estimated by the equation

$$D = 1 - \frac{m - V_u(1 - f)}{f} \tag{1}$$

where m is the hatchability of eggs from unmated females, V_u is the surviving proportion of unfertilized eggs (determined from unmated females), and f is the proportion fertilized. Hatchability is a simple and commonly used method for obtaining an estimate for dominant lethality since embryo lethality is a far higher proportion of total lethality than either larval or pupal lethality. To estimate D′, the frequency of total dominant lethal mutations, adult survival data can be inserted in equation (1); m′ and V'_u from adults correspond to m and V_u for the embryos.

The determination of the frequency of fertilized eggs, f, is taken from the experimental animal data by the equation

$$f = 1 - \frac{p}{V'_u} \tag{2}$$

where p is the ratio of surviving adult males from mated females to the total eggs from mated females. In irradiated eggs, the frequencies of fertilized eggs dying at any stage of development, s_f, can be computed by

$$s_f = \frac{s_m - s_u(1 - f)}{f} \tag{3}$$

where s_m is the proportion of embryos from mated females that die at a particular stage of development and s_u is the proportion of embryos from unmated females that die at the same developmental stage. When this equation is used for all stages of embryonic, larval, and pupal development, it is a stage-by-stage analysis in which the effect of fertilization can be compared between unfertilized and fertilized eggs.

Fertilized eggs that survive to adulthood are simply those without a dominant lethal mutation

$$V'_f = 1 - D' \quad . \tag{4}$$

Thus the frequency of genomes bearing at least one recessive lethal mutation, r′, can be obtained from the F_1 lethality data by

$$r' = 1 - (V'_u/V'_f) \, . \tag{5}$$

More commonly, as explained earlier, the F_1 daughters are collected as virgins and tested separately. If any one of these has a depressed hatchability, she is mated, and an analysis based on equations (1), (2), and (3) is performed on the data obtained from her offspring. From this it can be judged whether she was heterozygous for one or more recessive lethals, one or more translocations, or both.

Computer programs have been developed for data analysis, and these speed the analysis substantially [24].

Costs

We shall list the costs in person-days. The research is labor intensive and, after capital equipment has been purchased, the operating costs other than labor are trivial.

Two persons can test one compound quantitatively for its capacity to induce dominant lethality in Habrobracon in thirty days. This would include three different doses of the mutagen. If quantitative data are wanted on recessive lethals and chromosomal translocations as well, then two persons can test one compound in approximately 100 days.

References

1 Atwood, K.C., R.C. von Borstel and A.R. Whiting, An influence of ploidy on the time of expression of dominant lethal mutations in Habrobracon, Genetics, 41 (1956) 804—813.

2 Grosch, D.S., Environmental aspects: radiation, in The Physiology of Insecta, Vol. II, 2nd edition, Academic Press, New York, 1974, pp. 85—126.

3 Heidenthal, G., The occurrence of X-ray induced dominant lethal mutations in Habrobracon, Genetics, 30 (1945) 197—205.

4 Jaynes, H.A. and P.A. Godwin, Sterilization of the white-pine weevil (*Pissodes strobi* Peck) with gamma radiation, J. Econ. Entomol., 50 (1957) 393—395.

5 Kerschner, J., Dominant lethals induced by X-rays in sperm of the wasp Melittobia, Anat. Rec., 96 (1946) 556.

6 LaChance, L.E., The effect of chelation and X-rays on fecundity and induced dominant lethality in Habrobracon, Radiation Res., 11 (1959) 218—228.

7 LaChance, L.E. and S.B. Bruns, Oogenesis and radiosensitivity in *Cochliomyia hominivorax* (Diptera: Calliphoridae), Biol. Bull., 124 (1963) 65—83.

8 LaChance, L.E. and A.P. Leverich, Radiosensitivity of developing reproductive cells in a female *Cochliomyia hominivorax*, Genetics, 47 (1962) 721—735.

9 Lee, W.R., The dosage response curve for radiation induced dominant lethal mutations in the honeybee, Genetics, 43 (1958) 480—492.

10 Lefevre, G., Jr. and U.B. Jonsson, Sperm transfer, storage, displacement, and utilization in *Drosophila melanogaster*, Genetics, 47 (1962) 1719—1736.

11 Löbbecke, E.-A. and R.C. von Borstel, Mutational response of Habrobracon oocytes in metaphase and prophase to ethyl methanesulfonate and nitrogen mustard, Genetics, 47 (1962) 853—864.

12 Löbbecke, E.-A. and R.C. von Borstel, Genetically nontransmissible nuclear damage induced by ultraviolet radiation in the wasp Habrobracon, Genetics, 48 (1963) 1313—1322.

13 Maxwell, J., Inactivation of sperm by X-radiation in Habrobracon, Biol. Bull., 74 (1938) 253—255.
14 Monesi, V., Relation between X-ray sensitivity and stages of the cell cycle in spermatogonia of the mouse, Radiation Res., 17 (1962) 809—838.
15 Mortimer, R. and R.C. von Borstel, Radiation-induced dominant lethality in haploid and diploid sperm of the wasp Mormoniella, Genetics, 48 (1963) 1545—1549.
16 Oakberg, E.F., Initial depletion and subsequent recovery of spermatogonia of the mouse after 20 R of gamma rays and 100, 300, and 600 R of X-rays, Radiation Res., 11 (1959) 700—719.
17 Smith, R.H., Temperature-sensitive mutations in Habrobracon, Genet. Res., 23 (1974) 63—74.
18 Smith, R.H. and R.C. von Borstel, Inducing mutations with chemicals in *Habrobracon*, in A. Hollaender (ed.), Chemical Mutagens, Principles and Methods for their Detection, Vol. 2, Plenum Press, New York, 1971, pp. 445—459.
19 Smith, R.H. and R.C. von Borstel, Genetic control of insect populations, Science, 178 (1972) 1164—1174.
20 von Borstel, R.C., Feulgen-negative nuclear division in Habrobracon eggs after lethal exposure to X-rays or nitrogen mustard, Nature, 175 (1955) 342.
21 von Borstel, R.C., Sulla natura della lethalità dominante indotta dalle radiazione, Atti Assoc. Genet. Ital., 5 (1960) 35—50.
22 von Borstel, R.C., Induction of nuclear damage by ionizing and ultraviolet radiation, in B.Chr. Christensen and B. Buchmann (eds.), Progress in Photobiology, Proc. 3rd Int. Congr. Photobiol., Elsevier, Amsterdam, 1961, pp. 243—250.
23 von Borstel, R.C., Effects of radiation on germ cells of insects: dominant lethals, gamete inactivation, and gonial cell killing, in Radiation and Radioisotopes Applied to Insects of Agricultural Importance, Proc. Symp. on Use and Application of Radioisotopes and Radiation in the Control of Plant and Animal Insect Pests, (Athens, 1963), International Atomic Energy Agency, Vienna, 1963, pp. 367—385.
24 von Borstel, R.C., Quarterly Progress Report to the National Aeronautics and Space Administration for the Period Covering July 1 to September 30, 1966. Oak Ridge National Laboratory Technical Memorandum — 1732, 1967, pp. 42.
25 von Borstel, R.C., Quarterly Progress Report to the National Aeronautics and Space Administration for the Period Covering October 1 to December 31, 1966. Oak Ridge National Laboratory Technical Memorandum — 1957, 1967, pp. 250.
26 von Borstel, R.C. and E.-A. Löbbecke, Cell killing and the problem of nuclear reactivation, Abstracts of the Sessions of the Second International Congress of Radiation Research (Harrogate, 1962), 1962, p. 240.
27 von Borstel, R.C. and M.L. Rekemeyer, Radiation-induced and genetically contrived dominant lethality in Habrobracon and *Drosophila*, Genetics, 44 (1959) 1053—1074.
28 von Borstel, R.C. and W.St. Amand, Stage sensitivity to X-radiation during meiosis and mitosis in the egg of the wasp Habrobracon, in F.H. Sobels (ed.), Symposium on Repair from Genetic Radiation Damage and Differential Radiosensitivity in Germ Cells (Leiden, The Netherlands, 1962), Pergamon Press, London, 1963, pp. 87—100.
29 von Borstel, R.C. and S. Wolff, Photoreactivation experiments on the nucleus and cytoplasm of the Habrobracon egg, Proc. Natl. Acad. Sci. (U.S.), 41 (1955) 1004—1009.
30 Wallace, B. and Th. Dobzhansky, Radiation, Genes, and Man, Holt. New York, 1959.
31 Whiting, A.R., R.H. Smith and R.C. von Borstel, Methods for radiation studies during oogenesis in *Habrobracon juglandis* (Ashmead), in Effects of Radiation on Meiotic Systems, International Atomic Energy Agency, Vienna, 1968, pp. 201—208.
32 Whiting, A.R. and R.C. von Borstel, Dominant lethal and inactivation effects of nitrogen mustard on Habrobracon sperm, Genetics, 39 (1954) 317—325.
33 Yanders, A.F., The effects of X-rays on sperm activity in Drosophila, Genetics, 44 (1959) 545—546.

ROOT TIPS OF *VICIA FABA* FOR THE STUDY OF THE INDUCTION OF CHROMOSOMAL ABERRATIONS

B. A. KIHLMAN

Department of Genetics and Plant Breeding, Agricultural College, S-750 07 Uppsala (Sweden)

INTRODUCTION

One of the oldest, simplest and least expensive methods for studying the induction of chromosomal aberrations utilizes plant root tips as experimental material. The merits of this material were already realized by radiobiologists in the 1930's. The root tip technique is particularly well suited for the study of chemically induced aberrations. However, the method is no exception to the rule that reliable and valuable results can only be obtained when the experiments are correctly and carefully carried out. This requires a good knowledge of the material and its reactions towards different types of treatment.

THE MATERIAL

Root tips from several plant species have been used for the study of induced chromosomal aberrations. However, most common as experimental material are probably root tips of *Vicia faba* (horse bean, broad bean, field bean) or of various *Allium* species (*e.g.* the common onion, *Allium cepa*, the tree onion, *Allium proliferum*, and the Welsh leek *Allium fistulosum*). Because of their low chromosome numbers and large chromosomes, all these are favourable cytological materials, but they also have the advantage of being available all the year round and easily handled and cultivated. In this paper only one material will be described, *i.e.* root tips of the field bean, *Vicia faba*, var. *minor*. For information about the cultivation and handling of *Allium* root tips, the reader may consult ref. 2.

Vicia faba is a popular material which has been used extensively, not only in cytological, but also in physiological and radiobiological studies. An additional advantage, of course, is that its physiological properties and reactions towards external agents are so well known.

As shown in Fig. 1, a root tip cell of *Vicia faba* contains 5 pairs of nearly equally long chromosomes with sub-terminal centromeres (S-chromosomes) and one pair with median centromere (M-chromosome), the diploid chromosome number (2n) thus being

Abbreviations: maleic hydrazide, 1.2-dihydro-3.6-pyridazinedione; tepa, Tris(1-aziridinyl)-phosphine oxide; thiotepa, Tris(1 aziridinyl)-phosphine sulphide.

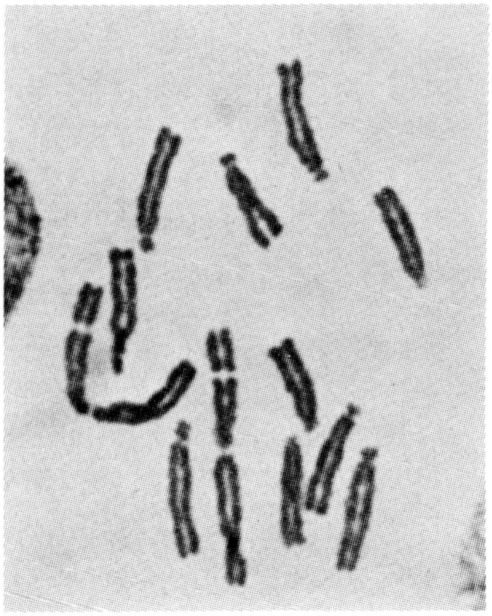

Fig. 1. The chromosome complement of the field bean, *Vicia faba*, var *minor* (2n = 12). From KIHLMAN[2]. (By courtesy of Plenum Press, New York.)

12. The M-chromosome contains a secondary (nucleolar) constriction which separates a large satellite from the rest of the chromosome. The ratio of the length of the M-chromosome to the mean length of the S-chromosomes is approx. 2.3:1. If chemically induced aberrations were randomly distributed between the chromosomes, 2.3 times as many aberrations should occur in S- as in M-chromosomes. However, since aberrations induced by chemical substances are seldom randomly distributed, the S/M aberration ratio usually diverges considerably from 2.3:1.

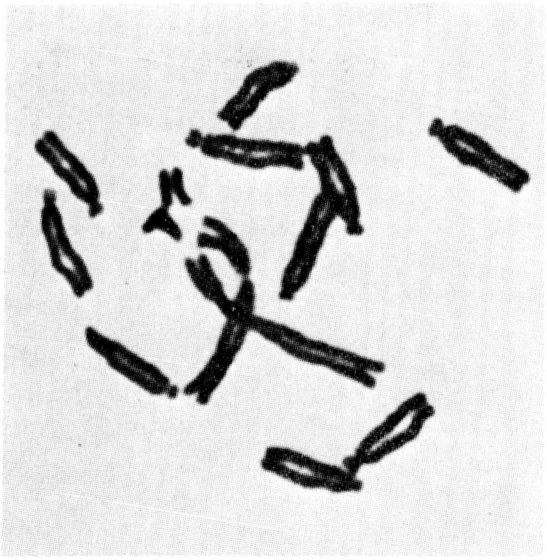

Fig. 2. Chromatid exchange localized in the nucleolar constriction after treatment with 8-ethoxy-caffeine. From KIHLMAN[1]. (By courtesy of Prentice-Hall, Englewood Cliffs.)

After treatment with certain chemicals, chromosomal aberrations may be localized almost exclusively in some particular region of one of the chromosome pairs[1] (Fig. 2).

At 20° the mean duration of the mitotic cycle in lateral roots of the field bean has been estimated to be 19.7 h. The interphase time was 17.7 h, 8.1 of which were spent in S (the period of DNA synthesis), 4.1 in G_1 (the period between telophase and S), and 5.5 in G_2 (the period between S and prophase). Mitosis itself lasted 2 h.

METHOD FOR GROWING ROOTS FROM SEEDS

Soaking

Seeds are soaked for 6–12 h in tap water. Before soaking, seeds may be disinfected by short immersion (3–10 min) in a decanted 5% calcium hypochlorite solution. However, as a rule, disinfection is not necessary.

Germination

After soaking, the seeds are allowed to germinate in moist Vermiculite or in Perlite (Deutsch. Perlite AB, Dortmund) for 4 days at 20°. At the end of this 4-day period, the seedlings have grown 3–5 cm long primary roots and are ready to be transferred to a water tank. However, before this is done, the seed coat should be removed and the shoot cut off. If lateral roots are to be used in the experiment, the tip of the primary root should be cut off as well, since the removal of the primary root meristem stimulates the growth of the lateral roots.

Growth period in water

The water in the tank should be fresh, well aerated and of constant temperature (*e.g.* 20°) during the growth period. These requirements can usually be met by growing the roots in thermo-regulated, running tap water. When tap water cannot be used (for instance when it is too heavily chlorinated), Hoaglands salt solution is recommended. This solution has the following composition: 0.005 M KNO_3, 0.005 M $Ca(NO_3)_2$,

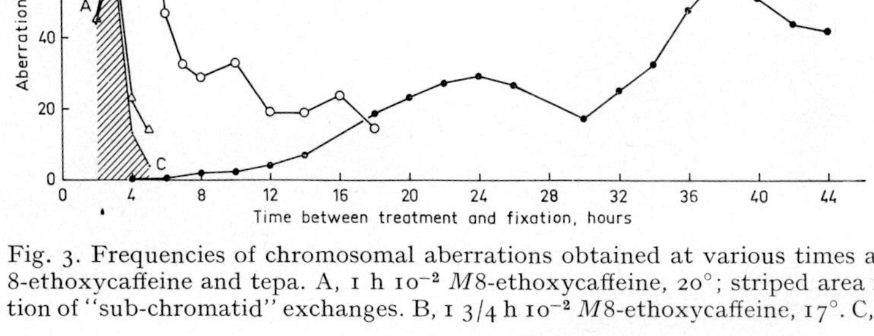

Fig. 3. Frequencies of chromosomal aberrations obtained at various times after treatments with 8-ethoxycaffeine and tepa. A, 1 h 10^{-2} M 8-ethoxycaffeine, 20°; striped area indicates the proportion of "sub-chromatid" exchanges. B, 1 3/4 h 10^{-2} M 8-ethoxycaffeine, 17°. C, 2 h $7.5 \cdot 10^{-4}$ M tepa.

0.005 M $MgSO_4$, 0.001 M KH_2PO_4, and 0.5% iron tartrate (1 ml/l). Primary roots can be used after 24 h in the water tank. To obtain lateral roots of suitable length (1–2 cm), the seedlings usually have to be grown for 4 days.

TREATMENT OF ROOT TIPS WITH CHEMICALS

When chemical substances are being tested for their chromosome-breaking activity, it should be realized that the effect is dependent not only on the structure and dose of the chemical, but also on factors such as the stage of the cell cycle at the time of treatment and the pH, temperature and oxygen tension of the treatment solution. Thus, for instance, there are substances in plant materials that act mainly or

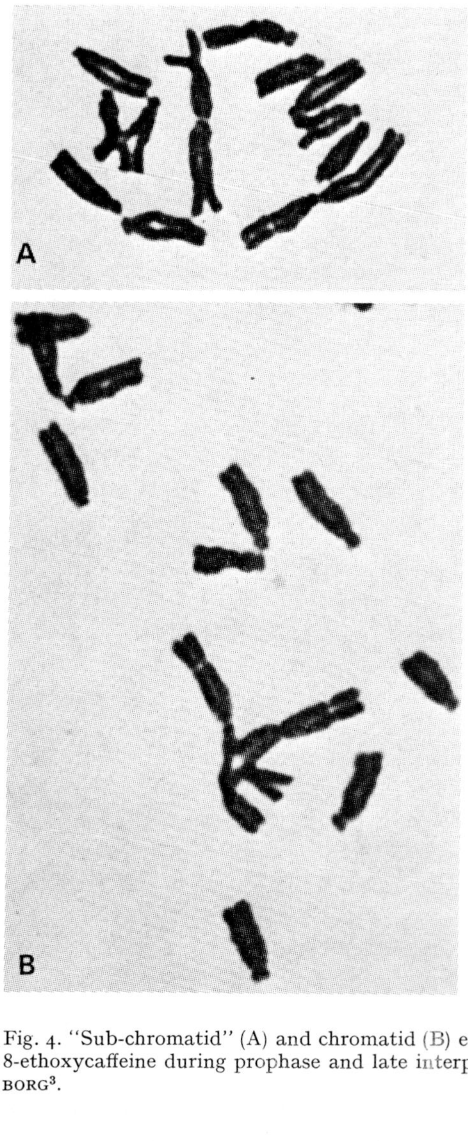

Fig. 4. "Sub-chromatid" (A) and chromatid (B) exchanges in metaphase after exposure of cells to 8-ethoxycaffeine during prophase and late interphase, respectively. From KIHLMAN AND KRONBORG[3].

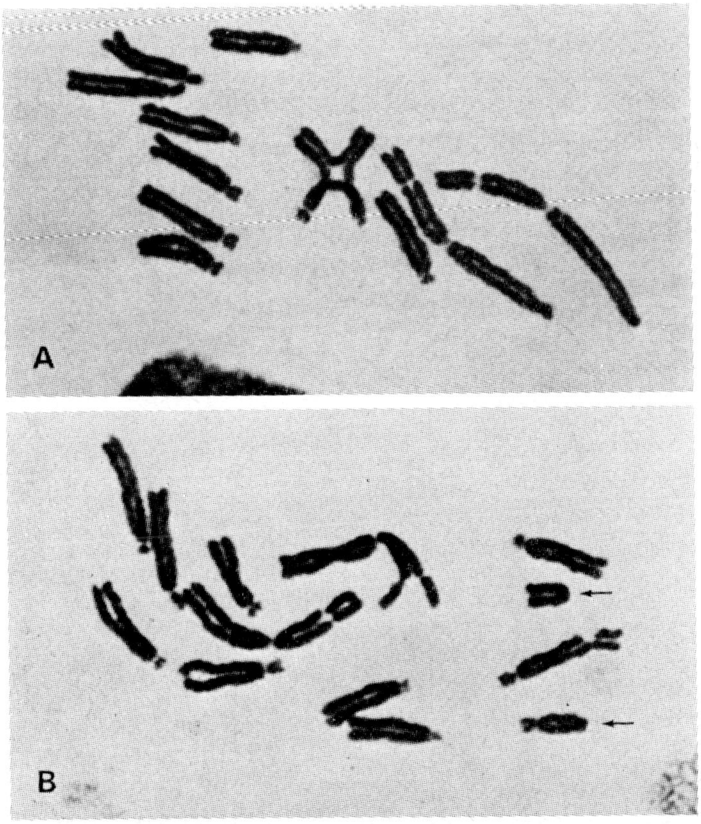

Fig. 5. Chromatid-type aberrations induced by the trifunctional alkylating agent tepa. A, Chromatid exchange (from STURELID[5]); B, chromatid break and isochromatid break (arrows) with proximal and distal sister union.

exclusively during late interphase and mitosis. Methylated oxypurines, such as caffeine, 8-methoxycaffeine, 8-ethoxycaffeine, and 1,3,7,9-tetramethyluric acid are substances of this type[1]. These agents produce "sub-chromatid" exchanges in prophase and chromatid aberrations in G_2, but very few, if any, aberrations at other stages of the cell cycle (Figs. 3A, B; 4A, B). Other substances appear to be active during all stages of the cell cycle, but since the formation of the aberrations is dependent on

TABLE I

TYPES AND FREQUENCIES OF CHROMOSOMAL ABERRATIONS OBTAINED AT 5 AND 18 h AFTER 1-h TREATMENTS WITH BLEOMYCIN.

Concentration of Bleomycin (mg/l)	*Time between treatment and fixation (h)*	*Abnormal metaphases (%)*	*Aberrations per 100 cells*					*Total aberrations per 100 cells*
			Breaks		*Exchanges*			
			Chromatid	*Isochromatid*	*"Subchromatid"*	*Chromatid*	*Chromosome*	
2	5	43	4	4	58	12	0	78
2.5	18	15	1	10	0	5	3	19

DNA synthesis, only cells exposed during or before S contain aberrations in the first mitosis after treatment. Such S-dependent effects are produced by alkylating agents (Fig. 3C). The aberrations produced by these agents are exclusively of the chromatid type (Fig. 5A, B). Finally, there are substances such as the antibiotics Phleomycin and Bleomycin that are active during all stages of the cell cycle, apparently independently of DNA synthesis, and that produce "sub-chromatid-", chromatid-, and chromosome-type aberrations (Table I).

The effect of substances with acid or basic properties is influenced by the pH of the treatment solution. Thus, the herbicide maleic hydrazide has a 6-fold stronger effect at pH 4.7 than at pH 7.3 (Fig. 6).

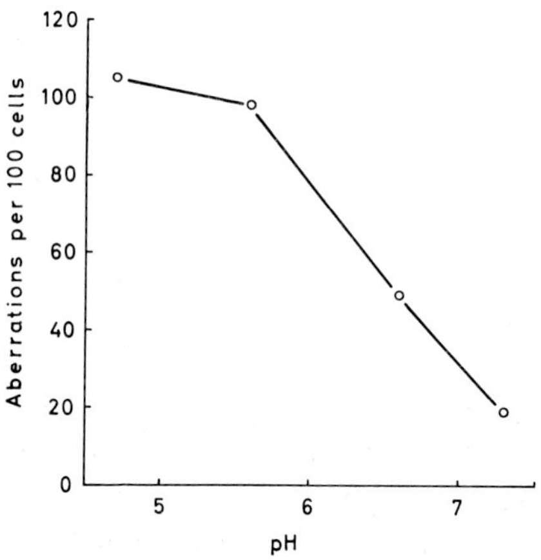

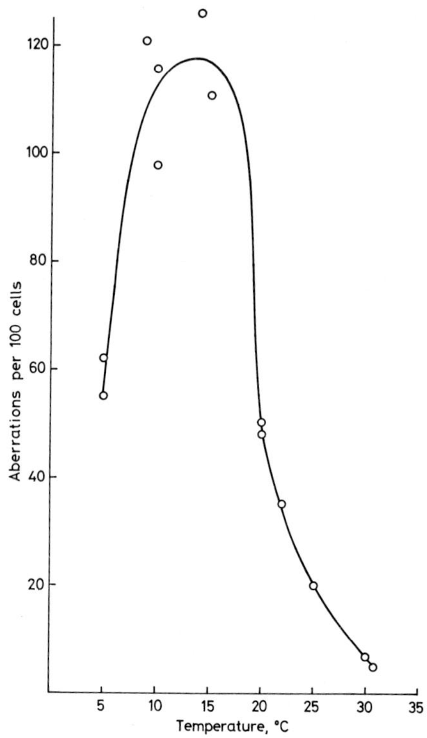

Fig. 6. The influence of the pH of the treatment solution on the frequency of aberrations obtained after treatments with the herbicide maleic hydrazide. From KIHLMAN[2]. (By courtesy of Plenum Press, New York.)

Fig. 7. The influence of treatment temperature on the frequency of aberrations obtained after treatments with 8-methoxycaffeine. From KIHLMAN AND KRONBORG[3].

Most chemical substances are more active at high than at low temperatures. Methylated oxypurines are exceptions to this rule. These substances appear to be most active at temperatures between 10° and 15°, and practically inactive at temperatures above 25° (Fig. 7).

The frequency of chromosomal aberrations produced by chemical substances may also be dependent on the oxygen tension during treatment. As a rule, a reduction of the oxygen tension results in a lower frequency of aberrations (Fig. 8).

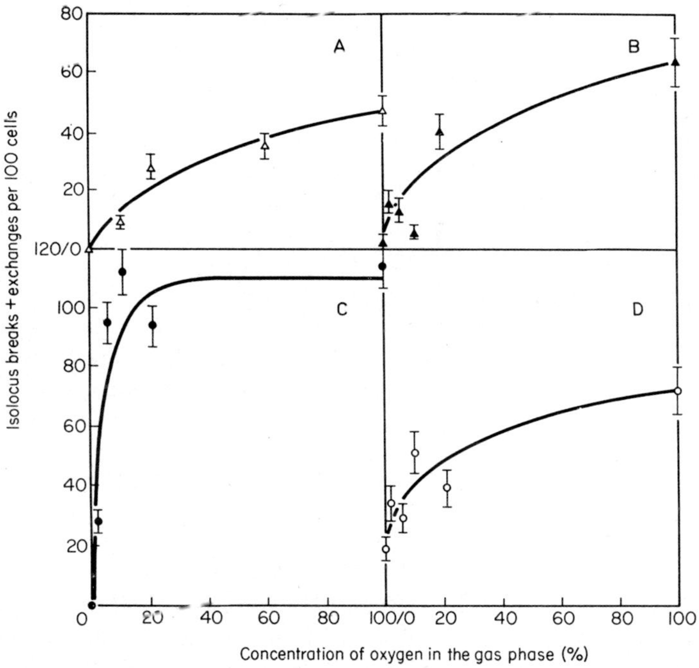

Fig. 8. The influence of oxygen concentration on the frequencies of aberrations obtained after treatments with potassium cyanide (A), methylphenylnitrosamine (B), 8-ethoxycaffeine (C), and maleic hydrazide (D). From KIHLMAN[1]. (By courtesy of Prentice-Hall, Englewood Cliffs.)

All these modifying factors should be considered when the experiment is planned. I should like to suggest the following procedure.

Preparation and volume of treatment solution. Size of experiment

Freshly prepared solutions should always be used, since aqueous solutions of chemical substances are often unstable. For the same reasons the solutions should not be exposed to strong light or high temperatures. If the substance does not dissolve easily in cold water, the water should be heated to the required temperature before the substance is added. As soon as the substance has dissolved, the solution should be chilled to room temperature. The treatments may be carried out in vials or tubes containing at least 25 ml of solution. Primary roots of three seedlings or lateral roots of one seedling can be treated in such vials. The number of seedlings to be used per treatment and fixation time depends on the type of root. In experiments with lateral roots, at least two seedlings should be used and in experiments with primary roots, at least six.

Treatment conditions

Treatment as well as recovery should take place in the dark. Well aerated, buffered solutions should be used and the temperature should be kept constant, since it is important that the treatments are carried out under controlled conditions.

Concentration of the substance to be tested

The chemical substances should be tested over a wide range of concentrations.

The upper limit of concentration is set by the water solubility of the substance as well as by its toxicity. The lethal concentration, if any, is of course dependent on the duration of the treatment. Plainly, there is also an upper limit of concentration worth testing. If the purpose is to find out whether the substance is a genetic risk, there is little point in testing concentrations more than 100 times higher than those likely to occur in the environment of man (after allowing for concentration along food chains, *etc.*).

Duration of the treatment period

It is advisable to keep the period of treatment as short as possible. If the treatment period is short, it is easier to keep the treatment conditions constant. It is also easier to determine the stage(s) of the mitotic cycle during which the aberrations are produced. However, since some substances are active only when given over an extended period, 24-h treatments should be included in the test. It should then be borne in mind that cells fixed after such long periods of treatment represent a very heterogeneous population with respect to the stage of the cell cycle at the time of treatment.

Fixation time

Since different chemical substances may produce aberrations at different stages of the cell cycle and affect the duration of the mitotic cycle differently, several fixation times should be used. When aberrations are produced in G_2 or mitosis, they appear in metaphase-anaphase from a few minutes to about 8 h after the beginning of the treatment. Methylated oxypurines produce this type of effect (Fig. 3A, B). On the other hand, when roots are exposed to alkylating agents, the appearance of aberrations in metaphase is delayed and the peak frequency of aberrations may not be obtained until the second mitosis after treatment, *i.e.* at about 40 h after the beginning of treatment (Fig. 3C). It is, therefore, advisable that the test should include at least one experiment in which roots are fixed at short intervals and over a period long enough to cover at least one, and preferably two, whole mitotic cycles.

Colchicine treatment

When aberrations are to be scored in metaphase, the roots are, as a rule, treated with colchicine before fixation. Colchicine prevents the formation of a functional mitotic spindle and thereby the arrangement of chromosomes on the metaphase plate. Furthermore, the splitting of the chromosomes into daughter chromosomes is delayed by colchicine, and, as a result, there is an accumulation of cells at a stage (c-metaphase) corresponding to metaphase in untreated cells. The analysis of the aberations is much facilitated by these effects. A suitable treatment is to expose the roots for 2–3 h at 19–20° to an aerated solution containing 0.02–0.05% colchicine.

FIXATION AND STAINING OF ROOT TIPS

Fixation

Roots are conveniently fixed in a mixture containing 3 parts of 100% alcohol and 1 part of glacial acetic acid. The alcohol can be either ethanol or methanol, the latter usually giving better results. The fixative should be freshly prepared and cool

(4–10°). Although fixation is completed within 15–20 min, the preparation of permanent slides is facilitated by leaving the roots in the fixative overnight at 4°. If kept in a freezer, roots can be stored in the fixative for several days, or even weeks.

Staining and preparation of slides

For preparation of slides, the Feulgen squash technique is recommended. A description of this well-known technique may seem superfluous but is nevertheless given here.

From the fixative the roots are transferred to distilled water and left there for a few minutes. They are then hydrolysed in 1 *M* hydrochloric acid (HCl) at 60° for 6–7 min. After hydrolysis, the roots are transferred to leukobasic fuchsin, *i.e.*, the Feulgen stain. Leucobasic fuchsin is prepared in the following way. One gram basic fuchsin (a suitable stain is "Diamantfuchsin für die Bakteriologie", Art. 1358, E. Merck AG, Darmstadt) is dissolved in 200 ml of distilled water that has been heated to boiling point. After the solution has been cooled to 50°, it is filtered into a dark bottle. To the filtered solution is added 20 ml of 1 *M* HCl and, when the temperature of the solution is 25° or less, 1.2 g potassium pyrosulphite (potassium metabisulphite), $K_2S_2O_5$. The solution is allowed to bleach at room temperature for 24 h in the dark. The bottle should be tightly stoppered. The decolorized stain is stored in the dark at 4°.

When staining is complete, which at 20° takes 45–60 min, the roots are transferred onto clean slides. The number of roots to be used for each slide depends on the type of root. The meristem from one primary root or three lateral roots usually gives a suitable number of cells. On the slide, the dark-stained tips are separated from the rest of the roots, crushed in a drop of 45% acetic acid with the flat end of an aluminium rod and squashed under a cover slip. The pressure is applied under several thicknesses of blotting paper. Sideways movements of cover slip must be avoided.

Preparations are made permanent according to the dry-ice procedure. The slides are placed on a block of dry ice and left there until they are thoroughly frozen which takes a couple of minutes. The cover slip is then popped off by slipping a scalpel under one of its corners. The material sticks to the slide, which is quickly passed to 100% ethanol (two changes). From the second change of ethanol the preparations can be mounted in Euparal, using a clean cover slip. If Canada balsam is used as a mounting medium, the slide is passed from ethanol to xylene (two changes) before mounting.

SCORING OF SLIDES

Chromosomal aberrations may be scored both in anaphase and in metaphase. However, scoring at metaphase is usually preferable, since only at this stage is it possible to distinguish the various types of aberration and to make a detailed analysis of the effect. As a rule, anaphase scoring should be used only in preliminary experiments to find out whether a chemical is active. The more rapid and simple anaphase method may also be preferable when the purpose of the experiment is to obtain an approximate idea of the concentrations which should be used or of the stage of the mitotic cycle during which the substance is most active. When anaphases are scored, the effect can usually be given only as the percentage of abnormal anaphases or as fragments and bridges per (100) anaphase(s). However, there is one type of aberration that can be scored more easily at anaphase than at metaphase, and that is the so-called

Fig. 9. Side-arm bridges at anaphase after treatment with 8-ethoxycaffeine during early prophase.

"sub-chromatid" exchange (Fig. 4A). During anaphase such exchanges give rise to characteristic side-arm bridges or "pseudochiasmata" (Fig. 9).

Metaphase scoring makes it possible to distinguish not only between "sub-chromatid-", chromatid- and chromosome-type aberrations, but also between the various types of breaks and exchanges that are produced by chemical substances. These aberrations have been described and schematically and photomicrographically illustrated in an earlier publication[2]. A "sub-chromatid" exchange is shown in Fig. 4A, chromatid exchanges in Figs 2, 4B and 5A, whereas Fig. 5B shows one chromatid break and one isochromatid break with sister chromatid union.

It is recommended that slides are masked and coded before analysis. The number of cells that should be analysed for each treatment and fixation time depends on the magnitude of the effect; the lower the frequency of aberrations, the more cells should be analysed. When the treatment produces more than 30 aberrations per 100 cells, it is usually sufficient to analyse 100–200 cells. The cells of untreated lateral roots contain less than 1% aberrations. Therefore, any marked increase above 1%, based on an analysis of 300 cells or more, indicates that the chemical substance in question produces chromosomal aberrations.

CONCLUSIONS

Advantages and disadvantages of Vicia root tips as a test material

Advantages. The material is easily available all the year round; it is inexpensive and easy to grow and handle. The method does not require sterile conditions or any expensive material or equipment.

The root meristem contains a high proportion of cells in mitosis. The chromosome number is low (2n = 12) and the chromosome very large, which makes scoring easy and accurate. An idea of the size of the *Vicia* chromosomes may be obtained by

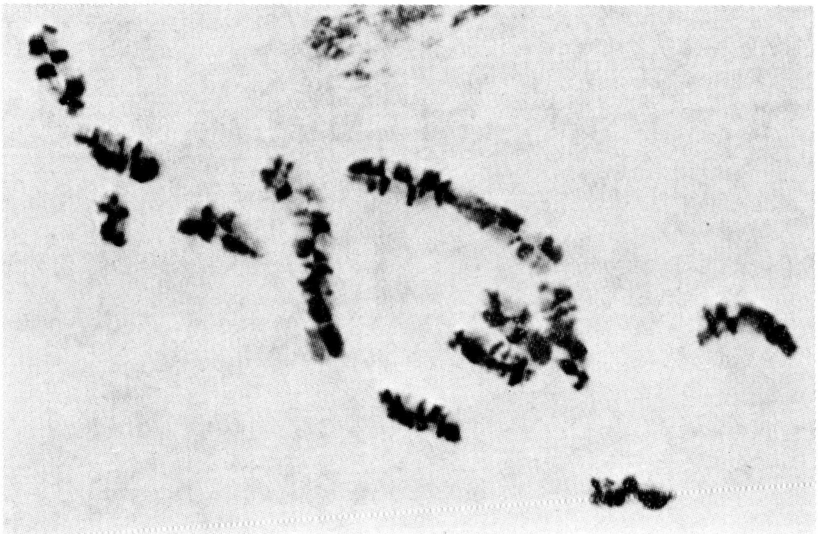

Fig. 10. High frequency of sister chromatid exchanges obtained after treatments with thiotepa and demonstrated by a modified FPG technique[4].

comparing the DNA content and chromosome number of a *Vicia* cell with those of Chinese hamster or human cells, two other materials commonly used for studying the production of chromosomal aberrations. A diploid root tip cell of *Vicia* contains about 46 pg of DNA, which is distributed among 12 chromosomes, a Chinese hamster cell 8.3 pg of DNA distributed among 22 chromosomes, and a human cell 7.3 pg of DNA distributed among 46 chromosomes.

The size of the chromosomes also makes *Vicia* root tips a favourable material for the study of the effects of chemical substances on the frequency of sister chromatid exchanges[4]. This phenomenon is of interest in connection with chemical mutagenesis, since like chromosomal aberrations, it appears to be an indication of damage to the genetic material.

A further advantage of root tips as a test material is that they can be exposed to known concentrations of chemical substances under controlled environmental conditions. Metabolic disturbances can be measured by biochemical methods or by autoradiography.

Finally, in contrast with artificial systems such as animal and plant cells cultured *in vitro*, root-tip cells have a very low spontaneous aberration frequency and a stable chromosome number. They also react more normally to treatment. Cells in tissue culture are very sensitive to changes in their environment and may react by fragmentation of chromosomes to treatments that have no effect on cells grown under normal conditions *in vivo*.

Disadvantages. The cell wall constitutes an obstacle to both biochemical and cytological studies on plant root tips. Thus, for squash preparations to be made and good separation of the cells in the root to be obtained, the pectic substances in the middle lamella of the cell wall have to be dissolved by hydrolysis with acids or enzymes. These procedures may change the morphology of the chromosomes and the composition of their chemical constituents, thereby making certain types of cytological analysis (*e.g.* analyses of the effects of chemicals on the frequency of sister chromatid exchanges[4]) difficult, if not impossible.

Another disadvantage of the material is that the root tip contains different types of cells which have mitotic cycles of different duration and, perhaps, different sensitivities to chemical treatment.

The impossibility of obtaining more than a very moderate synchrony of the cell divisions in the root meristem is a further disadvantage.

Finally, as a method for detecting chemical substances which may be a genetic risk to man, the root tip technique suffers from the serious disadvantage of not taking into account mammalian metabolism. Therefore, for this particular purpose it should be used only in conjunction with, and as a complement to, tests with mammalian systems *in vivo*.

REFERENCES

1 KIHLMAN, B. A., *Actions of Chemicals on Dividing Cells*, Prentice-Hall, Englewood Cliffs, N.J., 1966, pp. 260.

2 KIHLMAN, B. A., Root tips for studying the effects of chemicals on chromosomes, in A. HOLLAENDER (ed.), *Chemical Mutagens*, Vol. 2, Plenum, New York, 1971, pp. 489–514.

3 KIHLMAN, B. A. AND D. KRONBORG, Caffeine, caffeine derivatives and chromosomal aberrations, V. The influence of temperature and concentration on the induced aberration frequency in *Vicia faba*, *Hereditas*, 71 (1972) 101–118.

4 KIHLMAN, B. A., AND D. KRONBORG, Sister chromatid exchanges in *Vicia faba*, I. Demonstration by a modified fluorescent plus Giemsa (EPG) technique, *Chromosoma*, 51 (1975) 1–10

5 STURELID, S., Chromosome-breaking capacity of tepa and analogues in *Vicia faba* and Chinese hamster cells, *Hereditas*, 68 (1971) 255–276.

SAFETY PRECAUTIONS IN WORK WITH MUTAGENIC AND CARCINOGENIC CHEMICALS

LARS EHRENBERG and CARL AXEL WACHTMEISTER

Wallenberg Laboratory, Stockholm University, S-106 91 Stockholm 50 (Sweden)

Introduction

In biological laboratories, mutagenic and/or carcinogenic compounds (according to a suggestion by Druckrey [4] simply called genotoxic agents) will be encountered under a variety of circumstances. It is obvious that carefully specified safety rules should be applied not only when a few mutagens are in routine use, for example for mutation-induction in a plant-breeding programme or treatment of patients with chemotherapeutic agents, but especially when positive control compounds are used in small quantities. The most difficult situations from the standpoint of framing safety rules arise from the screening of potential mutagens. In this case we have no prior knowledge of which of the tested compounds will be found to exhibit mutagenic properties. In any particular case the safety procedure used will depend on the chemical used and the prevailing laboratory conditions. However, certain general precautions can be taken irrespective of the types and amounts of active compounds involved. The rules will have much in common with those developed for work with radioisotopes [2,5,9] *.

In addition to the rules for the experimenter, precautions must also be taken in the disposal of surplus active agents and in the cleaning of contaminated laboratory space and glassware.

One additional aspect deserves mentioning. Scientists in charge of programmes concerned with the testing of chemicals for genotoxic activity will

* It would have been desirable to base safety regulations for work with mutagenic and/or carcinogenic chemicals on more precise estimates of the risk at a given exposure dose [6,7]. It has been suggested to express the risk of chemicals in "rad-equivalents" [1,7], which would permit the application of radiation risk philosophy with more realistic permissible exposure levels in working environments and to the public. Although many complications are still at hand, especially concerning cancer risk and with regard to synergistic and additive effects, data indicate [7,11,12] that at present the risk of mutation might be expressed in such quantitative terms. These data show that the exposure doses which give the same risk as the maximum permissible dose to radiological workers (100 mrad γ-radiation per week) are very small indeed, corresponding, for example, to the uptake of 5 mg methanesulfonate (MMS) per week by a 60 kg person, or the exposure to ethylene oxide at 50 ppm (the present TLV in most countries) for totally 6 min during a working week.

often be local or national authorities in the fields of chemical mutagenicity, carcinogenicity and teratogenicity. As well as having a duty to protect their personnel, themselves and the public from unwanted consequences of their activity, they also have a responsibility to give information to others who are using genotoxic chemicals for other purposes. These scientists should also be aware that, because of the great number of unknowns in organic chemistry the rules presented here have to be provisional in many respects. It should be stressed that it is important to consult chemical expertise in cases where any doubt may be raised as to the handling or destruction of a specific chemical.

1. General rules (adapted in part from the booklet "Precautions for Laboratory Workers Who Handle Carcinogenic Aromatic Amines", ref. 3)

a. Labelling

All mutagenic/carcinogenic agents should be kept in bottles clearly marked "may be carcinogenic" and "not to be handled without the proper precautions".

b. Storage

The substances should always be kept in carefully closed bottles stored in closed containers, e.g. dessicators, and kept in the refrigerator.

c. Warning of staff

All persons likely to use these substances must be warned of the risks involved and told how to avoid them. The persons may sign a declaration of awareness when receiving standard safety rules.

d. Transport of dangerous material

Bottles or ampoules containing smaller amounts of mutagenic compounds which have to be transported to other laboratories, for example by mail, should be packed in parcels containing amounts of absorbing materials sufficient to absorb all the transported agent. Absorbing substances include Silica gel, Fuller's earth and activated charcoal.

Concerning the distribution and handling on a larger scale of toxic chemicals in general, considerable efforts have been made in recent years on the international level to improve the safety aspects [10].

It is our experience, however, that many well-known genotoxic agents are sometimes still delivered from commercial companies in glass bottles without adequate protective measures in case of breakage.

e. Ventilation

All operations with mutagens such as the opening of containers, preparation of solutions and treatment of contaminated glassware should, if possible, be done in a chemical hood with good ventilation. The ventilation can be checked with a burning match. Ensure that expelled air is not likely to enter other rooms of the same building or the ventilation system of any adjacent building (it may be tested, for example, with methyl formate, pyridine or Swedish "surströmming" or, under special precautions, with some radioactively labelled gas — Xe, Kr).

f. Working area

Limit all activities with mutagenic or carcinogenic compounds to a particular area in the laboratory (even when working in a hood). If possible use trays covered with filter paper. It is a good rule that no contaminated bottle or beaker is placed anywhere else than on the trays.

g. Protective clothes

When working with mutagens and carcinogens, protective eye glasses and a laboratory coat should always be used. All protective equipment, but especially rubber or plastic gloves may, however, give a false feeling of safety, as well as decreasing the ease of manipulations. Most low molecular weight, moderately polar compounds (e.g. ethyl methanesulfonate, ethylene oxide, ethylenimine) penetrate easily into rubber or plastic materials and pass rapidly at least through thin plastic films. The destruction of compounds absorbed into rubber also gives rise to quite special problems. It is therefore better to work with bare hands, *but with the utmost care, following every step by sight* to detect accidentally spilled drops (or particles).

It has to be borne in mind that laboratory clothes protect only for a limited time. They must therefore be changed often. Moreover, in order to avoid risks to washing personnel, contaminated clothes should be put into a special bag, treated separately and if possible washed by a procedure adapted to chemical properties of contaminants.

h. Pipetting and weighing

Note that pipetting a fluid from one vessel to another is far safer than pouring it; move pipettes slowly to prevent drops from falling. Use eye protectors.

When weighing fluids the density of which is known, a manually operated graduated pipette should be used. In case of fluids of only approximately known density, the weighing is made by difference. When the required volume of the solution is large, a suitable covered weighing bottle should be used, otherwise the final bottle (with cover) is carefully weighed and the solution prepared in this bottle. The bottles should always be filled inside the hood, covered and taken to the balance. The dilution to the required volume should likewise be made in the hood. In cases where a special weighing bottle is used the fluid is transported from this to the final bottle by means, for example, of a pasteur pipette, using parts of the pre-measured solvent to rinse weighing bottles and pipettes.

It is recommended that solutions of a desired strength should be prepared by adjusting the volume of solvent to the exact weight of the sample, to avoid prolonged handling of sample by weighing exactly, to a pre-determined weight. The error introduced by not using volumetric flasks should be negligible in case of dilute solutions.

Corresponding techniques should be used in case of solid compounds. Although such are often less volatile, it is essential to prevent formation of "dust" or even aerosols by making only slow, careful movements with the spatula in the transport operations. Use slightly moist filter paper on the tray.

i. General measures for preventing accidents

The leader of the research team should ensure that any of his workers who are sick or are subject to pronounced psychological stress are not requested to perform manipulations which require great care and concentration.

Eating and drinking in the laboratory should be forbidden and outsiders as well as the workers should be aware of this rule.

Since smoking (besides synergistic action) may lead to contamination of lips and in special cases to the formation of toxic compounds upon contact of aerosols with the glowing stub, restrictions are justified. Prohibition in a laboratory may have to be introduced depending of the type of premises, organisation and compounds studied.

j. Accidents

Contaminated skin

When skin is accidentally contaminated with a genotoxic agent (or by a reactive chemical of any type), two points are essential: to reduce the time of reaction with tissue components and to reduce the dose.

Do *not* rinse skin with organic solvents, especially not dimethyl sulfoxide, which will probably facilitate absorption into the skin of part of the active material. In the case of compounds with a high or even partial solubility in water, start rinsing immediately with large quantities of cold or faintly warm water, preferably under the tap but alternatively in a tray or by cautious rubbing with wet cloths which are often changed.

A compound which has already penetrated through the outer skin will of course never be completely removable, it will continue to spread through the skin by diffusion. When the skin surface is freed from the compound by rinsing with water, the absorbed compound will partly diffuse outwards, a fairly slow process. Therefore, it is necessary to continue the washing procedure for several hours (even during transport to hospital), especially in the case of compounds which react slowly.

In the case of markedly lipophilic compounds, such as hydrocarbons with a water solubility of less than 0.1%, washing with a water solution of soap or a detergent of the type used in modern wash-procedures should replace washing with water.

Contaminated eyes

Since many chemicals, especially reactive alkylating agents may lead to irreparable eye damage, all workers should carry eye protectors of a type which does not obstruct the vision. If the eye is contaminated, rinse with copious amounts of water and obtain medical aid as quickly as possible.

When institutions (companies etc.) have access to permanent medical service sections severe consequences of accidents may be counteracted if the medical staff are informed in advance about types of chemicals being used and thus are prepared to take the right measures. (But it should be added that getting, for example, ethylenimine into the eyes is a kind of accident that should not be permitted to happen).

It is of course essential to have on display the phone number of the nearest hospital or local doctor who might be needed in an emergency.

Contaminated bench areas, reagent flasks and other items

Clean carefully with dry absorbing material (cellulose, cotton etc.) followed by similar material slightly moistened with an organic solvent, e.g. acetone. Avoid contact of skin with contaminated part of the absorbing material, which should be collected for destruction.

In case of contamination with compounds of high reactivity, chemical destruction may sometimes be performed at once on the contaminated area (cf. Section 2c below).

2. Destruction of mutagenic agents

Mutagens and carcinogens must regularly be destroyed in connection with standard washing of used glassware. In addition, decontamination must also be carried out following accidental spills and in connection with the disposal of surplus active compounds.

a. General rules

The only general technique for the cleaning of glassware which can be safely recommended is its treatment with sodium dicromate in strong sulfuric acid. (Sodium dichromate ($Na_2Cr_2O_7$) 50 g) is dissolved in water (50 ml). To this solution 1000 ml of concentrated sulfuric acid is added slowly with stirring). This is the cleaning agent most commonly used in organic chemical laboratories. It is important to note that it is necessary to keep the content of Cr^{VI} and H_2SO_4 high in the mixture. Therefore, it is recommended to check the mixture periodically and to renew it when the liquid becomes greenish. Preferably the dichromate treatment is performed in two consecutive batches of acid, starting the treatment with a partly inactivated batch. The time necessary for inactivation of chemicals is seldom known but 1—2 days is generally considered sufficient when freshly prepared reagent is used. After this time, all materials should be dissolved in the reagent and can be easily rinsed away with water into the sewage system, probably in a largely degraded form (hydrolyzed, oxidized, sulfonated).

Non-destructive techniques involve the expensive rinsing of glassware with large volumes of organic solvents (acetone, methanol or ethanol) followed by water. Alternatively, cleaning with hot detergent solutions in laboratory washing machines is sometimes recommended. Both techniques imply that the final destruction of the active agents will occur gradually in the sewage treatment plants, an assumption which in most cases is probably true. However, both techniques involve obvious risks for the personnel employed in the laboratory. The sewage system in the laboratory may sometimes unexpectedly be a source of contamination of the air *inside* the laboratory. Furthermore, this air may also become contaminated by the volatilization of chemical substances at the high temperarures used in mechanical dish-washing equipment. Generally it is a safe rule that glassware contaminated with mutagenic and carcinogenic agents should not be mixed with ordinary dirty laboratory glass without adequate previous treatment.

b. Destruction by fire

Most organic compounds can be destroyed by burning at a sufficiently high temperature in the presence of excess oxygen, at least if a sufficiently long reaction time is allowed for. The conditions necessary for complete combustion of any particular compound are, however, rarely known. Therefore, a strong warning must be given against incineration of filter paper, cloths etc. contaminated with mutagenic or carcinogenic agents or against the disposal of unused samples in open fires, where combustion is often incomplete and, furthermore, persons are often exposed to the smoke. Incinerators designed for communal sewage disposal are better for this purpose, but it is best if samples containing larger amounts of genotoxic compounds are burned only in incinerators specially designed for the controlled destruction of, for example, used organic solvents.

c. Specific destruction of genetically active compounds

In the case of many types of genotoxic compounds it is possible to take advantage of their main chemical reactivity to effect a chemical destruction by agents less drastic than dichromate-sulfuric acid. This applies especially to easily oxidizable compounds and to alkylating, arylating and acylating compounds.

The former type of compounds may easily be destroyed by a sufficient excess of various milder oxidative agents. Thus, a saturated solution of potassium permanganate in acetone (15 g in 1000 ml) is a suitable reagent to destroy, for example, hydrazines, aromatic amines, or compounds containing isolated carbon-carbon double bonds. Alkylating compounds may, in principle, be destroyed by reactions with suitable nucleophiles, e.g. water, hydroxide ions, ammonia, sulfides, thiosulfate. The reactivities of various alkylating agents, however, vary within very wide limits, and the kinetic data required for calculating suitable reaction conditions in the case of a certain nucleophile are often not easily accessible [13]. Furthermore, the practical reaction rates are influenced not only by the structure of the compounds in question but also by their solubility in the reaction medium used.

One has to be aware of the possibility that a destruction method suitable for one type of compound may for other compounds either be ineffective or give rise to secondary products with a strong mutagenic or carcinogenic activity.

For these reasons general rules cannot be given, and when work with a new class of compound is initiated, it is important that an organic chemist be consulted who can look at the problem thoroughly before recommending the correct course for action.

Destruction methods suitable for use under laboratory conditions with 11 compounds often used as positive controls are shown in Table I. The list contains eight primary alkylating agents, one nitrosamine, one polycyclic hydrocarbon, and the complex compound mitomycin C.

Methyl- (MMS) and ethyl-methanesulfonates (EMS) are of moderate solubility in water and may be destroyed in 10% thiosulfate solution. These compounds demonstrate the large differences in reactivity which often exist between only slightly different compounds. (Table II) [13].
An attempt to destroy 99.9% residual MMS by treatment for 30 min (10 half lives) at room temperature with 10% thiosulfate, would hence be quite inadequate to destroy EMS which would not even be inactivated to 50% in the same period.

Special care should be taken when gram quantities or more of MMS, EMS or

TABLE I

SPECIFIC METHODS FOR DESTROYING MUTAGENIC CHEMICALS COMMONLY USED AS POSITIVE CONTROLS

It has to be noted that these methods, and especially the reaction times required, are not directly applicable to related compounds because of differences in reactivity. For additional data see Table I and II on pp. 412 and 414.

Compound	Method	Reagent	Estimated reaction time at room temperature
Methyl methanesulfonate	Nucleophilic substitution	$Na_2S_2O_3 \cdot 10\ H_2O$ 10% in water	~ 1 h
Ethyl methanesulfonate	Nucleophilic substitution	$Na_2S_2O_3 \cdot 10\ H_2O$ 10% in water	~20 h
Ethyleneimine	Nucleophilic substitution	$Na_2S_2O_3 \cdot 10\ H_2O$ 10% in 0.5 M acetate buffer, pH 5	~ 1 h
Triethylenemelamine	Acid-catalyzed hydrolysis	1M HCl	< 1 min
Trenimone	Acid-catalyzed hydrolysis	1M HCl	< 1 h
N-methyl-*N'*-nitro-*N*-nitrosoguanidine	Nucleophilic substitution	$Na_2S_2O_3 \cdot 10\ H_2O$ 2% in phosphate buffer, pH 8—9 [a]	< 1 h
ICR 170	Nucleophilic substitution	0.2M KOH in methanol	< 1 h
Cyclophosphamide	Nucleophilic substitution	0.2M KOH in methanol	< 1 h
Mitomycin C	Oxidation	$KMnO_4$ 1% in water	100°C, 0.5 h
N,N-dimethylnitrosamine	Oxidation, or (1) Reduction + (2) Oxidation	Dichromate-sulphuric acid e.g. Zn/acetic acid, then $KMnO_4$ 1% in water	< 1 day < 1 h stirring <10 min
Benzo(*a*)pyrene	Oxidation	Dichromate-sulfuric acid	1—2 days

[a] At higher pH values gaseous diazomethane may be formed.

TABLE II

Compound	Half-life in water		Half-life in 10% (w/v) $Na_2S_2O_3 \cdot 10\ H_2O$	
	20°C	37°C	20°C	37°C
MMS	77 h	9.5 h	3 min	50 sec
EMS	96 h	11.5 h	1.4 h	17 min

other highly reactive reagents have to be destroyed. Violent reactions may occur when 10% thiosulfate solutions are used; large volumes of aqueous bicarbonate solutions are rather recommended. Table I contains ethylenimine and the two ethylenimine derivatives, TEM and Trenimone. Ethylenimine (pK_a = 8.0) is of extremely low reactivity under basic conditions but is very reactive towards nucleophiles including water, at pH values lower than pH 9. A 10% solution of thiosulfate buffered to pH 5, therefore, is a suitable destruction agent. The two ethylenimine derivatives behave similarly.

The compound ICR 170, being a nitrogen mustard, is of high stability under acidic conditions but can be destroyed very easily with potassium hydroxide in methanol. The same applies to cyclophosphamide.

N-Methyl-*N'*-nitro-*N*-nitrosoguanidine is only very slowly hydrolyzed in water, but is rapidly destroyed by 10% thiosulfate solution. Treatment of MNNG and of the related nitrosoamides MNU and ENU with alkali to destroy them should be avoided because of the risk for exposure to diazomethane, a highly toxic gas produced in the reaction.

Dimethylnitrosamine is probably best destroyed by a dichromate-sulfuric acid mixture. Alternatively, it may be reduced (e.g. by zinc powder and acetic acid) to the (carcinogenic) hydrazine which is very easily oxidized to nitrogen.

Benzo(*a*)pyrene is readily destroyed by a dichromate-sulfuric acid mixture.

d. Detection of residual alkylating agents

Because of the very large differences in reactivity between various alkylating agents, attention should be given to analytical methods for detecting and even quantitating traces of alkylating agents and related compounds. Several reagents have been used for this, the most well-known being 4-(p-nitrobenzyl)-pyridine (I) [8], which reacts with an agent R-X to form a quaternary ion (II). This ion is readily transformed by bases into a coloured — often blue or purple — compound III (see Fig. 1).

The recommended procedure for detection differs somewhat depending on the main type of alkylating agent. The rate of reaction of *p*-nitrobenzylpyridine with aziridines (ethylenimines) is greatly increased under faintly acidic conditions. To some extent, the same applies to the reaction of epoxides. In the following test procedures, given by Sawicki and Sawicki [15], method A, below, is applicable in most cases with the exception of aziridines (ethylenimines), whereas method B, below, is designated for detection of aziridines and most probably also works with epoxides.

(A) A silica gel plate or a filter paper strip is impregnated with a 2% solution of 4-(p-nitrobenzyl)-pyridine in acetone, preferably by spraying. (The paper

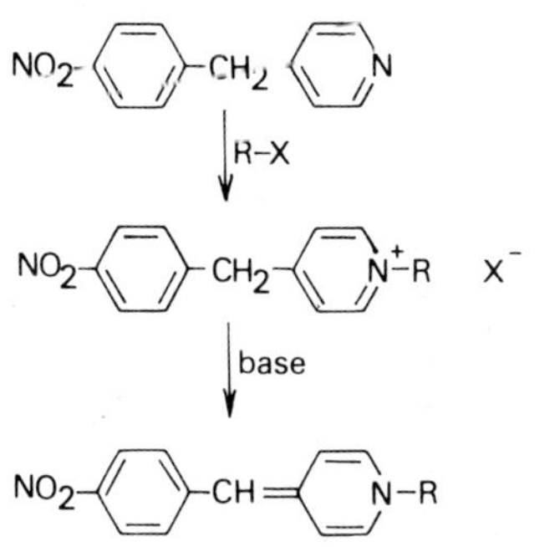

I

II

III

strip alternatively may be dipped in the reagent solution.) A drop of the test solution is applied to the air-dried plate or strip, which is then heated to 110°C for 10 min in an oven. Spray with a 10% solution of 1,3-propanediamine in acetone to obtain more intense colours, generally red, violet or blue.

(B) Spray the silica gel plate with a 0.1 M aqueous solution of potassium hydrogen phthalate, and then with a 5% solution of p-nitrobenzylpyridine in water. After application of a spot of the test solution, heat at 110°C for 30 min, cool and spray with 1 M aqueous potassium carbonate solution. Aziridines give blue colours.

Alternative modifications of the test, involving photometric measurement of the intensity of the colour produced, are given by Preussmann et.al. [14].

The detection limit of the methods given varies somewhat from compound to compound, but generally it is within the range of 1—50 μg. The original papers should be consulted for information on this point, on errors inherent in the methods and on applications for quantitative determinations.

Acknowledgement

This work was supported by the Swedish Council for Forestry and Agricultural Research.

References

1 Bridges, B.A., The three-tier approach to mutagenicity screening and the concept of radiation-equivalent dose, Mutation Res., 26 (1974) 335—340.

2 Chase, G.D. and J.L. Rabinowitz, Principles of Radioisotope Methodology, Burgess, Minneapolis, Minn., 2nd ed. 1962, pp. 5—8 and quoted papers.

3 Chester Beatty Research Institute, Royal Cancer Hospital, London: Precautions for Laboratory Workers who Handle Carcinogenic Amines. Hopkin and Williams, 1966.

4 Druckrey, H., in C. Ramel (ed.), Evaluation of Genetic Risks of Environmental Chemicals, Ambio Special Report No. 3, 1973, p. 7.

5 Ehrenberg, L., in K.B. Augustinsson (ed.), Experimentell Biokemi, Svenska Bokförlaget Bonniers, Stockholm, 1966, pp. 241—279 (in Swedish).

6 Ehreneberg, L., Genetic toxicity of environmental chemicals, Genetika (Belgrade), 6 (1974) 367—398.

7 Ehrenberg, L., K.D. Hiesche, S. Osterman-Golkar and I. Wennberg, Evaluation of genetic risks of alkylating agents: tissue doses in the mouse from air contaminated with ethylene oxide, Mutation Res., 24 (1974) 83—103.

8 Epstein, J., R.W. Rosenthal and R.J. Ess, Use of γ-(4-nitrobenzyl)pyridine as analytical reagent for ethylenimines and alkylating agents, Anal. Chem., 27 (1955) 1435—1439.

9 Kamen, M.D., Isotopic Tracers in Biology, Academic Press, New York, 3rd ed., 1957, pp. 407—410.

10 Meadowcraft, A.E., Environmental protection in the distribution of hazardous chemicals, Chem. Soc. Rev. (Lond.), 4 (1975) 99—106.
11 Osterman-Golkar, S., Studies on the Reaction Kinetics of Biologically Active Electrophilic Reagents as a Basis for Risk Estimates, Doctoral Thesis, Stockholm, October 1975.
12 Osterman-Golkar, S., L. Ehrenberg, D. Segerbäck and I. Hällström, Evaluation of genetic risks of alkylating agents. II. Haemoglobin as a dose monitor, Mutation Res., 34 (1976) 1—10.
13 Osterman-Golkar, S., L. Ehrenberg and C.A. Wachtmeister, Reaction kinetics and biological action in barley of mono-functional methanesulfonic esters, Radiation Bot., 10 (1970) 303—327.
14 Preussmann, R., H. Schneider und F. Epple, Untersuchungen zum Nachweis alkylierender Agentien. II. Der Nachweis verschiedener Klassen alkylierender Agentien mit einer Modifikation der Farbreaktion mit 4-(4-Nitrobenzyl)-pyridin (NBP), Arzneimittel-Forsch., 19 (1969) 1059—1073.
15 Sawicki, E. and C.R. Sawicki, Analysis of alkylating agents. Application to air pollution. Ann. N.Y. Acad. Sci., 163 (1969) 895—920.

HANDLING OF MUTAGENIC CHEMICALS: EXPERIMENTAL SAFETY

LARS EHRENBERG and CARL AXEL WACHTMEISTER

Wallenberg Laboratory, Stockholm University, S-106 91 Stockholm 50 (Sweden)

Introduction

The handling of mutagenic chemicals should not only aim at the safety of personnel. Measures have also to be taken to prevent misleading experimental results through the use of defective chemicals. Usually the biologist has to accept that analytical data and/or degree of purity correspond to the statement on the bottle of a purchased chemical (although, in the case of labile chemicals, he is certainly justified to request a *dated* analytical attest). Especially many electrophilic reagents (alkylating, arylating, acylating, etc., agents) are so reactive that improper handling may lead to their deterioration, with the effect that the intended chemical has partly or wholly disappeared when the test is done. Evidently, decay of a chemical may take place during storage, but, in addition, the decomposition in the prepared test solution before it is used has to be considered. For the latter reason, some knowledge of the stability of chemicals in solution is required, in order to safeguard that sufficiently fresh solutions of reactive chemicals are used.

Isotope-labelled mutagenic compounds represent, because of radiolytic decomposition, a special problem. Although this is outside the scope of the present chapter, we feel it justified to stress the necessity of extreme care to safeguard radiochemical purity at the time of delivery and at the time of use.

1. Prevention of decay

For alkylating (etc.) agents of the types mostly used as positive controls (Table I) the following conditions contribute to decay, with suggested general preventive measures in consequence:

(a) Chemical reaction with nucleophiles contaminating the stored samples. The greatest hazard in this respect is presented by the ubiquitous water, which may enter the sample from air humidity, from carelessly dried pipettes, etc. Due to its relatively low molecular weight (MW), some 10% water is sufficient to destroy completely an alkylating agent of MW about 200. If a reaction product of hydrolysis is able to solvolyze the compound further, an infinitely small amount of water will ultimately lead to complete destruction of the

TABLE I

A FEW CHEMICALS FREQUENTLY USED AS POSITIVE CONTROLS IN TESTING FOR MUTAGENICITY; AND PROCESSES LEADING TO DECOMPOSITION DURING STORAGE AND USE

Abbreviated name	Chemical name	Processes leading to decomposition (chemical change)
1. Alkylating agents		Generally: alkylation of water and other nucleophiles
MMS	Methyl methanesulfonate	= General
EMS	Ethyl methanesulfonate	= General
(EI)	Ethylenimine, aziridine	The protonated species, EIH^+, alkylating; hence protected by alkali; polymer formation.
TEM	Triethylenemelamine, 2,4,6-tris(1-aziridinyl)-1,3,5-triazine	Trifunctional alkylating agent; also the free base is alkylating, but decomposition is enhanced at low pH. Photosensitive?
Trenimone	[a]	—
MNNG	*N*-methyl-*N*′-nitro-*N*-nitrosoguanidine	Sensitive to light. Less unstable than other alkylnitrosamides, but like these very sensitive to alkali.
ICR 170	Quinacrine mustard [a]	Readily hydrolyzed. Probably sensitive to light.
Cyclophosphamide	[a]	Intramolecular alkylation with loss of chloride, accompanied by secondary reactions.
Mitomycin C	[a]	Relatively stable to hydrolysis at pH 6—9; solution may be stored for a few days in the refrigerator. Readily decomposed by light.
2. Nitrosamines		
DMNA	*N,N*-dimethylnitrosamine	Decomposed by various routes. Sensitive to heat and light.
3. Polycyclic hydrocarbons (PAH)		
	Benzo(*a*)pyrene	In light, rapidly oxidized by air.

[a] For complete formulae, see Fishbein et al. [6].

chemical (this seems to be the case with the acylating agent, diethyl pyrocarbonate, probably the reason why commercial samples sometimes contain as little as 20% of the stated product; cf. ref. 5).

Rule (a). Avoid any exposure to gaseous or liquid water. Pipettes, spoons, etc., should be carefully dried. In particular, expensive chemicals should be stored in a desiccator, with silica gel. If a chemical has to be stored in solution, dry non-reactive solvents should be used, e.g. hydrocarbons, acetone, acetonitrile; it has to be observed that alcohols are not generally safe. Possibly molecular sieves (pore size 3A) will offer a method of removing contaminating water (experiments under way).

(b) All reactions, including those leading to decay, proceed at a decreased rate at lowered temperatures.

Rule (b). Therefore, chemicals should be stored in the cold; e.g. desiccators according to (a) are preferently stored in the refrigerator.

Observe the possibly conflicting consequences of rules (a) and (b): *if a desiccator (or any container) is brought from the refrigerator to room temperature air moisture will precipitate on the colder material before it has assumed room temperature. If any bottle is not tightly closed, or if it is opened before it has become warm enough, its content will be contaminated with water.* This statement is italicized because violation of this rule is probably the most common cause of defective experimental data and loss of expensive chemicals; even in laboratories with good discipline in this respect, there are at times moments of hurry, such as Friday afternoons, when a bottle is opened too fast with the excuse that "only this time cannot matter". In critical cases it is recommended to warm desiccators or bottles to a temperature some 5°C above room temperature before they are opened.

(c) Some chemicals are destroyed photochemically by daylight. Examples of this, among the chemicals listed in Table I, are dimethylnitrosamine, MNNG and mitomycin C.

Rule (c). Dark bottles should be used, and chemicals should be protected from direct daylight. In cases of doubt it is better to assume light sensitivity.

(d) Some chemicals (alkylnitrosoureas?) may decay by unknown mechanisms, be it through inter- or intramolecular rearrangements of the proper compound, be it through oxidation by air oxygen (often catalyzed by traces of transition metals) or be it in reactions with unknown impurities.

Preventive measures. A low temperature (in critical cases deep-freezing) should be applied. In order to prevent free radical chain reactions, special measures such as dilution might be required; an organic chemist should be consulted about this.

2. Stability of solutions

a. General

A nucleophilic reagent, RX, dissolved in water, will be hydrolyzed at a rate

$$-\frac{d[RX]}{dt} = k'_{H_2O}[RX] \qquad (1)$$

where k'_{H_2O} is the first-order rate constant for hydrolysis. If the solution contains, in addition, a buffer and other dissolved compounds, RX may react also with these, with the effect that it disappears faster than in water alone.

Considering, for simplicity, a phosphate buffer, this will contain the nucleophiles, in order of increasing reactivity: $H_2PO_4^-$, HPO_4^{2-}. PO_4^{3-}. At pH about 7, the concentration of PO_4^{3-} is so low that the reaction with HPO_4^{2-} will predominate. The total rate of decomposition may hence be written:

$$-\frac{d[RX]}{dt} = k'_{H_2O}[RX] + k_{HPO_4^{2-}}\,[RX][HPO_4^{2-}] \qquad (2)$$

where $k_{HPO_4^{2-}}$ is the second-order rate constant for reaction of RX with hydrogen phosphate anion (cf. ref. 11).

Values of k'_{H_2O} and $k_{HPO_4^{2-}}$ are given in Table II, together with the time required for 10% and 50% decomposition at 37°C. The corresponding rates of decay at 25°C and 20°C may be estimated through division of rate constants by 3.5 and 7 respectively, and the corresponding times required for the given degrees of fractional decomposition by multiplying by 3.5 and 7, respectively.

Due to the enhanced decomposition at higher temperatures, sterilization of injection solutions through boiling is in most cases not possible.

Experimental error may also be caused by evaporation of chemicals from test solutions. This risk is high, especially in the case of relatively volatile com-

TABLE II

RATE CONSTANTS FOR HYDROLYSIS AND REACTION WITH HYDROGEN PHOSPHATE ANION AT 37°C OF SOME ALKYLATING AGENTS AND THEIR STABILITY IN WATER AND IN 0.1 M PHOSPHATE BUFFER pH 7

For abbreviations see Table I.

Compound	k'_{H_2O} (h^{-1})	$k_{HPO_4^{2-}}$ ($l\ mol^{-1}h^{-1}$)	Time for destruction			
			in water		in 0.1 M phosphate pH 7	
			10%	50%	10%	50%
MMS	0.072	1.5	1.5 h	9.5 h	40 min	—
EMS	0.060	0.28	1.8 h	11.5 h	1.4 h	9 h
iPMS	3.1	a	2 min	13 min	— a	— a
EI (pH 7)	0.013	1.5	8 h	53 h	1 h	7 h
TEM (pH 7)	0.024	0.6	4 h	29 h	1.7 h	12 h
(pH 3)	H^+ catalyzed 3000 $l\ mol^{-1}h^{-1}$	—	13 sec	1.4 min	—	—
MNNG	0.014 (pH 4.5—6)					
pH 7 (incl. reaction with OH^-)	0.09	0.002	1.2 h	7.5 h		
pH 8 (incl. reaction with OH^-)	0.74		8 min	1 h		
Cyclophosphamide				±140 h	—	—

a = water (because of S_N1 type reaction).

pounds of limited solubility in water, such as alkyl halides and hydrocarbons of intermediate molecular weight. This requires (i) that the concentration is determined in the stock solutions and its dilutions and, during the course of the experiment, in the test solution and (ii) that tests such as treatments of cell suspensions are done in closed vessels.

b. Remarks on specific compounds (see Tables I and II)

Many monofunctional alkyl alkanesulfonates (e.g. methyl methanesulfonate (MMS) and ethyl methanesulfonate (EMS)) are relatively stable in solutions. That this is not always so, however, is exemplified by iPMS (isopropyl methanesulfonate; added, for comparison, to Table II) which requires preparation of solutions within a minute or so before use [11]. This type of rapid hydrolysis is characteristic of compounds with a branching on the alkylating carbon, as in *sec*-butyl methanesulfonate and 1,4-dimethylmyleran.

Ethylenimine (aziridine) offers special problems. The alkylating reactivity resides in the ethylenimonium ion (EH^+) whereas the free base (EI) although little reactive seems to be responsible for the transport through membranes [10]. The protolytic equilibrium

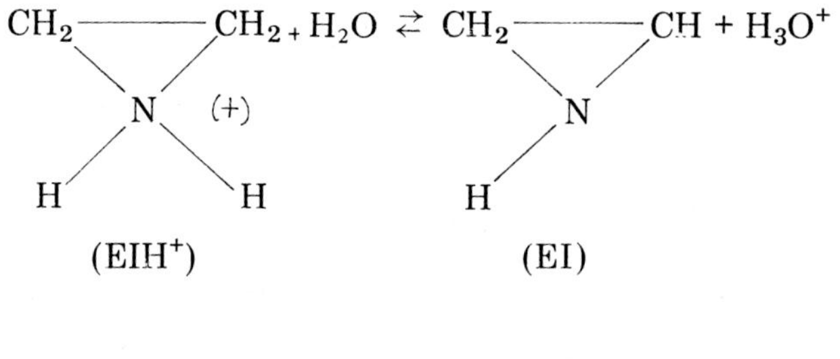

(3)

has pK_a = 8.0 (see ref. 1) with the effect that at pH 7 about 91% of the total concentration is available as the imonium cation, EIH^+. Whereas alkylating agents are usually destroyed by alkali, due to the high nucleophilic strength of OH^-, EI is instead protected by alkali. The EI liquid is therefore usually stored in the presence of some solid NaOH, from which it may be distilled (b.p. 57°C). Observe the necessity of protecting personnel in such handling.

Triethylenemelamine (TEM) is relatively stable in neutral solution but requires precautions to avoid acid conditions, because of the fast hydrogen ion-catalyzed reaction [13]. As illustrated by the rate of decay at pH 3 (Table II), an accidental lowering of pH may lead to rapid destruction.

Trenimone has apparently not been studied concerning its reaction rates in aqueous systems. From its structure, it must be assumed to react similarly to TEM.

N-Methyl-N'-nitro-N-nitrosoguanidine (*MNNG*) offers several problems of stability, although it seems to be more stable than the alkylnitrosoureas. Purchased preparations are sometimes partly decayed at delivery (probably due to moisture and light), and prepared solutions should therefore be checked, e.g. spectrophotometrically (ϵ_{400nm} = 163; ϵ_{280nm} = 17.600). Solutions are most stable in the range of pH 4.5—6. In more acidic solutions a H^+-catalyzed denitrosation occurs, and at higher pH a rapid reaction with hydroxyl ions (cf. ref. 12). Solutions are readily decomposed by light [15].

ICR 170. The difficulty with this compound is the rapid hydrolysis of the nitrogen mustard part of the DNA-intercalating molecule. The kinetics of the reactions of ICR 170 with water and other nucleophiles, however, is apparently unknown. There is reason to believe that investigations with ICR 170 are sometimes performed without due consideration to its partial or complete hydrolysis under the conditions used.

Cyclophosphamide (Endoxan). This compound, although like ICR 170 containing the 2-chloroethylamino function, has no alkylating activity as such and is relatively stable in the pH range 2—10. The fastest spontaneous decomposition process involves intramolecular alkylation of nitrogen with loss of chloride ($t_{1/2} \sim$ 6 days in water at 37°C) [7,8]. The postulated alkylating bicyclic primary intermediate rapidly undergoes secondary reactions [7]. The reactions of cyclophosphamide in vivo or in vitro apparently are quite different from the reactions in pure water [7].

Mitomycin C is most stable at pH 6—7 but especially labile under acidic conditions [14]. Thus it is rapidly destroyed, for example, by heating to 80°C for 40 min at pH $<$ 5. The compound is readily decomposed by visible light, especially in certain organic solvents (benzene, acetone, chloroform), but less rapidly in water. It is highly sensitive to oxidizing agents. The purity of mitomycin can be ascertained spectrophotometrically.

Polycyclic aromatic hydrocarbons are notoriously prone to undergo photooxidations of various types when exposed to light of wavelength $>$ 300 nm, e.g. as present in solar radiation at ground level [2]. When adsorbed on solid phases, for example particulate matter from the atmosphere, their reactivity is considerably increased [2].

3. Analysis and purification

In critical situations the purity, or degree of destruction, has to be verified by analysis. This may be called for: (a) in specially important investigations, e.g. when a compound is used as a positive control or for evaluation of the test system and for accurate dose-effect estimations; (b) when an old preparation is going to be used; (c) when a compound is known to have been accidentally exposed to factors (moisture, light, high temperature, etc.) known to enhance decomposition; and (d) when unexpected experimental results are obtained.

In their review, Fishbein et al. [6] refer to analytical procedures for many important genotoxic compounds (of those listed in Table I all but Trenimone).

For alkylating agents like the methanesulfonates, where decomposition follows a simple, easily grasped reaction scheme, standard titrimetric methods may be used. MMS, EMS and related compounds are hydrolyzed under formation of an equivalent amount of acid, for example

$$CH_3SO_2OCH_3 + 2H_2O \rightarrow CH_3SO_2O^- + H_3O^+ + HOCH_3 .$$

Through rapid titration with NaOH the amount of acid at hand in the sample is determined (due to continuous alkylation of OH^- the end-point may be diffuse; titration should therefore be *rapid*). By complete titration, suitably done by heating a dissolved sample with excess NaOH followed by back-titration with HCl, the total amount of alkylating agent plus decomposition products (if this, as in the example, gives rise to acid) is determined. In the general case the biologist is advised, however, to make contact with physical or organic chemists who can assist in suggesting or performing the required analysis by available techniques (NMR, IR, UV spectrometry; GLC, where a standard sample is required; mass spectroscopy, etc.). In certain cases (MNNG and mitomycin C) the test concentration of the compound is easily determined by UV spectrometry in water solution.

4. Reproducibility of the dose

A number of parameters which may easily be controlled may have a great influence on the reproducibility of experiments. To a certain extent the effect of mutagenic agents can be ascribed to the *dose*, D, defined as the concentration, C(t), integrated over the time of treatment (t_2—t_1), i.e. integrales of equations of the general form (1) and (2) [3,4,9].

$$D = \int_{t_1}^{t_2} C(t)\,dt \tag{4}$$

When a chemical may be considered stable, or if its decay is relatively slow, this expression may be approximated by

$$D = t \times C, \text{ or } D = t\frac{C_0 + C_t}{2},$$

respectively, where t is the duration and C_o and C_t the initial and final concentrations, respectively. If the decay of the compound is rapid compared to the duration of the treatment, the value of D according to equation (4), will be strongly dependent on the pseudo-first-order rate constant for the decay, k′. Now, k′ is the sum of the rates of reaction with water and with other nucleophiles such as buffer compounds (cf. equation (2)).

$$k' = k'_{H_2O} + k_{HPO_4^{2-}}\,[HPO_4^{2-}] + \ldots \tag{2a}$$

It may be shown that relatively small changes of pH and of the buffer concentrations, but above all, of the kind of buffer used, may have such drastic effects on the dose that an expected effect of a studied electrophilic reagent may be changed by one order of magnitude or more (cf. ref. 5).

An error of the experimental *temperature* will of course affect the dose as defined by equation (4), the uncertainty being about 15% per degree for a rapidly decaying compound. The biological consequences of this error will, however, to a great extent be eliminated by the influence of the temperature change on the rate of reaction with biological macromolecules.

In studies of a more slowly decaying compound the dose will be less affected by an error in temperature, but the influence on rates of reaction with macromolecules may give a considerable uncertainty of observed biological effects.

In the context of reproducibility the complicated situation with a compound like ethylenimine, which participates in an acid-base equilibrium, deserves mentioning (cf. equation (3) above)

$$\underset{\text{(acid)}}{EIH^+} + H_2O \rightleftarrows \underset{\text{(base)}}{EI} + H_3O^+ \tag{3a}$$

In this case the EIH^+ is the reactive principle whereas the base, EI, is the principle absorbed by cells. pK_a of EIH^+ is about 8. Therefore, at pH 7 about 10% of the total concentration of EI ($[EI + EIH^+]$) is available as the base. If pH is changed accidentally by 0.3 units this value will change to 5% (at pH 6.7) or 20% (at pH 7.3), with a corresponding change of the effectiveness of EI by a factor of two downwards or upwards, respectively. Because EIH^+ reacts relatively slowly, changes of pH will have less effect on the (extracellular) dose. In fact, a lowering of pH in the medium increases the concentration of EIH^+ according to equation (3) but the concomitant increase of the overall reactivity is partly counteracted by the decrease of concentrations of nucleophiles such as HPO_4^{2-} or non-protonated amines. Furthermore, the alkylation of EI by EIH^+ should not be forgotten. This reaction may occur in several steps and then lead to a complicated mixture of still alkylating polyamines. For these reasons it is very difficult to run acceptably reproducible mutation experiments with ethylenimines. Adding to this the volatility and toxicity of ethylenimine (cf. pp. 401), the use of a compound of this type as a positive standard should be restricted.

5. Cross-contamination of samples

High mutation frequencies are spuriously observed in control samples. In most cases such effects can be ascribed to contamination with mutagenic compounds from handling of solutions of strong mutagens in the proximity (e.g. in the same hood) of control media. Such effects may particularly be obtained from volatile, strong mutagens such as sodium azide (the hydrazoic acid, HN_3, in equilibrium with N_3^-, being volatile), or ethylenimine (the free base being volatile), but they may also be a consequence of using contaminated pipettes and other glassware.

Contamination of this kind, e.g. from positive controls, may have disastrous consequences in tests either by preventing detection of weak mutagens through increased mean value in the control or through increased variation in general, or by giving false-positive results in tests of non-mutagenic agents.

In work with volatile compounds the prevention of cross-contamination requires serious consideration, and may come into conflict with requirements on independence in comparisons of controls and tests. If for these reasons control samples have to be kept in a different room (hood, etc.), there should at the same time be an additional control sample in the absolute vicinity of the test sample and both controls should be considered in the statistical evaluation.

Acknowledgement

This work was supported by the Swedish Council for Forestry and Agricultural Research.

References

1 Buist, G.B. and H.J. Lucas, Basicity constants and rates of hydration of some imines, J. Amer. Chem. Soc., 79 (1957) 6157—6160.

2 Committee on Biologic Effects of Atmospheric Pollutants. Particulate Polycyclic Organic Matter, Natl. Acad. Sci., Washington, D.C., 1972, 361 pp.

3 Ehrenberg, L., Genetic toxicity of environmental chemicals, Genetika (Belgrade), 6 (1974) 367—398.

4 Ehrenberg, L., K.D. Hiesche, S. Osterman-Golkar and I. Wennberg, Evaluation of genetic risks of alkylating agents: tissue doses in the mouse from air contaminated with ethylene oxide. Mutation Res., 24 (1974) 83—103.

5 Ehrenberg, L., I. Fedorcsák and F. Solymosy, Diethyl pyrocarbonate in nucleic acid research, Proc. Nucleic Acid Res. Molec. Biol., 16 (1976) 189—262.

6 Fishbein, L., W.G. Flamm and H.L. Falk, Chemical Mutagens. Environmental Effects on Biological Systems, Academic Press, New York, 1970.

7 Friedman, O.M., Recent biologic and chemical studies of cyclophosphamide (NSC-26271), Cancer Chemother. Rep., 51 (1967) 327—333.

8 Hirata, M., H. Kagawa and M. Baba, Studies on cyclophosphamide. Part 1 Chemical determination and degradation kinetics in aqueous media. Annu. Rep. Shionogi Res. Lab., 17 (1967) 107—113.

9 Hussain, S. and L. Ehrenberg, Prophage inductive efficiency of alkylating agents and radiations, Int. J. Radiation Biol., 27 (1975) 355—362.

10 Konstantinov, K., M. Harms-Ringdahl, L. Ehrenberg, J. Dumanović and I. Božović, Influence of pH and temperature on the effects of ethylenimine (EI) in wheat and barley seeds, Radiation Bot., 10 (1970) 499—509.

11 Osterman-Golkar, S., L. Ehrenberg and C.A. Wachtmeister, Reaction kinetics and biological action in barley of monofunctional methanesulfonic esters, Radiation Bot., 10 (1970) 303—327.

12 Osterman-Golkar, S., Reaction kinetics of *N*-methyl-*N'*-nitro-*N*-nitrosoguanidine and *N*-ethyl-*N'*-nitro-*N*-nitrosoguanidine, Mutation Res., 24 (1974) 219—226.

13 Ross, W.C.J., The reactions of certain epoxides in aqueous solutions, J. Chem. Soc. (Lond.), (1950) 2257—2272.

14 Wakaki, S., H. Marumo, K. Tomioka, G. Shimizu, E. Kato, H. Kamada, S. Kudo and Y. Fujimoto, Isolation of new fractions of antitumor mitomycins, Antibiot. Chemother., 8 (1958) 228—240.

15 Veleminský, J. Porkorný and T. Gichner, The effect of visible light on mutagenic activity of 1-methyl-1-nitrosourea and 1-methyl-3-nitro-1-nitrosoguanidine, Biol. Plant. (Praha), 10 (1968) 85—94.

ASPECTS OF STATISTICAL INFERENCE IN TESTING FOR GENETIC TOXICITY

LARS EHRENBERG

University of Stockholm, Department of Radiobiology, Wallenberg Laboratory, Lilla Frescati, S-104 05 Stockholm 50, (Sweden)

Contents

1 Introduction

It would not be appropriate here to present all the various statistical methods which may be used to evaluate data obtained in tests for genetic toxicity. Testing programmes and the types of interfering experimental parameters are so manifold, that this would require a text-book of statistics. The aim of this chapter is rather to discuss, against the standard background of statistical knowledge any student of biology is assumed to possess, a few questions of statistical inference which have sometimes been neglected in the evaluation of results of tests and, more often, in the planning of such tests. It will become apparent that both the economy of work and validity of conclusions will benefit if the statistical aspects are considered *at the stage of planning the experiments.* If this is not done, important analytical possibilities may be lost. Not least, serious consideration must be given to the problem of randomization of the experimental material, in order to ensure that treated and control samples are comparable in all respects but the treatment.

Statistical inference deals with testing of hypotheses such as the Null hypothesis (H_0), which may be formulated to state that the difference between the mean value of a treated sample and the corresponding mean of the control sample is zero, or that the slope of an experimental dose-response curve does not differ from 0 or from that of a theoretical curve. For statistical treatment to be valid it is an indispensable condition that the samples of measurements can be considered representative of, i.e. drawn completely at random from, the infinitely large populations the mean values of which we want to estimate. Hence, the student of mutagenicity should not consider seemingly trivial problems of *sampling technique* to be irrelevant.

For test data to be evaluated adequately it is essential to *combine commonsense with statistical analysis in the process of accepting or rejecting hypotheses.* This exhortation is often ignored, sometimes with disastrous consequences. It must be stressed that the scientist involved in toxicological testing should, if he lacks special training in statistical methods, try to gain a feeling for the statistical dimension of his own measurements, and to become aware of the concept of variation as an ever present component of the biological experiment. He could then, in the closest possible way, enter into fruitful collaboration with the specialist statistician who knows how to allow for this variation and who, in his turn, will learn to appreciate biological problems also.

A few problems of statistical inference of specific importance to toxicological testing and risk estimates will be discussed here and, with the general aims given above in mind, these problems will be presented in the simplest possible way.

A few calculation methods are also presented in Section 3. and, in addition, the formulae used are summarized in Appendix 2.

2 Evaluation of test results

In any mutation test one or several samples of animals, plants, cells or microbes are exposed to an agent of unknown mutagenic properties, and the mutation frequencies observed are compared with a control value which, generally, should be assayed simultaneously. The use of control data of a well-established test system, accumulated during a long period of time, may sometimes be used in order to increase the statistical efficacy. However, one should avoid using the same control data again and again in comparison with test data. The best use of accumulated control data is the checking of the stability of the biological test system, through a homogeneity test of control samples (cf. Section 3.5.1).

A mean value ($\bar{x}$), determined on a sample — in our case the mutation frequency — is an estimate of a true mean value ($\bar{\xi}$) which would be obtained if we performed the test, under the same conditions, with an infinite number of individuals. Due to random variation, a sample of limited size is unable to pinpoint the true mean value; the word estimate used above, means that what the laboratory experiment gives is an indication of an upper and a lower limit, the "confidence limits", which, with a certain probability, enclose the true mean value, $\bar{\xi}$. Evidently, the higher we want to set this probability the wider we must make the confidence interval. Conversely, at a given probability the confidence limits can be narrowed only by increasing the sample size.

A number of functions computed from sample measurements have the same characteristics as estimates of true mean values and can also be provided with confidence intervals. Of special importance in mutagenicity testing and risk estimation are the difference between two mean values, e.g. between a treated sample and control sample, and the slope (the regression coefficient) of a linear regression curve (see Section 3.5.3). If, in the former of these two cases, the two mean values, the difference between which is being studied, are known with equal accuracy, the confidence interval of the difference will be approximately equal to $\sqrt{2} = 1.4$ times wider than that of the (average) confidence interval of the two measurements (cf. Section 3.2).

The statistical treatment of data consists, essentially, in the testing of hypotheses. The basic hypothesis tested is often referred to as the Null hypothesis (H_0). In the evaluation of the mutagenic activity of a chemical X, H_0 may be formulated: X is not mutagenic. If computations show the probability, *P*, to be smaller than the chosen level of significance, H_0 is rejected.

If we have chosen to reject H_0 at a certain significance level, α^*, this level obviously represents the probability of rejecting a true hypothesis. Conversely, we may happen to accept (or not reject [1]) a hypothesis that is actually false

[1] In the usual terminology of textbooks of statistics we reject the hypothesis (e.g. H_0) when the significance estimate gives a small P ($P \leqslant \alpha^*$) and, conversely, we accept the hypothesis when $P > \alpha^*$. In order to avoid any misinterpretation of the word "accept" as meaning "prove", we will in the following say that the hypothesis is not rejected when $P > \alpha^*$. This is more in line with the fact that a negative test result often calls for continued studies, e.g. when the power $(1 - \beta)$ of the test to detect a critical difference is too small.

TABLE I

POSITIVE AND NEGATIVE BIOLOGICAL TESTS: α-ERROR AND β-ERROR, RESPECTIVELY

(A numerical example is given in Table III and accompanying text of Section 3.2).

Result of statistical test	Decision	If H_0 is true	If H_0 is false
$P < \alpha^*$ (Biol. test positive)	Reject H_0	Decision incorrect (α-error)	Decision correct
$P > \alpha^*$ (Biol. test negative)	Do not reject H_0	Decision correct	Decision incorrect (β-error)

(i.e. whether a compound is or is not mutagenic). Therefore, using the result of the statistical test as a guidance for our decision to reject or not reject H_0, the four situations shown in Table I are possible.

Evidently our decision to reject or not reject H_0 is correct in only two out of the four possible cases. The error we make when we happen to reject a hypothesis that is actually true is called the α-error, or the Type I error. This case, dealt with in Section 2.1, would mean that we erroneously label an innocent compound as mutagenic. The non-rejection of H_0 when H_0 is actually false, referred to as the β-error or Type II error, is discussed in Section 2.2. This error, meaning that we conclude that a compound is non-mutagenic although it is in fact mutagenic, may have serious consequences if the compound is going to have a wide distribution in the human environment.

Strictly speaking, values of α and β connected with experimental data would mean, respectively, that in a great number of experiments with a *non-mutagenic* compound random variation of measurements would lead us in $100\alpha\%$ of the cases to erroneously reject the hypothesis of no effect and, similarly, that in a great number of tests of a certain *mutagenic* compound we would not detect its mutagenic activity in $100\beta\%$ of the experiments. Computed values of α and β hence give an indication of the reliability of the conclusions drawn.

In the following Sections 2.1 and 2.2 the cases of significance obtained and significance not obtained, respectively, are discussed. The interrelationships of the β-error and the resolving power of biological tests is discussed in 2.3. In Table II an idealized "flow-sheet" summarizes the various steps in the analysis, with resulting decisions and conclusions, a few comments being given in 2.5.

2.1 Significant effect obtained

In the qualitative evaluation of mutagenicity of a compound, i.e. of whether it is mutagenic or not, we are usually concerned with the analysis of the difference, d, between the experimental mean values of a sample treated with the mutagen ($\bar{x}_{Tr}$) and a control sample ($\bar{x}_C$)

$$d = \bar{x}_{Tr} - \bar{x}_C \qquad (1a)$$

These three parameters are estimates of the true difference, δ, of the true mean values, ξ_{Tr} and ξ_C (which would have been obtained in infinitely large populations studied under the same experimental conditions):

$$\delta = \bar{\xi}_{Tr} - \bar{\xi}_C \tag{1b}$$

The statement of the Null hypothesis (H_0) in the estimate of the significance of the observed mutagenic effect, d, may be formulated: the means of the treated and control populations do not differ, i.e.

$$\bar{\xi}_{Tr} - \bar{\xi}_C = 0 \text{ (or } \bar{\xi}_{Tr} = \bar{\xi}_C) \tag{2}$$

If we have estimated the variances σ^2_{Tr} and σ^2_C by s^2_{Tr} and s^2_C, the experimental difference (d) may be compared with any hypothetical difference (δ) using the t-distributed variable

$$\frac{(\bar{x}_{Tr} - \bar{x}_C) - (\bar{\xi}_{Tr} - \bar{\xi}_C)}{\sqrt{s^2_{\bar{x}_{Tr}} + s^2_{\bar{x}_C}}} = t \tag{3}$$

Inserting $\delta = 0$ for the hypothesis of no difference we obtain

$$\frac{\bar{x}_{Tr} - \bar{x}_C}{\sqrt{s^2_{\bar{x}_{Tr}} + s^2_{\bar{x}_C}}} = t \tag{3a}$$

where the probability (P) corresponding to t is taken from a t-table * at ($n_{Tr} + n_C - 2$) d.f. ($s_{\bar{x}}$ are the standard errors, calculated as $s/\sqrt{n}$). If n_C and n_{Tr} are not approximately equal this has to be allowed for by a weighting of the standard errors (see Section 3.2). If the true variances, σ^2, are known, for example from previous tests or through the application of certain distributions (cf. 3.1), these should be used and P read at ∞ d.f.

This may also be expressed by saying that the expression

$$\frac{\bar{x}_{Tr} - \bar{x}_C}{\sqrt{\sigma^2_{\bar{x}_{Tr}} + \sigma^2_{\bar{x}_C}}} = z \tag{3b}$$

is normally distributed, tables of the distribution of z being applicable.

The denominators of equations (3a) and (3b) are the standard deviations (standard errors) of the difference, s_d or σ_d, respectively (cf. Section 3.2).

$$\sigma_d = \sqrt{\sigma^2_{\bar{x}_{Tr}} + \sigma^2_{\bar{x}_C}} = \sqrt{\frac{\sigma^2_{Tr}}{n_{Tr}} + \frac{\sigma^2_C}{n_C}} \tag{4}$$

Expressions (3a) and (3b) may hence be written

$$\frac{d}{s_d} = t \text{ and } \frac{d}{\sigma_d} = z \tag{3c and d}$$

If we call α^* the predefined α error, and $P < \alpha^*$, which occurs when

$$|d| > t_{\alpha^*} s_d \text{ or } |d| > z_{\alpha^*} \sigma_d \tag{5}$$

---

* Throughout this Chapter, tables for two-sided tests of t and z are used (e.g. identified from the value 1.960 at $P = 0.05$ of z and of t at ∞ d.f.).

the decision made is to reject H_0, with the conclusion that we consider $\bar{\xi}_{Tr}$ and $\bar{\xi}_C$ to be different.

If the tested compound is in fact non-mutagenic (i.e. $\bar{\xi}_{Tr} = \bar{\xi}_C$), a highly improbable event (<100α*%) would have led us to draw the erroneous conclusion of mutagenicity.

If the purpose of the biological experiment is to provide a *quantitative* statement as to the mutagenic effectiveness of a compound, the difference d may be used (at least under certain conditions) as a measure of this effectiveness. In such situations it is of value to indicate the confidence interval which encloses the true difference, δ, with, for example, 95% probability.

$$d - t_{0.05}\, s_d < \delta < d + t_{0.05}\, s_d \qquad (6a)$$

or

$$d - z_{0.05}\, \sigma_d < \delta < d + z_{0.05}\, \sigma_d \qquad (6b)$$

Clearly, if we have rejected the hypothesis $\delta = 0$ at a significance level less than 5%, the 95% confidence interval will not enclose the value 0.

In biological experimentation significance is often expressed in terms of levels of increasing strength, depending on whether *P* falls above or below certain "critical" levels:

Level of *P*	Significance
$P > 0.05$	Not significant
$0.05 > P > 0.01$	Significance indicated (weak significance)
$0.01 > P > 0.001$	Significance clearly shown
$P < 0.001$	Strong significance

It is important to remember that none of these significance levels gives a definite proof of mutagenicity. The critical *P* values of the intervals give the probabilities, referred to as α-errors, that a difference as large as or larger than the one found in the experiment could have been obtained by chance even if the tested compound had no mutagenic action whatsoever.

In order to illustrate the importance of the α-error we may consider a laboratory testing some 4 compounds per day for mutagenicity. If we assume that, say, at least 3 of the tested compounds are non-mutagenic, approximately once a week one of these innocent compounds would be erroneously considered mutagenic at the 5% significance level. This would happen about 10 times per year and once per year, respectively, when non-mutagenic compounds are judged at the 1% and 0.1% significance levels, respectively, to be mutagenic (cf. Table I).

Besides the scope of the testing activity and the costs involved, the importance of a compound should also be considered. For instance, if a compound is of great value, or potential value, to human well-being and economy, the risk that it is erroneously judged as mutagenic should be kept small and the critical level of the probability set in advance at $P < 0.02$ or $P < 0.01$, or even $P < 0.001$. If then the outcome of the test gives, for example, $0.05 > P > 0.02$ additional test data as well as complementary chemical information are required before a final conclusion is drawn concerning the mutagenicity of the compound in the test system used (cf. Section 2.5). At this point a word of warning is justified. The statistical decision rules are valid only if the individuals within each sample are drawn at random from the population, and that the samples do not differ in factors other than the mutagenic treatment. If some other variable that is not identified or considered in the statistical evaluation is operating *within* the samples, the chance of drawing false conclusions about mutagenicity increases beyond the α-error of correctly performed experiments, with the consequence that the significance may be overestimated. Such variables may be introduced by, for example, a non-random subdivision of animals to test groups at the start of an experiment, or by a non-random positioning of animal cages (petri dishes, etc.) in the test laboratory with a possible influence of light, temperature, noise, etc. Such factors will decrease the reliability of tests and, conversely, even if they are identified and properly considered in the statistical analysis, they will decrease the resolving power of the test. Various homogeneity tests should therefore regularly be applied on positive and negative controls in order to demonstrate the absence of disturbing influences (cf. Section 3.5.1). Even slight indications of heterogeneity suggests the use of many replicates, or repetitions of experiments.

2.2 Significant effect not obtained

When the statistical evaluation gives a P higher than the critical level (α^*) at which the Null hypothesis (H_0) is to be rejected, the result is usually given as no significant difference or no significant effect. According to the habitual way of judging biological experiments a posteriori such a statement is made if the significance estimate gives $P > 0.05$.

Unfortunately, largely as a result of statistical ignorance, this statement is sometimes interpreted as meaning that the compound has been shown or proved to be non-mutagenic. *This is far from true.*

Like all other mean values the mean value of the difference between a test value and the proper control value has an uncertainty which may be expressed in terms of confidence limits which, with a certain probability, enclose the true mean value.

The same applies if we do not directly calculate the mean values compared and their difference but, because of the necessity of considering additional variables, perform statistical analysis according to some more complicated model, such as hierarchic variance analysis. In this case, the mutagenic action is evaluated by a variance component which, if non-significant, has an uncertainty which depends on the variance component for the subordinate level and its number of degrees of freedom (d.f.) (in principle, the size of the test). Even if a difference is found to be zero (or if, correspondingly, a variance component = 0 in the hierarchic analysis) it has confidence intervals that are the broader, the safer we want to be that they enclose the true mean value.

It is therefore not possible from a negative test (of limited size) to draw the *definite* conclusion that a compound is non-mutagenic, non-carcinogenic etc. When a test gives no difference at all between treated samples and the control sample, or when it exhibits no significant difference (i.e. $P > \alpha^*$), this may tell us that the mutagenic activity of the tested compound is not very great (a very relative measure!), *but we cannot present any statement about non-mutagenicity unless we also indicate the reliability of this statement.* The way of doing this depends on whether we are able or not to define beforehand a critical difference ($\delta = \Delta$) which we may consider a priori to signify a genetic risk. If we know the translation factor from the test system to man, as well as the exposure levels and the size of exposed groups, we may define a Δ signifying risk, as a function of the safety margin applied in the biological test. Δ may also be given an arbitrary value, corresponding to what the experimenter thinks is the resolving power (cf. Section 2.3) of his system, e.g. a doubling or a 20% increase of the habitual ("historical") control value. In cases in which the value of Δ is not defined, the observed difference, d, should be presented with its standard error σ_d or s_d or confidence interval:

$$d \pm z_\alpha \sigma_d \text{ or } d \pm t_\alpha s_d \qquad \text{(see Section 2.1)}$$

Confidence intervals of this type will cover an unknown true difference (for instance a value Δ any user of the test result may later want to apply) in approximately 100 $(1-\alpha)$% of the cases.

If a critical difference, Δ, has been defined, the β-error of a conclusion of non-mutagenicity should be estimated. This strictly refers to a comparison of the Null hypothesis, H_0: $\delta = 0$, with its contra-hypothesis, $H_1 : \delta = \Delta$, and involves a calculation of the risk of not rejecting H_0 when H_1 is in fact true (cf. Table I). As discussed in the preceding Section (2.1) the rule for the decision is to reject H_0 if $d > z_{\alpha*}\sigma_d$ and the α-error has therefore a probability corresponding to the area to the right of this limit under the curve of the sampling distribution of $d = \bar{x}_{Tr} - \bar{x}_C$ when $\delta = 0$, as illustrated in an example given in Section 3.3 and Fig. 4. The β-error is the probability of obtaining $d < z_{\alpha*}\sigma_d$ when $\delta = \Delta$. β corresponds, therefore, to the fraction of the area under the distribution curve of $\bar{x}_{Tr} - \bar{x}_C = \Delta$ which falls to the left of the value, $d = z_{\alpha*}\sigma_d$, where H_0 is rejected (cf. Fig. 4). This means that of we draw samples a great number of times from a population with the true mean value $\delta = \Delta$, random variation would lead us to conclude that $\delta = 0$ in 100β% of cases.

The sampling distribution of H_1 may be given as (cf. Section 3.3)

$$z = \frac{\Delta - d}{\sigma_d} \qquad (7)^1$$

We then obtain z_β (or when σ is not known, t_β) by inserting for d the rejection level $z_{\alpha*}\sigma_d$ (or $t_{\alpha*}s_d$):

$$z_\beta = \frac{\Delta - z_{\alpha*}\sigma_d}{\sigma_d} = \frac{\Delta}{\sigma_d} - z_{\alpha*} \qquad (8b)$$

[1] This is in fact the same as equation (3) although, for practical reasons, the sign is reversed.

and read the value of β from tables of the z (or t) distribution.

It follows that $1-\beta$ is a measure of the power of an experiment to detect a given difference, Δ, when testing H_0. The only way to decrease the β-error, i.e. increase the power of the test to detect a critical difference Δ at given α-error, is to increase the size of the test. Rules for calculating the required size of the test are given in Section 3.4.

Fig. 1 illustrates the interrelationship of the sample size and the β-error of accepting the hypothesis of no effect, for different values of Δ (i.e. values of δ in the contra-hypothesis). The data correspond to the rejection level of $H_0 : \alpha^* = 0.05$. The values of Δ are given as fractions of (historical) control values, i.e. $\Delta = 1$ means a doubling and $\Delta = 0.2$ a 20% increase, etc., of the mutation frequently. The sample size is given as the number, np, of mutants observed in the control sample, i.e. the number of individuals tested being $n = \frac{np}{p}$ where p is the proportion of mutants, the treated sample comprising the same number of individuals, n. The β-errors were computed from equation (23d) assuming np to be Poisson distributed (cf. Section 3.1).

The values of np at the points of intersection of the curves with the horizontal lines for $\beta = 0.01$, 0.05 and 0.20 indicate the sample sizes required for the determination of effects = Δ, at 5% significance level, with the efficiencies $(1-\beta)$ 99, 95 and 80%, respectively. A large β-error may call for a repeated

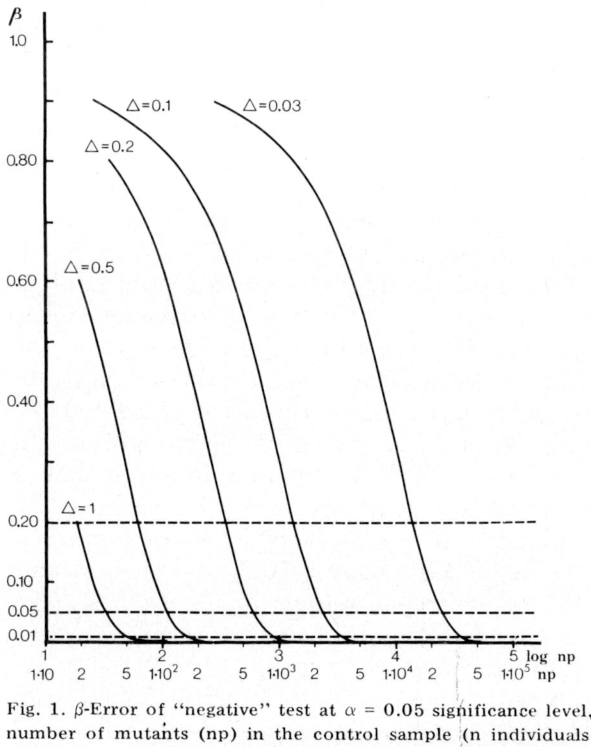

Fig. 1. β-Error of "negative" test at $\alpha = 0.05$ significance level, as a function of sample size expressed as number of mutants (np) in the control sample (n individuals in both control and test, proportion of mutants in control = p). The curves are computed for various hypothetical values of fractional increase of p or np compared to historical control, $\Delta = 1$: 100% increase, $\Delta = 0.5$: 50% increase etc. Intersection points of dashed lines with curves give the required sample sizes for the determination of given Δs at α-error = 0.05 and the respective β-errors.

test with samples of a size estimated for reasonable values of α^* and β^* (cf. Section 3.4).

2.3 β-error, Δ and resolving power of biological tests

The definition of a value Δ as "tolerable" or "safe" has a bearing on legislative decisions about permissible levels and is therefore primarily a matter for public health autorities and administrators. A primary requisite for such a definition of a safe Δ is that data from test systems used are qualitatively and quantitatively translatable in terms of risk in man. In the following deliberations this is assumed to be the case as a first approximation.

At present there is no evidence to support the view that dose—response curves of chemically induced mutation or cancer have a safe threshold [7]. Known facts rather indicate the existence of a linear dose—response from the dose zero [3,7]. The risk that exposure to chemicals will give rise to genetical damage has therefore to be considered stochastically, in the same way as radiation risks, with the consequence that the risk (e.g. the probability that any member of a population will be harmed), at low exposure levels, is proportional to the exposure dose and to the size of the exposed group. This requires special care in the evaluation of compounds considered for general use, including exposure of sensitive groups such as children and pregnant women. If, like Mantel and Bryan [22] we insist on only considering a compound safe if we can show with 99% probability (i.e. $\beta^* = 0.01$) that it gives less than one mutant (or one cancer case) in a 100 million population, we will find that this degree of safety is incompatible with the resolving power of most biological test systems. As appears from the examples given below (Section 4) the latter are able at moderate values of β^* (0.05—0.20) and α^* (0.01—0.05) to detect some 10—100% increase above the control frequencies.

A comparison with radiation risks may give some impression of what these figures mean: a 100% increase (i.e. a doubling) of the mutation frequency is caused by about 100 rad of γ-irradiation absorbed at low dose-rate [21]. A doubling of the mutation rate in man is expected to cause an increment of the order of 1% of heritable damage becoming manifest among live-born in the next few generations [27,28]. Hence, the resolving power of the majority of biological tests would permit the detection of Δ in the range of 10^{-3}—10^{-2}, whereas the Δ corresponding to "safety" in Mantel-Bryan's sense would amount to about 10^{-8}. If an environmental factor cannot be tested with a safety margin some 10^5 times the expected exposure level, this small Δ will lead to a requirement for unrealistically, if not ridiculously, large test series (cf. ref. 8). (The detection, at $\beta^* = \alpha^* = 0.01$, of $\Delta = 10^{-8}$, at a spontaneous mutation frequency = 0.01, would, for example, require about $1/f^2 \cdot 5 \cdot 10^{15}$ tested organisms in each of the test and the control series, f being the safety margin used in the experiment; cf. Section 3.3).

This point is raised as one example of the necessity of combining statistics with common-sense (Section 1) in the evaluation of test data. Additional information on the chemical properties of a compound and of its metabolites might help to bridge the apparent gap between the resolving power of tests and the degree of safety one may want to achieve.

2.4 Double-tail and single-tail test

Often the reliability of an experimental mean value is indicated by giving the upper and lower limits of a confidence interval $\bar{x} \pm z_\alpha \sigma_{\bar{x}}$ (or, if σ was estimated by s, confidence interval $\bar{x} \pm t_\alpha s_{\bar{x}}$), which may be said to have the 100 (1—α)% chance of including the true mean value $\bar{\xi}$ (equation 6; Section 3.2). Referring to a choice of giving the 95% confidence interval (α = 0.05), we may further illustrate what this interval means by saying that its lower limit, $\bar{x}—z_\alpha\sigma$, has $\frac{1}{2}\alpha$ = 2.5% chance of falling above a true mean, $\bar{\xi}$, and that its upper limit, $\bar{x} + z_\alpha\sigma$, in 2.5% of a number of tests, will be smaller than the true mean.

If the studied mean value is a difference $d = \bar{x}_1—\bar{x}_2$ estimating a true difference $\delta = \bar{\xi}_1—\bar{\xi}_2$, we should in the general case, present the experimental value of d together with a confidence interval, having the chance (1—α) of covering the true mean, δ :

$$d—z_\alpha\sigma_d < \delta < d + z_\alpha\sigma_d \qquad \text{(6a and b in Section 2.1)}$$

We illustrate this with a constructed example. In a treated and a control sample of 16 observations each we have estimated the mean values $\bar{\xi}_{Tr}$ and $\bar{\xi}_C$ of the two samples by $\bar{x}_{Tr}$ = 10 and $\bar{x}_C$ = 10, respectively, and we have estimated the standard deviations, σ, of both samples by s = 5. Hence d = 10—10 = 0, and the confidence interval may be given as

$$—t_\alpha s_d < \delta < + t_\alpha s_d\ ; \qquad \text{(6c)}$$

using t for ($n_{Tr} + n_c$ — 2) = 30 d.f., s_d will be computed

from $s_d = \sqrt{\frac{5^2}{16} + \frac{5^2}{16}} = 1.77$.

If we chose to present the 95% confidence interval we would have — $—t_{0.05} \cdot 1.77 < \delta < + t_{0.05} \cdot 1.77 = —3.61 < \delta < + 3.61$.

If the studied mean value is a difference between the mutation frequencies in a control and a treated sample, we are often in the situation that the only alternative of no mutagenic activity, i.e. $\delta = \bar{\xi}_{Tr}—\bar{\xi}_C = 0$, is a mutagenic activity, i.e.

$$\delta > 0; \bar{\xi}_{Tr} > \bar{\xi}_C.$$

We may then increase the sensitivity of tests, and decrease the errors, by applying a single-tailed (one-sided) statistical test.

Fig. 2 illustrates the confidence intervals of a difference in single-sided and double-sided tests. In the one-sided test we use z or t for the double probability limit. Since in this Chapter we have referred to the z- and t-tables designed for two-sided statistical tests, we should use the values of $z_{2\alpha}$ or $t_{2\alpha}$ in one-sided tests at the significance level α, e.g. for $t_{0.05}$ (single-sided) use $t_{0.10}$ (double-sided). In the constructed example given above we may then give the 95% confidence interval as

$$0 < \delta < t_{0.10}\ 1.77 \text{ or } 0 < \delta < 3.00 \qquad \text{(6d; cf. 6c)}$$

i.e. with a lower value of the upper limit than obtained in the two-sided test.

For reasons given above it should be stated in every report whether double-

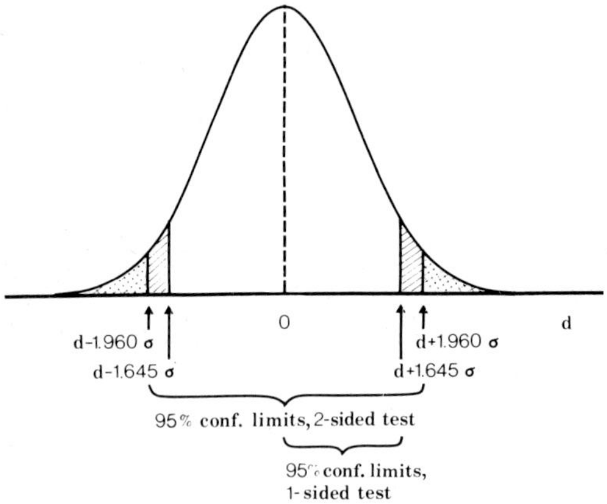

Fig. 2. Illustration of the 95% confidence interval in two-sided and one-sided tests. Distribution of $d = \bar{x}_{Tr} - \bar{x}_C$ when $\delta = 0$.

tailed or single-tailed estimates of confidence intervals, significances, etc. were performed. A warning is justified at this point: if it is considered possible that a studied chemical has some kind of "antimutagenic" action, i.e. giving room for the possible alternative $\bar{\xi}_{Tr} < \bar{\xi}_C$, two-sided confidence intervals and statistical tests should be used.

2.5 A suggested protocol for statistical evaluation

This section summarizes a few general points which, with the modifications called for in specific cases, should be considered in the evaluation of test data.

A test of a compound X of unknown mutagenicity is considered. Even if a critical difference, Δ (cf. Sections 2.2 and 2.3), had been defined in advance, together with acceptable α- and β-errors, the biological test is best economized if it is run in the conventional scale, using samples of standard size. It is then the goal of the evaluation to make a choice between the following two statements:

(1) X has to be considered mutagenic;

(2) the mutagenicity of X, if any, is so low that the compound may be considered non-mutagenic.

As appears from the foregoing the evaluation of a test may also lead to the conclusion that additional testing and/or complementary information is required before a statement can be made.

Table II summarizes in a flow sheet the statistical procedure, with the different decisions and conclusions that may occur. The following comments on the different steps in Table II are required.

Step 3. The search for errors, primarily by inspection of test data (e.g. from print-outs of original data and mean values for all pertinent causes of variation), followed by heterogeneity tests for such variations, including sampling errors, which could disturb the statistical evaluation, should be done at an initial stage of the evaluation.

TABLE II

FLOW SHEET OF PRECEDURE, DECISIONS AND CONCLUSIONS IN THE TEST OF A COMPOUND X FOR MUTAGENICITY

Comments in Section 2.5.

1. State goal of biological test
2. Perform biological test
3. Inspect test data, and data from negative and positive controls, with regard to normal behaviour (erroneous values, e.g. outliers, heterogeneity)
4. Statistical test: State the Null hypothesis, H_0 e.g.

   $$\delta = \bar{\xi}_{Tr} - \bar{\xi}_C = 0$$

   State the level of rejecting H_0, i.e. an acceptable α-error (α^*). If required, state a critical difference, Δ, indicating genetic risk. State an acceptable β-error (β^*). Decide whether two-sided or one-sided test should be applied. Test H_0 by appropriate statistical procedures. Decide whether reject or not reject H_0
5. Alternative situations

(i)	(ii)	(iii)	(iv)	(v)	(vi)
$P < \alpha^*$	$P \approx \alpha^*$		$P > \alpha^*$		
Decision: Reject H_0	Decision depends on whether the applied statistic has under- or overestimated the α-error (see text)	Decision: Do not reject H_0			
		Δ defined (or can be defined)			Δ not defined
		State H_1: $\delta = \Delta$ Test H_1 (equation 8, p. 439; 8b, p. 426)			
		$P \leqslant \beta^*$	$P > \beta^*$		
Conclusion X is mutagenic (α-error of conclusion = α^* or P; see text)		*Conclusion* X is non-mutagenic (β-error of conclusion = β^* or P, see text)	*Conclusion* Test was not able to decide whether X is mutagenic		*Conclusion* X is not mutagenic or of low mutagenicity as indicated by confidence interval of found d.
			X important	X not important	
			Decision: Repeat biologic test at appropriate size (n)	Decision: Report or publish (=alt. vi)	

Step 4. The choice of the rejection level (α^*) for the Null hypothesis and of an acceptable β-error (β^*) is quite arbitrary. Since both the conclusion that a non-mutagenic compound is mutagenic (α-error) and the conclusion of non-mutagenicity of a mutagenic compound (β-error) are undesirable events, low values of α^* and β^* should be chosen, especially in the case of economically important compounds with a wide distribution.

A statement of a critical difference, Δ, could either be referred to the risk one may want to exclude in human populations (cf. Section 2.3) or be referred to some routine practice, e.g. a doubling or a 20% increase of the control level (i.e. $\Delta = \bar{\xi}_C$ or $\Delta = 0.2\,\bar{\xi}_C$, respectively).

Step 5. Alternative (i). Although the rejection level, α^*, was stated (in Step 4), we are permitted to reject H_0 at the level of P given by the test of H_0. If, for example, α^* was beforehand stated to be 0.05, but the test of H_0 gives $P < 0.001$, we may reject H_0 at the 0.1% significance level, with the consequence that we would reach a conclusion of mutagenicity with the α-error 0.1%.

Alternative (ii). Knowing that most statistical models give an approximate estimate of α, it may be necessary, for scientific reasons or because of legal requirements, etc., to clarify whether $P < \alpha^*$ or $P > \alpha^*$. It should then be ascertained whether the statistical model applied is known to overestimate or underestimate α. If necessary, an alternative statistical test should be used, for example with transformed data, or assuming some other distribution of the variates. In critical cases, P should be calculated exactly.

Alternative (iii). If $P \approx \beta^*$, the statistical models applied should be looked into as in Alternative (ii) in critical cases.

Alternative (iv). If the compound X is so important that a statement of mutagenicity or non-mutagenicity should have the α-error $\leqslant \alpha^*$ or the β-error $\leqslant \beta^*$, respectively, the required size of the test should be computed (Section 3.4). If the required size appears to be very large, exceeding what can easily be handled in the laboratory, the application of sequential analysis (cf. ref. 4) could lead to large savings in time, space and money (see also Section 2.3). For a safer conclusion with regard to mutagenicity, additional information, for example from other biological test systems and from chemical structure, should be utilized.

Alternatives (v) and (vi). Depending on the circumstances, the data may be reported or published, with a statement that the test showed no or a low mutagenic activity, the statement being presented as the found difference, d, with a confidence interval (cf. Sections 2.6 and 3.2).

In the examples given below (para 4) this protocol is not followed strictly for the reason that α^* and β^* were not defined a priori. Instead, it is shown how the various points in this protocol could be handled.

2.6 *Presentation of data*

It appears from the foregoing that the evaluation of test results, particularly negative ones, is a relative matter. In order to be optimally useful to other laboratories or to administrators or other users of the data, it is important that publications or other reports contain all essential information required for interlaboratory comparison, safety evaluation, etc. This information should comprise:

(1) Layout of experiment, sampling technique, methods to avoid bias (coding, etc.);
(2) pertinent data on test organism, observed effects and recording errors, medium composition etc.;
(3) time variables, and stage of cell division cycle tested; and
(4) quantitative data from the reported test on
negative control,
exposed sample (tested concentrations), and
positive control (tested concentrations).

These data should comprise (in applicable cases or with modifications):

(1) number of replications;
(2) number of individuals in control series and treatment series, per replication; and
(3) number of mutants in control and treatment series in each replication, or in each treatment group but then provided with the standard deviation for replicates.

(Until the dominant lethal test, see Section 4.4, has been standardized it is advisable in critical cases to publish — or otherwise make available — individual data for total and dead implants, and, if investigated, of corpora lutea. Presentation restricted to mean values per treatment group, week, etc. should be provided with variances.) Established long-term values of the negative and positive controls should be given with their variation. If this has been published elsewhere it should be cited.

The statistical evaluation should comprise:

(1) values of α^* and, if defined, Δ;

(2) statistical method, reference to whether double- or single-sided tests were performed;

(3) result of positive tests: α-error; confidence interval of found effect; and

(4) result of negative tests: confidence interval of found difference (absence of effect). If Δ defined: β-error at specified α^*.

3 Some calculation rules

3.1 Variances

A. Normal distribution.

σ^2 of measurements is estimated by

$$s^2 = \frac{\Sigma(x - \bar{x})^2}{n - 1} \tag{9}$$

Often computation is facilitated by using the expression

$$\Sigma(x - \bar{x})^2 = \Sigma x^2 - \frac{(\Sigma x)^2}{n} \tag{9a}$$

The variance of a mean value of n measurements is obtained as

$$\sigma_{\bar{x}}^2 = \frac{\sigma^2}{n}, \text{ estimated by } s_{\bar{x}}^2 = \frac{s^2}{n} \tag{10}$$

The standard deviation of a mean (Standard Error, S.E.) is hence obtained as

$$\sigma_{\bar{x}} = \frac{\sigma}{\sqrt{n}}, \text{ estimated by } s_{\bar{x}} = \frac{s}{\sqrt{n}} \tag{10a}$$

B. Provided causes of variation other than the sampling error are small, and if the events are independent, proportions (p) follow the binominal distribution, which has to be applied for exact calculation of significances. If the sample size, n, is sufficiently large, the distribution of proportions will approximately follow the normal distribution, the variance of observations (e.g. mutation = 1, not mutation = 0) being estimated through

$$\sigma^2 = p\,(1 - p) \tag{11}$$

Here, the proportion is the mean value, with variance

$$\sigma_{\bar{x}}^2 = \frac{p(1 - p)}{n} \text{ and S.E.} = \sigma_{\bar{x}} = \sqrt{\frac{p(1 - p)}{n}} \tag{11a}$$

The validity of this approximation increases with increasing values of n · p(1-p), and Dixon and Massey [4] suggest that it is permissible when n · p(1—p) > 5, but in critical cases ($\alpha \approx \alpha^*$ or $\beta \approx \beta^*$) great prudence is called for if the number of mutants is small, say <30. This uncertainty is decreased if the variance is stabilized through angular transformation of p or square-root transformation of np (equations 18 and 19; see Section 3.2 and Example 1).

C. In mutation tests the proportion p of mutants is often ver small, i.e. (1—p) ≈ 1. In this case the number of mutants scored, np, will follow the Poisson distribution,

$$\sigma^2 = \text{np and } \sigma = \sqrt{\text{np}}. \tag{12}$$

These values are mean values, and σ therefore is at the same time S.E. Of course, if a test comprises m petri dishes with the mean $\overline{\text{np}}$ of observed mutant colonies per petri dish, $\sigma_{\overline{\text{np}}} = \sqrt{\frac{\text{np}}{\text{m}}}$ if np is Poisson distributed.

3.2 Confidence intervals of mean values and differences; test of H_0

Confidence intervals of means ($\overline{x}$) are calculated from

$$\overline{x} - z_\alpha \sigma_{\overline{x}} < \xi < \overline{x} + z_\alpha \sigma_{\overline{x}} \tag{13a}$$

where $\sigma_{\overline{x}}$ is the standard deviation of the mean (standard error, S.E.), obtained from equation (10a). If σ is not known but estimated in the test, and if n is small (<200), the *t* distribution applies, i.e.

$$\overline{x} - t_\alpha s_{\overline{x}} < \xi < \overline{x} + t_\alpha s_{\overline{x}} \tag{13b}$$

As for the use of $z_{2\alpha}$ or $t_{2\alpha}$ in the single-sided tests, see Section 2.4.

Essentially the same calculation rules are used in the statistical test of the hypothesis of no difference of two mean values (e.g. $H_0 : \xi_{Tr} = \xi_C$) and in the computation of the confidence interval of the found difference, $d = \overline{x}_{Tr} - \overline{x}_C$. In the case where we use the normally distributed statistic

$$z = \frac{\overline{x}_{Tr} - \overline{x}_C}{\sqrt{\frac{\sigma_{Tr}^2}{n_{Tr}} + \frac{\sigma_C^2}{n_C}}} \tag{3b, Section 2.1}$$

to test the hypothesis that $\xi_{Tr} - \xi_C = 0$, with the decision to reject the hypothesis if z is larger than $z_{\alpha*}$, the denominator of equation (3b) is the standard deviation (S.E.) of d,

$$\sigma_d = \sqrt{\frac{\sigma_{Tr}^2}{n_{Tr}} + \frac{\sigma_C^2}{n_C}} \tag{4, Section 2.1}$$

the 100 (1—α)% confidence interval being given by equations (6a and b).

If σ is not known, but estimated in the experiment by s, the corresponding *t*-statistics are used. If $\sigma_{Tr} = \sigma_C = \sigma$ and if n_{Tr} and n_C are different, a pooled value, s_p, of s_{Tr} and s_C should be used to estimate σ :

$$s_p^2 = \frac{(n_{Tr} - 1)s_{Tr}^2 + (n_C - 1)s_C^2}{n_{Tr} + n_C - 2} \tag{14a}$$

and H_0 tested by the statistic

$$z = \frac{\bar{x}_{Tr} - \bar{x}_C}{s_p \sqrt{\frac{1}{n_{Tr}} + \frac{1}{n_C}}}, \tag{15}$$

H_0 being rejected if $t \geqslant t_{\alpha*}$ at $(n_{Tr} + n_C - 2)$ d.f. Likewise, the $100(1-\alpha)\%$ confidence interval of d should be given by equation (6a) using t_α for $(n_{Tr} + n_C - 2)$ d.f. and estimating s_d through

$$s_d = s_p \sqrt{\frac{1}{n_{Tr}} + \frac{1}{n_C}}. \tag{14b}$$

If the two samples compared have different σ, the statistic z of formula (3b) will approximately follow the normal distribution at large sample sizes. If, however, σ_{Tr} and σ_C are not known but estimated by s_{Tr} and s_C the corresponding statistic

$$t = \frac{\bar{x}_{Tr} - \bar{x}_C}{\sqrt{\frac{s_{Tr}^2}{n_{Tr}} + \frac{s_C^2}{n_C}}} \qquad (16) = (3a)$$

will approximately follow the t-distribution, however, with a reduced d.f. (cf. ref. 4, p. 119), leading to a risk of overestimating significances.

The variances of two samples may be compared by an F-test of $H_0 : \sigma_{Tr}^2 = \sigma_C^2$, rejecting H_0 at the level $\alpha*$ if

$$F = \frac{s_{Tr}^2}{s_C^2} \left(\text{or } F = \frac{s_C^2}{s_{Tr}^2}\right) \tag{17}$$

at $(n_{Tr}-1)/(n_C-1)$ d.f. (or vice versa) is sufficiently large to give $P < \alpha*$.

A difference in variances between compared samples is frequently encountered in tests for mutagenicity. In the ideal case (cf. p. 434), the variable: the number of mutants per petri dish (as in the bacterial assay; cf. Section 4.2) or per female (as in the dominant lethal test; cf. Section 4.4) will be approximately Poisson distributed, with σ^2 proportional to the mean value (equation 12). In such cases it is preferable to stabilize the variance by suitable *transformations* (see ref. 4, p. 324) such as the square-root transformation

$$a = \sqrt{x} + \sqrt{x+1} \tag{18}$$

of the original variable x. The variable a has $\sigma^2 \approx 1$, and this transformation may be applied if $x = np \geqslant 0.8$. For similar reasons, the use of low proportions as the variable (e.g. the "mutation index", $\frac{\text{number of dead implants}}{\text{total number of implants}}$, in the dominant lethal test), will lead to variances approximately proportional to the mean value (equation 11). In this case a stabilization of variances is suitably obtained through an angular transformation of the proportions $p = \frac{x}{n}$. The transformation

$$\phi = \arcsin \sqrt{\frac{x}{n+1}} + \arcsin \sqrt{\frac{x+1}{n+1}} \tag{19}$$

may be applied in $np \geqslant 0.8$ [4], and has the variance

$$\sigma_\phi^2 = \frac{1}{n + 1/2}. \tag{19a}$$

Example 3.2.1. Testing a mutation frequency in 1600 individuals, $100p = 1.0\%$ was found. Give the 95% confidence interval of the frequency.
$np(1-p) \approx 16$ (>5). Hence σ^2 may be estimated from equation (11).

$$\bar{\xi} = 0.01 \pm z_{0.05} \cdot \sigma$$

$$z_{0.05} = 1.96; \sigma = \sqrt{\frac{0.01 \cdot 0.99}{1600}} = \frac{0.0995}{40} = 0.00249.$$

This gives the confidence interval

$$0.01 - 1.96 \cdot 0.00249 < \bar{\xi} < 0.01 + 1.96 \cdot 0.0249$$

$$0.00512 < \bar{\xi} < 0.01488 \tag{20a}$$

Since $(1-p) \approx 1$ we might as well apply the Poisson distribution on the number of mutants scored
$np = 1600 \cdot 0.01 = 16$

$\sigma = \sqrt{16} = 4$ (cf. equation 12)

Confidence interval:

$$16 - 1.96 \cdot 4 < n\bar{\xi} < 16 + 1.96 \cdot 4$$

$$8.16 < n\bar{\xi} < 23.84$$

i.e., for the primarily interesting proportion,

$$\frac{8.16}{1600} < \bar{\xi} < \frac{23.84}{1600}$$

$$0.00510 < \bar{\xi} < 0.01490 \tag{20b}$$

i.e. essentially the same result as the one obtained in the estimate (20a) above.

Since the Poisson and binomial distributions are not symmetrical the confidence intervals given by (20a) and (20b) will underestimate the lower and upper limits of the interval. The interval calculated from the binomial distribution will be (see statistical tables, for example in ref. 10, Table VIII : I; see also Fig. 3 which illustrates the confidence intervals of Poisson distributed mean values and of binominally distributed small proportions, p):

$$16 - 1.96\sqrt{16} + 0.97 < n\bar{\xi} < 16 + 1.96\sqrt{16} + 2.08$$

$$8.16 + 0.97 < n\bar{\xi} < 23.84 + 2.08$$

$$9.13 < n\bar{\xi} < 25.92$$

$$0.0057 < \bar{\xi} < 0.0162 \tag{20c}$$

At higher values of np the approximative method will become more correct. At lower frequences (tabulated values of) exactly calculated limits should be used.

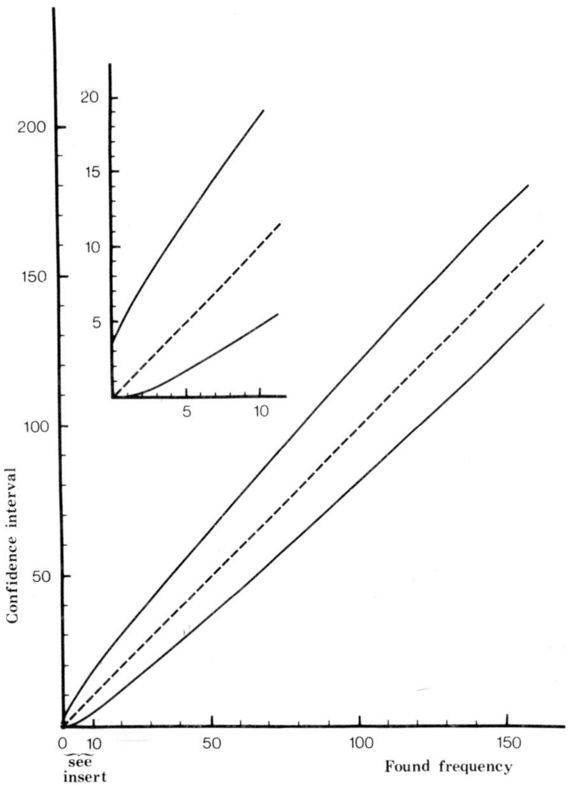

Fig. 3. 95% confidence limits (two-sided) of Poisson distributed frequencies, i.e. numbers of mutants, np. The upper and lower confidence limits are read from the ordinate axis at a point on the abscissa corresponding to a given value. The confidence interval of proportions (p) are obtained by dividing with number of individuals, n. The curves are satisfactorily valid for proportions, p, up to 0.05 (cf. ref. 10, Table VIII_I).

Application of the square-root transformation on np ($\overline{np}$ = 16) gives the confidence interval (cf. equation 18)

$$a = \underset{4.0000}{\sqrt{16}} + \underset{4.1231}{\sqrt{17}} \pm 1.96 \cdot 1$$

(The value of np for any value of a is easily obtained from

$$np = \frac{(a^2 - 1)^2}{4a^2} \tag{21}$$

Variable	Lower limit of conf. interval	Upper limit of conf. interval	
a	6.163	10.083	
np	9.002	24.919	(20d)
$p(=\overline{\xi})$	0.0056	0.0156	

i.e., in better agreement with the limits calculated exactly. The arc sine transformation (equation 19) applied on p gives the same result.

Example 3.2.2. Calculate the 95% confidence interval of the proportion when 0 mutants were found in 1000 tested organisms.

The confidence interval of 0 has only an upper limit. The meaning of the 95% confidence intervals of Example 1 is that the upper limit has the chance 2.5% of being equal to or smaller than the true mean value $\bar{\xi}$ (and corresponding for the lower limit). In order to bring the "95% confidence interval" of zero into conformity with that of any value >0, we should then compute a limit such that the upper limit has 2.5% probability of being $\leqslant\bar{\xi}$. This is the same as computing a proportion of mutants, p, with the property that, in a test of 1000 organisms, zero mutants will appear with 2.5% probability. The probability of non-mutant in one tested organism is 1—p. The probability of no mutant in 1000 organisms is $(1-p)^{1000}$. This gives

$$(1-p)^{1000} = 0.025; \quad p = 1 - 0.025^{1/1000}$$

$$\text{(general formula: } (1-p)^n = P) \tag{22}$$

p = 0.00369; and the confidence interval $0 < \bar{\xi} < 0.00369$.

This illustrates the important fact that an experimental finding of zero events might very well be obtained if the true mean value is rather great:

Probability (single-tail)	Confidence limit
0.10	2.30
0.05	2.99
0.025	3.69 (Example 1)
0.005	5.30

See also insert of Fig. 3 for the value 0.

3.3 Determination of β for an alternative hypothesis, i.e. the power of the test (1—β) to detect a critical difference, Δ

There are two samples with the following statistical parameters:

	Sample 1	Sample 2
Means	$\bar{x}_C$	$\bar{x}_{Tr}$
Numbers of individuals	n_C	$n_{Tr}(\simeq n_C)$
Variances	σ^2_C	σ^2_{Tr}

In mutation tests we are mostly investigating if the mean of a treated sample is greater than that of the control sample. We therefore restrict ourselves to cases where one-sided statistical tests should be used. For calculations of β in two-sided tests the reader is referred to textbooks of statistics (cf. ref. 4; Chapter 14).

We test at the significance level α^* the hypothesis (H_0) that $\bar{\xi}_{Tr} = \bar{\xi}_C$.

$\bar{x}_{Tr} - \bar{x}_C = d$ (could be negative or positive)

The mode of reasoning in the evaluation of the β-error is illustrated by a hypothetical example.

Two samples having the following parameters are compared

Treated	Control	Difference
$\bar{x}_{Tr} = 50$	$\bar{x}_C = 49$	$d = 1$
$\sigma^2_{Tr} = 50$	$\sigma^2_C = 49$	$\sigma^2_d = 0.495\ (0.704^2)$
$n_{Tr} = 200$	$n_C = 200$	

We are interested in the power of this test in detecting a difference amounting to (i) 1% of the historical control value ($\bar{\xi}_{hist} = 50$), i.e. $\delta = 0.5$ and (ii) 5% of $\bar{\xi}_{hist}$, i.e. $\delta = 2.5$. (The equality of s^2 and $\bar{x}$ indicates that we are dealing with Poisson distributed values, for instance samples of 200 petri dishes with mutant colonies of bacteria (cf. Section 4.1), or with the determination of a low radioactivity at a historical background of 50 cpm. The fact that by estimating the variances by s^2 equality with $\bar{x}$ was obtained, shows further that we might consider σ to be known and are allowed to use z-statistics. However, even for a variable following the normal distribution, z-statistics would have been permissible, the *t*-distribution being close to the z-distribution at the large n_{Tr} and n_C of the test.)

We have decided to reject the Null hypothesis in the one-sided test at the 5% significance level. Testing H_0 we obtain

$$z = \frac{d}{\sigma_d} = \frac{50 - 49}{\sqrt{\frac{50}{200} + \frac{49}{200}}} = \frac{1}{0.704} = 1.420$$

Since we reject H_0 if $z > z_{2\alpha*} = 1.645$, we will in this case not reject H_0, in other words the data do not permit us to conclude that $\bar{\xi}_{Tr} > \bar{\xi}_C$. Our conclusion is therefore:

$\delta = 0$, or $\bar{\xi}_{Tr} = \bar{\xi}_C$.

We first investigate the β-error of this conclusion if the true difference is $\delta = 0.5$ and at the same time the power of the test $(1-\beta)$ to detect this effect of the treatment. We therefore state the contra-hypothesis, $H_1' : \delta = 0.5$. Rejecting H_0 when $\frac{d}{\sigma_d} \geqslant z_{2\alpha*}$ means rejection when $d \geqslant z_{2\alpha*} \cdot \sigma_d = 1.645 \cdot 0.704 = 1.16$. The hypothetical difference with mean $\delta = 0.5$ has the same variance as the one of the experiment (see Fig. 4, which illustrates the sampling distributions of the alternatives). Comparing this distribution with the level at which H_0 is rejected we obtain a z-distributed variable (equation 7) from which $z_{2\beta}$ is calculated according to equation (8) (see Section 2.2):

$$z_{2\beta} = \frac{\Delta - z_{2\alpha*} \cdot \sigma_d}{\sigma_d} = \frac{\Delta}{\sigma_d} - z_{2\alpha*} \qquad (8)$$

$$z_{2\beta} = \frac{0.50 - 1.16}{0.704} = -0.94$$

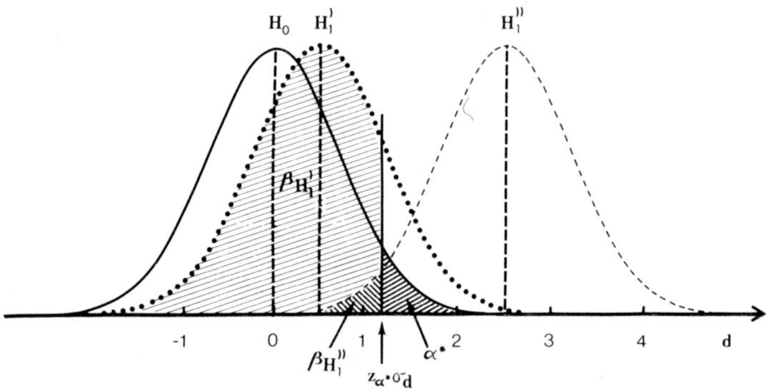

Fig. 4. Sampling distributions of d - $\bar{x}_{Tr}-\bar{x}_C$ for three alternative hypotheses: H_0 : $\delta = \bar{\xi}_{Tr}-\bar{\xi}_C = 0$; H_1' : $\delta = 0.5$; H_1'' : $\delta = 2.5$. The figures refer to examples in Sections 3.3 and Table III, where $\sigma_d = 0.704$. H_0 is rejected when $d > z_{\alpha*}\sigma_d$. Using $\alpha^* = 0.05$ in the one-sided test, the limit of rejection is therefore $d > 1.645 \cdot 0.704 = 1.16$, shown in the figure, and the α-error is the shaded area to the right of this limit of the H_0 curve. The β-errors of not rejecting H_0 when H_1' is true or when H_1'' is true correspond to the shaded areas to the left of $z_{\alpha*}\sigma_d$ under the respective curves. See text for details.

Understanding the β-error as the risk of not rejecting H_0 when H_1' is true corresponds numerically to the area under the distribution curve of H_1' falling to the left of the value, $z_{2\alpha*} \cdot \sigma_d = 1.16$, at which H_0 is rejected. If the mean value of H_1' is equal to the rejection level, equation (8) gives $z_{2\beta} = 0$, corresponding to $2P = 1$ and $P = 0.5$, i.e. the β-error will be 50%. If, as in the present case, a negative value of $z_{2\beta}$ is obtained, the probability for the corresponding positive $|z_{2\beta}|$ read from a z-table (for two-sided tests) means $P(-z_{2\beta}) = 2(1-\beta)$
We obtain: $P(-z_{2\beta}) = 0.35$
Hence $(1-\beta) = 0.175$ and $\beta = 0.825$

Conclusion. The large β-error does not allow us to conclude that $\bar{\xi}_{Tr}-\bar{\xi}_C = 0$. Further measurements (or additional information) are required for a conclusion of no effect of the treatment (cf. Table II, alternative 5(iv)).
Testing the second contra-hypothesis, H_1'' : $\delta = 2.5$, we obtain in a corresponding manner:

$$z_{2\beta} = \frac{2.50 - 1.16}{0.704} = 1.903$$

$$P(z_{2\beta}) = 0.059$$

$$\beta = \frac{0.059}{2} = 0.0285 \approx 0.03$$

Conclusion. In this case $\beta < 0.05$, and we feel entitled to conclude that the treatment had no effect (N.B. if by "effect" we mean a true difference $\delta = 2.5$). The errors of not rejecting H_0 when either H_1' or H_1'' is true, or of rejecting H_0 when H_0 is true, are summarized in Table III which thus gives flesh to the bones of Table I (Section 2).

3.4 Estimate of the required size of a test

The size of a mutation test required for the detection of a critical effect, Δ, with a given power $(1-\beta^*)$ of the statistical test and at a given significance

TABLE III

ILLUSTRATION OF α- AND β-ERROS

(cf. Table I and see example in Section 3.3)

Decision	Null hypothesis H_0 true	Contra-hypothesis	
		(a) H_1' true	(b) H_1'' true
Reject H_0	Incorrect in $100\alpha\% = 5\%$ of the cases	Correct in $100(1-\beta)\% = 17.5\%$ of the cases (power = 17.5%)	Correct, in 97% of the cases (power = 97%)
Do not reject H_0	Correct, in $100(1-\alpha)\% = 95\%$ of the cases	Incorrect, in $100\beta\%$ = 82.5% of the cases	Incorrect, in 3% of the cases

level, α^*, may be estimated from the expression

$$\frac{\Delta}{\sqrt{\frac{\sigma_{Tr}^2}{n_{Tr}} + \frac{\sigma_C^2}{n_C}}} \geqslant z_{\alpha*} + z_{\beta*} \tag{23}$$

When σ^2 has been estimated from s^2 in sufficiently large samples in preceding work, n is easily determined from a t-test using t for ∞ d.f.

$$n \geqslant \frac{(t_{\alpha*} + t_{\beta*})^2 \cdot (s_{Tr}^2 + s_C^2)}{\Delta^2} \tag{23a}$$

If we assume that $\sigma_{Tr}^2 = \sigma_C^2 = \sigma^2$, or if we use a pooled average to estimate σ^2 of the difference (equation 14), we obtain

$$n \geqslant \frac{(t_{\alpha*} + t_{\beta*})^2 \cdot 2}{(\Delta/\sigma)^2} \tag{23b}$$

In mutation testing we are often dealing with low proportions, p, where the number of mutation cases, np = x, are approximately Poisson distributed with the variance $\sigma^2 = \overline{np} = \bar{x}$ (equation 12). Using expressions derived from equation (23) we may then directly obtain sufficiently accurate estimates of the required size (n) of tests

$$n \geqslant \frac{(z_{\alpha*} + z_{\beta*})^2 (2p + \Delta)}{\Delta^2} \tag{23c}$$

or the required number of mutants

$$x(=np) \geqslant \frac{(z_{\alpha*} + z_{\beta*})^2 (2 + \Delta/p)}{(\Delta/p)^2} \tag{23d}$$

In the latter expression, Δ/p is the fractional increase of p we want to detect. If Δ is small compared to p, the last factor of the numerators of equations (23c) and (23d) approaches 2p or 2, respectively.

Example (constructed).

A test system has a spontaneous mutation frequency 1% ("historical control" value). Estimate the (equal) sizes of a control and a treated sample required for the detection, with power 95% ($\beta^* = 0.05$) and at significance level 0.05 (α^*) an increase of the mutation frequency amounting to: (a) 1% (i.e. a doubling of the historical control value); (b) 0.2%; (c) 0.01% (i.e. 1% of the control value).

Since we are dealing with an increase of p, we apply the one-sided test, and use $z_{2\alpha^*} = z_{2\beta^*} = 1.645$ (cf. Section 2.4), and $(z_{2\alpha^*} + z_{2\beta^*})^2 = 4 \cdot 1.645^2 = 10.82$.

Application of equations (23c) and (23d) gives for the three values of Δ (cf. Fig. 1, values of np at $\beta = 0.05$):

TABLE IV

	Δ/p_C $\left(\frac{p_{Tr} - p_C}{p_C}\right)$	Δ $(=p_{Tr} - p_C)$	Estimate of n	Estimate of x(=np)
(a)	1	0.01	$n \geqslant \frac{10.82 \cdot 0.03}{0.01^2} = 3250$	$x \geqslant \frac{10.82 \cdot 3}{1^2} = 33$
(b)	0.2	0.002	$n \geqslant \frac{10.82 \cdot 0.022}{0.002^2} = 59{,}500$	$x \geqslant \frac{10.82 \cdot 2.2}{0.2^2} = 595$
(c)	0.01	0.0001	$n \geqslant \frac{10.82 \cdot 2 \cdot 0.01}{10^{-8}} \approx 22 \cdot 10^6$	$x \geqslant \frac{10.82 \cdot 2}{0.01^2} \approx 220{,}000$

If we are studying an event of a kind that is expected not to occur in the control material, that is, which has never been observed in control series (i.e. assuming $p_C = 0$) appreciably smaller sample sizes are required, as is illustrated in Fig. 3. In this case we calculate the number of individuals (n) required to make the probability $\leqslant \beta^*$ that no mutant will be found, if the true mutation frequency were $= \Delta$ $(=p_{Tr})$. This is the same as saying that n should be sufficiently large to permit, with probability $1-\beta^*$, at least one event to be observed if the true frequency is equal to Δ. This gives the equation (cf. equation 22)

$$(1 - \Delta)^n \leqslant \beta^* \tag{22a}$$

$$n \geqslant \frac{\log \beta^*}{\log(1 - \Delta)} \approx \frac{\log \beta^*}{-0.434\Delta} \tag{22b}$$

For the three alternative values of Δ in the preceding constructed example this would give

	$\Delta(=p_{Tr})$	n
(a)	0.01	300
(b)	0.002	1,500
(c)	0.0001	30,000

Above (p. 438) we have shown that the 95% confidence interval, in the one-sided test, of the observed value 0 is 3.0 (at low proportions). In agreement with this the proportion 0.01 would be obtained as the upper 5% value, if we tested 300 individuals without observing any mutant.

3.5 Special considerations

3.5.1 Homogeneity of test

This problem has several aspects, two of which will be discussed briefly.

Homogeneity of replications. If mean values of replications within a test can be shown not to vary significantly more than expected from a theoretical distribution (e.g. Poisson or binomial distribution), the efficiency of the test can be increased by basing the number of degrees of freedom (d.f.) on individual observations. In addition, a general agreement with theory facilitates the detection of accidentally deviating values ("outliers", cf. Section 3.5.2). χ^2 and variance analytic methods may be used in testing hypotheses of homogeneity. One suitable test of inter-replication or inter-occasion variation for a test repeated on different days is the factorial variance analysis (a):

(a) Factorial analysis

Cause of variation (number)	d.f.
Occasion (m)	m − 1
Treatment (control + treated = 2)	1
Interaction occasion × treatment	m − 1
Error (n measurements per treatment and occasion)	2m(n − 1)

(b) Hierarchic analysis

Cause of variation (number)	d.f.
Occasion (m)	m − 1
Treatment within occasion (2)	m
Error (n measurements per treatment and occasion)	2m(n − 1)

In the absence of a significant interaction effect, the square sum of the treatment effect may then be compared in the F-test at the d.f. of the pooled error. If an interaction effect is indicated, its mean square should constitute the error, with ensuing low efficacy of the test. The treatment effect is then better estimated in the hierarchic analysis (b).

If a significant between-replicate variation occurs, the reasons for this should be investigated. If improvements do not help, comparisons have to be based on means with d.f. equal to the number of petri dishes (in bacterial mutation tests) or males (in the dominant lethal test), with a decreased power of the test in consequence. This is exemplified in Section 4.

Long-term mean values and pooling of data. A test laboratory should continuously investigate the stability of its test systems by statistical tests of homogeneity of scored values in positive and negative controls over periods of suitable length as well as cumulatively over longer periods. Besides standard methods, regression analysis should be used to identify trends (cf. ref. 25). A demonstrated absence (or more strictly a low probability) of heterogeneity will give a safer basis for the evaluation of critical cases.

If a test system responds homogeneously, (negative) control data from many experiments may be pooled, allowing a reduction of the standard error of the difference (equations 4 and 14) and a corresponding increase of the resolving power in critical cases. If in the same fashion a positive control is regularly applied, homogeneity in the response to this compound will justify the pooling of data more generally, e.g. of great numbers of treatments of related compounds which each appear to give a negative result.

At this point a word of warning is required. Pooled control means or historical control values should be used with care and restriction, and only as an auxiliary tool in cases of special importance. A recurrent use of the same control data in many comparisons might easily violate the available numbers of degrees of freedom and, further, lead to systematic errors in conclusions partly because variation of known or unknown origin may occur in the control (cf. Section 2).

Variation between laboratories. The reasons for differences between results obtained in different laboratories have too rarely been seriously analyzed, although they might give important information on synergisms, on heritable variation in sensitivity between strains of test organisms, on defective techniques, etc. (cf. ref. 17).

Interlaboratory standardization of testing procedures will prevent the publishing — which in the case of important compounds may have serious consequences — of conclusions that are erroneous because of defective experiments (cf. Section 2.1).

3.5.2 Outliers

The last-mentioned case exemplifies the general problem of identifying spurious deviating values. This is a problem in all sciences which deal with quantitative measurements. Hence, rules for discarding extreme values, based on the small probability that they belong to the population (usually represented by a sample of at least three values), are dealt with not only in statistics (cf. ref. 4; Chapter 16: 3—5, Table A-8e) but also, for example, in analytical chemistry (cf. ref. 9).

The importance should be pointed out of *not* discarding values which are apparently deviant if they can be shown with sufficiently high probability to belong to the studied population, as well as of discarding values if this probability is small. If there are no a priori observations of illness of an animal, defective medium, or similar data as a cause of the deviating value, it would endanger the validity of the test if the value were discarded at, say, $P > 0.10$ and in any case at $P > 0.20$. On the other hand, an outlier should be discarded below some low P, which depends on the number of simultaneous tests run by the laboratory and on other factors. In most cases discarding below $P < 0.05$ does not seem wrong.

Evidently, outliers may occur on any level, from individual observations to whole experiments (cf. preceding subparagraph).

The occurrence of *clusters* are in certain cases detectable as outliers (cf. Example 4.4).

If a test system permits the scoring of >1 mutant originating from the same mutational event, the observed numbers of mutants are not independent, and the statistical treatment of data have to be based on frequencies of mutant "clusters" with independent origin, with the corresponding requirement on scoring offspring families separately for mutants. This is a routine in the analysis of mutants in experiments with Drosophila [23] and barley [6].

If the clustering of mutants is disregarded in the significance test, i.e. if mutants are considered to have arisen independently, it is obvious that a spurious occurrence of very large clusters involves a risk that wrong conclusions will be drawn. This is exemplified by the dominant lethal test (Section 4.4). The use of tables like those of Kastenbaum and Bowman [15] for significance estimates, on data uncorrected for large clusters, might lead to erroneous conclusions. Large clusters may be identified and, if necessary, discarded by applying the methods for detection of outliers. (It should be remarked that the similarity between the chlorophyll mutation test in barley and the dominant lethal test in rodents is only apparent. In the barley test the mutants of a cluster usually have a common origin, a mutational event in a precursor cell in the embryo giving rise to a mutated sector of the spike or the plant [14,20] whereas in the dominant lethal test the few dead implants in a female seem normally to originate from independent events. "Large clusters" in the former case are usually due to spontaneous mutations in previous generations or to a large size of the mutated sector, whereas in the latter case abnormal data seem to have a non-genetic cause ("pseudo-pregnancy").)

In the barley test the now wide-spread habit of threshing together all plants of a treated sample, and to express mutant frequencies as per cent of second-generation seedlings (cf. ref. 12), makes any statistical treatment of differences observed impossible and what is still worse, it makes mean values highly questionable. There is only one way of escaping the separate analysis of offspring of spikes-within-plants, and that is to pick (in a random fashion!) one seed from each plant of a large treated sample, a method which might be applied when testing cheap chemicals.

3.5.3 Shape of dose—response curves

In a majority of cases available data for induced mutation indicate the existence of a linear component in dose—response relationships, i.e. that the mutation frequency, Y, may be written

$$Y = C + bD + cD^2 + \ldots\ldots \qquad (24)$$

where C is the spontaneous frequency, D the dose of the applied factor, and b and c constants. Some investigators are however prepared, on rather unsatisfactory evidence, to deny the existence of a linear term. This is especially true in the cases of the incidence of cancer and of terata (for a review see ref. 7).

For radiation-induced cancer, considerations of the influence of linear energy transfer have led to the idenitification of a predominance of the dose-square component (cf. ref. 16), probably because the studies did not include very low doses, and the dose response of terata is generally (but on the basis of very incomplete information) considered to have a dose threshold (D_{thr}) below which causative agents are inefficient; in other words the incidence

(E_T) would approximately follow the relationship

$$E_T = k(D - D_{thr})^n \tag{25}$$

where k and n are constants; $n > 1$.

The existence or non-existence of a safe dose-threshold is of utmost importance to risk estimates at low doses. Therefore, in studies of dose—response curves, the coefficient b of the linear term in equation (24) has to be evaluated carefully. b has to be treated as a mean value, provided with a confidence interval with the same meaning as discussed above (Section 3.2). Thus, if b is not significantly different from zero this is far from a proof that no linear term exists.

In formulations of confidence intervals of b, enclosing with given probability the true value, B

$$b - t_\alpha s_b < B < b + t_\alpha s_b \tag{26}$$

or in tests of $H_0 : B = 0$, σ_b may be estimated by

$$s_b = \sqrt{\frac{1}{n-2} \cdot \frac{\Sigma(y - Y)^2}{\Sigma(D - \bar{D})^2}} \tag{27}$$

where (y—Y) are the vertical distances of measurements from the regression line. In statistical tests s_b has n—2 d.f.

In the evaluation of b and of the β-error when the hypothesis $B = 0$ is not rejected, a weighting of points with regard to their contribution is essential (cf. ref. 5). This leads to complicated calculations best performed on the computer.

4 Examples

4.1 *Recessive lethals in Drosophila*

(Treatment of rare events by 2 × 2 contingency test and direct calculation of probability.)

4.1.a. Clark [2] tested the mutagenicity of technical grade DDT in different systems. In *D. melanogaster* samples of 350 chromosomes II were tested for recessive lethal mutations. The control contained 1, and the treated sample 6 recessive lethal mutations. This gives us the following 2 × 2 contingency table:

Control	x = 1	y = 349	a′ = 350
Treated	z = 6	u = 344	b′ = 350
Sum	a = 7	b = 693	n = 700

From the $\chi^2_C = 2.31$ (C denoting Yates's correction) i.e.

$$\chi^2_C = \frac{[(z - \frac{1}{2})(y - \frac{1}{2}) - (x + \frac{1}{2})(u + \frac{1}{2})]^2 n}{a \cdot b \cdot a' \cdot b'} = \frac{(|z \cdot y - x \cdot y| - n/2)^2 u}{a \cdot b \cdot a' \cdot b'} = 2.31 \tag{28}$$

the double-tailed test gives $0.20 > P > 0.10$. Clark, who does not say how he calculated χ^2, obtains $\chi^2 = 2.47$, with the same probability interval. From this, he draws the "tentative" conclusion that DDT does not give rise to recessive lethal mutations in Drosophila.

One cannot object to this conclusion — although the observed mutant frequency was 6 times higher in the test than in the control (for which reason there is little point in calculating a β-error). The author was able to indicate the ability of DDT to cause genetic injury of other kinds not only in Drosophila but also in mouse and Neurospora, and he is hence not concerned with the decision whether DDT is mutagenic or not. The question is only: does DDT also induce recessive lethal mutations in Drosophila? His "tentative" conclusion is therefore the right one, since it suggests further experiments.

Two things should be noted in this case:
(a) the χ^2 test is normally a double-tailed one but, in view of positive evidence of mutagenicity in other systems, the single-tailed χ^2 test is justified. Hence, $\chi^2_{2\alpha}$ gives $0.10 > \alpha > 0.05$, with α closer to the lower limit; and
(b) despite corrections for continuity, the χ^2 analysis may, especially at low expected values, lead to appreciable errors in the estimated probability (see Section 4.2). With low numbers, and in any case if one value is zero, the exact calculation of P is not only more correct but may even be simpler.

The exact calculation comprises the determination of the probability of the observed or a stronger deviation from expectation and gives, hence, the single-tailed probability

$$P = \frac{a!b!a'!b'!}{x!y!z!u!n!} + \frac{a!b!a'!b'!}{(x-1)!(y+1)!(z+1)!(u-1)!n!} + \ldots \qquad (29)$$

$$\frac{a!b!a'!b'!}{0!(y+x)!(z+x)!(u-x)!n!}$$

The operation may be described as the summation of the probabilities, at constant margin totals, of obtaining x (where x is the smallest entry), x—1 (where the other three values have to be adjusted), x—2, etc., until x = 0.

In the present case, where x = 1, the calculation comprises two terms, with x =1 and x = 0:

$$P = \frac{7!693!350!350!}{1!6!349!344!700!} + \frac{7!693!350!350!}{0!7!350!343!700!} = 0.054 + 0.008 = 0.062 \qquad (29a)$$

This probability is in agreement with the one obtained in the (single-tailed) χ^2 test above. It is in fact so close to 0.05, the limit usually set for low significance, that the description of the result, alongside with others, as a "certain indication" of mutagenicity might have been justified; of course then together with the α-error (= 0.062) of this conclusion.

4.1.b. In his study Clark [2] finds, in mouse spermatocytes, significant increases in the frequencies of breakage, univalents and stickiness; however frequencies of translocations, gaps and despiralization are not greater. The basic data for chromosome breakage are as follows:

	With breaks	Without breaks	Total scored
Control	0	684	684
Treated	3	304	307
Sum	3	988	991

The high χ^2 given by Clark ($\chi^2 = 6.7; P \approx 0.01$) was evidently, by mistake, calculated without using the discontinuity correction. With this correction (equation 28), $\chi_C^2 = 3.86$ is obtained, corresponding to $2\alpha \simeq 0.05$, i.e. $\alpha \simeq 0.025$ (single-tailed test). Since, as a general rule, the smallest theoretical frequency of the 2×2 χ^2 test should not be less than 5, the exact calculation should be applied in this case (equation 29), giving

$$\alpha = P = \frac{3!988!684!307!}{0!3!684!304!991!} = \frac{305 \cdot 306 \cdot 307}{989 \cdot 990 \cdot 991} \simeq \left(\frac{306}{990}\right)^3 = 0.030 \qquad (29b)$$

Clark's conclusion that an effect was obtained is thus sustained.

4.2 Back-mutations in S. typhimurium his⁻ strains

In this back-mutation test a little histidine is added to the medium to permit a couple of cell divisions. For this reason, a constant number of colonies develop in the control even if the survival is reduced considerably through treatment with a non-mutagenic agent. Therefore, the variate has to be *number of colonies per plate*, not number of colonies per unit number of surviving cells. Usually 7—30 colonies are obtained in control samples comprising five plates each (U. Rannug, personal communication). Examples will be given of one positive (a) and one negative test (b).

(a) Test of the mutagenic action of ethylene dichloride-tar (EDC-tar, waste-product from vinylchloride industry), applied in DMSO solution, in the presence of the active microsomal system (U. Rannug, personal communication). Mutant numbers per plate were as follows at a concentration (100 μg/plate) giving 95—100% survival:

Control	Treated
15	24
23	25
13	31
17	37
18	41
$\Sigma x = 86$	$\Sigma x = 158$
$\bar{x} = 17.2$	$\bar{x} = 31.6$
$s_{\bar{x}}^2 = 2.84$	$s_{\bar{x}}^2 = 10.96$
S.E. = $s_{\bar{x}} = 1.69$	S.E. = $s_{\bar{x}} = 3.31$

Variance of the mean and the standard error were calculated assuming normal distribution. A *t*-test of the difference leads to

$$t = \frac{31.6 - 17.2}{\sqrt{2.84 + 10.96}} = \frac{14.4}{\sqrt{13.80}} = 3.88$$

The corresponding probability at (5 + 5—2) = 8 d.f. gives $0.01 > P > 0.001$ (double-tailed test). In the single-tailed test this gives $0.005 > P > 0.0005$. The conclusion that the EDC-tar is mutagenic has thus an α-error of less than 0.5%.

(b) Testing mutagenicity of the herbicide, "Karmex 80". The test, with the same layout as the one under (a) above, gives the following data (mutants/plate).

Control	Treated
41	24
26	17
21	39
25	40
25	31
$\Sigma x = 138$	$\Sigma x = 151$
$\bar{x} = 27.6$	$\bar{x} = 30.2$
$s_{\bar{x}}^2 = 11.96$	$s_{\bar{x}}^2 = 18.34$
$s_{\bar{x}} = 3.46$	$s_{\bar{x}} = 4.28$

$$t = \frac{30.2 - 27.6}{\sqrt{11.96 + 18.34}} = \frac{2.6}{\sqrt{30.3}} = 0.47.$$

The computed t, with 8 d.f., corresponds to $P \approx 0.35$ (one-sided test). In this case H_0 is not rejected, i.e. data do not permit the conclusion that Karmex 80 is mutagenic.

The one-sided confidence interval of the found (insignificant) difference may be computed as: $0 < \delta < d + t_{2\alpha} \cdot s_d$, where $s_d = 5.504$ (see above). For $\alpha = 0.05$ and 0.01, using t-values for 8 d.f., the upper limit of the interval would be

α	$t_{2\alpha}$	Upper limit	Corresponding increase (% of control mean)
0.05	1.860	2.6 + 10.2 = 12.8	46.3
0.01	2.896	2.6 + 15.9 = 18.5	67.0

Defining, as done by Rannug (personal communication), the practical limit where a test is to be considered positive as "the doubling of the control value", the β-error of not rejecting H_0 when this alternative is true could be calculated. Assuming that the control value $\bar{x}_C = 27.6$ of the test is representative of a fixed value, e.g. a historical control, the contra-hypothesis $H_1' : \delta = (\Delta) = 27.6$ should be tested according to equation (8), giving

$$t_{2\beta} = \frac{\Delta}{s_d} - t_{2\alpha*} = \frac{27.6}{5.504} - t_{2\alpha*}$$

However, with regard to the fact that the control value varies between experiments (compare examples (a) and (b)), the correct hypothesis, if a "doubling" is specifically considered as an effect, would be

$$H_1'': \frac{\bar{\xi}_{Tr}}{\bar{\xi}_C} = 2, \text{ i.e. } \frac{\delta}{\bar{\xi}_C} = 1. \quad (30a)$$

We have then to consider the contribution of the uncertainty of the denominator in the estimate of the variance of

$$\frac{d}{\bar{x}_C} = \frac{\bar{x}_{Tr} - \bar{x}_C}{\bar{x}_C} \quad \left(= \frac{\bar{x}_{Tr}}{\bar{x}_C} - 1\right) \quad (30b)$$

which may be obtained from

$$s^2_{d(rel.)} = s^2_{d/\bar{x}_C} s^2_{\bar{x}_{Tr}/\bar{x}_C} = \left(\frac{\bar{x}_{Tr}}{\bar{x}_C}\right)^2 \left(\frac{s^2_{\bar{x}_C}}{\bar{x}^2_C} + \frac{s^2_{\bar{x}_{Tr}}}{\bar{x}^2_{Tr}}\right) \quad (31)$$

The test pf H_1'' according to equation 8 then gives

$$t_{2\beta} = \frac{1}{s_{d(rel.)}} - t_{2\alpha*} = \frac{1}{0.207} - t_{2\alpha*}$$

For these different ways of looking at the problem we obtain the following values of β (in one-sided tests):

Contra-hypothesis	α*	β-error
H_1': δ = 27.6	0.05	0.007
	0.01	0.03
H_1'': $\frac{\delta}{\xi_C} = 1$	0.05	0.01
	0.01	0.05

We see that if H_1' is true, the risk level of erroneously concluding non-mutagenicity is satisfactorily small. If H_1'' is true, the β-errors are higher due to the greater variance of the quotient, but still acceptably small at both rejection levels.

Comments. The variance (σ^2) of the number of mutant colonies per plate ($\bar{x}$) is approximately equal to $\bar{x}$, in agreement with a Poisson distribution (unpublished data); cf. equation (12). This has the following consequences.

Equation (23d) applies for the calculation of the sample size required for the detection of a critical absolute effect, Δ, with a given power ($1-\beta$*) at the significance level α*. For instance, values of np for α* = 0.05 and for various values of β* and Δ may be obtained from Fig. 1. (If, however, a relative increase — equation (30b) — is considered, a greater value of σ^2_d has to be used in this calculation.) It appears that if the control values are greater than 25 colonies per plate, five plates per test is sufficient for the determination of a doubling at α- and β-errors $\approx$ 0.01.

The fact that σ^2 varies with the mean value may invalidate significance estimates in critical cases ($P \approx \alpha^*$; cf. Table II) due to over-estimation of the d.f. (see Section 3.2). The stabilitization of σ^2 through the square-root transformation (equation 18) will then give safer significance estimates.

4.3 Micronuclei in young erythrocytes

Evaluation of data from this kind of test encounters the special problems of very low proportions.

(a) Positive test: The effect of 10 rad of X-radiation (D. Jenssen, personal communication). The following data were obtained:

Treatment	Mouse number	Cells with micronuclei	Cells investigated	Proportion (%)
Control	1	2	876	0.23
	2	1	807	0.12
	3	3	729	0.41
	Sum (1—3)	6	2412	0.25
10 rad	4	13	842	1.54
	5	12	716	1.68
	6	6	710	0.85
	Sum (4—6)	31	2268	1.37

None of the treatment series shows heterogeneity. Therefore significance could be estimated by χ^2 on the sum frequencies; $\chi^2_c = 17.22$; at $P = 0.001 : \chi^2 = 10.827$. Hence, $P < 0.001$ (double- or single-tailed test), and $H_0 : \bar{\xi}_{Tr} = \bar{\xi}_C$ is rejected. A t-test of the found difference between the proportions leads to the same conclusion.

Comments: generally significance testing of 2×2 tables is best done by exact calculation of P (equation 29), easily performed on a computer.

(b) Negative test: Effect of 2,4-D (100 mg/kg b.w.) (D. Jenssen, personal communication). Values obtained:

Treatment	Mouse number	Cells with micronuclei	Cells investigated	Proportion (100p%)
Control	1	5	1284	0.39
	2	4	1641	0.24
	3	4	1256	0.34
	Sum (1—3)	13	4181	0.31
Treated with 2,4-D	4	2	1134	0.18
	5	3	893	0.34
	6	5	1145	0.44
	Sum (4—6)	10	3172	0.32

A significance estimate by χ^2 suggests non-rejection of H_0 ($\chi^2 = 0.0011$; $0.98 > P > 0.95$).

The calculation of the confidence interval of the found difference ($\simeq 0$) is best done by an exact calculation applying the binominal distribution, preferably done by computer. A rapid estimate of the confidence interval may be done by using the simplified form of the standard deviation

$$\sigma_d = \sqrt{\frac{p_C}{n_C} + \frac{p_{Tr}}{n_{Tr}}} \qquad (32)$$

(1—p being close to one.)
This gives σ_d = 0.132% and the upper 95 and 99% confidence limits in the one-sided test are estimated from:

α-error	z	Upper limit of d $(0.01 + z_\alpha \sigma_d)$
0.05	1.645	0.22%
0.01	2.326	0.31%

The confidence intervals are thus about as large as the control values. The preceding example offers an opportunity of comparing the efficacy of the biological test with the radiation doses causing the same effect. In this very sensitive test the effect of 10 rad of X-rays induced 1.12% of cells with micronuclei. We may, for example, choose the effect of 1 rad (0.11% cells with micronuclei) or of 5 rad (0.55%) as a critical level signifying risk, and calculate the β-error of not rejecting H_0 for either of these contra-hypotheses. Applying equation (8), and estimating σ_d by $\sqrt{\frac{p_C}{n_C} + \frac{p_{Tr}}{n_{Tr}}} = 0.132\%$, we obtain, at the rejection levels α^* = 5 or 1%:

	H_1' $\delta = 0.11\%$	H_1'' $\delta = 0.55\%$
$\alpha^* = 0.05$	$\beta = 0.79$	$\beta = 0.006$
$\alpha^* = 0.01$	—	$\beta = 0.03$

If an effect corresponding to that induced by 1 rad of X-rays is assumed to signify a risk, a statement of no effect in consequence of the non-rejection of H_0 would hardly be justified, and additional information would be required for a decision about the mutagenicity of 2,4-D. In fact the power of the test, 1—β, to detect this difference at 5% significance level is only 21%. The size of the test required to reach 95% power may be estimated from equation (23c)

$$n \geqslant \frac{(t_{2\alpha^*} + t_{2\beta^*})^2 (2p + \Delta)}{\Delta^2} = \frac{(2 \cdot 1.645)^2 (2 \cdot 0.0031 + 0.0011)}{0.0011^2}$$

$$n > 50{,}000$$

From Fig. 1 it may be inferred (by interpolation between curves for the relative increase of number of mutants Δ = 0.5 and Δ = 0.2) that the control would have to comprise about $2 \cdot 10^2$ mutants, i.e., at p = 0.3%, the same number of cells tested as computed above. However if Δ equivalent to the effect of 5 rad (H_1'' : δ = 0.55% cells with micronuclei) is the level considered to signify risk, a conclusion of non-mutagenicity on the basis of non-rejection of H_0 has a satisfactorily great chance of being correct.

4.4 Dominant lethal test (DLT)

4.4.1 General remarks

In his chapter of this book, Bateman [1] points out the difficulties inherent in the statistical evaluation of test data from DLT. It is a general experience that the statistical treatment is often more time-consuming than the test itself [13]. In agreement with the Editor, this author's viewpoint on the DLT will be published elsewhere. In this article we only present certain details from a DLT test in order to show the importance of identifying and eliminating outliers (clusters) (cf. Section 3.5.2). This will have to be seen in the context of the following general viewpoints with special regard to critical decisions ($P \geqslant \alpha^*$; see Section 2.5).

(a) Choice of variable (cf. Bateman, ref. 1)

The probability (p) that a fertilized egg will be implanted and the probability (r) that an implant will die give rise to two superimposed binomial distributions which lead to a binomial distribution (m,pr). Since pr is a small fraction of the total number of implantation sites, m, which may be considered constant [25], the average number of dead implants per female, x(DI) = prm, follows approximately the Poisson distribution (to be published). x(DI) is therefore a useful variable both for the identification of outliers and for significance testing.

The use of the mutation index, $\frac{\mathrm{x(DI)}}{\mathrm{x(TI)}}$, where x(TI) is the total number of implants per female, has the disadvantage of having a variable denominator, but may be used if females with abnormally low x(TI) are discarded (cf. example below).

Salsburg's suggestion of using $pr = \frac{\mathrm{x(DI)}}{m}$ gives essentially the same result as the direct use of x(DI).

(b) Transformations

The variance of x(DI) will be approximately proportional to x(DI). Therefore a stabilization of σ^2, e.g., by the square root transformation (equation 18) $a = \sqrt{\mathrm{x(DI)}} + \sqrt{\mathrm{x(DI)} + 1}$, should be used in order to avoid overestimation of significances in critical cases (cf. Section 3.2 and equation 16). For similar reasons the arc sine transformation should be applied on the proportion variables (equation 19).

Using these transformations, the z-test will be allowed, giving a higher resolving power.

(c) Required size of a test

Assuming the number of dead implants per female, x(DI), to be Poisson distributed, equation (23d) applies for estimating the required size of tests for detecting a difference Δ at wanted power ($1-\beta^*$) and significance level (α^*). The sample sizes given first as number of mutants, np, required to determine a "doubling" ($\frac{\Delta}{p_C} = 1$) and a 20% increase over the control level ($\frac{\Delta}{p_C} = 0.2$) at $\alpha^* = \beta^* = 0.05$ are estimated in Table IV as np = 33 and 595, respectively (see also Section 3.4 and Fig. 1). This means that at least 33 and 595 dead implants should be determined in the control. If, for example, the number of implants

TABLE V

LAYOUT OF VARIANCE ANALYSIS OF A DOMINANT LETHAL TEST ACCORDING TO A MIXED FACTORIAL—NESTED DESIGN

n_j refers to number of entried with regard to treatment (t), week (w), males (m) and females (f).

Level cause of variation	Degrees of freedom (d.f.)
1. Treatment (n_t)	$n_t - 1$
2. Week (n_w)	$n_w - 1$
3. Interaction treatment × week	$(n_t - 1)(n_w - 1)$
4. Males (n_m per treatment) within treatment (n_t)	$n_t(n_m - 1)$
5. Interaction male × week within treatment	$n_t(n_m - 1)(n_w - 1)$
6. Error: between females within treatment × week × male group	$(n_f - 1) - n_t \cdot n_w \cdot n_m$
Total	$n_f - 1$

per female is 10, and 8% of the implants are dead, we would require $\frac{33}{0.08 \cdot 10}$ = 42 and $\frac{595}{0.08 \cdot 10}$ = 750 females, respectively, in each of the control and the test. If outliers are not eliminated leading to, for example, a two-fold larger σ^2, these figures should be multiplied by two. Elimination of outliers may hence considerably increase the resolving power (cf. Section 2.3) of the DLT.

(d) Variance analysis

In this author's view a DLT test should be evaluated initially according to a *mixed factorial-hierarchic variance analysis* of a properly transformed variable, as indicated in Table V. If the error variance agrees with theory (tested by χ^2; see ref. 4, p. 101), this strengthens the reliability of the test. Indications of a male effect (Table V, level 4) or a male interaction effect (level 5) would then allow the detection of inter-male variations [24] as a guidance to a decision whether the mean square of level 4 data should be used as the error variance in the F-test, or whether the experiment should be repeated with new males (cf. Section 2.5). Following a "chronic" treatment (i.e. of all stages of spermatogenesis) a significant week effect (level 2) or treatment × week interaction (level 3) will likewise indicate heterogeneity, usually due to experimental defects. Following acute treatment, a mutagenic action will appear at levels 1 or 3 or both.

4.4.2 Example: outliers

A DLT test in mice of the mutagenic action of irradiated potatoes [19] comprised three groups of five males, treated for one week with an alcoholic extract of irradiated potatoes (Group III), or with an extract from unirradiated potatoes (Group II, specific control) and an absolute control (Group I). For each of 5 weeks each male was caged with three females, giving ⩽15 pregnancies per treatment × week cell. No significant effects were obtained; however, some difference between Groups I and II in the third week was said to be indicated.

Disregarding here the purpose of the study, the data from week 3, Groups I and II are used to illustrate the importance of discarding clusters. The Group × week cells, in all 15, regularly gave variances of x(TI), s^2(TI) = about 3, with three exceptions, which could be ascribed to outliers. Two of these values fell in the analysis considered here. After exclusion of the low x(TI), the variances became normalized.

TABLE VIa

MEAN VALUES, NUMBER OF FEMALES AND VARIANCE ESTIMATES OF TOTAL IMPLANTS (x(TI)) AND DEAD IMPLANTS (x(DI)) IN THE COMPARISON OF GROUPS I AND II, WEEK 3, BEFORE AND AFTER DISCARDING EXTREMES.

(Data from ref. 19.)

Variable		Group I		Group II	
		All females	Extremes discarded	All females	Extremes discarded
Total implants x(TI)	$\bar{x}$(TI)	11.2	11.9	10.8	11.3
	n	14	13	15	14
	s^2(TI)	10.12	3.41	6.53	3.30
Dead implants x(DI)	$\bar{x}$(DI)	0.92	—	1.71	1.08
	n	14	—	15	14
	s^2(DI)	0.91	—	6.68	1.08

In a similar fashion all but three of the 15 variances of x(DI) were found to be close to the respective mean values, $\bar{x}$(DI), as expected in the case of Poisson distribution of x(DI). The three abnormally high s^2(DI) became "normalized" following exclusion of outliers. One of these cases fell in the groups considered here (Table VIa). For these values, the rules of excluding outliers are given in Table VIb.

Table VIc shows significance estimates, by means of different statistics, of the difference of the means of dead implants of Groups I and II, without and with exclusion of the big cluster of 10 DI ("pseudo-pregnancy") in Group II. It is evident that all rejection rules (e.g. the χ-test, cf. ref. 18) which presuppose independence of individual observations of dead implants give significance at the 5% level for the difference. The parallel analysis of the material after exclusion of outliers shows that the significance was caused by the presence of one single big cluster in the material.

The *t*-test does not give significance in this case, because of the strong

TABLE VIb

DEVIATING FEMALES IN GROUPS I AND II, WEEK 3

	Abnormal	*P* of female belonging to population			Decision
		Poisson or binomial distribution [a]		Found distribution [c]	
		P(a priori) [b]	*P*(a posteriori) [b]	*P*	
Group I, female 1	x(TI) = 2 (x(DI) = 0)	$1.0 \cdot 10^{-6}$	$2 \cdot 10^{-4}$	<0.005	Discard
Group II, female 1	x(TI) = 4 (x(DI) = 1)	$1.0 \cdot 10^{-5}$	0.002	≃0.3	Discard
Group II, female 2	x(DI) = 10	$6.3 \cdot 10^{-9}$	$1.3 \cdot 10^{-6}$	<0.005	Discard

[a] Probabilities calculated assuming binomial distribution of x(TI) and Poisson distribution of x(DI).

[b] A priori denotes the probability that *a certain* female exhibits the found or a more deviating value. A posteriori denotes the probability that *any one* out of 200 females (the number in the whole test) will show the found or a larger deviation from the mean value. The values were estimated from: P (a post.) $= 1 - (1 - P(\text{a priori}))^n \simeq n \cdot P(\text{a priori})$.

[c] *P* estimated according to conventional rules of identifying extremes (cf. Section 3.5.2.).

TABLE VIc

COMPARISON OF MEANS OF x(DI) OF GROUPS I AND II, WEEK 3, FROM TABLE VIa, WITHOUT AND WITH DISCARDING OF ONE OUTLIER

Probabilities refer to double-tailed tests.

Statistical test	Significance estimate	
	All values	Cluster excluded
χ^2 (2 × 2 contingency)	$\chi^2 = 4.71$ $0.05 > P > 0.02$	$\chi^2 = 0.285$ $0.7 > P > 0.5$
χ^2 (2 × 2 contingency, with Yates's correction	$\chi_c^2 = 3.99$ $0.05 > P > 0.02$	$\chi_c^2 = 0.108$ $0.8 > P > 0.7$
z-test of proportions x(DI)/x(TI) = p; variances estimated by p(1 − p)	$t = 2.20$ $0.05 > P > 0.02$	0.533 $P \simeq 0.7$
z-test of difference between arc sine transformed proportions (in radians); σ^2 estimated by 1/(n + ½)	z = 2.18 $0.05 > P > 0.02$	
t-test of $\bar{x}_{II}(DI) - \bar{x}_I(DI)$ (variances = s^2 in Table VIa)	$t = 1.07$ (25 d.f.) $0.3 > P > 0.2$	$t = 0.41$ (24 d.f.) $0.7 > P > 0.6$

effect on s_d of the presence of the cluster. If clusters are present in samples compared, non-parametric methods such as the sensitive Smirnov-Kolmogorov test [26] should be used. This test may also be used to compare a found with a theoretical distribution.

Appendix 1: Symbols used

x,	Value of measurement (variate) in general.
n,	Number of observations
p,	Proportion; hence, np = frequency, e.g. number of mutants in a test of size n individuals.
d.f.,	Number of degrees of freedom.
$\bar{x}_C$ and $\bar{x}_{Tr}$,	Mean values of control and treated samples, respectively.
ξ_C and ξ_{Tr},	True mean values of control and treated series (in the experiment estimated by x_C and x_{Tr} respectively)
$d = \bar{x}_{Tr} - \bar{x}_C$,	Found difference (between mean values of control and a treated sample).
$\delta = \bar{\xi}_{Tr} - \bar{\xi}_C$,	True value of the difference estimated by d.
b and B,	Estimated and true value of regression coefficient.
Δ,	Preset permissible value of the mutagenic activity of a compound. (If the probability (β) that $\delta \geqslant \Delta$ is acceptably small, a compound may be considered "practically" non-mutagenic.)
s and s^2,	Standard deviation and variance of samples.
$s_{\bar{x}}$ or S.E.,	Standard error ($=\frac{s}{\sqrt{n}}$)
σ and σ^2,	True standard deviation and variance, estimated by s and s^2.
P,	Probability.
α,	Observed probability of rejecting a true hypothesis (α-error); also used for confidence intervals of found mean values.

β,	Observed probability that not rejecting a hypothesis is incorrect (β-error).
α^* and β^*,	Pre-defined upper limits of α- and β-errors, in testing of hypotheses.
$1-\beta$,	Power of a test, i.e. the probability that a difference Δ is detected at a given significance level α^*.
$t(8)$,	t at d.f. = 8.

Appendix 2: References to equations

Variance:	*Equation(s)*
Normal distribution, calculation	(9 and 9a)
Of mean value	(10)
Comparison of variances (F-test)	(17)
Binomial distribution	(11)
Poisson distribution	(12)
Ratio	(31)
Mean value:	
Standard error	(10a,11a and 12)
Confidence interval	(13a,b and 32)
of proportions and frequencies	(20a–d) Fig. 3
of $\bar{x} = 0$	(22, 22a and b) Fig. 3
Sample size for detecting hypothetical mean (Δ)	(22a and b)
Difference of two means	
Estimated	(1a)
True	(1b)
Null hypothesis	(2)
Significance test	(3a–d, 15 and 16)
Standard error	(4)
weighted	(14a and b)
if unknown	(16)
Rejection rule	(5)
Confidence interval	(6a–c)
of difference = 0	(6c and d)
Confidence interval of proportions, different estimates	Section 3.2, (32)
Double-tail and single tail test	Section 2.4
Test of hypothetical difference ($\delta = \Delta$)	(7)
β-error	(8,8b) Fig. 1
Sample size for detection of hypothetical difference (Δ)	(23a–d) Fig. 1
Expressed as fraction of control mean variance of the same	(31)
Binomial distribution	
Calculation of P	(29)

Applications	(29a,b)
χ^2 *test*, 2×2 contingency test	(28)
Probability that deviating value found a posteriori belongs to population	Section 4.5
Transformations	
Square root	(18)
Confidence interval of transformed mean	(20d and 21)
Arc sine	(19)
Variance	(19a)
Dose—response	
Non-shouldered	(24)
Shouldered	(25)
Confidence interval of linear term	(26)
Standard error of slope	(27)
Variance analysis (models)	
Inter-replicate variation	Section 3.5.1
Mixed hierarchic-factorial (dominant lethal test)	Section 4.4

Acknowledgements

Valuable suggestions from Professor Bengt von Bahr, Stockholm; Dr. Gunnar Ekman, Stockholm and Dr. Brian Kilbey, Edinburgh, are acknowledged. The paper is based on work supported by the Swedish Council for Forestry and Agricultural Research and the Swedish Atomic Research Council.

References

1 Bateman, A.J., The dominant lethal assay in the male mouse, *this volume* pp. 323—332.

2 Clark, J.M., Mutagenicity of DDT in mice, *Drosophila melanogaster* and *Neurospora crassa*, Austr. J. Biol. Sci., 27 (1974) 427—440.

3 Crump, K.S., D.G. Hoel, C.H. Langley and R. Peto, Fundamental carcinogenic processes and their implications for low dose risk assessment, Cancer Res., 36 (1976) 2973—2979.

4 Dixon, W.J. and F.J. Massey, Introduction to Statistical Analysis, 3rd ed., McGraw-Hill, New York, 1969. (See especially pp. 263—282.)

5 Draper, N.R. and H. Smith, Applied Regression Analysis, Wiley, New York, 1966.

6 Ehrenberg, L., Higher plants, in A. Hollaender (ed.), Chemical Mutagens: Principles and Methods for Their Detection, Vol. 2, Plenum Press, New York, 1971, pp. 365—386.

7 Ehrenberg, L., Genetic toxicity of environmental chemicals, Genetika (Belgrade), 6 (1974) 367—398.

8 FDA Advisory Committee on Protocols for Safety Evaluation: Panel on Carcinogenesis Report on Cancer Testing in the Safety Evaluation on Food Additives and Pesticides, Toxicol. Appl. Pharmacol., 20 (1971) 419—438.

9 Fisher, R.B., Quantitative Chemical Analysis, 2nd ed., Saunders, Philadelphia, Pa., 1961, pp. 86—87.

10 Fisher, R.A. and F. Yates, Statistical Tables for Biological, Agricultural and Medical Research, 6th ed., Oliver and Boyd, Edinburgh, 1963.

11 Friedman, L., A Proposed Procedure for the Assessment of Health Hazards of Carcinogens at Very Low Levels of Exposure, WHO Tech. Rep. Series No. 546, WHO, Geneva, 1974, pp. 14—19.

12 Gaul, H., Critical analysis of the methods for determining the mutation frequency after seed treatment with mutagens, Genet. Agrar., 12 (1960) 297—318.

13 Green, S. and J.A. Springer, The dominant-lethal test: potential limitations and statistical considerations for safety evaluations, Environ. Health Perspect., 6 (1973) 37—46.

14 Jacobsen, P., Demarcation of mutant-carrying regions in barley plants after ethylmethanesulfonate seed treatment, Radiat. Botany, 6 (1966) 313—328.

15 Kastenbaum, M.A. and K.O. Bowman, Tables for determining the statistical significance of mutation frequences, Mutation Res., 9 (1970) 527—549.

16 Kellerer, A.M. and H.H. Rossi, The theory of dual radiation action, Curr. Top. Radiat. Res. Q., 8 (1972) 85--158.

17 Kilian, D.J., F.M. Moreland, M.C. Benge, M.S. Legator, E.B. Whorton, Jr., A collaborative study to measure interlaboratory variation with in vivo bone marrow metaphase procedure, *this volume* pp. 243—260.

18 Krüger, J., Statistical Methods in Mutation Research, Appendix in F. Vogel and G. Röhrborn (eds), Chemical Mutagenesis in Mammals and Man, Springer-Verlag, Berlin, 1971, pp. 460—502.

19 Levinsky, H.V., M. Wilson and H.N. MacFarland, A study of the mutagenic effect of an alcoholic extract of irradiated potatoes in the mouse, IFIP Techn. Report Ser., R9, 1973.

20 Lindgren, D., G. Eriksson and K. Šulovská, The mutagenic effect of ^{90}Sr in barley, Hereditas, 65 (1970) 107—132.

21 Lüning, K.G. and A.G. Searle, Estimates of the genetic risks from ionising irradiation, Mutation Res., 12 (1971) 291—304.

22 Mantel, N. and W.R. Bryan, Safety testing of carcinogenic agents, J. Natl. Cancer Inst., 27 (1961) 455—470.

23 Muller, H.J., The standard error of the frequency of mutants some of which are of common origin, Genetics, 37 (1952) 608.

24 Ray, V.A. and M.L. Hyneck, Some primary considerations in the interpretation of the dominant-lethal assay, Environ. Health Perspect., 6 (1973) 27—35.

25 Salsburg, D.S., Statistical considerations for dominant-lethal mutagenic trials, Environ. Health Perspect., 6 (1973) 51—58.

26 Tate, M.W. and R.C. Clelland, Nonparametric and Shortcut Statistics, Interstate Printers and Publishers, Danville, Ill., 1957.

27 UNSCEAR, United Nations Scientific Committee on the Effects on Atomic Radiation, Report to the General Assembly, Vol. II, United Nations, New York, 1972.

28 UNSCEAR, to be published in 1977.

METHODS OF OBTAINING EPIDEMIOLOGICAL DATA FOR MUTATION RISK ESTIMATION

HOWARD B. NEWCOMBE

Biology and Health Physics Division, Atomic Energy of Canada Limited, Chalk River, Ontario (Canada)

Introduction

Unfortunately, little is known about the extent of the effect on health which a given increase in mutation rate would cause in man. This is true because of a lack of certainty about the amount of ill health that is maintained in the population by the pressure of recurrent natural mutations. Without such knowledge, the importance of exposures to known mutagens is difficult to assess quantitatively with a view to setting reasonable standards for the protection of the human gene pool. The purpose of this chapter is to consider some epidemiological methods which may be used to obtain relevant human data that are in short supply, which are needed to bring the problem of environmental mutagenesis into perspective with other public health problems.

Kinds of data needed

Attempts to measure by direct observation the amount of additional genetic ill health in man resulting from exposure to a known potent mutagen, i.e. ionizing radiation, have been disappointing even where the doses were large. Also, experiments with laboratory mammals give little reason to hope that this direct approach will be particularly rewarding. Relevant epidemiological studies will thus be profitably directed towards identifying and measuring the social burden imposed by the mutation-maintained component of the hereditary diseases already present in a normal population. Without such information, even precise knowledge of the extent of an induced increase in the mutation rate of man would be virtually uninterpretable in terms of the magnitude of the resulting public health problem.

It is commonly assumed by human geneticists that the rarer and more severe dominantly inherited diseases are maintained in the population by the "pressure" of recurrent mutations. However, adequate evidence of the correctness of the assumption exists only for a very few diseases. Moreover, information on the collective frequency of the severe dominant traits is fragmentary. Frequencies of a few of these diseases are known, but the vast majority are individually too rare to have been the subjects of appropriate surveys. It is even less certain that common or mild conditions, or those in which the symptoms emerge beyond reproductive age, are mutation-maintained.

Without further evidence for the various plausible assumptions that are often made when estimating genetic risks, it remains difficult to say with certainty whether even a sustained doubling of the mutation rate of man would be important or trivial in relation to other public health problems. For this reason, better human data are needed, on (a) the collective frequencies of the different categories of genetic disease, i.e. dominant, recessive etc., (b) the extents of the social burdens which each of these represents, and (c) the distinction between the mutation-maintained and the selection-maintained diseases.

Conventional procedures and their products

Geneticists interested in the frequencies and modes of inheritance of particular diseases have traditionally used interview methods to obtain pedigrees, and have sometimes searched manually through the records of various health services and the vital registration system, on an ad hoc basis, to find new cases for follow-up. These manual methods need not to be described here, but some of their products will be of interest as indicating current gaps in knowledge that might be filled if more extensive use could be made of existing sources.

The first serious attempt to assess the magnitude of the total load of hereditary disease was made about 18 years ago [10,13]. The work was based on the population of Northern Ireland. The procedures employed are not described in the reports, but data on a number of diseases in that population, such as deaf-mutism and muscular dystrophy, had already been obtained by Stevenson using conventional methods. The result of the survey was a list of hereditary conditions with their frequencies among liveborn individuals and in the living population, broken down by the mode of inheritance. Totals, as modified later in official reports [14,15] are given in Table I. In the period since the original lists were prepared, however, ideas have changed regarding the mode of inheritance of certain of the diseases. In particular, some of the more common conditions assigned a dominant aetiology then would now be regarded as irregularly inherited and as probably multifactorial in origin. As a result the frequency of the dominant category may have been overestimated.

TABLE I

FREQUENCIES OF HEREDIATARY DISEASE IN NORTHERN IRELAND

Based on refs. 10 and 13 as updated in refs. 14 and 15.

Disease aetiology	Frequency per 100 liveborn
Dominant	0.95
Recessive	0.21
Sex-linked	0.04
Chromosomal	0.42
Congenital malformation	2.50
Other irregularly inherited	1.50
TOTAL	5.62

An alternative procedure for obtaining relevant data involves the use of the discharge records for the paediatric service of a single hospital [4]. The products from this approach, although useful, suffer from three serious limitations for genetic risk estimation:

(a) the numbers of cases cannot be related with assurance to a birth population of known size;

(b) the relative numbers of cases of different diseases seen in a given year may not represent the relative numbers in a birth population, and may under-represent conditions that require infrequent hospitalization, and

(c) diseases with onsets late in life will be under-represented in the records of a paediatric service.

Such limitations might be largely avoided if hospital discharge records for a whole area, and for all ages, were integrated with each other and with the vital registration system for that area. Only by the use of unconventional methods, however, would such an extensive integration become feasible.

For the dominantly inherited class of diseases, an alternative assessment of their collective impact may be derived by pooling results from conventional studies of specific conditions. Data on the mutation rates for dominant diseases have been reviewed by various authors [3,11,17,18] (Table II). However, knowledge of mutation rates is less helpful in predicting the rise to a new equilibrium frequency of disease, which could result from a given increase in mutation rate, than is knowledge of the current frequency of dominant diseases.

TABLE II

DOMINANT MUTATION RATES IN MAN

Adapted from refs. 3, 11 and 17.

Disease	Rate/10^6
Neurofibromatosis	100
Polycystic kidney disease	90
Dominant deafness (all kinds)	47
Apert syndrome	30
Intestinal polyposis	20
Achondroplasia (chondrodystrophy)	12
Myotonia dystrophica	12
Osteogenesis imperfecta	10
Pelger anomaly	9
Multiple exostoses (diaphyseal aclasis)	8
Tuberous sclerosis (epiloia)	8
Retinoblastoma	7
Marfan syndrome	5
Microphthalmus	5
Facioscapular muscular dystrophy	5
Waardenburg syndrome	4
Myotonia congenita	4
Aniridia	3
Acrocephalosyndactyly	3
Amyotrophic lateral sclerosis	3
Huntington chorea	2
TOTAL	387/10^6

Little attempt has been made in the past to review exhaustively what data are available in the literature on the incidence and prevalence of genetic disease. However, a compendium on birth defects has been published which contains some 842 articles on specific conditions [2]. These articles report in a uniform fashion the incidence and/or prevalence of each condition where the information is available; this information has been extracted for some 39 dominant diseases (Tables III and IV). The severe conditions likely to be maintained by mutation have a combined frequency of about 0.1 per 100 liveborn, and if the less severe conditions that might also be mutation maintained are added, the combined frequency becomes about 0.3 per 100. If the mutation rate were to increase to a new level these disease frequencies would presumably rise to a new equilibrium in proportion to the new mutation rate.

What needs to be determined is the collective frequency for those dominant diseases that are severe and individually too rare for their separate frequencies to have been determined. The above compendium contains articles on an additional 160 dominant conditions for which frequencies are not given. Many of these are said either to be very rare, or to be mild in their expressions. Thus the missing information on the collective frequency of the remaining serious dominant conditions is only likely to be obtained from an exhaustive survey of a birth cohort, in numbers sufficient to show whether this figure approaches or exceeds 0.1 additional seriously affected cases per 100 liveborn.

The data on severity are likewise limited, as obtained from published accounts of studies of individual diseases based on conventional methods. Data

TABLE III

FREQUENCIES OF SEVERE DOMINANT DISEASES

From ref. 2; references refer to article numbers in the book.

	Ref. (article)	Disease	Frequency per 10^6	
1.	(395)	Hypercholestremia (type II)	2000	selection maintained (?)
2.	(667)	Polycystic kidney (adult type)	1000	
3.	(586)	Neurofibromatosis	300	
4.	(831)	Waardenburg syndrome (20% deaf)	250	
5.	(16)	Achondroplasia	100	
6.	(440)	Intestinal polyposis (types I and II)	60	
7.	(638)	Osteogenesis imperfecta	60	
8.	(387)	Huntington chorea	50	
9.	(723)	Retinoblastoma	50	
10.	(37)	Amelogenesis imperfecta	25	
11.	(222)	Dentin dysplasia	20	
12.	(25)	Acute intermittent porphyria	15	
13.	(515)	Marfan syndrome	15	
14.	(46)	Aniridia	10	
15.	(140)	Cleft lip or palate with lip pits	10	
16.	(807)	Tuberous sclerosis (epiloia)	10	
17.	(237)	Dominant osteopetrosis	10	
18.	(115)	Cervicoocculoacoustic syndrome	2	
19.	(789)	Torsion dystonia	2	
		Total of 1 and 2	3000	
		Total of 3—19	989	

TABLE IV

FREQUENCIES OF DOMINANT DISEASES OF LESS SEVERE KINDS

From ref. 2; references refer to article numbers.

	Ref. (article)	Disease	Frequency per 10^6	
1.	(179)	Congenital ptosis (80% unilateral)	common	Selection maintained (?)
2.	(828)	Vitiligo (patchy depigmentation)	10 000	
3.	(109)	Cephalopathic epilepsy (declines with age)	5000	
4.	(642)	Otosclerosis (penetrance 30%)	3000	
5.	(456)	Keloid	1500	
6.	(636)	Osteochondritis dissecans	1000	
7.	(668)	Polydactyly (variable severity)	700	
8.	(678)	Primary basilar impression	300	
9.	(772)	Syndactyly (types I to IV)	300	
10.	(93)	Blepharochalasis and thyroid enlargment	250	
11.	(496)	Lop ear (dom. with some deafness)	200	
12.	(748)	Sphaerocytosis	200	
13.	(418)	Hypodontia and nail dysgenesis	140	
14.	(223)	Dentinogenesis imperfecta	125	
15.	(640)	Osteopoikilosis	50	
16.	(821)	Variegated porphyria	15	
17.	(280)	Erythropoietric protoporphyria	10	
18.	(202)	Cyclic neutropenia	10	
19.	(562)	Multiple exostoses (majority mild)	10	
20.	(260)	Ehlers-Danlos syndrome	5	
		Total of 1—6	20 500 +	
		Total of 7—20	2315	

TABLE V

TIME OF ONSET OF SEVERE DOMINANT DISEASES

From ref. 2.

Disease	Time of onset of troublesome symptoms (decade of life)	
Waardenburg syndrome	1st	(congenital for deafness and clefts)
Achondroplasia	1st	(usually recognized at birth)
Osteogenesis imperfecta	1st	(congenital for 1/3 of cases)
Amelogenesis imperfecta	1st	(early loss of teeth)
Aniridia	1st	(congenital)
Cleft lip or palate with lip pits	1st	(congenital)
Cervicoocculo acoustic syndrome	1st	(congenital deafness)
Intestinal polyposis	1st—2nd	(death in 3rd—4th)
Marfan syndrome	1st—2nd	(death in 4th)
Torsion dystonia	2nd	(progressive)
Acute intermittent porphyria	2nd—3rd	
Dom. osteopetrosis	2nd—3rd	(50% asymptomatic)
Tuberous sclerosis	3rd	(slowly progressive, death in 3rd)
Hypercholestremia	3rd—4th	
Huntington chorea	3rd—4th	(death 10—20 years after onset)
Dentin dysplasia	3rd—4th	(for complete loss of teeth)
Polycystic kidney disease	4th—5th	
Retinoblastoma	—	(birth to old age; 20% death in U.S.A.)
Neurofibromatosis	—	(slowly progressive)

TABLE VI

TIME OF ONSET OF DOMINANT DISEASES OF LESS SEVERE KINDS

From ref. 2.

Disease	Time of onset of noticeable symptoms (decade of life)	
Congenital ptosis	1st	(80% unilateral, drooping eyelid)
Polydactyly	1st	(congenital)
Syndactyly	1st	(congenital)
Blepharochalasis etc.	1st	(infancy)
Lop ear	1st	(congenital)
Dentinogenesis imperfecta	1st	(early loss of teeth)
Erythropoietic protoporphyria	1st	(mild photosensitivity of skin)
Cyclic neutropenia	1st	(first seen as gingivitis and oral ulceration)
Ehlers-Danlos syndrome	1st	
Centralopathic epilepsy	1st—2nd	(but declines thereafter)
Osteochondritis dissecans	1st—2nd	(early for femoral head, later for knee)
Spherocytosis	1st—2nd	(seen as haemolytic anaemia)
Hypodontia and nail dysgenesis	1st—2nd	
Multiple exotoses	1st—2nd	(1st onward)
Vitiligo	2nd	(patchy depigmentation of skin)
Variegated porphyria	2nd	(usually after puberty)
Otosclerosis	2nd—3rd	(100% show hearing loss by age 40)
Keloid	2nd—3rd	(becoming rare in old age)
Primary basilar impression	3rd—4th	(for neurological signs)
Osteopoikilosis	—	(variable)

have been extracted from the compendium on the times of onset of the 39 dominant diseases mentioned earlier, and on the times of death for some of them (Tables V and VI). But such information is descriptive rather than statistical. For a better assessment of the collective impact of all dominant diseases quantitative measures are required, such as age-specific and cumulative rates of onset, rates of hospital utilization, and rates of death.

Methods for integrating source files

More extensive use of existing health records, for a well-defined population, may serve to provide certain of the needed epidemiological data that are currently lacking in the results from ad hoc studies of particular genetic diseases. However, because of the enormous sizes of some of the routine files, substantial exploitation of these potential sources is only possible with the aid of a computer and of special methods for processing the data. Fortunately, many of the important files are already available in machine-readable form for some major populations. Also, the kinds of procedures that would permit extraction of data bearing on mutation risk assessment could serve the purposes of investigators and public health authorities concerned with broader problems, including carcinogenesis and teratogenesis. Thus it is not too difficult to justify some investment of effort in the development of methods that will make the information content of the files more accessible for all such purposes.

In essence what is needed is a method for bringing together from a diversity of source files all relevant facts pertaining to particular individuals or families. It is not only the diagnostic information that would be thus extracted, but also the chronological histories of sibship formation, together with various biological and social data concerning either the parents or the offspring or both. The computer methods that render this possible are almost as easy to apply to the whole of a given magnetic tape file as to a part of it. The methods thus lend themselves to a demographic approach in which it is a whole population which is studied.

Many of the routine health-related records contain adequate personal identifying information in the form of names, birthdates and such. Much therefore depends on the availability of computer techniques which will bring together and "link" facts contained in various independently prepared records of important health events pertaining to the same individual. It is also necessary that such "record linkages" be carried out rapidly, accurately and cheaply.

The techniques that have been developed for this type of epidemiological work over the past 18 years will be described along with certain of their uses relating to the assessment of mutational risks. The unique features of the approach are (a) that it uses a computer to perform the kind of interfiling of personal records usually regarded as requiring human judgment, and (b) that it leads progressively to an integration of health records, from diverse sources, around the individual and the family as the important units, on a scale that would not otherwise be possible.

The health files

The methods required for this epidemiological approach are best illustrated by referring to work done in British Columbia (population 2.2 million in 1971), and the files of health-related records available for that population. In this work the vital registration system, certain specialized disease registers, and records from a universal hospital insurance scheme are all tapped for information.

To identify sibship groupings of individuals, use is made of the magnetic tape versions of:

(a) all *marriage records* back to 1946 (i.e. between 10,000 and 20,000 per year), and

(b) all *birth records* over the same period (i.e. between 30,000 and 35,000 per year),

with the birth registrations being linked by computer to the appropriate parental marriage records, or to the senior birth registration whenever the parental marriage record is missing.

Similarly, to compile individual case histories, ill health records of various kinds are linked by computer to the appropriate birth records, or to the senior ill health record whenever a child's birth record is missing. These ill health records are also on magnetic tape, and include:

(c) all *death records* for individuals born from 1946 onward and dying in British Columbia (i.e. about 30,000 up to 1970),

(d) records of *handicapping conditions* of children and adults, from the British Columbia Health Surveillance Registry (about 48,000 registered between 1952 and 1972), and

(e) *hospital admission-separation* records from a system of universal hospital insurance (currently in use are 235,000 records for children 10 years of age and

TABLE VII

IDENTIFYING INFORMATION AVAILABLE IN THE MAGNETIC TAPE FILES OF VITAL AND HEALTH RECORDS

Data for British Columbia.

Event and its date and place			Father				Mother				Child							
	Date (or yr*)	City & Prov.	Surname	Initials	Age	Birth Prov.	Maid. surn.	Initials	Age	Birth Prov.	Surname	Given names	Sex	Birth order	Full birth date	City of birth	Prov. of birth	OTHER (see below)
Marriage	√	√	√	√	√	√	√	√	√	√								(1)
Livebirth	√	√	√	√	√	√	√	√	√	√	√	√	√	√	√	√	√	(2)
Stillbirth	√	√	√	√	√	√	√	√	√	√	√	√	√	√	√	√	√	(2)
Death	√	√	√	√		√	√	√		√	√	√	√	√	√	√	√	(3)
Handicap	√*										√	√	√		√	√		(4)
Congen. malf. surveillance	√*										√	√	√		√	√		(4)
Hospital separation	√	√									√	√	√		√			(5)

(1) Marital status of groom and bride.

(2) Legitimate or illegitimate status; singleton, twin, triplet, etc.

(3) If married woman, own maiden surname instead of mother's; marital status.

(4) Coded address representing city, school district and provincial health unit area; whether born in home province.

(5) Hospital code and admission number; maiden surname if different; marital status; address as city, street and number; religion; three initials of head of family, or spouse or guardian; plus codes for relationship and whether address is the same.

under at the time of discharge in the period 1961—70, excluding only those with certain common diseases of no genetic interest).
The records of the Health Surveillance Registry are themselves derived from multiple sources, and autopsy records are made available to the Registry to provide final diagnoses for deceased individuals. Also, an aetiology code is included on the handicap records, in addition to the particular disease code, to indicate whether the disease is hereditary, and if so its mode of inheritance, or whether it has a known environmental cause.

The personal identifying information contained in the magnetic tape versions of the various records, which makes all of these computer linkages possible, is shown in Table VII. Thus, linkages of the vital registrations into brother-sister groupings are enormously simplified by the presence on the records of identifying information relating to both members of a married pair, normally first as groom and bride on a marriage registration and later as father and mother on a birth registration. Linkages of ill health records to the birth records are based mainly on names, birth dates, and birth places where available. For hospital records, which lack birth place, the three initials of the head of family are normally available and may be used for linkage.

Computer strategy

With such identifying information, it would clearly be possible to perform the linkages into individual and sibship groupings manually on a small scale. If a computer could be programmed to follow the rules of logic employed by a human when performing such a task, the scale of the operation could be enormously increased. It is thus necessary, when devising procedures for integrating and linking diverse kinds of health records by computer, to consciously arrange for the machine to employ a few of the strategems that a filing clerk would use intuitively.

In particular, two devices are especially valuable (a) to get around file sequencing troubles arising from variations in the spellings of surnames, which create difficulty when searching for potentially linkable pairs of records, and (b) to aid in deciding whether a particular pair of records relates to the same individual or sibship. The first of these devices is a standard system for phonetically coding surnames, known as the Russell Soundex code. The second is a way of weighting the separate agreements and disagreements of the various items of identifying information on a pair of records, to reflect accurately their discriminating powers. The weights are called "binit" weights.

The Soundex surname phonetic codes may be added to the magnetic tape records by the computer, using the following rules:

(1) the first letter of the surname remains uncoded;
(2) W and H are ignored entirely;
(3) A, E, I, O, U, Y are not coded but serve as separators (see item 5 below);
(4) other letters are coded as follows until three digits are used up (the remaining letters are ignored),

B, P, F, V	coded 1
D, T	coded 3

L		coded 4
M, N		coded 5
R		coded 6
All other consonants	(C, G, K, Q, S, X, Z)	coded 2

(5) Exceptions are letters which follow prefix letters which would, if coded, have the same code. These are ignored in all cases unless a separator (see item 3 above) precedes them.

The name Anderson, for example, is coded A536. Similar sounding names tend to have the same codes; e.g. Bergmans and Brigham both have the code B625.

The "binit" weighting system for various agreements and disagreements is based quite simply on the likelihood that the agreement or disagreement could have occurred by chance. Thus if the initial J has a frequency of 1 in 32 for males, the odds against a record which has an initial J being paired by pure chance with another which carries the same initial would be 32 : 1. If the initial were Z, with a frequency of one in 1000, the odds would be 1000 : 1. Similarly, if sex is entered wrongly into the tape records only once in 4000 times, the odds against a pair of records being correctly matched, when one is said to relate to a male and the other to a female, will be reduced by 1 in 4000.

It is convenient in practice to convert these odds to logarithms to the base 2, so that they can be added. Thus, agreements of the initials J and Z will carry weights of +5 (i.e. $\log_2$ of 32) and +10 (i.e. $\log_2$ of 1000). Similarly, disagreement with respect to sex on a pair of records will carry a negative weight, because it argues against a correct linkage, and the value will be —12 (i.e. $\log_2$ of 1/4000).

The raw data are obtained in almost any convenient fashion. The files themselves will provide the frequencies of all initials from A to Z, separately for males and for females. Telephone directories can provide reasonably accurate frequencies for surnames. A table of populations will give a sufficiently accurate indication of the odds in favour of being born in a particular place. The probability of being born on a particular day of the month is 1/30, and in a particular month 1/12, and so on. Only when determining the weights to be attached to various disagreements is it necessary to carry out a few hundred, or more, linkages by hand to see how often the various items disagree in genuinely linked pairs.

When the weights have been calculated, the computer is given a set of look-up tables of weights, and from these it extracts the weights for the particular agreements and disagreements represented by a pair of records undergoing comparison. The procedure needs to be tested in actual use, as a check against errors in the logic, before it is finalized. But generally a weight of +5 or more indicates a genuine linkage, and a weight of —5 or less indicates a definite non-link, with relatively few cases of doubtful linkage falling in between.

Refinements in the weighting system have been discussed in detail elsewhere [5,8,9]. However, this weighting system should be tested first in its simplest form on a small scale before attempting to use it for a large linkage operation. Refinement can then be made after familiarity has been gained with the simpler version. The chief requirement for the development of adequate methods for computer integration of health record files is that the methods be developed gradually and that they be well tested at each stage.

File organization

Two kinds of linked files may be envisaged, one containing sibship groupings of birth records with the parental marriage records, the other containing the individual histories as represented by the linked ill health records and including the birth and death records where available. The two files are differently sequenced.

The sibship file is most conveniently arranged in order of the husband's surname followed by the wife's maiden surname, preferably in phonetically coded form. The two surnames are together highly discriminating and we found only sixteen cases of a Mr. Brown who had married a Miss Smith in a file of 114,000 marriages. Searching in such a file, by computer, is straightforward when starting with records that have been linked to the birth registrations to obtain the double surname pairs.

The individual health histories are best sequenced by the single surname phonetic code, followed either by the birth date or the first given name, whichever is most readily available or most reliable. Other sequences of this file or of parts of it, for example in double surname sequence or in birth year and registration number sequence, may be required for special purposes.

To obtain sibship histories of disease, the individual histories, condensed as necessary, are re-sorted in order of the parental surname pairs obtained from the birth registrations. In this sequence, the individual ill health histories may be merged with the sibship histories of marriages and births. The order within a sibship group of records will then be

(a) the parental marriage registration,

(b) the live birth and stillbirth registrations in chronological order, and

(c) interfiled and immediately following each birth registration, the ill health history of the child in chronological order.

The form of this organization is illustrated in Table VIII.

Computer speeds and costs

The development of a computer program for linking records requires careful planning and testing, but once developed it can be used to process large files with rapidity and at small cost.

A typical linkage program which is now being used at Chalk River to link British Columbia records of deaths, registered handicaps, congenital anomaly surveillance reports and hospital discharges (back to the appropriate birth registrations) carries out these operations at a rate of about 20,000 incoming health records per minute.

Development costs need not be great, provided the work is undertaken methodically, with ample testing at each stage using small sample files. Large crash programs probably would be expensive if not preceded by such trial runs on a modest scale.

TABLE VIII

EXAMPLE OF THE FILE ORGANIZATION FOR A SIBSHIP HISTORY OF DISEASE (FICTITIOUS)

This is obtained by merging the individual histories of birth, ill health and death, with the sibship histories of parental marriages and births.

Kind of event (birth order in parentheses)	Event Year	Father				Mother				Child					
		Surname	Initials	Birth Prov. *	Age	Maid. surn.	Initials	Birth Prov. *	Age	Given names	Sex	Birth yr	mo	dy	Disease
Marriage	1948	Cox	JW	07	25	Bell	MA	09	21						
Live birth (1)	1950	Cox	JW	07	27	Bell	MA	09	23	Annie May	F	50	01	09	
Live birth (2)	1952	Cox	JW	07	29	Bell	MA	09	25	Brian John	M	52	04	14	
Reg. handicap	1952	Cox	—	—	—	—	—	—	—	Brian John	M	52	04	14	Spina bifida
Hospitalization	1955	Cox	—	—	—	—	—	—	—	Brian J-	M	52	04	14	Spina bifida
Death	1956	Cox	JW	07	—	Bell	MA	09	—	Brian John	M	52	04	14	Spina bifida
Stillbirth (3)	1954	Cox	JW	07	31	Bell	MA	09	27	Mary Jane	F	54	07	26	Anencephaly

* Coded from a two-digit geographic code for all Candain provinces and all major countries; 07 = Saskatchewan and 09 = British Columbia.

Epidemiological products for risk assessment

A number of uses of the family linked files for purposes relating to the estimation of genetic risks have already been demonstrated, and other such uses have been planned in some detail. They include:

(a) a search for effects of naturally occurring mutations,

(b) an attempt to detect favourable selective forces that may be maintaining the prevalence of some of the otherwise troublesome genetic conditions,

(c) an attempt to gain further information on the genetic aetiology of various diseases from more extensive data on risks to cases' siblings,

(d) a reassessment of the total load of genetic disease detectable in children,

(e) a quantification of the social burden from genetic diseases once they have occurred,

(f) a test for possible admixtures of other aetiologies among cases that have been nominally assigned a simple dominant cause.

The first three of the above uses have been described in a review article [6] and the fourth in a recent paper [12]; preparations are currently being made for the remaining two. The methods employed, or to be employed, in these six kinds of study will be mentioned briefly.

Considerable importance would be attached to any direct demonstration of a harmful effect of an increased number of mutations. For this reason possible correlations between the health of children and the ages of the parents at the times of their births are regarded as relevant. Linkage of British Columbia health records with the corresponding birth registrations has served to bring together, on a substantial scale, diagnostic data pertaining to children with the ages of their parents. What has been found, in addition to the expected increase in Down's syndrome among offspring of older mothers, is a variety of correlations including an increased risk of certain groups of congenital malformations in the children of older fathers, independent of maternal age.

More important, the study using the linked records has demonstrated the reliability of the method and its potential for further use. For example, heterogeneity with respect to racial origin was found to give rise to at least one spurious correlation with parental age. Information on race may be extracted readily by computer from the parental marriage registrations, which are linked routinely with the birth registrations of the offspring. Only where the data are abundant will such additional breakdowns be useful. Thus it is unlikely that any but the present approach would lend itself to refinements of this sort.

Similarly, studies of favourable selective forces which may maintain the prevalence of some hereditary diseases, and show them to be independent of recurrent mutations, require an abundance of family fertility and mortality data for a whole population. Erythroblastosis, for example, is due to gene differences which are presumably maintained by natural selection. However, the generally accepted theory that mothers compensate reproductively for the loss of an erythroblastotic child is not supported by data from the linked record files. For this study, birth records over a given period were used to initiate searches by computer for the births of later siblings. This was done not only for erythroblastotic children but for some 30,000 control children as well. Such a large control enabled the data to be broken down by birth order and mother's age simultaneously, and thus resulted in a degree of precision which would not have been possible in a more limited study of conventional design.

Substantial studies of the risks to siblings of "affected" children have likewise been possible, and could be extended in the future to supply one of the kinds of information needed for assessing genetic aetiology. With the computer linkage methods it is not difficult to link routinely all the birth registrations for a population of two million into sibship groupings. With relevant ill health records also linked to the birth records, it is relatively simple to manipulate the files to determine the risks to later brothers and sisters of affected children. For the future, it will be of value to determine whether, for whole groups of affected individuals for whom the causal diseases have been assigned a similar genetic aetiology (e.g. dominant, or recessive, or other), the sibling risks confirm the correctness of the assignment. This test will be applied first to registrations of handicapping conditions.

These handicap registrations, and their genetic aetiology codes, have also been used recently to prepare lists of all dominant, recessive, sex-linked, chromosomal and irregularly inherited diseases, with their frequencies, in a young population. The study has shown (a) that the irregularly inherited congenital malformations and other multifactorial diseases are considerably more common than had been supposed from the results of earlier studies, and (b) that the serious dominant diseases are much less common, at least as judged from the numbers reported in children (see Table IX). The various reasons for the discrepancies, as compared with the data from Northern Ireland, are still under discussion. However, it is the value of such information for genetic risk estimation that is providing much of the incentive for additional work.

For the future, there will be a need to quantify further the social burden from the various categories of genetic diseases, in terms of age at the time of serious manifestation, age-specific mortality, extent of the use of hospital services, etc. Such statistics can be extracted for virtually all cases identified in the handicap registrations, simply by using the integrated files to follow up these known individuals in a systematic fashion.

Such epidemiological products should not be regarded as taking the place of data obtained from ad hoc studies of particular genetic diseases carried out in a

TABLE IX

FREQUENCY OF SERIOUS HEREDITARY DISEASE. COMPARISON OF BRITISH COLUMBIA DATA WITH NORTHERN IRELAND DATA

From ref. 12; comparison with Table I of this paper

Disease aetiology	Frequency per 100 liveborn	
	B.C. children	N.I. all ages
Dominant	0.08	0.95
Recessive	0.11	0.21
Sex-linked	0.04	0.04
Chromosomal	0.20	0.42
Congenital malformations	4.28	2.50
Other irregularly inherited	4.73	1.50
TOTAL	9.44	5.62

conventional manner. Rather, the purpose is to provide population data, relevant to genetic risks, of kinds that are currently unobtainable in sufficient quantity by the methods usually employed by human geneticists.

Some practical considerations

Questions of a practical nature may be asked concerning (a) an appropriate size of population for study in the manner described, (b) the pros and cons of using, and parasitizing, registers developed for purposes other than genetic, (c) the cost of card punching, and (d) the availability of software.

Our own experience relates to a population of roughly two million people, in which there are about 30,000 births per year. For the purpose of determining the collective frequency of serious dominant diseases, this is more than ample in size. It does not matter that individual rare conditions may not be found in such a population, because it is the combined frequency rather than the individual frequencies that are important. Where extensive use is made of a vital registration system there is little point in choosing to study only a part of the registration area, since whole vital registration magnetic tapes will have to be scanned anyway. Also, for studies of selective forces which may be maintaining genetic diseases, the sensitivity of detection will be strongly correlated with population size. Since direct observations of the harm caused by mutagens are unlikely, this consideration should not influence the choice of the size of the study population.

Geneticists may hesitate to use existing registers, for example of congenital anomalies, because of under-registration, lack of diagnostic specificity, and the fear that genetic uses of the register may not be given sufficient priority by health authorities. These defects can be corrected, however, where health departments have been persuaded that hereditary and environmental causes of congenital anomalies need to be studied together. Human geneticists have not in the past been notably successful in ensuring continuity and adequate use of registers maintained solely for genetic purposes, of which there have been several (e.g. that at the University Institute of Human Genetics in Copenhagen, and the Eugenics Records Office at Cold Spring Harbor in New York). Also, integration with the vital registration system is unlikely to be workable outside public health auspices.

The cost of entering the required information into punchcards or magnetic tape may be taken as in the vicinity of ten cents per record. This operation is best carried out in a public health statistical organization where it will represent a small addition to existing related operations. From there on the software, i.e. the record linkage programs, have to be developed separately for each operation. One full time person with programming skills and a flexible approach to new problems is required, along with a supervisor who understands both the biological problems and the computerized approach to record linkage.

Much will depend on whether geneticists regard environmental mutagenesis and teratogenesis as having comparable social and public health importance. If they do, then the developments described above will not be regarded as unduly demanding on their own time or on society's resources.

Discussion

There are many influences that are known to substantially affect human health and well-being. Whether the environmental mutagens constitute an important or a trivial public health problem, in this context, still needs to be demonstrated in some quantitative fashion.

The present account has described methods which emphasize the demographic and epidemiological nature of some of the data needed to assess genetic risks. These procedures do not create the raw data. They are designed to manipulate any or all of the available information for a whole population, and to make more readily accessible what is contained in a diversity of files.

The effort involved in this demographic approach may seem large if genetic data are the only products. But the integrated health files may be used to study a multiplicity of problems including those associated not only with mutagenesis, but with carcinogenesis and teratogenesis as well. It is in this context that the epidemiological approaches which have been described are most likely to be developed in the future.

References

1 BEIR, The Effects on Populations of Exposures to Low Levels of Ionizing Radiation, Report of the Advisory Committee on the Biological Effects of Ionizing Radiations, National Academy of Sciences, National Research Council, Washington, D.C., 1972.

2 Bergsma, D., Birth Defects Atlas and Compendium, Williams and Wilkins, Baltimore, Md., 1973.

3 Cavalli-Sforza, L.L. and W.F. Bodmer, The Genetics of Human Populations, W.H. Freeman and Co., San Francisco, Calif., 1971.

4 Childs, B., S.M. Miller and A.G. Bearn, Gene mutation as a cause of human disease, in H.E. Sutton and M.I. Harris (eds.), Mutagenic Effects of Environmental Contaminants, Academic Press, New York, 1972, pp. 3—14.

5 Newcombe, H.B., Record linking: the design of efficient systems for linking records into individual and family histories, Amer. J. Human Genetics, 19 (1967) 335—359.

6 Newcombe, H.B., Present state and long-term objectives of the British Columbia population study, in J.F. Crow and J.V. Neel (eds.), Proc. Third Int. Congr. of Human Genetics, Plenary Sessions and Symposia, Johns Hopkins Press, Baltimore, Md., 1967, pp. 291—313.

7 Newcombe, H.B., Mutation and the amount of human ill health, in O.F. Nygaard, H.I. Adler and W.K. Sinclair (eds.), Radiation Research, Biochemical, Chemical and Physical Perspectives, Academic Press, New York, 1975, pp. 939—946.

8 Newcombe, H.B., J.M. Kennedy, S.J. Axford and A.P. James, Automatic linkage of vital records, Science, 130 (1959) 954—959.

9 Smith, M.E. and H.B. Newcombe, Methods for computer linkage of hospital admission-separation records into cumulative health histories, Meth. Information Med., 14 (1975) 118—125.

10 Stevenson, A.C., The load of hereditary defects in human populations, in D.E. Smith (ed.), Proc. Int. Congr. of Radiation Res., Burlington, Vermont, Radiation Res., Suppl. 1 (1959) 306—325.

11 Sutton, H.E., Impact of induced mutations on human populations, Mutation Res., 33 (1975) 17—24.

12 Trimble, B.K. and J.H. Doughty, The amount of hereditary disease in human populations, Ann. Human Genet., 38 (1974) 199—223.

13 UNSCEAR, Report of the United Nations Scientific Committee on the Effects of Atomic Radiation, General Assembly Official Records, Thirteenth Session, Supplement No. 17 (A/3838), United Nations, New York, 1958.

14 UNSCEAR, Seventeenth Session, Supplement No. 16 (A/5216), United Nations, New York, 1962, pp. 84—88.

15 UNSCEAR, Twenty First Session, Supplement No. 14 (A/6314), United Nations, New York, 1966, pp. 98—99, 126—127.

16 UNSCEAR, Ionizing Radiation: Levels and Effects, Vols. I and II, United Nations, New York, 1972.

17 Vogel, F., Approaches to an evaluation of the genetic load due to mutagenic agents in the human population, Mutation Res., 29 (1975) 263—269.

18 Vogel, F. and R. Rathenberg, Spontaneous mutation in man, in Advances in Human Genetics, Vol. 5, Plenum, New York, 1975, pp. 223—318.

SUBJECT INDEX